80 Years Of Cadillac LaSalle

By Walter M.P. McCall

Editing and Design

By George H. Dammann

Crestline Publishing

1251 NORTH JEFFERSON AVE. SARASOTA, FLA. 34237

CRESTLINE AGRICULTURAL SERIES

AMERICAN GASOLINE ENGINES SINCE 1872
(584 Pages - 2,350 Illustrations)

150 YEARS OF INTERNATIONAL HARVESTER
(416 Pages - 1,900 Illustrations)

ENCYCLOPEDIA OF AMERICAN FARM TRACTORS
(352 Pages - 1,500 Illustrations)

ENCYCLOPEDIA OF AMERICAN STEAM TRACTION ENGINES
(320 Pages - 1,250 Illustrations)

NEBRASKA TRACTOR TESTS SINCE 1920
(584 Pages - 2,350 Illustrations)

80 YEARS OF CADILLAC - LaSALLE
By Walter M. P. McCall

Copyright©1982 by Crestline Publishing Co., Inc. in the name of Walter McCall

Library of Congress Catalog Number 79-50804

ISBN Number - 0-912612-17-7

Photographics by Crestline Publishing Co.

Cover Design by William J. Hentges, Warren, Mich.

Typesetting by Serbin Printing, Sarasota, Fla.

Printed and Bound by Walsworth Publishing Co., Marceline, Missouri

CRESTLINE AUTOMOTIVE SERIES

U.S. MILITARY WHEELED VEHICLES
(472 Pages - 2,100 Illustrations)

FORD TRUCKS SINCE 1905
(416 Pages - 2,000 Illustrations)

ILLUSTRATED HISTORY OF FORD
(320 Pages - 1,400 Illustrations)

THE HISTORY OF HUDSON
(336 Pages - 1,650 Illustrations)

60 YEARS OF CHEVROLET
(320 Pages - 1,650 Illustrations)

75 YEARS OF PONTIAC-OAKLAND
(528 Pages - 2,500 Photos)

THE CARS OF OLDSMOBILE
(416 Pages - 2,000 Illustrations)

80 YEARS OF CADILLAC-LaSALLE
(448 Pages - 2,200 Illustrations)

THE PLYMOUTH AND DeSOTO STORY
(416 Pages - 2,000 Illustrations)

THE DODGE STORY
(320 Pages - 1,600 Illustrations)

70 YEARS OF CHRYSLER
(384 Pages - 1,950 Illustrations)

ENCYCLOPEDIA OF AMERICAN CARS, 1930-1942
(384 Pages - 2,000 Illustrations)

ENCYCLOPEDIA OF AMERICAN CARS, 1946-1959
(416 Pages - 1,800 Illustrations)

AMERICAN FIRE ENGINES SINCE 1900
(384 Pages - 2,000 Illustrations)

AMERICAN FUNERAL CARS & AMBULANCES SINCE 1900
(352 Pages - 1,900 Illustrations)

CRESTLINE AVIATION SERIES

AIRLINES OF NORTH AMERICA
(240 Pages - 1,000 Illustrations)

Published in 1988 by: Crestline Publishing Co.
1251 N. Jefferson
Sarasota, Florida 34237

All rights to this book are reserved, and no part may be reproduced in any manner whatsoever without the express written permission of the author or the publisher. For further information, contact Walter McCall, 2297 Hall Ave., Windsor, Ontario, Canada, N8W 2L8, or Crestline Publishing Co.

Many Thanks

Publication of this book is the realization of a long-held ambition to provide for all admirers of the marque a single, authoratative reference source that presents, in chronological sequence, a photographic record of each and every standard production model and body style produced over the fourscore years the Cadillac Motor Car Division has been in existence. An undertaking of this magnitude, of course, requires much planning, research and the co-operation of many people. My sincere gratitude, therefore, is extended to the following:

To the many fine people at Cadillac who cheerfully accommodated my frequent research visits to 2860 Clark Avenue in Detroit since 1976, especially former Cadillac Motor Car Division Public Relations Director William J. Knight and his capable and helpful assistant, Norbert Bartos, as well as the PR secretaries who were endlessly opening the archives room for me and fetching various bulging file folders:

The Automobile History Collection of the Detroit Public Library, the late James J. Bradley and his staff for access to the trove of early Cadillac material in the superb Nathan Lazernick Collection, as well as early auto industry trade journals:

My valued, longtime friend and fellow Cadillac enthusiast Thomas A. McPherson for his generous contributions of photos and material for this book, and for making it all possible by introducing the undersigned to Crestline Publisher George H. Dammann:

To all others who freely offered photos, information and ideas, including: Albert J. Burch, who was there at Fisher Body when all this was happening; Rick Lenz; Kevin McCabe, David C. Beveridge, Steven B. Loftin, Al Judge, James K. Wagner, Don Butler, Joel C. Woods, Steven Hagy, Dan G. Martin, Kenneth H. Arksey, John G. Linhardt, The Professional Car Society and PCS members Tom Parkinson and Dr. Rodger D. White; the Craven Foundation; Eureka Coach Co. Ltd., Mel Stein and his AHA Mfg. Co., Clarence C. Woodard and American Custom Coachworks.

Most of the photographic and statistical data used in this publication came directly from the Cadillac Motor Car Division's own archives and records. Model numbers, prices and weights were culled from the Branham Automobile Reference Books in the Detroit Public Library Auto History Collection. While the emphasis in this text is on production models actually offered to the public through the Cadillac-LaSalle dealer organization, we have endeavored to include a selection of prototypes, customs and semi-customs, one-offs, specials and conversions. Similarly, because of the significant aspect of Cadillac's business represented by commercial chassis production since the mid-1930's, we have included at the end of each chapter a cross-section of Cadillac-chassised funeral cars and ambulances. These professional cars are covered in greater detail in another Crestline publication, "American Funeral Cars and Ambulances" by Thomas A. McPherson.

Finally, my heartfelt thanks to my good wife Denise, a longtime General Motors of Canada Ltd. employee who finish-typed much of the original manuscript.

Walter M. P. McCall
Windsor, Ontario, Canada
June 1, 1982

FOREWORD

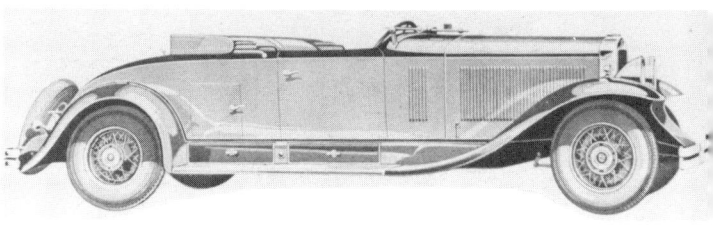

Of the thousands of passenger car nameplates that have appeared since the dawn of the Motor Age in America, none has enjoyed the continued success and aura of prestige long associated with the products of the Cadillac Motor Car Division of General Motors. Fourscore years after the first single-cylinder prototype chugged out onto the streets of Detroit, the stately Cadillac car remains a universally recognized and respected symbol of the best the United States has to offer.

To be the finest of anything — be it a bicycle, a food blender, a ballpoint pen or a motor home — is to be called the "Cadillac" of its kind. Ownership of a Cadillac, or even to be seen riding in one, is irrefutable evidence that one has indeed arrived. No rationalizing its solid engineering, remarkable resale value or even its velvety ride can deter the popular notion that this is, more than anything else, the ultimate symbol of material success. Cadillac's golden grip on the North American luxury car market appears invincible, although the division's traditional markets are being shaken and transformed as never before.

What is know today as the "Cadillac mystique" did not come automatically, or overnight, It all began with a tall, bearded Vermonter with a fetish for thousandsths-of-an-inch accuracy. Henry Martyn Leland in 1902 was associated with a prominent machine tool business in Detroit. He had been approached by a group of men who wanted him to evaluate the machinery owned by a faltering horseless carriage company they were anxious to liquidate. The astute Leland sensed an opportunity to get into the infant automobile business: he had an engine that had been turned down by Oldsmobile, and here was a company with a car but no motor! The rest is one of the longest and most colorful chapters in U.S. auto industry history.

Cadillac didn't even try to court the luxury car buyer for many years. The first Cadillacs were priced at about half the cost of most cars of their day and well below the prestigious Pierce-Arrows, Wintons and Packards.

Leland was content to build fewer, but better, cars than his competitors and success came to the fledgling Cadillac Autombile Co. largely by reputation. Leland's insistence on close manufacturing tolerances and the principle of parts interchangeability he had learned years earlier in the small-arms industry brought Cadillac the coverted Dewar Trophy in 1908. The following year the company became part of Billy Durant's ambitious General Motors, but retained its operating autonomy. In 1910 Cadillac placed the first volume order for closed car bodies, launching the Fisher Bros. fortune, and in 1912 the company was awarded another Dewar Trophy for the self-starter. In 1914, Cadillac introduced the industry's first mass-produced V-8 engine, incurring the scorn of other manufacturers and resulting in one of the most famous ads ever written — "The Penalty of Leadership." Leland left Cadillac in a patriotic huff in 1917 and went on to organize another company — Lincoln.

The 1920's brought numerous mechanical improvements including the inherently-balanced engine, improved lighting, safety glass, and the synchromesh transmission. Under the leadership of the legendary Lawrence P. Fisher, Cadillac mounted a determined bid for supremacy in the American fine car field. Fisher brought in Harley J. Earl to "style" the division's cars. Earl's first assignment, and triumph, was the 1927 LaSalle, a new lower-priced Cadillac companion. It was the age of conspicuous automotive luxury and style and the exclusive Automobile Salons. Cadillac offered 50 body styles in a dazzling array of 500 color combinations. But the Jazz Age ended with the Crash of 1929 and the ensuing Depression.

Cadillac astounded the industry with the introduction of its V-16 and V-12 supercars. The division had acquired Fleetwood, which now executed custom and semi-custom body creations for Cadillac exclusively. While Cadillac styling individuality reached its peak during the golden classic era, production tumbled to miniscule levels. Had it not been part of the General Motors empire with its vast resources spread over many divisions, Cadillac might not have survived the Depression.

One by one the independent luxury car makers fell by the wayside....Pierce-Arrow, Marmon, Peerless and even mighty Duesenberg. The custom-body business was all but wiped out. Streamlining was the rage of the 1930s. Packard continued to dominate the luxury car market even after it came out with its popular-priced One-Twenty. Cadillac surprised everyone with a second-generation V-16 for 1938 and William J. Mitchell's style-setting Sixty Special.

By the end of 1940, the V-16 and even the LaSalle were gone. Cadillac put its magic name on its lowest-priced 1941 cars, and sales doubled. For the first time in more than a decade all Cadillacs were powered by the same V-8 engine. When the nation was plunged into the Second World War, Cadillac switched from passenger car to tank production. The division's sturdy 346 CID engine teamed with the GM Hydra-Matic automatic transmission proved a tough, dependable armored vehicle powerplant. It was a seller's market when civilian car production haltingly resumed in 1946. Cadillac set the automobile styling world on its ear with its bold fishtail fins for 1948, and in 1949 introduced the shortstroke, high-compression V-8 engine that set the standard for the U.S. industry for the next 30 years. That memorable year also saw the one-millionth Cadillac come off the line and the first Coupe de Ville hardtop. Cadillac shifted into high and swept past longtime rival Packard for good.

Sales topped the 100,000 mark for the first time in 1950. In 1952 the division observed its 50th anniversary as the oldest continuous manufacturer of motor cars in the City of Detroit. Fifty-three saw the debut of the flashy Eldorado convertible, and the GM Motorama took to the road. The Fabulous Fifties brought new production and sales records, new styling, the glitter of chrome and Cadillac's famous "dollar grin". The hyper-expensive Eldorado Brougham arrived to challenge Ford's Continental Mark II and Cadillac's 1959 tail fins soared to absurd heights. The Sixties saw more restrained styling and continued engineering advances. Government began to intervene in vehicle design with ever more stringent standards and regulations.

The front-wheel-drive Fleetwood Eldorado was introduced for 1967, Cadillac's high-styled entry into the new luxury-personal car market. The venerable Sixty Special evolved into the Fleetwood Brougham. Cadillac's engines continued to grow, to 429 cubic inches in 1964, to 472 for 1968 and to a monstrous 500 CID by 1970. The Cadillacs of the Seventies boasted myriad comfort and convenience options and extensive electronics. In 1975 the division responded to the inroads being made by imported luxury sedans with its new, smaller "international-sized" Seville. Cadillac's — and the U.S. industry's — "last" convertible was driven off the Clark Avenue assembly line with appropriate fanfare in 1976.

The next generation of Cadillacs, for 1977, were smaller, lighter and significantly more fuel-efficient — new kinds of Cadillacs for a changing world. There was an all-new, downsized Eldorado for 1979 and a stunningly-styled, bustle-back FWD Seville for 1980. Diesel engines were now available in Cadillacs and computers controlled various engine functions. Cadillac's first subcompact, the 1982 Cimarron, was the smallest Cadillac offered in 77 years. As the Eighties began, steel was rising for two huge new Cadillac assembly plants in the Detroit area and the curtain began to come down on the aging Clark Ave. and Fisher-Fleetwood facilities, Cadillac's "home" for six decades.

Cadillac's long and proud history is one of continous engineering and styling innovations, brilliant marketing strategies and customer loyalty that is the envy of the industry. In a viciously competitive, fast-changing industry, Cadillac stands uniquely alone as an unparalleled success story...truly an American Standard for the World, and likely to remain so.

Walter M. P. McCall
2297 Hall Ave.
Windsor, Ontario, Canada
June 1, 1982

1902

The phenomenonally successful commercial enterprise known around the world today as the Cadillac Motor Car Division of General Motors Corporation was born on the hot afternoon of August 22, 1902 when an enthusiastic group of Detroit businessmen sat down to dissolve a small automobile manufacturing company and organize a new one.

Several days earlier, principal officers of the failing Henry Ford Company had approached Henry M. Leland for an appraisal of the equipment and buildings owned by the company they wished to liquidate. The Henry Ford Company had been organized three years earlier, in 1899, as the Detroit Automobile Company. Its chief engineer had been a bright, young man named Henry Ford who had designed, built and successfully driven a motor quadricycle in Detroit in 1896. The Detroit Automobile Company failed in November, 1900 but was revived under the new name of the Henry Ford Company late in 1901. But Mr. Ford soon left, and after only a few months the faltering firm's owners decided to wind up the business. In its three years of existence the company had produced only a handful of cars, most of them racing machines.

Henry Martyn Leland and his son, Wilfred, operated a large and highly-regarded machine tool business. Born in Vermont in 1843, Henry M. Leland learned the precision toolmaking trade in the arms factories of Colonel Samuel Colt. Totally devoted to the principles of accuracy and uncompromising high quality, Mr. Leland was a leading proponent of the principle of interchangeability of parts.

Mr. Leland's company, Leland, Faulconer and Norton, was established in Detroit in the 1890's. Mr. Leland had already acquired a widespread reputation as a true master of precision.

At the dawn of the new Motor Age, Ransom Eli Olds had founded his Olds Motor Vehicle Company in Lansing, Michigan in 1897. Two years later he moved his operation to Detroit and reorganized it as the Olds Motor Works. The first mass-producer of automobiles in America, his company was soon turning out thousands of curved-dash Oldsmobiles annually. But Oldsmobile's manufacturing methods evidently left something to be desired, so R. E. Olds went to Leland and Faulconer for an improved, quieter transmission. And he got it. In 1901, Leland and Faulconer was awarded a contract to supply Oldsmobile with 2,000 engines. Oldsmobile's only other engine supplier at the time were the Dodge brothers, Horace and John.

With its precision machining and Leland manufacturing quality, the Leland-built engine was markedly better and more efficient than its competitor. Encouraged by the possibilities, H. M. Leland went to work on a much-improved Oldsmobile engine. But Oldsmobile turned it down, citing time delays and the additional engineering costs necessary to adapt it to the Olds chassis. Mr. Leland installed his superior engine in his personal Oldsmobile and returned to the machine tool business.

As he studied the Henry Ford Company's assets, Mr. Leland recognized a golden opportunity to get into the budding motor vehicle industry.

He advised the Henry Ford Co. officers against liquidating their company. Instead, he offered his proven engine for installation in their car. A meeting was convened on August 22 and the firm reorganized. But what about a name? Detroit had just celebrated the 200th anniversary of its founding. What better choice for a name, then, than that of the noble French explorer who had first set foot on the spot — Le Sieur Antoine de la Mothe Cadillac? Thus was born the Cadillac Automobile Company.

Under terms of the agreement, Leland and Faulconer would supply engines, transmissions and steering gear for the new car which would be built in the former Henry Ford Co. plant at Cass and Amsterdam Streets in Detroit. The first three prototype vehicles would be built at Leland and Faulconer. Work on the first car commenced in September, and the first automobile to bear the Cadillac name was completed on October 17, 1902. A vigorous program of testing and development began on the streets around L. & F. that fall.

The company's first officers included Lemuel W. Bowen, president, and William E. Metzger, sales manager. Henry M. Leland was appointed a director and held a small block of stock. The newly-formed Cadillac Automobile Company ran its first advertising in November of 1902. "Wise agents" were encouraged to pick up the new auto agency. In retrospect, it would be difficult to count the millionaires the Cadillac motor car franchise has made over the past fourscore years.

Cadillac did not start out in the luxury car market. Indeed, the company's single product, although of admittedly high quality, was moderately priced when compared with other cars of its day.

The first automobile to bear the Cadillac name is seen here as it underwent its first outdoor tests in Detroit late in the summer of 1902. The single-cylinder car was running at the time this historic photograph was taken, but the vehicle was not driveable. Alanson P. Brush, the Cadillac's principal designer, grips the steering wheel while Wilfred C. Leland peers over the side. These first, halting tests were conducted outside the Leland and Faulconer Manufacturing Co. plant on Trombley Ave. on Detroit's east side. Visible behind the left rear wheel are Ernest E. Sweet, Henry M. Leland's engineering consultant, and factory superintendent Walter H. Phipps. Such amenities as fenders and lamps had not yet been attached to the prototype Cadillac.

1902

The first Cadillac was designated the Model "A". It was a buggylike two-seater built on a 76-inch wheelbase and weighed 1,395 pounds in standard trim. It stood 57.5 inches high, was 111 inches long and 67 inches wide. The car was powered by a one-cylinder engine with five-inch bore and stroke. The horizontally-mounted engine was rated at 10 NACC horsepower. The car had a planetary transmission with two forward speeds and reverse, and was chain-driven. Top speed was about 30 miles an hour.

The retail price, F.O.B. Detroit, was $750. Accessories included a buggy-type top, lamps for night driving and a detachable tonneau which could be added to enable the vehicle to carry four passengers. The tonneau cost an extra $100. Smart patent leather fenders were standard.

Late in 1902, the first few Cadillac passenger cars were turned over to Mr. Metzger for display at the 1903 National Automobile Show in New York. Volume production was scheduled to begin in March of the new year.

In other auto industry news that year, the Thomas B. Jeffery Co. introduced a new car called the Rambler and built 1,500 of them. Oldsmobile production passed the 2,000 mark, and the Ohio Automobile Co. changed its name to the Packard Motor Car Company. Cadillac and Packard were destined to become close rivals in the luxury car field.

There is a very strong resemblance between the first Cadillac and the first production Ford. Both, in fact, were designated Model "A"s. The first Cadillacs were built in the latter part of 1902, with full production beginning in March of the following year. Work on the first Ford Model "A" did not begin until very late in 1902, and the car did not go into production until the middle of 1903. The new Ford Motor Company was not organized until June, 1903. Both cars were powered by horizontally-mounted engines, but where the Cadillac utilized a single-cylinder design, the Ford featured a parallel twin powerplant. Historians appear to agree that the external similarities between the two vehicles are largely coincidental.

From the very beginning, the fledgling Cadillac Automobile Company was justifiably proud of its product's outstanding performance. Early in the prototype car's road test program, test driver Alanson P. Brush put the first Model "A" Cadillac to a most convincing test by driving it up the front steps of the Wayne County Building in Cadillac Square in downtown Detroit — to the cheers of a large crowd of bystanders. This remarkable feat was later duplicated all over the country by dozens of Cadillac salesmen and owners, and such convincing demonstrations of the little one-lunger's tractive capabilities were discreetly mentioned in some of Cadillac's early advertising. Alanson P. Brush went on to build a runabout bearing his own name and later designed the first Oakland, in addition to contributing much to the design of the first Cadillac.

Factory historical records reveal that the first Cadillac was completed on October 17, 1902. The first three prototype cars were built in the Leland and Faulconer Manufacturing Co. plant on Trombley Ave. in Detroit rather than in the former Henry Ford Co. factory at Cass and Amsterdam Streets. Production did not begin until March of the following year. The first three vehicles were constructed for extensive testing and for public exhibition at the forthcoming 1903 National Automobile Show in New York. The first three Cadillacs produced in 1902 were designated Model "A"s and were powered by a horizontally-mounted, single-cylinder engine with five-inch bore and stroke. The first complete car, shown here, has been fitted with patent leather fenders.

1903

A distinguished new automotive nameplate — Cadillac — entered the rapidly expanding American automobile market this year. The all-new Cadillac had its public premiere at the Third Annual National Automobile Show held in Madison Square Garden in New York City from January 17th to 24th, 1903. Sponsored jointly by the Automobile Club of America and the National Association of Automobile Manufacturers, the prestigious New York show was by far the largest of its kind in the country. It was first held in 1900.

Cadillac's first public showing was a spectacular success. An elated William E. Metzger, sales manager for the fledgling Cadillac Automobile Company, returned to Detroit with $10 deposits for 2,286 cars — virtually all of the company's planned production for the year. Not a single production car had yet been built. The first would not come off the line until March. The first Cadillac sold was delivered to a customer in Buffalo, N.Y. And the rush was on...

The company's sole product was a sturdy little two-passenger runabout known as the Cadillac Model "A". The first Cadillac ads, which appeared in November, 1902, heralded the new car as... "The Winner of 1903 ... It's Just Good All Over!" Technical specifications were generally identical to those of the first three prototype cars assembled in 1902. The Model "A" rode on a diminutive 76-inch wheelbase. In stock form, the standard Model "A" weighed just over 1,000 pounds. With the single-cylinder, horizontally-mounted engine wide open, the little Cadillac was capable of speeds up to 30 miles an hour.

The precision-made Leland engine put out 9.7 brake horsepower, but had a NACC rating of 10 HP. The chain-driven runabout was equipped with a planetary transmission with two forward speeds and reverse. The Cadillac rode on 3 X 28-inch pneumatic tires mounted on spoked wheels of selected hickory.

The retail price of the standard Model "A" Runabout with two seats was a flat $750. For another $100 the customer could have an optional rear entrance tonneau that increased the Cadillac's seating capacity to four. Lamps — two headlights and a taillight — and a buggy-type top for the two-seat runabout, were the only other accessories available.

Production of the Cadillac began in the Detroit plant in March, 1903. From the beginning, the factory was hard-pressed to fill the orders taken at the New York show, but 2,497 Model "A" Cadillacs were built during 1903. In all, the new company was off to a very strong start in a fiercely competitive industry.

Cadillac's initial success was not limited to America's shores. In February, 1903 the Anglo-American Motor Car Co. of London, England ordered, sight unseen, a Model "A" rear-entrance tonneau. The car's importer, Frederick S. Bennett, immediately entered the little American runabout in a major hill climb at Sunrising Hill where it turned in a most impressive performance competing against much more powerful cars. In September of the same year, the Model "A" took first place in its class in the gruelling Thousand Miles Trial in England.

In other U.S. automobile industry news during Cadillac's first year on the market, the Ford Motor Company was incorporated in June and was soon turning out Model "A" Fords that were startlingly similar in appearance to the 1903 Cadillac. There were substantial mechanical differences, however.

The Packard Motor Car Company moved from Ohio to Detroit and later that year introduced the soon-to-be-famous slogan, "Ask the Man Who Owns One." The Buick Motor Company was organized, and the Association of Licenced Automobile Manufacturers was formed with 10 member companies including Winton and the Electric Vehicle Co.

This is the standard Cadillac Model "A" Runabout as it appeared at the time of its public introduction early in 1903. The retail price as shown was $750, F.O.B. Detroit. Patent leather fenders were standard, but that was all. Cadillac customers could specify an optional rear tonneau to carry two additional passengers, a buggy-type top or a set of lamps for night driving. The first Cadillacs bore a strong external resemblance to the Model "A" Ford which was introduced later this same year, although there were numerous mechanical differences between the two cars.

This is the 1903 Cadillac Model "A" Runabout complete with optional buggy top. The top was a $30 accessory, and a popular one. Forward visibility was unlimited but the driver couldn't see much through those tiny oval side windows and to the rear. These early Cadillacs rode on 3 x 28-inch pneumatic tires mounted on hickory spoked wheels.

1903

For an extra $100, the 1903 Cadillac could be ordered with an optional rear tonneau. This increased seating capacity from the standard two to four. Thus equipped, the 1903 Model "A" carried a retail price of $850 F.O.B. Detroit. The massive flywheel is visible below the front seat.

Thousands of visitors to the Cadillac Motor Car Division's "home" on Clark Avenue in Detroit have paused to admire this nicely restored 1903 Model "A" Cadillac with rear entrance tonneau that occupies a position of honor in the lobby of Cadillac's Visitor Reception Center, where plant tours begin. The script-type nameplate on the radiator did not appear for another two years. This car was restored twice, the last refurbishing taking place in 1976. Cadillac's Model "A" sports optional lamps and has an accessory hand-rail.

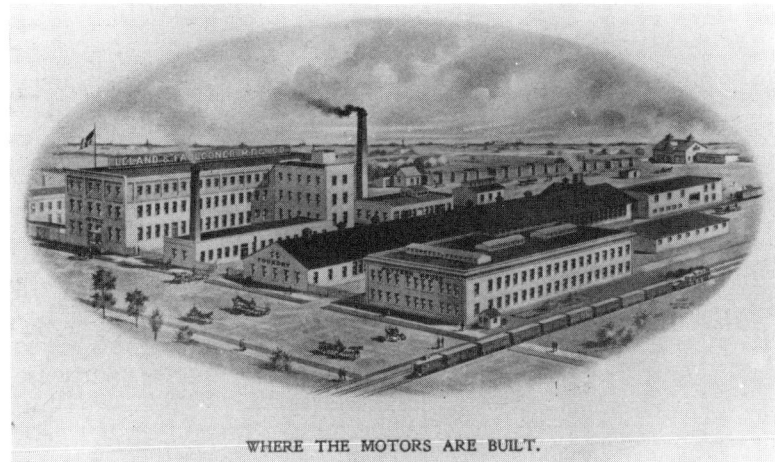

WHERE THE MOTORS ARE BUILT.

This is the sprawling Leland and Faulconer Manufacturing Co. foundry and machine shop complex on Trombley Ave. in Detroit, where the first three Cadillac motor cars were assembled in the latter part of 1902. Under the agreement signed in August, 1902, Leland and Faulconer was to supply engines, transmissions and steering gear to the newly-formed Cadillac Automobile Co., successor to the Henry Ford Company. Cadillac chassis and bodies would be made in the former Henry Ford Co. plant at Cass and Amsterdam Streets in Detroit, and components supplied by L & F assembled there. Production began in March, 1903. Leland, Faulconer and Norton dated back to the early 1890's, and Henry Martyn Leland's machine tool business had acquired a formidable reputation for high quality.

THE HOME OF THE "CADILLAC."

This illustration of the Cadillac Automobile Company factory in Detroit was taken from the new firm's first catalog published late in 1902. The plant was that of the now-defunct Henry Ford Company, successor to the Detroit Automobile Company which had been organized in 1899. In August, 1902 the directors of the failing Henry Ford Co. approached Henry M. Leland and asked him to evaluate the company's assets prior to liquidation. Mr. Leland, who had a promising single-cylinder automobile engine but no car to put it in, immediately sensed an opportunity. He talked the Henry Ford Company's officers into installing the Leland and Faulconer engine in their car, and the rest is automotive industry history. Within a few weeks the Henry Ford Company had been reorganized as the Cadillac Automobile company, named in honor of the French adventurer who had founded Detroit two centuries earlier.

1903

No, this is not Cadillac's lowest-priced economy model. It is a 1903 Model "A" chassis that has been stripped down to the bare frame for racing. Factory records show that a single-cylinder Cadillac won a race that year, but the date and place are not indicated. Automobile racing was a popular spectator sport in the Detroit area at this time and was a real test of a new car's mettle. Inasmuch as there is minimal bodywork available for display of the competitor's number, the young driver displays it on his armband.

A Cadillac is a Cadillac, but this is primitive motoring! These two intrepid employees of the Allen Service Station in Murfreesboro, Tenn. drove this one-cylinder Cadillac chassis, sans body or any other creature comforts, from Murfreesboro to Detroit. The year was not indicated but it was likely about 1940. About the only "frill" on this breezy runabout is a set of electric lights.

The famous Henry Ford Museum in Dearborn, Michigan has this beautifully restored 1903 Model "A" Cadillac in its vast vehicle collection. The car is a four-passenger tonneau model with optional headlamps. It has been posed beside one of the historic buildings in the adjacent Greenfield Village. Note the white rubber pneumatic tires and dash handrail.

Attired in typical motoring garb of the era, this couple poses with a 1903 Model "A" Cadillac two-seater runabout at Greenfield Village in Dearborn, Michigan. The little 1903 Cadillac Model "A" is in stock form, just as it might have been when shipped from the factory at a complete price of $750.

1903

Bundled up against the weather in fur coat and goggles, this privileged lady poses proudly behind the steering wheel of her 1903 Model "A" Cadillac. The large wicker baskets over the rear fenders fulfilled the function of what later became the trunk. Note the bulb horn attached to the steering column.

The 1953 Glidden Tour passed through Detroit, Michigan and Cadillac was well represented. Here, Wilfred C. Leland, son of the legendary Henry M. Leland, admires a fully-equipped 1903 Cadillac Model "A" with top and rear tonneau. The big, single headlamp, sidelamps, horn and wicker picnic basket are accessories. Parked next to the 1903 model is a spanking new 1954 Cadillac convertible. It's hard to believe that the late-model ragtop is now a collector's item too!

This cutaway view of the Model "A" Cadillac chassis appeared in the new company's first product catalog, which was issued to agents and prospective purchasers late in 1902. The horizontally-mounted, single-cylinder engine had a displacement of 98.2 cubic inches and carried a NACC rating of 10 horsepower. Brake horsepower was shown as 9.7. Wheelbase was a diminutive 76 inches. The cooling system had a capacity of three gallons. Cadillac's sporty little Model "A" boasted a top speed of 25 to 30 miles an hour in standard form.

Ten horsepower overtakes one-horse power. Scenes such as this were not uncommon in the early days of motoring in the United States. The car is a 1903 Cadillac Model "A" Rear-Entrance Tonneau. The make of the buggy blocking the motor car's progress is anybody's guess. Yes, that's a lady at the wheel.

1904

The year 1904 was one of continued growth and prosperity for Cadillac. Sales and production had exceeded all expectations. In just one year on the market, the sturdy little single-cylinder Cadillac had acquired a worldwide reputation for quality and performance. The bustling Detroit plant produced 2,418 cars in 1904.

But not all the news that year was good. Disaster struck the firm on April 13, 1904. Fire broke out in the Cadillac Automobile Company, and before being brought under control destroyed the main vehicle assembly plant and the south wing. The company was suddenly out of business. Deposits on 1,500 cars had to be returned. This interruption, however, lasted only four weeks. Thirty days after the fire, Cadillacs were again being produced. To keep up with a massive backlog of orders, the plant was forced to work six-day weeks and night shifts all through 1904.

For the 1904 model year, Cadillac introduced a refined, much improved new line of cars designated the Cadillac Model "B". The Model "B" cars were powered by the same proven Leland-built horizontally mounted engine used in the Model "A". The extensively restyled Model "B" Cadillac was offered in two-seat runabout, four-passenger surrey, four-passenger touring and light delivery models. The Model "B" sported an imitation engine hood up front that was actually a luggage compartment.

The popular Model "A" was carried into its second year with virtually no changes. Models included the two-passenger runabout, a detachable four-passenger tonneau and a new-for-1904 Model "A" light delivery car.

Leather or rubber buggy tops were still optional equipment on the Model "A" Runabout. Illumination was also still an extra-cost option, as was a warning horn. The optional lighting set included two headlamps and a small taillight.

Stunts and showmanship were still an important element in selling motor cars in 1904. In April, Detroit newspapers carried an impressive photo of a Model "A" Cadillac hauling 14 passengers up the Shelby Street hill. In an unanticipated testimonial to the Cadillac's climbing ability, a slightly inebriated gentleman drove one up the front steps of the Capitol in Washington, D.C. This feat was alluded to in some of Cadillac's advertising, although the company was

Cadillac's highly successful little Model "A" Runabout was carried over into the 1904 model year with no significant changes. A top and lights were still extra-cost options. With its gracefully curved front and modestly ornamented sides, the Model "A" presented a tidy, rather handsome appearance. Standard color was maroon. The price remained at $750 Net F.O.B. Detroit.

This is the 1904 Cadillac Model "A" Light Runabout equipped with the optional buggy-type top. With a leather top complete with sides and storm apron, this car carried a price of $800. With rubber top, the price was $780 complete. The horizontal tube radiator was canted to conform with the small vehicle's frontal styling. Upholstery was of practical leather. Decorative scroll moldings added character to the car's otherwise plain sides.

Shown here is the 1904 Cadillac Model "A" Detachable Tonneau. Rear seat passengers climbed into the car through a small door at the rear. As the name implied, the entire rear tonneau could be lifted off to convert the car into a sporty two-seat runabout. Standard color was maroon with black trim, and the base price was $850. The car shown sports optional lighting equipment and has a bulb horn clamped to the steering column.

1904

careful to point out that the fellow concerned... "had to pay for his fun." And a Canton, Ohio Cadillac owner wrote in to report that he had successfully pulled two truckloads of railroad iron weighing five tons up a four per cent grade from a standing start.

In auto industry news, Studebaker sold its first gasoline-driven automobile, moving away from less efficient electrics, and Ranson E. Olds sold his interest in the Olds Motor Works and left to organize the Reo Motor Car Company.

But all was not running smoothly at Cadillac by the end of the year. The factory was still hard-pressed to keep up with orders, and problems were cropping up in the cars. Cadillac business was taking up more and more of Henry M. and Wilfred Leland's time. On Christmas Eve, 1904 William H. Murphy and Lemuel C. Bowen again called on the Lelands. The next day they spent in the automobile plant. Another reorganization was in order.

This reorganization took the form of a consolidation of the Leland and Faulconer Manufacturing Co. and the Cadillac Automobile Company. The consolidation permitted Leland and Faulconer to concentrate on their automotive business and to supply components that up to now had been purchased outside. With the consolidation came a new name: the Cadillac Motor Car Company. Henry Martyn Leland was named General Manager of the new company on December 27, 1904, a position he would hold until 1917.

Late in the year, Cadillac announced its first four-cylinder car, the Cadillac Model "D". But the company's highly-regarded little one-lunger would continue to occupy a prominent position in Cadillac's proliferating model lineup for some years to come.

Here is the one-lung "heart" of Cadillac's first cars. The Cadillac Automobile Co. had no reservations when it came to promoting the performance of this small but potent powerplant, which it boastfully advertised as the "Little Hercules". Now downrated at 8-¼ horsepower, the Leland-built single cylinder engine was used in both Cadillac's Model "A" and "B" cars in 1904.

This is the new Cadillac Model "B" Surrey. A four-passenger car, it carried a price of $900. With this body style, passengers could step right up into the car. However, a rear entrance tonneau body was also offered in the new "B" series. This was called the Model "B" Touring Car and had an identical price of $900. Standard color was maroon with black trim.

In profile, the new Model "B" Cadillac Surrey presented a larger, more massive appearance when compared with the lighter Model "A". The Model "B" Surrey was advertised as a variation of the touring car for those who chose to get in and out from the sides, rather than from the rear. Standard tires on the 1904 Cadillacs were Goodrich double tube three-inch detachables.

For the 1904 model year, Cadillac came out with an improved, more refined car designated the Cadillac Model "B". This car's most noticeable exterior change was a dummy engine hood that served as a small luggage compartment. To make room for the dummy hood, the Model "B"'s radiator was repositioned into a vertical position. The 1904 Cadillac Model "B" Runabout shown here was priced at $800, or $50 more than the companion Model "A" two-seater.

1904

The brash, young Cadillac Automobile Co. never failed to exploit any opportunity to publicly demonstrate the strength and stamina of its solid little single-cylinder cars. In this convincing display of the car's useable power, a Model "A" Cadillac hauls a load of 14 passengers up a ten percent grade. This photograph was taken in April, 1904 on the Shelby St. hill near downtown Detroit. The driver is industry pioneer Alanson P. Brush who was associated with Cadillac at the time of its creation in 1902.

This advertising cut shows clearly the mechanical details of the simple 1904 Cadillac Model "A" light passenger car chassis. This chassis, with angle iron frame, was used under all three Model "A" bodies, which could be quickly removed and changed without disturbing working parts. Standard running gear included Whitney roller drive chains of one-inch pitch.

Long, arduous cross-country motor tours were important in proving the dependability of the newfangled horseless carriage, and for gaining public acceptance for this revolutionary new mode of personal travel. Early automobile clubs staged such runs as effective means of lobbying for more and better public roads in the U.S. This 1904 Cadillac Model "B" Touring Car pauses during the 1904 World's Fair Tour from New York to St. Louis. Cadillac took first place and a first-class certificate in this competition, for cars of 10 to 80 horsepower and ranging from $1,000 to $6,000 in price.

For 1904, the Cadillac Automobile Company broadened its product range with the addition of a light delivery car, available on either the Model "A" or the new Model "B" chassis. The Model "A" commercial car shown here carried a price of $850. Cadillac found a good market for its well-engineered delivery car. This one is equipped with lamps and a C-type semi-open cab with large oval side windows.

Cadillac also offered a light delivery car in its new "B" Model series. This car varied considerably in appearance from the companion Model "A" commercial car. Price was $900. Cadillac advertised the new Model "B" Delivery as.. "Just meets the requirements of the merchant who is tired of using three men and as many horses for the work that can be done better by one man and a Cadillac".

At first glance, the Model "B" chassis appears identical to that of the lower-priced Cadillac Model "A". But there were significant differences. For one, the radiator of the Model "B" was repositioned to a new, vertical position under an extended front platform that carried an imitation engine hood that served as a luggage compartment.

1905

Late in 1904, Cadillac announced its entry into the upper end of the U.S. automobile market, a segment of the industry the company was destined to eventually dominate. In December, 1904 Cadillac released details of its long-rumored new four-cylinder car. This impressive new model was called the Cadillac Model "D", and it would be the company's premium product for the 1905 model year.

Introduction of the new Model "D" marked a sharp and significant swing by Cadillac away from its own well-established vehicle market, namely that for moderately priced single-cylinder cars. The introduction of the luxurious new Model "D", however, would in no way cut into Cadillac's sales of single-cylinder cars. For 1905, Cadillac would offer its customers a choice of no less than three separate series of one-cylinder automobiles. These were designated Models "B", "E" and "F". Cadillac's tremendously successful single-cylinder cars were all powered by the same horizontally-mounted engine of "square" dimensions (five inch bore and stroke), which for 1905 was rated at nine horsepower.

The 1905 Cadillac Model "B" was a carryover of the highly successful car introduced the previous year. The new Model "E" and "F" passenger cars were much improved successors to the very first Cadillacs which had been introduced three years earlier. The pleasingly styled new "E" and "F" models had conventional type louvered engine hoods. The "B" retained its dummy hood. But the spotlight at the 1905 National Automobile Show in New York was on the impressive new Cadillac Model "D".

The big Model "D" Cadillac for 1905 was offered in only one body style, a superbly engineered five-passenger, side entrance touring car. Built on a 100-inch wheelbase, Cadillac's new flagship car was powered by a Leland-built in-line four-cylinder engine of 301 cubic inch displacement. It was rated at 30 horsepower — more than three times as powerful as the company's popular single-cylinder models. The Model "D" boasted a top speed of 50 miles an hour. The wood-framed body sported small doors for elegant entry into the rear seats. For an extra $250, buyers could specify a body with some weight-saving aluminum panels. A top was still an extra-cost luxury.

The all-new Model "D" Cadillac carried a base price of $2,800 F.O.B. the Detroit factory. Although much higher than the company's standard products, this price did not immediately place Cadillac into the true luxury market, which consisted of vehicles in the $5,000 and up range, like the Winton, Thomas, Locomobile, Pierce and imported makes.

For the first time, Le Sieur Antoine de la Mothe Cadillac's coat of arms appeared as an insignia on Cadillac's cars. This would not become a registered trademark until the following year.

Cadillac was now the largest single producer of motor vehicles in the world. In August the busy Detroit plant turned out its 8,000th car. During 1905, Cadillac built 3,712 cars. This total included 3,556 single-cylinder cars and 156 Model "D"'s. Late in 1905, Henry M. Leland ordered for himself a special closed car on a single-cylinder chassis. This historic vehicle was actually a 1906 model and is described in the following chapter.

In October, 1905 the complex consolidation of the former Cadillac Automobile Company and the Leland and Faulconer Manufacturing Co. was finally completed. Henry M. Leland was named the first general manager of the Cadillac Motor Car Company at a salary of $750 per month. Mr. C. A. Black held the title of president. With approximately 600 employees at the Cadillac plant at Cass and Amsterdam Streets and another 400 at Leland and Faulconer, the company now had a work force of more than 1,000.

At the Fifth Annual National Automobile Show, still the largest and most important annual exhibition of its kind in the nation, various auto manufacturers displayed 177 gasoline-powered cars, 31 electrics and four steam cars. The American automobile industry that year produced more than 24,000 cars bearing 76 nameplates. The Society of Automotive Engineers was also formed this year.

Detroit's picturesque Belle Isle was a popular place for motor outings during the first decade of this century. This photograph was taken there in 1906. The car is a 1905 Cadillac Model "B" Touring Car. The little one-cylinder tourer is hauling five passengers. Not bad for nine horsepower! Note the optional driving lamps.

With close to 2,000 sales in its first year on the market, Cadillac wisely carried its popular Model "B" into 1905 with minimal changes. Wheelbase of this single-cylinder car was 76 inches. This is the 1905 Cadillac Model "B" Touring Car with detachable tonneau. A rear-entrance model, this car carried a price of $900. It was available in a standard Brewster Green finish, with an optional choice of maroon. Designated a "convertible", the Model "B" could be easily converted to a two-passenger runabout by lifting off the rear body and changing the drive sprockets.

1905

Now known as the Cadillac Motor Car Co., Cadillac entered its third model year with a greatly expanded range of products. New for 1905 was the Cadillac Model "E", a thoroughly improved single-cylinder light runabout. The price, however, remained at a bargain $750. Standard finish was Brewster Green with Primrose running gear. The individual seats were upholstered in hand-buffed black leather. Unlike its direct predecessor, the little Model "A", the Model "E" Runabout was not convertible into a four-passenger car.

This 1905 Cadillac Model "B" Detachable Tonneau Touring Car at the side of the road appears to be in a park setting. Presumably, this couple is just checking directions rather than experiencing any mechanical trouble. For 1905, the single-cylinder Model "B" continued to sport a dummy hood that doubled as a small luggage compartment. The radiator is visible below the hood. The four-passenger "B" tourer could be converted to a two-seat runabout simply by undoing four screws and sliding the rear tonneau off. This car had a standard weight of 1,450 pounds. The lamps were optional.

Cadillac's Model "E" Runabout, when equipped with a buggy type top, was aimed at the professional market. In fact, Cadillac actually advertised this car as the Physician's Light Runabout, or the "Doctor's Delight". Equipped with a rubber top, sides and storm apron the Physician's Light Runabout carried a basic price of $780 F.O.B. Detroit. Similarly equipped with a leather top, sides and apron the price went up to $800. Even today, it would appear that Cadillac retains the favor of the American medical profession, judging from the number of Clark Avenue products visible in any hospital doctors' parking lot.

All decked out for a Sunday drive, this gowned and capped young lady looks right at home behind the wheel of her light-colored 1905 Cadillac Model "E" Runabout. Cadillac promoted the Model "E" Runabout as a light, powerful machine of "the semi-racer type". Wheelbase was 74 inches. The new "E"-model featured a conventional style hood with louvers. Top and lights were still optional extra-cost equipment.

This 1905 Cadillac Light Runabout, in concours condition, is the prized possession of Cadillac fancier George Pickering, of West Bloomfield, Mich. The 65-year-old single-cylinder Cadillac is still capable of its original performance. Note the acetylene generator for the optional gas headlights. Mr. Pickering also owns a 1912 Cadillac in original condition.

1905

Idling at the curb, a Cadillac Model "F" side-entrance touring car prepares to go out for a run. Notice the ornamental brass handrail on the dash, and the horn mounted between the newly-styled engine hood and the flared front fender. Lamps were still considered a luxury. This Lazarnick photograph was taken in New York City in 1905, when the car was still brand new.

Although small in physical size, Cadillac's superbly engineered little one-cylinder cars quickly acquired an enviable reputation for strength and reliability. Consequently, single-cylinder Cadillacs were soon performing all sorts of jobs all over the country. The U. S. Government was an early user of Cadillac motor cars. This 1905 Model "F" Cadillac chassis has been equipped with a light duty wagon body for service with the U.S. Army Signal Corps at Fort Leavenworth, Kansas. Note the brass swivel spotlight mounted on the dash and the hang-on tail lamp.

Here is how the 1905 Cadillac Model "F" Runabout looked when it was equipped with the optional buggy top. This photograph was taken in New York City in 1906. The snout of a Model "B" Cadillac is visible just behind this car. These tops were rather bulky affairs, but they fulfilled their intended purpose well. This car sports a set of optional dash lamps.

Posed in a park somewhere in Vermont in 1907, this is a four-passenger Cadillac touring car in full road trim. This car carries both head and cowl lamps. By now, license plates were required just about everywhere. Headlamps of this period burned oil or acetylene and were still considered something of a luxury. Electric lights were still a few years away.

This is how the optional cape cart top looked when folded down. This photograph was taken in a New York City park. The top design utilized three tubular metal top bows. These tops were offered in a choice of rubber or leather, complete with apron and side curtains. The optional horn is mounted between the hood and right front fender with the bulb clamped to the steering column. Note the tire chains, even though the weather appears relatively warm and dry.

1905

With the introduction of its new Model "F" for 1905, Cadillac now offered its customers no less than three series of single-cylinder cars in Models "E", "F" and "B". The "E" and "F" Series Cadillacs were totally new for 1905. This is the Model "F" Runabout, which is distinguishable from the Model "E" Runabout in that it has a single seat instead of the bucket-type seats standard on the "E" Runabout. Lamps and top were available at extra cost. The new Model "F" Cadillac sported a conventional engine hood. The Model "F" rode on a 76-inch wheelbase, two inches longer than that of the companion Model "E".

Product endorsement, it would appear, begins at home. In its 1905 product brochure, Cadillac boasted that it had had five Cadillac commercial cars in service at its Detroit factory for the previous year doing the work of 15 horses. When the company announced its all-new Model "F" for 1905, a delivery body was included in the model lineup. Standard color was maroon with black trim. Price was $950 F.O.B. Detroit. The Model "F" Commercial Car weighed about 1,400 pounds and rode on the same 106-inch wheelbase as all other Model "F" Cadillacs. The semi-open "C" cab was still standard on light duty delivery trucks of this era.

Cadillac advertising at this time emphasized mechanical and construction features. Automobile buyers at the dawn of the auto age were both demanding, and knowledgeable. This is the new Model "F" Cadillac chassis, which was available in the stripped-down form shown here for $800 F.O.B. the Detroit factory. All 1905 Cadillacs were powered by the same horizontally-mounted, single-cylinder Leland-built engine of "square" five inch bore and stroke rated at nine horsepower. The Model "F" chassis featured a new pressed steel frame and tubular type radiator. Gasoline capacity was seven gallons, water capacity three gallons.

The 1905 Cadillac Model "F" Touring Car shown here was the result of strong demand for the popular Model "B" Surrey the company introduced the previous year. The convenient side-entrance body style was gaining in popularity on the old standby rear-entrance design. The fixed-body Model "F" Tourer had rear seat doors only. Price was $950 F.O.B. Detroit. Weight was 1,350 pounds. Standard color was deep Brewster Green with primrose running gear. The Model "F" Touring Car had an over-all length of nine feet, four inches and stood five feet, four inches high. Upholstery was hand-buffed black leather.

For those who still preferred this body style, Cadillac for 1905 offered a rear-entrance tonneau type body for its new Model "F". The detachable tonneau was designated a "convertible", inasmuch as this car could be quickly and easily converted into a two-seat runabout simply by unfastening a few screws and sliding off the rear body. This car was a companion to the non-convertible Model "F" Touring Car. Wheelbase for both models was 76 inches.

1905

Late in 1904, Cadillac released details of its long-rumored four-cylinder car. It was the 1905 Cadillac Model "D", a premium model five-passenger touring car powered by a 30 horsepower engine with 4-3/8 bore and five-inch stroke. The standard Model "D" tipped the scales at a ponderous 2,600 pounds — twice that of its single-cylinder companions — but was available in stripped-down form at 2,000 pounds. Running boards were standard. The impressive Model "D" Cadillac rolled quietly along on 34-inch wood artillery wheels. Standard tires were 34 x 4½ Dunlop pneumatics.

At first glance, this looks like a British or European-built car of the early 1900's. But it is a 1905 Cadillac Model "D" five-passenger touring car. The spare tire lashed to the side is no frill: blowouts were still a fact of life to anyone driving a car with pneumatic tires. Note the broad radiator and the relatively low hood line, and the flared front fenders. The Model "D" was powered by a four-cylinder, L-head engine rated at 30 horsepower — more than three times the output of the little one-lunger that powered Cadillac's other models.

Cadillac's luxurious Model "D" was a big, heavy passenger car. The $2,800 five-passenger tourer all but dwarfed the company's single-cylinder cars in comparison. Stripped of its front fenders, this 1905 Model "D" was a participant in a Triple "A" tour that year. Note the lofty perch of the rear seat passenger. Doors were supplied only for the rear seat. The side-entrance feature was considered a major advance over the rear-entrance tonneau body style.

Cadillac's most important announcement for 1905 was its dramatic entry into the four-cylinder automobile market. The company continued to sell no less than three separate lines of its popular single-cylinder cars, but with the introduction of the four-cylinder Model "D" in December, 1904 Cadillac entered the upper end of the market, for good. The big Model "D" Cadillac was available in only one body style, a five-passenger side-entrance touring car. Wheelbase was 100 inches, and the base price was $2,800 F.O.B. Detroit.

As expected, the all-new Model "D" stole the spotlight at the Cadillac Motor Car Company's exhibit at the Fifth Annual National Automobile Show in New York early in 1905. In the foreground is a sectioned chassis showing off the new Model "D"'s four-cylinder engine. Behind it, Models "D", "F" and "B" are visible in the crowded Cadillac display.

1906

This was to be Cadillac's biggest year ever for sales of single-cylinder cars. The Cadillac Motor Car Co. boasted that there were nearly 14,000 one-cylinder Cadillacs in use throughout the world, and that production of single-cylinder Cadillacs alone exceeded that of any three other makes combined.

But after three years, the little single-cylinder Cadillac was beginning to show its age. Four-cylinder cars were steadily increasing in popularity and the public was demanding more convenience and sophistication in its automobiles.

For 1906, Cadillac went to market with four series of cars, two each in single and four-cylinder models. These included the one-cylinder Model "K" and "M", and considerably more expensive offerings in the four-cylinder Models "H" and "L".

Also in 1906, Cadillac for the first time offered its customers closed body styles. From 1906 until 1910, the Cadillac Motor Car Co. purchased its closed bodies from Seavers and Erdman of Detroit. The first Cadillac to carry a closed body was a special single-cylinder car designed and built especially for Henry M. Leland's personal use. Actually constructed late in 1905, this towering two-seater soon became a familiar sight on Detroit streets. It was five inches taller than it was long. Nicknamed "Osceola" after a Seminole Indian chief, it was driven by Mr. Leland for many years before being retired to the family's summer home near Kingsville, Ontario. This car was exhibited at the 1933-34 Chicago World's Fairs and was on display in Chicago's Museum of Science and Industry for 20 years.

In 1953, the Leland family donated this historic vehicle to the Detroit Historical Museum where it is still on display today.

Cadillac's 1906 single-cylinder cars continued on a 76-inch wheelbase. These cars were officially rated at 10 horsepower, but the Leland-built single-cylinder engine of five-inch bore and stroke consistently produced more that that. It was Cadillac policy to deliberately underrate engine output. Single-cylinder car prices began at $750 for the standard two-seat runabout and ranged up to $950 for the Model "M" light touring car and delivery car.

The company's advertising that year included yet another testimonial to the one-cylinder Cadillac's strength. A Los Angeles doctor originated a stunt called "Leaping the Gap" in which a stock model "K" runabout jumped from ramp to ramp without damage to axles or suspension. The sporty little Model "K" was available as a two-seat runabout with or without top. The companion Model "M" was a nicely styled four-passenger light tourer.

Cadillac's new Model "H" passenger cars were powered by a four-cylinder engine rated at an advertised 30 horsepower. The top-line Model "L" Cadillacs were advertised as 40-horsepower cars. The Model "H" series included a two-passenger runabout at $2,400 ($2,450 with top); a five-passenger touring car at $2,500 ($2,625 with optional top), and a stylish two-passenger coupe that sold for $3,000. The regal Cadillac Model "L" was offered in only two body styles, as a five or seven-passenger touring car with a base price of $3,750 ($3,900 with optional top) or a formal open-front limousine which sold for a hefty $5,000.

The Cadillac Motor Car Company proudly boasted that its Detroit factory was the largest and best-equipped automobile factory in the world. The plant produced 3,559 cars in calendar year 1906.

This was also a milestone year for Cadillac corporate identity. The flowing Cadillac script began to appear on the radiators of some of the company's 1906 models. The previous year the company had applied for registration of a trademark incorporating La Mothe Cadillac's family coat of arms. This was granted on August 7, 1906. The coat of arms was divided into two parts — a four-panel shield topped by a seven-spiked coronet. These symbols were enclosed in a graceful laurel wreath. Cadillac had used a decal emblem of this design as early as 1904. In modified form, these emblems are still used on Cadillacs to this day.

Cadillac's competitors were not standing still, either. The talk of the shows this year was the six-cylinder engine. Pierce-Arrow and National exhibited luxurious six-cylinder cars. Even the Ford Motor Co. came out with a premium-priced six-cylinder car. Ford's new Model "K" was priced at $2,500 and posed direct competition for Cadillac's Model "H" cars. Reo had introduced a new four-cylinder car, and Maxwell built a 12-cylinder racer.

The first completely enclosed car to bear the Cadillac name was this truly historic two-passenger coupe, which was designed and built for Henry M. Leland's personal use and to determine the production feasibility of closed cars. Constructed late in 1905 on a 1906 Cadillac single-cylinder chassis, it was nicknamed "Osceola" by its proud owner after a Seminole indian chief. Finished in black and blue, it was a familiar sight on Detroit streets for many years. Resembling a telephone booth on wheels, "Osceola" stood a towering 87 inches high — five inches taller than it was long. More than once factory personnel were called to right this vehicle after it had tipped over, usually in snow. At the ripe old age of 87, Mr. Leland poses with his beloved "Osceola" in this photograph taken in 1930.

1906

Cadillac's staple product was still the light, single-cylinder car. This is the 1906 Cadillac Model "K" Light Runabout, a two-seater that carried a price of $750 in standard form. The seat is of the new "tulip" style and has a center divider. Lamp mounting brackets are visible on the dash, but lights were still considered an extra-cost luxury.

After many years of faithful service to its owner, Henry M. Leland retired "Osceola" to the family's summer residence near Kingsville, Ontario where it was used to run errands. The first closed-body Cadillac was exhibited at 1933-34 Chicago World's Fairs and was later on display at the Chicago Museum of Science and Industry for 20 years. In 1953 the Leland family presented this car to the Detroit Historical Museum, where it is still on display today. Seated in "Osceola" in this 1928 photograph taken at Kingsville is Mr. Leland's granddaughter, Miriam Leland Woodbridge.

This is the 1906 Cadillac Model "K" Light Runabout as it appeared when equipped with the optional buggy top. When ordered with a rubber top, the base price of the two-seater runabout went up $30 to $780. A leather top was a $50 option, pushing the base price of the Model "K" Runabout when thus equipped up to $800.

These gentlemen are showing off a new 1906 Cadillac Model "K" Light Runabout. This car carried a standard price of $750 — precisely the same price as the very first Cadillac runabout which had been introduced three years earlier. The dash lamps were not standard equipment. The sloping rear deck included a small compartment for tools.

Single-cylinder Cadillacs were rugged, superbly-engineered cars. It was no surprise that some of these little one-lungers outlasted luxury makes costing four or five times as much. This 1906 Cadillac Model "K" chassis with two-seat runabout body was photographed sometime in the mid-1920's. Note the Cadillac script on the radiator. This was the first year the script nameplate appeared on the company's cars.

1906

Factory brass didn't hesitate to pose for advertising photos. That's old Henry M. Leland himself perched in the rear seat of this 1906 Cadillac Model "M" Light Touring Car. This was perhaps the most stylish of all of the company's single-cylinder cars.

This is a catalog illustration of the 1906 Cadillac single-cylinder chassis as used under the company's Models "K" and "M". Wheelbase was 76 inches. The engine was essentially the same as that used in the very first Cadillacs built four years earlier. For 1906, output was rated at 10 horsepower. A new cylinder cost only $4.50 and a replacement piston $3.50. Bore and stroke remained at five inches. Cadillac advertising stressed the principle of interchangeability and the company boasted that all of its engine components were manufactured using limit gauges. These parts were rejected if they exceeded prescribed limits specified to one-thousandth of an inch.

For 1906, Cadillac also offered a side-entrance touring car in its new Model "M" series. This handsome tourer carries a gracefully styled body of the new "tulip" design. Formally identified as the Model "M" Light Touring Car, this four-passenger vehicle carried a price of $950 F.O.B. the Detroit factory, minus lamps. Wheelbase was 76 inches and the single-cylinder car was rated at 10 horsepower.

This is another view of a 1906 Cadillac Model "M" Light Touring Car photographed on an outing. The driver appears to be a liveried chauffer. This Model "M" Cadillac is equipped with head and cowl lamps, still optional equipment on most Cadillacs. The Model "M" Tourer seated four passengers and was of the popular side-entrance style.

Proudly posed in front of the camera is a spanking, new 1906 Cadillac Model "M" Light Touring Car. This car is about to set out on a tour, a popular Sunday pastime in this era. This was a lot of car for a single-cylinder, 10-horsepower engine to haul around but the Leland-built one-lunger did it with ease. Cadillac engines were always underrated, consistently delivering considerably more power than advertised.

1906

Whoever lettered the sign carried on this car got a little ahead of themselves. The 1895 date shown precedes the organization of the Cadillac Automobile Co. by at least seven years and this model did not appear until 1906. The row of shiny new Fords in the background would indicate that this photograph was taken just after the Second World War. Nevertheless, it is a tribute to the dependability of the single-cylinder Cadillac. This two-seat runabout carries a locally improvised utility box on its rear deck.

Chugging up a mild grade on the 1906 Glidden Tour is a 1906 Cadillac Model "M" Touring Car. This car has been equipped with a surrey top, head and cowl lamps. Such tours did much to illustrate the need for more and better roads to accommodate a rapidly expanding vehicle population in America.

The Cadillac Motor Car Company continued to offer a light delivery car. This is the 1906 Model "M" commercial car, which carried a standard price of $950 F.O.B. Detroit. The driver was well protected from the weather in a "C"-type cab, but the small circular side windows afforded little visibility. These light delivery cars had double rear doors and a padded bench-type seat.

As the American automobile market continued to grow, Cadillac's product offerings continued to proliferate. For 1906, the Cadillac Motor Car Co. introduced a new premium series that carried the designation of Cadillac Model "H". This expensive new series included both open and closed body styles. This is the 1906 Cadillac Model "H" Runabout which carried a basic price of $2,400. This hefty price tag still did not include lamps.

Included in the all-new Cadillac Model "H" lineup was a large touring car. The 1906 Cadillac Model "H" Side Entrance Tourer carried a standard price of $2,500 which did not include the lamps shown. This same Model "H" chassis was available with a runabout body for $100 less. All prices were still quoted F.O.B. the Detroit factory.

1906

The most expensive car in Cadillac's new Model "H" series was a closed car, a rather aristocratic-looking coupe. The stately Model "H" Coupe came complete with bevelled window glass and rear window draperies but no lamps. At $3,000 F.O.B. Detroit, this was an expensive car in its day. This was the first year that Cadillac would offer its customers a choice of both open and closed cars.

Cadillac's most expensive 1906 models were to be found in yet another completely new series designated the Cadillac Model "L". The new top-line "L" models were powered by an improved four-cylinder engine that had been uprated to 40 horsepower. This engine had a five-inch bore and stroke, the same as the company's now aging single-cylinder cars. This is the 1906 Cadillac Model "L" Touring Car for seven passengers. The big "L" tourer was also available in a five-passenger configuration without the rear facing auxiliary seats. Built on a 110-inch wheelbase, this luxuriously appointed touring car had a base price of $3,750 not including lamps. With a top, the price went up to $3,900.

The name of Cadillac has long been synonymous with chauffer-driven limousines. The company's sleek Fleetwood Limousines and seven-passenger sedans continue to dominate this relatively small but highly prestigious segment of the market to this day. Cadillac introduced its very first limousine in 1906. It was the tall, oppulently appointed and very expensive four-cylinder Model "L" Limousine. Five passengers could ride in coddled comfort inside and two more could brave the elements on the open front seat. Upholstery was of best Morocco leather and the ceiling was finished in satin. Equipment included electric lights, an electric signal bell and speaking tube. The French plate glass windows could be lowered. This impressive limousine carried an equally impressive price of $5,000. Wheelbase was 110 inches.

Photographed at the company's exhibit at the 1906 National Automobile Show in New York, this is a Model "L" four-cylinder Cadillac chassis. At its big exhibit at this show, Cadillac showed four series of cars — Models "K" and "M" one-cylinder models and the four-cylinder Models "H" and "L". This would be the company's biggest year for single-cylinder cars. Six-cylinder cars were the rage by now and were offered by such makes as Pierce-Arrow, Ford and National.

1907

Long before he entered the infant automobile industry, Henry Martyn Leland had acquired an almost legendary reputation within the American machine tool industry as a true master of precision. Trained in the highly disciplined and demanding requirements of the firearms industry, the Yankee toolmaker was convinced that complete interchangeability of components was essential to product dependability and was the real standard of production quality.

In 1907, the far-sighted and progressive Mr. Leland imported the famous Johannsen, or "Jo-block" gauges from Sweden. These gauges, used in the setup of production machinery, eventually made possible the complete interchangeability of parts which Mr. Leland saw as the key to success in the young automobile industry. The gauges were the invention of a Swedish-American toolmaker named Carl Edward Johannsen. Dissatisfied with the gauges in use in Sweden, Johannsen developed his own. He publicly announced his system in 1906 and was seeking a sponsor in the U.S. Mr. Leland was years ahead of the rest of the auto industry in adopting them for use in his plants. Five years later Mr. Johannsen set up a plant in the U.S. to commercially produce his "Jo-block" gauges.

The famed "Jo-blocks" are not to be confused with the "Go-Not Go" gauges which had been in use at Leland and Faulconer for some time. Limit gauges used in the L & F machine shops were constantly checked against highly accurate master gauges which were correct to one one-hundred thousandth of an inch.

To minimize production delays, these gauges were designed to permit plant workers to make quick, precise dimensional checks without having to actually measure the part. It was simply "Go" or "No Go" when the part was put to the gauge test.

For 1907, Cadillac again offered four series of cars, two single-cylinder lines and two four-cylinder series.

While the little one-cylinder Cadillac had enjoyed its best year ever in 1906, the design was obviously becoming dated and was quickly being eclipsed by competitive multi-cylinder cars. The one-lunger would be around for another two years but the writing was already on the wall. Cadillac's engineering and design staffs were already hard at work on a successor.

The single-cylinder Model "K" was available in only one body style this year, a two-seat runabout that was priced at $800. The price had increased $50 for the first time since Cadillac introduced its first one-cylinder runabout in 1903. The companion Model "M" was available in no less than three touring car variations — a straight-lined standard tourer, a Victoria and as a light touring car with folding rear tonneau. There was also a Model "M" coupe and light delivery car.

Cadillac came out with its new Model "G" four-cylinder line to meet a demand for what it called "a thoroughly high grade, medium-powered and medium-sized automobile at a price somewhat lower than the large touring cars".

The all-new Model "G" was offered in runabout and touring car body styles, with or without a top. The sporty "G" runabout and touring car were both priced at $2,000. The Model "G" was a direct shaft-drive car, without complex chains and sprockets, and was conservatively advertised as a 20-horsepower car. Wheelbase was 100 inches.

The company's most prestigious products for 1907 were the refined Model "H"'s. These were luxurious four-cylinder cars rated at 30 horsepower. Only two body styles were offered, a big touring car and a formal open-front limousine. The touring car was priced at $2,500 while the stately limousine carried a base price of $3,600. The premium Model "H" Cadillacs were built on a 102-inch wheelbase.

In another unsolicited testimonial to Cadillac engineering, a Seattle owner named F. G. Plummer set a new nonstop driving record of 1,000 miles in his single-cylinder car. The run took 71 hours and 32 minutes through Northwestern rain and snow.

Six-cylinder cars continued to gain in popularity at the Seventh Annual National Automobile Show. The man credited with much of the design of the first Cadillac in 1902, Alanson P. Brush, had left to organize a new company in Pontiac, Mich. to build a two-cylinder car also of his own design called the Oakland. Buick had developed its own four-cylinder engine and a Hewitt limousine became the first production U. S. car powered by a V-8 engine.

But all was not well at the Cadillac Motor Car Co. The company's sales had plunged 37 per cent. Something had to be done. The solution, a totally new four-cylinder car, was still a year away. Work on this new model began in earnest in mid-1907.

The single-cylinder runabout remained one of Cadillac's most popular models. This is the 1907 Cadillac Model "K" Two-Seat Runabout, which was priced at $800 F.O.B. Detroit, or $50 higher than the previous year. A stripped Model "K" chassis with dash and hood was available for $750. Lamps were still not included in the base price. Three choices of tops were optionally available. The runabout body could also be purchased separately, for $75.

1907

This 1907 Model "K" Cadillac Runabout is equipped with the stylish optional full-leather "Victoria" top. This top, with outside bows and small side windows, cost $100 extra. A rubber buggy type top cost $40 and a leather buggy top $70. The large headlamps and small cowl lamps were also extra-cost options obtained through the dealer.

In this scene clipped from a 1950's movie, a 1907 single-cylinder Cadillac beats a hasty retreat from the broom-wielding lady in the background. The car is a Model "K" two-seat runabout with optional folding top. It carries 1914 New York license plates and is in excellent original condition. Note the Cadillac script on the lower radiator.

This unique Cadillac model was a true "convertible". As the name implied, the Folding Tonneau model featured a rear seat that could be neatly collapsed and folded away when four-passenger carrying capacity was not needed. The 1907 Cadillac Model "M" Folding Tonneau was priced at $1,000. In the cut at the upper right, this car is illustrated in its two-passenger configuration.

For 1907, Cadillac offered two distinctly different touring car body styles. This is the elegantly-styled Model "M" Victoria Light Touring Car, which carried a price of $950. A cape cart top added $100 to the price. Cadillac's Victoria bodies were finished in a rich wine color called Purple Lake with fine red pinstriping. This was the finest of Cadillac's single-cylinder models.

Here is an interesting Leland family portrait. That's old Henry M. Leland himself behind the steering wheel. Seated on the front seat with him is his son, Wilfred C. Leland. Henry M.'s grandson Wilfred C. Leland, Jr. is seated in the rear. The car is a 1907 Cadillac Model "M" Victoria Tourer for four passengers. The photo was taken on a winter's day sometime in the late 1920's.

1907

This is the other touring car body style offered on the 1907 Cadillac Model "M" single-cylinder chassis. This style was referred to as the "Straight" Light Touring Car. The price was $950 — the same as for the gracefully curved Victoria Light Tourer. A top was $100 extra. Standard one-cylinder Cadillac bodies were finished in Brewster Green with what was described as a hairline red stripe. Running gear was red with fine black striping.

The most expensive single-cylinder Cadillac was the two-passenger coupe. This is the 1907 Model "M" Coupe, which was priced at $1,200. The prototype for this model was "Osceola", a custom-built coupe constructed especially for Henry M. Leland on a 1906 Cadillac single-cylinder chassis. The lamps shown cost extra. Standard tires on these cars were Dunlop or Clincher types supplied by the Hartford Rubber Works, Hartford, Conn.; Morgan and Wright of Detroit, or G. & J. Clinchers made by the G. & J. Tire Company of Indianapolis. Model "K" Cadillacs rode on 30 x 3-inch tires while the Model "M" used 30 x 3-½-inch tires.

For 1907, the Cadillac Motor Car Company introduced an all-new series of four-cylinder cars identified as the Cadillac Model "G". One of the most popular body styles in this new series was the two-seat runabout shown here. This sporty Cadillac carried a price of $2,000. Wheelbase was 100 inches. The moderately-priced Model "G" was rated at 20 horsepower. Lamps and a top were extra-cost options.

This profile view gives a good indication of the Model "M" Coupe's towering dimensions. Despite its ungainly shape, the coupe body has rather graceful lines. This model was to prove very popular with physicians of its day, who still made house calls.

This is the 1907 Cadillac Model "G" Roadster, a three-passenger car with "dickey" seat in the rear. The hardy soul who rode back there had to brave the elements while more fortunate front seat passengers were afforded some protection from the elements under the three-bow buggy top. The Model "G" transmission was of the selective sliding gear type with three forward speeds. It was encased in an aluminum housing.

1907

This is the 1907 Cadillac Model "G" Touring Car. In standard form shown here, the Model "G" Tourer was priced at $2,000. Addition of a Cadillac cape cart top raised the price to $2,120. Doors were still not used for the front seat, and headlamps remained on the options list. This was a five-passenger car built on a 100-inch wheelbase. Standard wheels are of the artillery type of best grade seasoned hickory.

Cadillac sales literature for 1907 described this car as a Model "G" Roadster with Victoria Top. The Model "G" was a new, medium-priced series that bridged the gap between Cadillac's popular single-cylinder cars and the costly Cadillac Model "H". Standard exterior finish was Brewster Green with lighter green running gear trimmed with white hairline striping. Wheelbase was 100 inches.

Illustrated here is the 1907 Cadillac Model "G" Touring Car with the optional Cadillac cape cart top in the raised position. All Cadillac Model "G"'s were powered by a smooth four-cylinder engine which was conservatively rated by the factory at 20 horsepower. As shown, this car carried a base price of $2,120. All lamps were extra.

The all new Cadillac Model "G" was a contemporary, pleasingly styled car that bore a strikingly different appearance from the company's still-popular single-cylinder model. The modern Model "G" featured direct shaft drive rather than sprockets and chains. This 1907 Cadillac Model "G" Touring Car took part in a New York-to-Detroit Endurance Run. Note the mud-caked wheels and fenders. The Cadillac script is embossed in the upper front portion of the radiator shell.

Cadillac's most prestigious models for 1907 were called the Model "H". Only two models were offered, a big touring car and an open-front limousine. The Cadillac Model "H" was powered by a four-cylinder, 30-horsepower engine. This is the 1907 Cadillac Model "H" Limousine which was priced at $3,600 — considerably less than the Model "L" Cadillac Limousine of 1906. The regal Model "H" was built on a 102-inch wheelbase, two inches longer than the lower priced Model "G". The Model "H" Tourer was priced at $2,500. As was the case on all other Cadillac models, lamps were not included in standard Model "H" prices.

1907

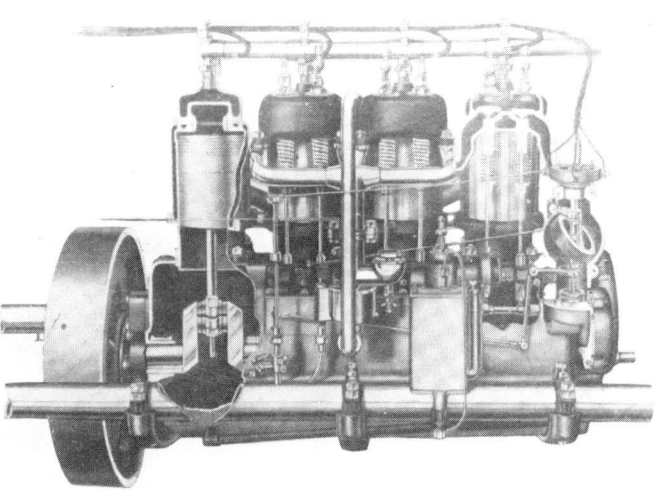

This is a cutaway view of the 1907 Cadillac Model "H" four-cylinder engine. Conservatively rated at 30 horsepower, this engine had a 4-1/8-inch bore and five-inch stroke. The water-cooled engine had a copper water jacket. Note the massive flywheel. This engine was mounted in a pressed steel, channel frame with four-spring suspension. Cadillacs of this series had double-acting drum-type brakes and planetary transmissions.

Racing was still a popular weekend sport. This is a single-cylinder Cadillac chassis stripped to the frame for competition. The driver peers intently ahead while his mechanic provides ballast on the left-hand side of the car. A spare tire is lashed to the rear. There is no indication of how this entry did.

Endurance runs were popular weekend outings in the early years of this century. This 1907 Cadillac Model "M" Victoria Touring Car is taking part in a mid-winter run sponsored by the Long Island Automobile Club. The hardy four-man crew checks in with one of the run organizers. That spare tire is no ornament — blowouts were still a fact of life for all automobile owners.

Automobile rides were still a novelty to much of the population. It was considered a real treat to be invited along for a horseless carriage ride in the early years of our century. Here, a high-spirited group of children fill the rear of a 1907 Cadillac Model "M" Straight Touring Car somewhere in New Jersey. Wonder if the two little girls in the background got to go along.

1908

The year 1908 was destined to be the most important in Cadillac's history. It was during this year that the slogan "Standard of the World" first appeared in Cadillac's advertising. The seven Fisher brothers formed a new company to make automobile bodies, and the legendary William Crapo Durant consolidated a loose group of struggling automobile companies into a new entity which he named General Motors.

But the major event affecting the Cadillac Motor Car Co. that year took place not in the United States but thousands of miles away, in England. Cadillac achieved worldwide recognition when it was awarded the coveted Dewar Trophy in a unique standardization test that was to establish Cadillac as a quality automobile nameplate for evermore.

Cadillac's spectacular achievement in winning this trophy, considered the Nobel Prize of the automotive industry, can be traced directly to the efforts of Frederick S. Bennett of the Anglo-American Motor Car Co. of London. Mr. Bennett, the ambitious young salesman who had imported the first Cadillac into Britain in 1903, sought some form of convincing public demonstration to prove the Cadillac's mechanical merits. Parts interchangeability was unheard of in England and on the Continent at this time. The accepted method of assembly and repair was "file and fit". Bennett was keenly aware of Leland's contempt of hand-fitting of any kind. He prevailed upon the prestigious Royal Automobile Club to oversee a unique standardization test. This test took place between Feb. 29 and March 13, 1908.

Three single-cylinder Cadillac runabouts were selected at random from an Anglo-American shipment on a London dock. The cars were driven under supervision to the new Brooklands Motordrome 23 miles away. There they were taken for 10 quick laps of the new track before being locked up in a garage on the grounds.

Between March 2 and 4, and under careful scrutiny, the three cars were completely disassembled. The 2,163 components were then thoroughly mixed up by a R.A.C. technical committee and piled up in three separate heaps. In a further test of interchangeability, 89 parts were selected and replaced from Anglo-American's dealer stocks. The cars were then put back together again. All three cars started and ran perfectly. The vehicles were then immediately taken out onto the Brooklands track and driven at top speeds for 500 miles. Their average speed was 34 miles an hour. At the conclusion of the run the Cadillacs were carefully examined again. On the basis of this remarkable test, Cadillac became the first American automaker ever to win the Dewar Trophy.

Sir Thomas Dewar had annually awarded this trophy since 1904 to the manufacturer deemed to have made the most significant advance in the state of the art for the previous year. In February, 1909 the large, silver cup was brought to Detroit and proudly displayed in Cadillac's main lobby. Mr. Leland personally shook the hands of more than 1,000 of his employees whom he credited with making this great honor possible. Cadillac had firmly established itself as a world-renowned builder of quality cars.

Cadillac garnered even more international honors that summer. In June, one of the three cars that had taken part in the R.A.C.'s standardization test won its class in the Royal Automobile Club's gruelling 2,000-Mile Trial.

Already something of an auto industry legend in his own right, Billy Durant in September, 1908 had brought together a group of struggling automobile companies in a new enterprise which he named General Motors. The nucleus of this venture, which Durant envisaged as eventually encompassing every segment of the U.S. automobile market from top to bottom, was made up of Buick and Oldsmobile. The first president of General Motors was William Eaton. Big Bill Durant was already casting envious glances at the Cadillac Motor Car Co.

In other industry news, Henry Ford on October 1, 1908 had launched production of a new low-priced, four-cylinder car he named the Ford Model "T". And an old, established carriage maker, the James Cunningham, Sons & Co. of Rochester, N.Y. embarked on the production of fine motor cars designed for the carriage trade.

Frederic J. Fisher, eldest of the seven Fisher brothers, organized the Fisher Body Co. in Detroit. Fisher built automobile bodies for a number of independent auto manufacturers, including Chandler. This company's association with Cadillac, however, was still several years away.

Cadillac built only 2,377 cars in 1908. The plant was busy tooling up for a completely new car. Announced in August, 1908 it was to be called the Cadillac Model "Thirty".

Production of the single-cylinder Cadillac entered its sixth and final year in 1908. The least expensive model in Cadillac's 1908 product lineup was the two-seat Model "S" Runabout, which was available in the standard "straight line" style illustrated here, or with a gracefully curved "Victoria" body. In either style the basic price was $850 — $50 higher than the previous year. The Model "S" was built on a new 82-inch wheelbase, six inches longer than comparable 1907 models.

1908

Economy runs are nothing new. Such competitions were routinely held 70 years ago. This 1908 Cadillac Model "S" Straight Line Runabout is taking part in what was billed as a "One Gallon Test" in the New York City area. Note the Cadillac script nameplate on the radiator and the rail around the rear deck. This car appears to be equipped with a third seat. The Cadillac Script is also visible on the apron that covers the lap of the dickey seat passenger.

Shown here is the 1908 Cadillac Model "S" Two-Seat Runabout with "Victoria" style body. Standard color scheme for the Model "S" Runabout was a dark blue body and siderails with lighter blue running gear highlighted with a complementing blue pinstripe treatment. Lamps continued as optional equipment. Cadillac now had its own topmaking department. Advertised horsepower remained at 10.

Cadillac's most expensive single-cylinder car was the ungainly-looking Model "T" Coupe. This two-passenger car with completely enclosed body carried a price of $1,350. Just like the prototype "Osceola", which had been constructed to Henry M. Leland's order to explore the feasibility of production closed bodies, the Model "T" Coupe was finished in dark blue and black with blue running gear. Cadillac advertised the 1908 Model "T" Coupe as "Ideal for the Physician, for shopping or for the opera". The tall coupe body could be removed and a light runabout body substituted for summer use.

Two members of the Long Island, N.Y., Automobile Club take a breather during a club run. Their car is a 1908 Cadillac Model "T" Tourer with the standard straight-line four-passenger body. A large, brass Cadillac script nameplate adorns the radiator. The car parked behind it appears to be a rival Packard.

The Cadillac Motor Car Co. continued to offer a light delivery car in its single-cylinder product line. This 1908 model would be the last. Unlike the one-cylinder Model "S" and "T", the Model "M" Delivery Car retained its 76-inch wheelbase. All other single-cylinder Cadillacs this year had 82-inch wheelbases. The Model "M" Commercial Car was priced at $1,000. Standard color was Brewster Green with red running gear.

1908

Cadillacs continued to do well in competition. This is a stripped-down single-cylinder Cadillac chassis in racing trim. The entire rear body has been removed to reduce weight and thus increase top speed. This Lazarnick photograph was marked "Darnstadt Cadillac, Motor Parkway."

This is how the three single-cylinder Cadillac runabouts looked after the mechanics had finished taking them apart on Wednesday, March 4, 1908. All of the components were then thoroughly mixed up and piled in three heaps for reassembly. More than 80 parts were selected for replacement from dealer parts stocks. Frederick S. Bennett of the Anglo-American Motor Co. of London, which had been importing Cadillacs since 1903, cheerfully complied. Cadillac passed the Royal Automobile Club's standardization test with flying colors. The prize was the coveted Dewar Trophy. Named after its donor, Sir Thomas Dewar, this award had been made annually to the company deemed to have made the most significant advancement in automotive engineering. It was recognized everywhere as the Nobel Prize of the auto world. The actual presentation of the trophy did not take place until the following year.

One of the most important single events in the entire history of Cadillac took place in England in 1908. In a unique test of the principle of parts interchangeability, three standard single-cylinder Cadillac runabouts were selected at random from a shipment on a London dock. The cars were driven to a garage at the new Brooklands Motordrome 23 miles away and locked up. Two days later they were carefully inspected by a committee of the prestigious Royal Automobile Club. The cars were then completely dismantled. The more than 2,000 parts were piled in three mixed heaps. The vehicles were then reassembled using the deliberately mixed-up parts. The cars were then driven out onto the Brooklands track and run for 500 miles at top speeds. After the uneventful high speed run, the vehicles were closely examined again. On the basis of this remarkable performance, Cadillac was awarded the R.A.C.'s coveted Dewar Trophy, the first time such an honor had ever fallen to an American-made car. The test was convincing proof of the merit of Henry M. Leland's theory of production accuracy and complete interchangeability of components. Here the three Cadillac runabouts are seen on the Brooklands track prior to beginning the 500-mile run.

Cadillac's largest one-cylinder car for 1908 was the four-passenger Model "T" Touring, which was also offered in "Straight Line" and "Victoria" body styles. In each case, the price F.O.B. Detroit was an even $1,000 without lamps or top. The Model "T" rode on the same new 82-inch wheelbase as the Cadillac Model "S". Standard Model "T" paint was a Brewster Green body with light red striping and red running gear with dark green stripe. Model "T" bodies could be purchased separately for $175 in either style. A bare Model "T" chassis cost $850, the same price as a standard Model "S" runabout.

1908

This is the standard 1908 Cadillac Model "T" four-passenger touring car with "Straight Line" body. Its $1,000 price was identical to that charged for the more stylish "tulip" Victoria tourer. Cadillac's touring cars were of the two-door, side-entrance style. Lamps and a top were optional at extra cost. Model "T" Cadillacs rode on 30 x 3-½-inch tires.

Cadillac's four-cylinder Model "G" entered its second model year in 1908. Changes were minimal. This is the standard Model "G" Runabout which was priced at $2,000. Note the third "dickey" seat on the rear deck. Lights and a top were extra-cost options. The Model "G" cut a dashing image in its standard form. The standard paint job was French Gray with fine red pinstriping and red leather upholstery.

Perhaps the sportiest of all of Cadillac's 1908 models was the racy-looking Model "G" Runabout. Standard color was a light French Gray with red pinstriping and red leather upholstery. Basic price was $2,000. The four-cylinder Model "G" was rated at 20 horsepower and was built on a 100-inch wheelbase. The Cadillac script is visible at the top of the radiator. All lamps were accessories. Sweeping fender lines enhanced the Model "G" Runabout's long, low appearance.

It didn't take much to prepare a Cadillac for competition. Just get out the tools and remove all unnecessary bodywork. This is a 1908 Model "G" Cadillac chassis stripped down to the essentials for dirt track competition. Even the hood has been removed from this car. Fender braces have been left on, however. This car carries a Cadillac script nameplate on its radiator core. The Cadillac script is also embossed into the top of the radiator shell. The license plate is 1908 New Jersey.

This is the standard Cadillac Model "G" Touring Car for 1908. This side-entrance touring car carried a price of $2,000 and came without lamps or a top. Standard finish was dark blue body and frame siderails touched off with a fine primrose stripe. Wheels and axles were cream yellow with dark blue striping. Brewster Green paint with red running gear was an optional color choice for this model.

1908

The most costly of the three Model "G" Cadillacs for 1908 was the Limousine. This heavy car was priced at $3,000. The four-cylinder, 20-horsepower Model "G" Limousine came complete with an electric dome light and speaking tube. Standard color scheme was dark blue and black with dark blue running gear accented with light pinstriping. Other Model "G" Cadillac models included a runabout and touring car.

This was one of the more intriguing mysteries to show up in Cadillac's archives. The information with this photo identified it as a 1907 or '08 Cadillac Model "G" Limousine. The rear bodywork is identical to that of the standard Model "G" Limousine, but the chassis is definitely something else. This car has a collapsible landaulet rear roof area and appears to have been outfitted for use as a taxi. Note the taxi meter and roof luggage rack. The car was photographed on the same Detroit rooftop that appears in many Cadillac factory photographs taken in the late 1920's and 1930's.

Topping Cadillac's product line for the third straight year was the luxurious four-cylinder Model "H". Again, this series was available in only two body styles, a heavy touring car and a formal limousine. This is the 1908 Model "H" Limousine, which had a basic price of $3,600 F.O.B. the Detroit factory. The 30-horsepower engine was capable of propelling the big Model "H" at speeds up to 50 miles an hour.

For the third consecutive model year, the heaviest and most expensive cars in Cadillac's product lineup were the luxurious Model "H"'s. Again, only two body styles were offered in this series — a heavy touring car and a companion limousine. This is the big Model "H" Touring Car. The 30-horsepower, four-cylinder Model "H" tourer was priced at $2,500. The standard color scheme was Brewster green with matching running gear. Red running gear with dark green striping was available on special order. This big Model "H" touring car has been equipped with a folding windshield and is a rolling billboard for the Healy removable wheel rims on which it rides.

Parked on a New York City street, this is a huge 1908 Cadillac Model "H" Limousine. The Model "H" Limousine was built on a 102-inch wheelbase. Bore was 4⅜ inches with a five-inch stroke. Standard color for the regal Cadillac limousine was a deep Brewster green with black trim and complementing Brewster green running gear with light pinstriping. Cadillac offered 11 models in 1908 on three different chassis with three engine choices.

1909

Introduction of the all-new Cadillac Model "Thirty" in August, 1908 marked the beginning of an important chapter in Cadillac Motor Car Co. history. The meticulously engineered, moderately priced Model Thirty would be Cadillac's bread-and-butter car for the next six years, bridging the technological gap between the comparatively primitive single-cylinder Cadillac and the industry's first mass-produced V-8.

This was also the year that Cadillac became a division of General Motors.

Design and development work on this important new model had begun in earnest in mid-1907, at a time when sales of Cadillac's highly successful but now dated one-cylinder cars had gone into a steep decline. The company badly needed a new product with which to capture the public's attention and to restore Cadillac to its former position of prominence among U.S. motor vehicle manufacturers. The new Cadillac Thirty was to be that car.

In a single, bold stroke, Cadillac replaced all three of its previous car lines — the popular single-cylinder "S" and "T"; the middle-range four-cylinder Model "G" and the premium, limited-production Model "H" — with a single model available in only three body styles. All three models were built on the same 106-inch wheelbase, and all were powered by the same refined four-cylinder engine.

The engine used in the new 1909 Cadillacs was officially rated at 25.6 horsepower. On the dynamometer, however, this engine easily produced in excess of 30 horsepower. This is where the Model Thirty got its name.

The Cadillac Model Thirty was a moderately-priced car, based on the 20-horsepower, four-cylinder Model "G" of 1908-1908. It was initially offered in only three body styles: a sporty roadster, a "convertible" four-passenger demi-tonneau and a standard five-passenger touring car. All three cars were priced exactly the same in base form — $1,400 F.O.B. Detroit. A pair of oil side lamps, a taillamp and horn were standard equipment. Model Thirty buyers could choose from a long list of optional equipment ranging from tops and windshields, various styles of headlights, Gabriel horns and speedometers.

There was no limousine in the initial 1909 Cadillac Model Thirty lineup. Customers could, however, order a Model Thirty chassis. Priced at $1,350, or only $50 less than a complete Model Thirty roadster, demi-tonneau or tourer, this chassis came with fenders, hood and dash suitable for the installation of a custom body.

The Model Thirty roadster and demi-tonneau featured a cowl extension of the car's hood line. This cowl housed a recessed instrument board and marked a significant step toward streamlining of the automobile body into a more harmonious unit.

The new "Thirty" was an astounding success. Offering exceptional value for the money, 4,500 were sold in the first six months on the market — more than the company's best previous year — and the entire year's production was virtually sold out within a few months of its introduction. A total of 5,900 were built during the 1909 model year.

After two unsuccessful earlier attempts, W. C. Durant succeeded in bringing Cadillac into his recently-organized General Motors organization. After the initial offer and the well-publicized Dewar Trophy win, the company's price had gone up $1 million. Durant's Buick division paid $5.6 million for Cadillac — $500,000 in cash and the balance in General Motors preferred stock. One of the conditions of the sale, imposed by Henry M. and Wilfred C. Leland, was that the existing management be allowed to run Cadillac as if they still owned it. Durant unhesitatingly agreed.

In July, 1909 a brilliant young inventor named Charles F. Kettering signed an order with Cadillac for 8,000 auto ignition sets. Mr. Kettering had formed his own company to develop and make the auto ignition system he had designed. This company, based in Dayton, Ohio, was called the Dayton Engineering Laboratories Co. Within a few years the company's acronym — Delco — would be known around the world and the company would become an important research and scientific component of General Motors.

Also that year, on April 1, Harry C. Urich founded the Fleetwood Metal Body Co. in Fleetwood, Pa. Successor to the former Reading Metal Body Co. which had been formed in 1905, Fleetwood would specialize in building high quality automobile bodies. This company was destined to later form a close association with Cadillac.

For the next two decades, Fleetwood produced bodies for many independent automobile manufacturers, among them Alco, Packard, Simplex, Locomobile, Chadwick and Rolls-Royce to name a few. In later years, Fleetwood would become a highly respected designer and builder of classic custom car bodies.

In other automotive industry news that year, the George N. Pierce Co. of Buffalo, N.Y. became the Pierce-Arrow Motor Car Co. The famed "three P's" — Packard, Peerless and Pierce-Arrow — dominated the luxury end of the American automobile market. Cadillac had yet to make its move up into the top echelons of the fine car business.

> In an extremely bold move, Cadillac late in the summer of 1908 replaced all three of its former car lines with a single new series. This new car, patterned after the Model "G", was called the Cadillac Model "Thirty". It was initially offered in only three body styles, a roadster, demi-tonneau and standard touring car. All three models were built on a 106-inch wheelbase and utilized the same four-cylinder engine. This powerplant was officially rated at just over 25 horsepower but it consistently produced over 30 on the dynamometer, hence the Model Thirty name. This is the 1909 Cadillac Model Thirty Roadster which sold in basic form for $1,400. The price included two oil side lamps, a taillamp and horn. It was designed to carry three passengers.

1909

This beautifully-restored 1909 Cadillac Model Thirty Touring Car is in the Craven Foundation collection in Toronto, Ont., Canada. Model Thirty buyers could choose from a long list of optional equipment ranging from Metzger folding windshields and Gabriel horns to folding tops and Stewart and Clark speedometers. While this professionally restored example is finished in a striking white with black trim, standard color on all 1909 Model Thirty Cadillacs was a Royal Blue body with black fenders and radiator shell with light blue pinstriping. Price of the popular Model Thirty touring car was $1,400, the same as for the roadster and demi-tonneau.

Quite similar in appearance to the standard touring car, this is the 1909 Cadillac Model Thirty Demi-Tonneau. By detaching the tonneau from the rear deck, this versatile car could be quickly converted into a sporty runabout. Like the Model Thirty roadster, the demi-tonneau differed in appearance from the touring car in that its instrument board was recessed into a cowl extension of the hood. This was a significant first step toward what would eventually become more streamlined "torpedo" body styling. All three 1909 Cadillac Model Thirties carried the same basic price: $1,400 F.O.B. the Detroit factory. Side lamps, a taillamp and horn were now standard equipment. This was a four-passenger car.

It was not surprising that the well-engineered, moderately-priced Model Thirty Cadillac soon became the darling of the motor touring set. Pennants flying, this Model Thirty touring car with optional windshield heads off on an organized run. Note the Cadillac script name badge on the radiator. This car is being followed by a rival Packard. By this time, Packard had already adopted its classic radiator shape.

Cadillac's new Model Thirty was powered by this four-cylinder, in-line engine of 226 cubic inch displacement. Bore was four inches and stroke four-and-a-half inches. Although advertised at 25.6 horsepower, this Leland-designed and built four easily delivered over 30 horsepower on the dynamometer stand. This is where the Model "Thirty" got its name. The shaft-drive power train utilized in the new Thirty was essentially similar to that of its predecessor, the Model "G" Cadillac of 1908.

1909

1909

The Cadillac Motor Car Co. did not market a Model Thirty Limousine until the 1910 model year. However, customers could order a Model Thirty chassis complete with hood, dash and fenders for $1,350. This was only $50 less than the price of any one of the three 1909 Model Thirties in standard form. It is not known who built this handsome limousine body, but it seems to suit the Model Thirty chassis well. The small semi-doors on the open chauffer's compartment are rather unique.

This is the standard 1909 Cadillac Model Thirty Touring Car as shown in base form in the company's 1909 catalog. This was the only Model Thirty that retained the old buckboard-style dash: the companion roadster and demi-tonneau sported cowls. Headlights were still optional, but a pair of oil sidelamps, a taillamp and horn were made standard this year.

The superbly-engineered Cadillac Model Thirty was an immediate success. This model was soon a familiar sight on roads all over the United States. The bowler-hatted gentleman in this photo is obviously proud of the Model Thirty touring car he is driving. By now the side-entrance tourer had all but replaced rear-entrance cars. The Model Thirty tourer was a five-passenger car.

Here's an interesting unsolicited testimonial to Cadillac dependability. Presumably attired as Le Sieur Antoine de la Mothe Cadillac himself, this gentleman is posing with what looks to be a well-travelled 1909 Cadillac Model Thirty touring car. The standard he bears read: "An Example of Cadillac Dependability". Unfortunately, the exact date and location of this endorsement is not known. One of the basic elements in Cadillac's success has been a large and unusually loyal owner body.

1910

Cadillac's new Model Thirty was a spectacular success in its first year on the market. Despite a much more limited range of models than previously offered, sales of the four-cylinder "Thirty" in its first few months on the market exceeded those of the company's best previous year. The sales slump which had begun two years earlier had been dramatically reversed.

For the 1910 model year, the Cadillac Model Thirty was carried forward with relatively few changes. Wheelbase was increased four inches, to 110 inches on the roadster, demi-tonneau and touring car. Engine output was uprated from 25.6 horsepower as advertised in 1909 to 28.9, although the engine consistently delivered 33 horsepower on the dynamometer. Displacement was increased to 255.3 cubic inches. Bore and stroke were now 4-1/4 by 4-1/2 inches.

The refined 1910 Cadillac Model Thirty was equipped with a Delco ignition system. Supplied to Cadillac by Charles F. Kettering's Dayton Engineering Laboratories Co., the new Delco ignition was used in conjunction with a conventional magneto. True self-starting was still two years away.

At long last, a pair of acetylene headlights and gas generator were made standard equipment on all Cadillacs. Also standard now were a pair of side oil lamps, a taillamp, horn, a set of basic tools, a tire pump, tire repair kit and tire holder.

Model Thirty base prices were increased from $1,400 the previous year to $1,600 in 1910.

The standard 1910 Cadillac Model Thirty lineup included a sporty two-seat roadster which was officially designated the "Gentleman's Roadster"; a four-passenger demi-tonneau which could be quickly converted to a two-seat runabout simply by detatching and removing the entire rear body; a popular five-passenger touring car and an all-new Model Thirty Limousine. Late in the year, an inside-drive coupe was added to the line. This handsome closed car carried a base price of $2,200.

With the introduction of the Model Thirty Limousine, the Cadillac Motor Car Co. re-entered a market it had vacated for the previous year. The big Model Thirty Limousine rode on its own 120-inch wheelbase, ten inches longer than the rest of the line. It carried a standard price of $3,000.

Probably the most significant event for Cadillac this year was the company's placement with the Fisher brothers of an order for 150 Model Thirty coupe bodies. Henry M. Leland had visited the Fisher Body Co., which had been formed in Detroit two years earlier, and he evidently liked what he saw. Mr. Leland gave the Fishers an order for 150 closed bodies for Cadillac. This was the first such quantity order for closed automobile bodies in the United States and is generally credited with giving Fisher Body its real start. In December, 1910 the Fisher Bros. operation was renamed the Fisher Closed Body Co.

Since 1906, Cadillac had purchased its relatively limited closed-body requirements from the Seavers and Erdman plant on Jefferson Ave. in Detroit.

While Cadillac continued to prosper, all was not well with its new corporate parent, the recently organized General Motors. In his enthusiasm to build GM into a self-contained, full-line automobile company, William C. Durant had dangerously overextended himself. On September 10 a meeting was called in New York to discuss the dissolution of the company. Speaking on behalf of Cadillac, Wilfred Leland cautioned against this. Consequently, the company was reorganized and Durant was forced out. But this would not be the last General Motors would hear from the irrepressible William Crapo Durant.

The 10th annual National Automobile Show was again held in New York City. A bold, new passenger car body style — the torpedo — made its debut this year. This open-style body featured high sides and a semi-enclosed look that imparted an impression of length and speed.

Most American automobile manufacturers, including Cadillac, were members of the Association of Licensed Automobile Manufacturers, which required them to pay royalties under the famed Selden Patent of Nov. 5, 1895. George B. Selden had obtained this patent for his "road machine" which never went into production.

The Cadillac Motor Car Division built 10,039 cars in 1910, of which 8,008 were 1910 models.

Cadillac's spectacularly successful Model Thirty went into its second model year with an expanded range of models, but only minor mechanical and equipment changes. Base price had gone up $200, from $1,400 to $1,600 in standard form. This is the 1910 Cadillac Model Thirty "Gentleman's Roadster" with deck. A rumble seat was available as an extra cost option.

Shown here is the 1910 Cadillac Model Thirty "Gentleman's Roadster" equipped with a rumble seat. This seat could be easily removed to convert the car into a two-passenger runabout with plain rear deck. A pair of gas headlights had finally become standard equipment. All 1910 Cadillac Model Thirties carried at $1,600 base price, except for the much more expensive new limousine.

As in 1909, buyers of the Cadillac Model Thirty could choose from a wide range of extra-cost options. This is a well-equipped Model Thirty "Gentleman's Roadster". The poor rumble seat passenger, however, was left out in the wet when it rained. Cadillac now had its own top-making department. Standard finish was still Royal Blue with black-enameled radiator shell and fenders.

1910

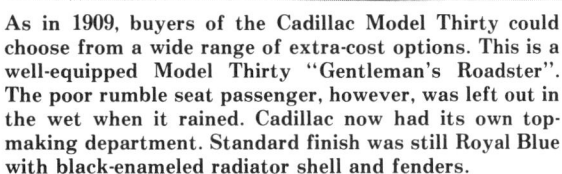

The versatile demi-tonneau continued to appeal to many of Cadillac's customers. As shown here, the Model Thirty Demi-Tonneau served as a comfortable four-passenger car. By simply detaching and removing the rear tonneau it could be quickly converted into a sporty two-passenger runabout or roadster. The windshield and folding top shown on this car were extra-cost options, but lamps had finally been added to the standard equipment list of all 1910 Cadillac models.

The big five-passenger touring car remained one of Cadillac's most popular models. This is the 1910 Model Thirty Touring Car in full road trim, complete with optional folding top. Only the touring car and new limousine retained the old style flat dash. All other Cadillac Thirties featured a more streamlined cowl with recessed instrument board. In standard form the Model Thirty Touring Car represented excellent value for $1,600 F.O.B. the factory in Detroit.

The open rear door affords a glimpse into the posh interior of a 1910 Cadillac Model Thirty Limousine. Cadillac's standard exterior color was still Royal Blue with black fenders and radiator shell, but this car appears to have been fitted with a rear body finished in gray. The high headroom was necessary to permit the occupants to wear top hats to or from work or social events.

After a one-year absence, Cadillac got back into the limousine market in 1910. The stately new 1910 Cadillac Model Thirty Limousine carried a base price of $3,000 — nearly twice as much as the other standard models in the Cadillac Thirty series. Wheelbase of the big limousine was 120 inches, or ten inches longer than that of the roadster, demi-tonneau and touring car. These high, boxy limousines were elegant cars designed primarily for city use. Standard equipment including lights and a speaking tube so the passengers could communicate with the chauffer, who still sat out in the cold.

This is a Cadillac Model Thirty chassis as exhibited at the 10th Annual National Automobile Show in New York City. Wheelbase of the standard Model Thirty had been increased four inches for 1910, to 110 inches. All of the 1910 Cadillacs were equipped with a Delco electric ignition system which was used in conjunction with a magneto. Self-starting was still two years away but adoption of Kettering's Delco system was a significant advance.

1910

Although Cadillac had discontinued the manufacture of light duty commercial cars at the end of the 1908 model year, the sturdy, well-engineered Model Thirty chassis quickly proved to be highly suitable for any number of special purposes. A good example is this special Fire Chief's Runabout. Note the big warning bell mounted on the cowl, the spare tire lashed to the right-hand side and the special equipment compartment that has been installed on the rear deck of this otherwise stock Model Thirty Roadster. It is believed that this speedy vehicle saw service with the New York City Fire Department.

The Cadillac Motor Car Division of General Motors still offered a bare chassis to custom body builders or individuals who wished to install their own. Some of these stripped chassis were exported to overseas customers. This very attractive little runabout carries an Italian-made custom body. Note the louvers in the hood, the complete enclosure of the cockpit and the two spare tires on the rear deck.

Organized meets and durability runs were still popular pastimes for those fortunate enough to own automobiles. The banner affixed to the hood of this Cadillac Model Thirty Demi-Tonneau reads "Montauk Light or Bust! Reliability Contest". This run took the contestants from New York City all the way out to the most easterly point on Long Island and back. The lady in the left front seat is appropriately attired for this long, dusty journey.

Only the words "Management Motor Contest Association" are visible on the banner that covers the hood of this Model Thirty Cadillac Touring Car. Note the pennant tied to the outside of the front seat. Brand loyalties were quite strong, and the owner of this Cadillac obviously wanted to be sure everyone along the way knew he was driving the "Standard of the World".

Cadillac's rugged Model Thirty was at home everywhere, from the paved streets of our sophisticated Eastern cities to the dusty trails and primitive roads of the American west, and overseas. This is a 1910 Cadillac Model Thirty Roadster out on a run. This car has been equipped with an optional folding windshield and carries a holder for two spare tires on its left-hand running board. Pennants were popular accoutrements on organized runs.

This Cadillac Model Thirty Roadster carries some "options" that were not to be found in the company's catalog. The machine gun mounted on the cowl was Government issue. These two Model Thirty Cadillacs were modified for service in the U. S. Army. Note the Cadillac "30" emblem and script on the radiator. An auxiliary seat has been attached to the rear deck. These soldiers look like they mean business.

1911

The 1911 model year marked the third and final year for the Cadillac Model Thirty. The "Thirty" series designation was dropped when the 1912 models were introduced, although essentially the same car, with many improvements, was produced for another three years until the advent of the historic V-8.

For 1911, the Cadillac Model Thirty product lineup was broadened with the addition of two important new models. These included a four-door touring car and a rakish new Cadillac Torpedo. A handsome Coupe had been added late in the 1910 model year, giving Cadillac an impressive range of seven body styles for 1911.

At the lower end of the line was the two-passenger Cadillac Model Thirty Roadster which was available with deck or an optional rumble seat for a third passenger. Next came the versatile Demi-Tonneau, which would be offered this year for the last time. A four-passenger car, the Demi-Tonneau could be easily converted into a light two-seat runabout by detaching and removing the rear tonneau. Next came the standard five-passenger Touring Car. The popular tourer was complemented for 1911 by a new touring car that had doors for both the front and rear seats. In the previous two years, all Cadillac Model Thirty touring cars had rear doors only. This new car was called the Model Thirty "Fore-Door" (meaning front, rather than number of doors) Touring Car.

Cadillac's most impressive new model, however, was the racy-looking Torpedo. The torpedo body style was the hit of the 1910 auto shows and the introduction of this semi-streamlined body style marked an important step forward in automobile styling.

Although similar in concept to the new Fore-Door Tourer, the Cadillac Torpedo featured a continuous, flowing bodyline that ran from the radiator to the rear of the car. Hood and body flowed pleasingly into one another via a swelled, pressed sheet-metal cowl. The resulting "bathtub" effect was both clean and attractive.

Also for 1911, the wheelbase on standard Model Thirties was increased another six inches, to 116 inches. Engine displacement went up again to 286.3 cubic inches, and the design was now "squared" with equal four-and-a-half inch bore and stroke. The four-cylinder engine had been again uprated, this time to 32.4 A.L.A.M. horsepower.

In its advertising, Cadillac proudly boasted that... "Every Cadillac is a Dewar Trophy Cadillac."

Cadillac's 1911 prices began at $1,700 for the Roadster, Demi-Tonneau and standard Touring Car. The new Fore-Door Tourer was priced at $1,800 and the Cadillac Torpedo at $1,850. The elegant new closed coupe carried a base price of $2,250. The luxurious Model Thirty Limousine continued at $3,000.

A Bosch magneto used in conjunction with the Delco ignition continued as standard equipment on all 1911 Cadillacs. A pair of gas headlamps and gas generator, one pair of side oil lamps, a taillamp, horn, tool set and tire repair kit were also standard. A Prest-O-Lite tank was substituted for the gas generator on the coupe and limousine.

Model Thirty customers could choose from a wide range of comfort and convenience accessories. These included seat slip covers ($40 to $60); an "automatic folding windshield" for $40, and a mohair cape top for the new Cadillac Torpedo at $95 extra. A complete Model Thirty chassis could be purchased for $1,605. Complete, finished bodies were offered starting at $145 for a rear tonneau to $475 for a Torpedo body and $1,350 for a fully-trimmed limousine body. In addition, the Fleetwood Metal Body Co. of Fleetwood, Pa. started building custom bodies for the four-cylinder Cadillac chassis.

Cadillacs continued to break records, and not just in sales. T. J. Beaudet drove a stripped-down Model Thirty Cadillac in a 24-hour race on a Los Angeles track. He covered 1,448 miles at an average speed of 60.33 miles an hour.

This was a milestone year for the industry. The contentious Selden Patent was successfully challenged in the United States Court of Appeals, which ruled that the patent was "valid, but not infringed" by Ford and other manufacturers. The burdensome payment of royalties that had infuriated the American automobile industry since its infancy finally came to an end, and the Association of Licenced Automobile Manufacturers, of which Cadillac had been a member, was disbanded. Its successor was the Automobile Board of Trade.

The Chevrolet Motor Company of Michigan was organized this year to build a six-cylinder car developed two years earlier by Louis Chevrolet, and the Studebaker Corp. finally abandoned production of electric cars in favor of gasoline-propelled vehicles only.

Cadillac built 10,071 cars in calendar year 1911 and 10,018 in model year 1911.

The Cadillac Motor Car Co. late in 1911 proudly announced its 1912 models: the world was ready for self-starting.

Cadillac's 1911 product line included seven models, two more than the previous year. The lightest of these was the Model Thirty Roadster with Deck, shown here. A rumble seat for carrying a third passenger was available at extra cost and could be attached to the rear deck. This roadster is equipped with an optional two-piece folding windshield. Base price with standard equipment was $1,700.

1911

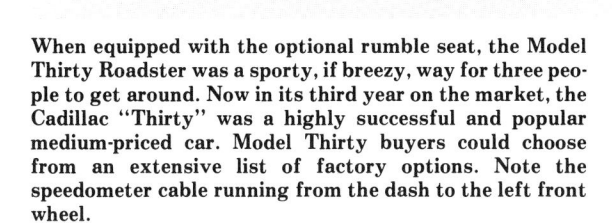

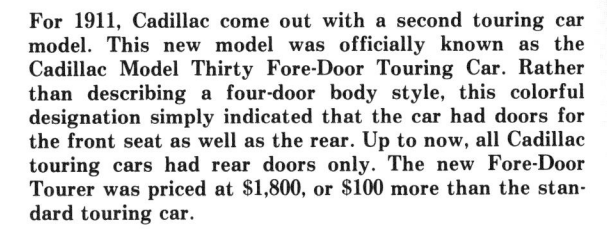

When equipped with the optional rumble seat, the Model Thirty Roadster was a sporty, if breezy, way for three people to get around. Now in its third year on the market, the Cadillac "Thirty" was a highly successful and popular medium-priced car. Model Thirty buyers could choose from an extensive list of factory options. Note the speedometer cable running from the dash to the left front wheel.

The 1911 model year would be the last for the versatile demi-tonneau model. This is an advertising cut of the 1911 Cadillac Model Thirty Demi-Tonneau, with optional folding top and windshield. For quick conversion to a runabout, the rear tonneau could be detached and removed. With the tonneau in place, this model was advertised as a four-passenger car. Price was identical to that of the standard touring car and the roadster — $1,700 F.O.B. Detroit.

For 1911, Cadillac come out with a second touring car model. This new model was officially known as the Cadillac Model Thirty Fore-Door Touring Car. Rather than describing a four-door body style, this colorful designation simply indicated that the car had doors for the front seat as well as the rear. Up to now, all Cadillac touring cars had rear doors only. The new Fore-Door Tourer was priced at $1,800, or $100 more than the standard touring car.

Pit stop! The clean, uncluttered lines of the new Fore-Door Touring Car are evident in this photograph, snapped on one of many stops during the 1911 Glidden Tour. The Model Thirty Cadillac also sported a handsome front-end ensemble. This car is flying Jacksonville and New York pennants.

This mud-splattered touring car is a veteran of the 1911 Glidden Tour. It is a Cadillac Model Thirty Fore-Door Touring Car. The optional top and windshield were probably well worth the extra cost on this run. Note the canvas-covered luggage lashed to the rear.

1911

Totally restored and posing proudly is this beautiful 1911 Model Thirty Touring Car, owned by Robert Adams of Mineola, N.Y. The car has been fitted with the accessory side-mounted spare rim, this one equipped with the same diamond tread pattern as used on the rear wheels. The front wheels are equipped with "No Skid" tread. A common practice, that extended into the mid-twenties, was to put heavy-tread tires on the rear for gripping in the mud or slush while equipping the front tires with a relatively smooth or ribbed tread design for better riding. This car is finished in traditional Royal Blue body and wheels, with black fenders, chassis, and radiator. Both cowl and runningboards are highly varnished natural wood. Note the interesting rear window design, of eight small panes of celluloid.

This photograph was simply too pretty to leave out. The car is a spanking new 1911 Cadillac Model Thirty Touring Car, the standard model as opposed to the new companion Fore-Door Tourer. It is finished in the standard Model Thirty color scheme — Royal Blue body and chassis with light decorative pinstriping. The wheels were light cream with black striping. Blue-painted wheels could be specified as an option. The inviting seats were upholstered in hand-buffed black leather.

Posed proudly at the side of the road, its front fenders festooned with Cadillac pennants, this is a 1911 Cadillac Model Thirty Touring Car that participated in an international run this year. This endurance run took the cars north from Detroit across the Canadian border and on in to Winnipeg, Manitoba. Note the long windshield braces which are attached to the front of the chassis frame. This would be Cadillac's last year for gas and oil lamps.

Cadillac's standard 1911 touring car was similar in appearance to the popular Model Thirty tourers offered in the previous two model years. It had no doors for front-seat passenger or driver. This is the standard 1911 Cadillac Model Thirty Touring Car about to set out on a durability run from Detroit to Milwaukee. The folding top was an extra-cost option. Note the dashboard, which follows the contour of the hood. Only the touring car and limousine still used the flat, buggy-style dash.

The bold, new "torpedo" body style was a big hit with both manufacturers and the public at the 1910 National Automobile Show. It was no surprise, then, when Cadillac added a four-door torpedo to its product line for 1911. The Cadillac Model Thirty Torpedo was priced at $1,850. A mohair cape top cost $95 extra.

1911

The new Cadillac Model Thirty Torpedo marked an important advance in vehicle styling. Gone was the cluttered "buggy look". The handsome Torpedo presented a clean, fleet appearance in a semi-enclosed design that featured an unbroken body line from the hood to the rear of the car. The sporty new Cadillac Torpedo was an immediate success. This light-colored Model Thirty Torpedo took part in the 1911 Glidden Tour. The spare tire and holder, folding windshield and top were extra-cost options.

Late in 1910, the Cadillac Motor Car Co. added an elegant enclosed-drive coupe to its Model Thirty line. Henry M. Leland had given the newly-formed Fisher Body Co. an order for 150 closed Cadillac bodies, a contract that is credited with giving Fisher Body its real start. The towering, two-passenger Cadillac "Thirty" Coupe was the company's second most expensive model. At $2,200 the coupe was priced at only $800 less than the big Model Thirty Limousine. These coupes sported functional rear decks.

At the top of the Model Thirty product line was the impressive Limousine. This formal car carried a base price of $3,000. The chauffer still rode out in the open while his passengers travelled in coddled luxury in the rear compartment. This model was sold as a seven-passenger car. Standard equipment included electric interior lights and a speaking tube.

This is the no-frills instrument board on the standard Cadillac Model Thirty Touring Car. The dash is rounded to follow the contour of the hood. The steering column has been "sectioned" to show the location of various switches. Note the speedometer mounted to the right. This model was equipped with an 18-inch diameter steering wheel with corrugated rubber rim and an aluminum spider. Delco ignition and a Bosch magneto were standard.

Despite a wide selection of body styles, some of Cadillac's customers preferred true individuality in the automobiles in which they rode, or drove. Here is a real custom creation. About the only recognizable Cadillac Model Thirty components are the hood and front fenders. The chassis has been greatly elongated to accommodate this formal town car body, which looks like a roadster and limousine body mounted end-to-end. The rear body looks as if it might be a transplant from a horse-drawn carriage. This was likely a one-off conversion.

1912

Cadillac's introduction of electric self-starting on its refined 1912 models marked a major advance in automobile design. In itself, automatic electric self-starting constituted an unprecedented breakthrough in automotive engineering. But when combined with improved ignition and electric lighting, all controlled from a centralized Delco electrical system, these remarkably advanced features immediately catapulted Cadillac far ahead of every other make of car in existence in sophistication and convenience.

The electric self-starter did away with the necessity of hand-cranking the engine. This was an onerous and often dangerous chore which was made even more inconvenient and unpleasant in bad weather. Up to now, because of the sheer muscle power often required just to turn the engine over, automobile operation was largely the province of men. With true self-starting, the world of motoring was opened to women as never before.

Soon Cadillac was advertising its products as . . . "The Car That Has No Crank."

The man generally credited with the perfection of the electric self-starter was Charles F. Kettering. A talented inventor, Mr. Kettering left the National Cash Register Co. in the early 1900's to form his own company, the Dayton Engineering Laboratories Co. in Dayton, O. The infant automobile industry offered inventors like Mr. Kettering unlimited opportunity to develop and market all kinds of new technology.

Three years earlier, in 1909, Mr. Kettering had succeeded in landing an order from Mr. Leland and his associates for 8,000 ignition sets for Cadillac's 1910-model cars. This engine ignition system operated in conjunction with a conventional magneto. The concept of a completely automatic means of starting automobile engines intrigued Mr. Kettering and his staff, who worked for the better part of two years to design and perfect it. When it was announced on the new 1912 Cadillacs in the fall of 1911, electric self-starting took the industry by storm. Suddenly virtually every other car seemed old-fashioned by comparison.

As described in Cadillac's 1912 advertising, the electric self-starter was essentially a miniature dynamo operated by the vehicle's engine. The dynamo was activated by the storage battery. To start the engine, the driver retarded the spark lever and pushed in the clutch pedal automatically engaging a gear on the electric motor with teeth in the engine flywheel. This resulted in the engine turning over, duplicating the effort formerly expended in hand-cranking. Once the engine started, the clutch was let back out and the electric motor gear disengaged from the flywheel. The electric motor again reverted to its role as a dynamo, or generator, supplying ignition current and charging the battery.

Within a few years, self-starters were standard on most makes. International recognition of Cadillac's latest engineering triumph was not long in coming.

During the summer of 1912, the Royal Automobile Club of Great Britain conducted extensive tests on the new Cadillac. The engine was subjected to more than a thousand stop/start cycles. This resulted in the Cadillac Motor Car Co. being awarded its second Dewar Trophy the following year. No other automobile manufacturer, let alone an American one, had ever received such a prestigious award twice.

Standard equipment on all 1912 Cadillacs included a pair of big Gray and Davis electric headlamps with adjustable focusing. The Delco electrical system also included two electric cowl lights, a taillight and even a light for the speedometer.

For 1912, Cadillac discontinued the "Model Thirty" designation that had been used on its four-cylinder cars since 1909. With the exception of the revolutionary self-starter, ignition and lighting system, there were no significant mechanical changes. All Cadillacs now rode on larger 36-inch by four-inch tires.

Only one new model was added this year — a handsome four-passenger Phaeton. The old, reliable Demi-Tonneau had been dropped along with the front-doorless standard touring car. All of Cadillac's big, open cars — the Tourer, Phaeton and sleek Torpedo — now had four doors. The two-door Roadster had been extensively restyled with an attractive new body, and the company at last offered the carriage trade a completely-enclosed Berline Limousine.

The 1912 Cadillac model lineup consisted of six distinctive body styles. Wheelbase remained at 116 inches for all standard models. Prices were adjusted upward slightly.

In industry news, Boyce came out with its first Moto-Meter and Edward G. Budd developed all-steel car bodies for Hupmobile and Oakland. The new Automobile Board of Trade took over where the A.L.A.M. had left off the previous year.

Cadillac produced 13,995 cars in the 1912 model year, with calendar year production at 12,708.

Equipped with the revolutionary new Delco self-starter, Cadillac advertised its 1912 models as. . ."The Car Without a Crank". Introduction of the Delco automatic electric starter, ignition and electric lights marked a major breakthrough in motoring convenience. This is the thoroughly-redesigned 1912 Cadillac Two-Passenger Roadster. Equipped as illustrated with a mohair cape top, this car sold for $1,860.

1912

This is the all-new 1912 Cadillac Roadster in full road trim with its folding top in place. The top was a $60 option. The Cadillac Roadster's sleek new all-metal body featured a pressed-steel cowl and side doors, though the right door was almost impossible to use. The two-piece folding windshield has its lower pane raised for some flow-through ventilation. The long, sloping rear deck contained a luggage compartment.

The five-passenger touring car continued as one of Cadillac's most popular models. The roomy tourer still sported a flat, square-cut dashboard between the engine and passenger compartments. All Cadillac touring cars now had four doors. This model carried a base price of $1,800. The lofty placement of the license plate atop the radiator kept it clear of mud and readable. This 1912 Cadillac Touring Car is taking part in a club run.

Still in is original condition after more than 68 years, this 1912 Cadillac Touring Car is owned by George Pickering, of West Bloomfield, Michigan. Mr. Pickering also owns a 1905 single-cylinder Cadillac. The mohair cape top shown on this beautifully-preserved example was a $90 extra cost option. Cadillac's 1912 roadster, touring car and phaeton were all priced at $1,800 F.O.B. Detroit in standard form.

Standard equipment on all 1912 Cadillacs was a pair of specially-designed Gray and Davis electric headlights. These handsome headlamps are shown to advantage on this 1912 Cadillac touring car. Cadillac's revolutionary electric illumination system also included a pair of cowl lamps, a taillight and a speedometer light. Prior to this, lights were cumbersome gas lamps which required a generator and frequent charging.

1912

This gang is part of the shipping department at Cadillac's big plant at Cass and Amsterdam Streets in Detroit. In the background is a long line of 1912 Cadillac touring cars waiting to be loaded into boxcars. The hood of this standard touring car is open, affording a glimpse of the four-cylinder engine. This intriguing photograph was marked "Boat Shippers", possibly in reference to this model's tublike body, but probably because this crew was responsible for getting the cars to a Detroit dock and onto the many Great Lakes freighters that were extensively used in this era.

All of Cadillac's new 1912 models rode on larger 36 by 4-inch tires, supplied by Hartford or Morgan & Wright. Transmission was of the selective, sliding-gear type with three forward speeds plus reverse. Axles were equipped with Timken bearings. The hood top and side panels featured an embossed rectangular design which is visible in this photo of a 1912 Cadillac touring car which has paused for a rest on a club run.

Except for the addition of such refinements as electric self-starting and electric headlights, Cadillac's handsome four-passenger Torpedo went into its second model year with little change. The speedy-looking Torpedo was a fairly popular model. It was priced $100 higher than the standard touring car and phaeton. This example has been equipped with a windshield and top. It is a participant in this year's Glidden Tour.

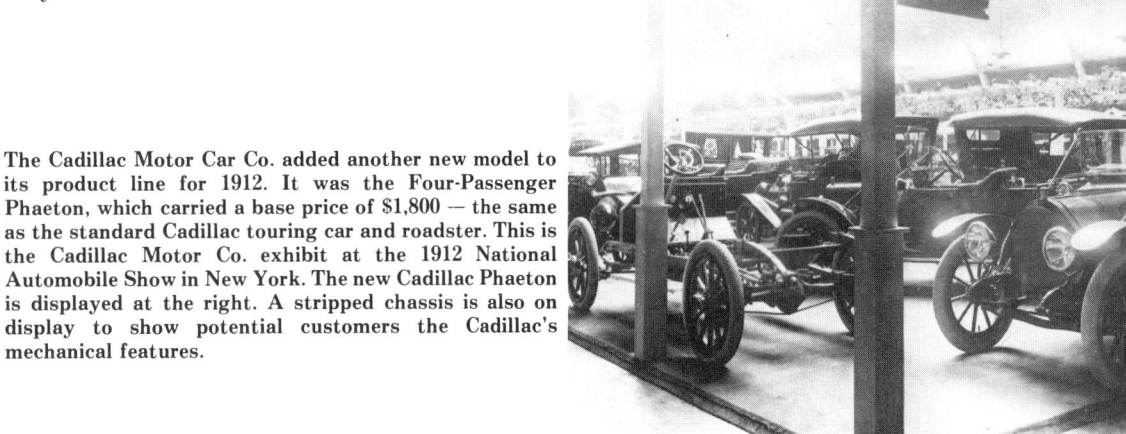

The Cadillac Motor Car Co. added another new model to its product line for 1912. It was the Four-Passenger Phaeton, which carried a base price of $1,800 — the same as the standard Cadillac touring car and roadster. This is the Cadillac Motor Co. exhibit at the 1912 National Automobile Show in New York. The new Cadillac Phaeton is displayed at the right. A stripped chassis is also on display to show potential customers the Cadillac's mechanical features.

1912

The exquisitely-finished and appointed Cadillac Coupe appealed to an everincreasing number of Cadillac's customers as closed bodies continued to gain in popularity. Enclosed cars, however, remained relatively expensive. The 1912 Cadillac Coupe for four passengers carried a base price of $2,250. But the price gap between the large, open cars and closed models was gradually narrowing. Like all other standard 1912 Cadillac models, the high, angular Coupe rode on a 116-inch wheelbase.

Cadillac's most expensive car, the big seven-passenger Limousine, underwent a significant change this year. For the first time, the chauffeur rode inside, completely out of the weather. The chauffeur's compartment was still separated from the luxuriously-appointed passenger compartment by a fixed partition. This style of Cadillac Limousine was designated a Berline and featured aluminum body panels. The price was increased for the first time since this model had been introduced in 1910, from $3,000 to $3,250 in 1912.

Introduction of the Delco-designed automatic electric starter, ignition and electric lighting system marked a major step forward in automobile design. This important innovation was credited to Charles F. Kettering of the Dayton Engineering Laboratories Co., of Dayton, O. To demonstrate the strength of Cadillac's new self-starter, the company staged this interesting demonstration. Weighted down with a dozen "passengers", this Cadillac four-cylinder chassis was propelled around the room solely by the Delco self-starting device.

A gifted and resourceful inventor, Charles F. Kettering had been working on his self-starting device for automobiles for several years prior to its introduction on the 1912 Cadillacs. Kettering's Dayton Engineering Laboratories Co. had been supplying Cadillac with electric ignition components since 1910. These were used in conjunction with a standard magneto system until 1912. The self-starter was a major breakthrough in American motor vehicle design and brought the Cadillac Motor Car Co. its second Dewar Trophy. Later known around General Motors as "Boss Kett", this is Mr. Kettering at work in his Dayton, O. shop.

1913

Cadillac Motor Car Company production and sales reached new record levels during the 1913 model year, as the company ended its first decade.

During the 1913 model year, the Detroit factory built an all-time record 15,018 cars. Virtually all of this production was snapped up by Cadillac's dealers before the first 1913-model Cadillac rolled off the assembly line. Calendar year production hit 17,284.

In 1913, Cadillac received its second Dewar Trophy. This important international honor was presented to Cadillac for the second time in five years on the basis of the Royal Automobile club of Great Britain's extensive testing of the revolutionary new Delco electric self-starter in the summer of 1912. No other manufacturer ever won the coveted Dewar Trophy twice.

Cadillac's improved 1913 models included some significant mechanical changes. First the wheelbase on all standard models was increased another four inches, to 120 inches.

But the most noticeable change was under the gabled Cadillac hood. Rated horsepower of the Cadillac four-cylinder engine was boosted by more than 25 per cent, from 32.4 in 1911 and 1912 to an astonishing 48.7 horsepower for 1913. To accomplish this, displacement was bored out to 365.8 cubic inches. Bore and stroke were again altered, to 4-¼ by 5-⅜ inches.

The 1913 Cadillac model lineup offered a choice of seven body styles. Of these, four were large, open touring-type passenger cars.

The 1913 Cadillac lineup started with the sporty two-passenger Cadillac Roadster. This popular model featured a restyled rear end with a gracefuly curved rear deck.

Next came the standard five-passenger Touring Car. At last this model had discarded the old-style, flat dash and featured a handsome, new curved cowl similar to those used on all other open Cadillacs. The four-passenger Cadillac Torpedo and the new Phaeton introduced the previous year were essentially unchanged in appearance. Cadillac added a new open model to its line for this year. It was a six passenger touring car which was advertised simply as the "Cadillac Six-Passenger Car." This new model carried a basic price that was $100 higher than the five-passenger tourer, the Torpedo and Phaeton.

Closed bodies were still limited to a four-passenger coupe and the formal limousine. Coupe and limousine bodies were constructed of aluminum. All other models had full-metal bodies built over wood frames.

The big Cadillac Limousine was extensively restyled. The chauffer performed his driving duties from what was described by the company as a "three-quarter closed" chauffer's compartment. Five passengers rode in luxurious comfort in the rear passenger compartment. This heavy limousine was sold as a seven-passenger car, but the sixth passenger was obliged to ride up front with the liveried driver, separated from the rear compartment by a permanent partition. Cadillac customers could also specify an Inside-Drive Limousine.

In addition to major engine changes, improvements were made in the Delco electric self-starter, ignition and lighting system introduced the previous year in what was hailed as one of the auto industry's most important advances.

With two Dewar Trophies under its belt, Cadillac proudly hailed its products as "The Standard of the World". This would become one of the best-known advertising slogans in the world and is still universally recognized today.

For well over a year, Cadillac's engineering and design staff had been hard at work on a new engine. This powerplant was one of the best-kept secrets in American auto industry history. But introduction of this engine was targeted for Cadillac's 1915 models. Cadillac had some other tricks up its corporate sleeve to hold public interest, and booming sales, until the new engine was ready.

With two Dewar Trophies to its credit, the Cadillac Motor Car Co. had no reservations about promoting its products under a new slogan: "Standard of the World". One of the most famous of all automobile advertising phrases, it is still recognized all over the world today. This is the 1913 Cadillac Two-Passenger Roadster. For 1913, this sporty car was given a new, gracefully curved rear deck. Base price was $1,975.

Of Cadillac's seven 1913 models, four were large, open cars of the touring type. This is the standard five-passenger touring car, which finally got a curved cowl replacing the old style, flat buggy dash. This was the only one of the four-door open models with a dropped sill below the rear entrance doors. All other models featured a conventional straight sill. Price started at $1,975 F.O.B. Detroit.

1913

Cadillac introduced two new large, open cars for 1913. This is one of them. It was simply designated the "Cadillac Six-Passenger Car". A variation of the standard touring car, it had a pair of folding auxiliary seats the tops of which are just visible in this photo. The six-passenger car was priced $100 higher than the standard tourer, at $2,075. Note the robe rail attached to the back of the front seat.

For 1913, Cadillac came out with yet another new open model. It was called the Cadillac Phaeton and was a four-passenger car. This model is distinguishable from the new Six-Passenger Car and Standard Tourer by the decorative molding that encircled the upper edge of the body from cowl to rear. Like the regular touring car and the Torpedo, this car carried a base price of $1,975.

The popular Cadillac Torpedo was now in its third model year. This was a four-passenger car. Its price started at $1,975 — the same as the two-passenger roadster, standard touring car and the new Phaeton. Cadillac's 1913 models were built on a new 120-inch wheelbase, four inches longer than the previous year.

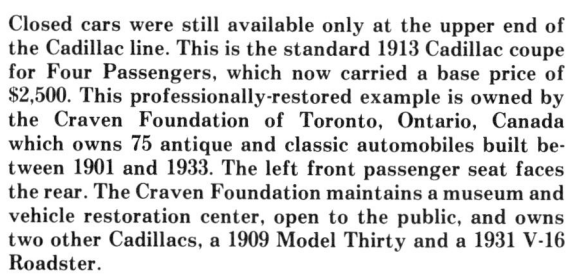

Closed cars were still available only at the upper end of the Cadillac line. This is the standard 1913 Cadillac coupe for Four Passengers, which now carried a base price of $2,500. This professionally-restored example is owned by the Craven Foundation of Toronto, Ontario, Canada which owns 75 antique and classic automobiles built between 1901 and 1933. The left front passenger seat faces the rear. The Craven Foundation maintains a museum and vehicle restoration center, open to the public, and owns two other Cadillacs, a 1909 Model Thirty and a 1931 V-16 Roadster.

1913

The most prestigious model in the entire Cadillac line, the heavy seven-passenger Limousine, was extensively restyled this year. The redesigned limousine body featured straighter, more harmonious lines. The chauffer's compartment was semi-open but did have some side glass. The regal Cadillac Limousine, however, was unchanged in price: $3,250 with standard equipment. The limousine body was made of aluminum.

The interior of the majestic Cadillac Limousine was an extension of the sumptuously-appointed drawing rooms of the fine homes of the well-to-do who could afford this car. The passenger compartment was upholstered in best quality blue broadcloth trimmed in matching broad and narrow lace. The passenger compartment of the 1913 Cadillac Limousine was designed to transport five passengers in conspicuous luxury. Two "revolving", or folding, seats were standard. The driver's compartment was three-quarters enclosed and was upholstered in hand-buffed, deep grain black leather.

The cowl design of this open-front limousine would indicate that it is a 1913 Cadillac. Body details, window reveals, etc. vary from the standard Cadillac Limosine, so this was probably a custom-built job. Note the flat roof and high, narrow doors to the passenger compartment. This limousine, however, does have doors for the chauffer's compartment.

A number of custom body builders were now supplying specially-designed custom bodies for the highly-regarded four-cylinder Cadillac chassis. The builder of this custom-bodied Cadillac Coupe is not identified, but it could be a Columbia. All those tiny windowpanes must have played havoc with the driver's vision. Note the location of the elegant sidelamps at the front of this angular body.

1913

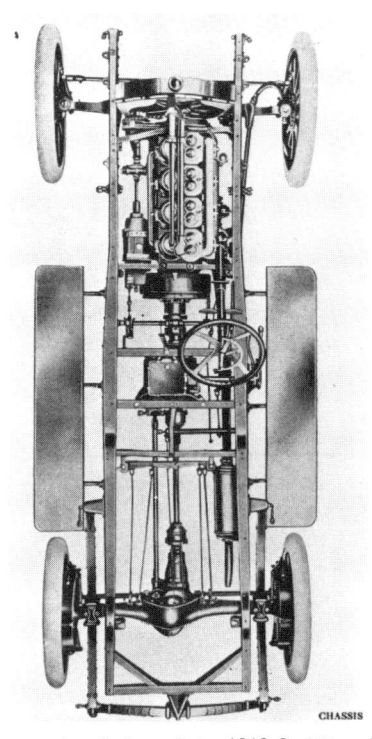

This is an overhead view of the 1913 Cadillac chassis. For 1913, horsepower of the four-cylinder engine was dramatically increased, from 32.4 in 1912 to 48.7 horsepower. Displacement was increased to 365.8 cubic inches and bore and stroke were now 4-¼ by 5-⅜ inches. Wheelbase had also been increased another four inches to 120 inches. This four-cylinder design had started out as the Cadillac Model Thirty for 1909. With many engineering improvements and refinements, this series was carried through the 1914 model year until the Cadillac V-8 burst upon the scene.

The Cadillac nameplate has been closely associated with the American funeral car and ambulance industry for many years. Indeed, Cadillac has enjoyed a near-monopoly in this specialized field for the past 30 years. Cadillac actually marketed hearses and ambulances on its own in the mid-1920's. In 1937 the company began to supply funeral car and ambulance manufacturers with a specially designed commercial chassis engineered for hearse and ambulance work. Because of Cadillac's reputation for quality and durability, Cadillacs were popular with the undertaking trade almost from the beginning. Good, used Cadillacs were eagerly sought for conversion to hearses. The Hoover Wagon Co. of York, Pa. mounted this eight-column, carved panel hearse body on a lengthened 1913 Cadillac touring car chassis in 1920.

This was certainly the smallest 1913 Cadillac built. The fully-operable scale replica of a Cadillac Roadster was built by Frederick S. Bennett, of London, England and was presented to Wilfred Leland, Jr. on his fifth birthday. The little roadster was powered by the revolutionary Cadillac self-starter. Here, young Wilfred takes a young lady out for a spin in his new toy. Frederick S. Bennett, of Anglo-American Motor Co. fame, was Cadillac's principal importer in Great Britain and was the man behind the 1908 standardization test that brought Cadillac its first Dewar Trophy.

After years of faithful family service, more than a few outdated but still serviceable old four-cylinder Cadillacs began second careers as utility vehicles, hearses and even tow trucks. This is the original Holmes Wrecker. It was built in 1916 utilizing a 1913 Cadillac touring car chassis. This historic vehicle was constructed by Ernest W. Holmes who operated a busy garage in Chattanooga, Tenn. Holmes' highly successful wrecker design was patented in 1918. Holmes is still one of the nation's largest manufacturers of specialized towing equipment today. A duplicate of the original Holmes Wrecker was built some years later. The moldings at the top of the front door and just below the seat indicate that this was a conversion of a Cadillac Phaeton.

1914

Now in its sixth model year, Cadillac's extremely successful four-cylinder car was finally nearing the end of the line. This series had commenced its existence as the highly-regarded Cadillac Model Thirty, a 1909 model that had been introduced in the latter part of 1908. Despite continued improvement and such pace-setting advances as self-starting, the popular four-cylinder Cadillac was by now definitely showing its age. Sales were still good, but something new was needed if the Cadillac Motor Car Company was to retain its very strong position in the American automobile industry.

This fact was certainly not lost on Cadillac's management. Under a formidable blanket of secrecy and security, a totally new Cadillac was under development. Testing was already in the advanced stages, and plans were being formulated to introduce the all-new car in mid-1914 as a 1915 model. Cadillac's marketing and sales staff knew that if they could keep the lid of security in place, their new car would take the industry by storm.

But there was still an entire model year to go. Outwardly, the 1914 Cadillacs showed little change from those sold in 1913. The most significant technical innovation incorporated into the 1914 Cadillacs was the introduction of a two-speed, direct-drive rear axle. By simply flicking an electric switch on the dash, the driver of a 1914 Cadillac could switch from low into high gear. The result, according to Cadillac's 1914 introductory advertisng, was an immediate 42-per cent boost in speed. The two-speed axle was a short-lived feature at Cadillac, however. The company was sued by the Austin Automobile Company of Grand Rapids, Michigan and the two-speed axle was quietly discontinued the following year.

Another noteworthy change this year was Cadillac's adoption of left-hand drive. Prior to 1914, virtually all Cadillacs produced were right-hand drive cars. The 1914 Cadillacs also boasted another steering innovation — a hinged steering wheel that swung up and out of the way when the driver entered the car. This was the same basic principle embodied in the tilt steering columns that were heralded as major comfort and convenience options when they were introduced more than half a century later.

Also on the list of mechanical refinements, the gasoline tank on Cadillac's 1914 cars was relocated to the rear of the vehicle. The proven, old reliable four-cylinder Cadillac engine was still rated at a respectable 48 horsepower. Wheelbase on all of Cadillac's standard passenger car models remained at 120 inches.

Cadillac ambitiously announced that it planned to build an unprecedented 18,000 cars in the 1914 model year. Actual 1914 production reached 14,002 cars.

The 1914 Cadillac model lineup included a comprehensive range of seven body styles. These included the sporty Roadster for Two Passengers, which carried a base price of $1,975 F.O.B. the factory in Detroit. The perenially popular Touring Car for Five Passengers carried the same price, as did the Cadillac Phaeton for Four Passengers. Cadillac's 1914 offerings also included a handsome Landaulet Coupe for Three Passengers that sold for $2,500. Then came the big Seven-Passenger Car, at $2,075. Rounding out the line was a new Inside-Drive Limousine, priced at $2,800, and the formal Seven-Passenger Cadillac Limousine, with prices starting at $3,250.

All of Cadillac's basic prices now included a top, windshield and demountable wheel rims.

In U.S. automotive industry news, Cadillac's sister General Motors division, Chevrolet, discontinued the production of six-cylinder cars concentrating instead on a new line of popularly-priced fours. Another GM division, Buick, switched over to six-cylinder engines. And the Dodge Brothers, Horace and John, started the manufacture of a new car that bore the Dodge name.

In far-away Europe, war had broken out. Within a very few years the United States would find itself enmeshed in what would eventually become known as the Great War.

As the 1914 model year wound down, preparations were being made for the startup of production of the next generation of Cadillacs. The development and test program was proceeding smoothly at an accelerated pace. Much of the critical development work and testing was being done in a nondescript building in suburban Mount Clemens, Mich., miles away from Cadillac's main plants and offices. Only those directly involved in the project had any inkling that something important was going on.

In September, 1914, the Cadillac Motor Car Company succeeded in catching the industry by surprise with the announcement of its sophisticated, new 1915 models. In general appearance, the new Cadillacs looked much like the 1914's they replaced. But under the Cadillac hood throbbed the auto industry's first mass-produced, V-type eight-cylinder engine. The milestone Cadillac Type 51 had arrived.

Throughout the industry, demand for closed cars was steadily increasing. For 1914, Cadillac expanded its range of closed body offerings with the introduction of this nicely styled Cadillac Landaulet Coupe. Advertised as a three-passenger car, the new Landaulet Coupe was priced at $2,500. The "bent" front window glass, curved beltline and modestly rounded rear deck give this car an elegant look. The angular, gabled hood, however, clashes with the soft curves of the car's body styling.

1914

Now in its sixth consecutive model year, the highly-successful four-cylinder Cadillac that had originally made its debut as the 1909 Cadillac Model Thirty was nearing the end of the line. This is the 1914 Cadillac Roadster for Two Passengers, which carried a factory price of $1,975. The most significant mechanical change on the 1914 Cadillacs was a new two-speed, direct drive rear axle.

"Firemen Choose the Cadillac", read the eye-catching headline in this 1914 Cadillac ad. A standard Cadillac Two-Passenger Roadster had been converted into a speedy Fire Chief's Runabout. Special equipment includes a pair of big spotlights attached to the cowl, an electric siren horn and an equipment compartment on the rear deck. Note the chief's speaking trumpet and the portable fire extinguisher mounted on the running board. The hood is lettered "Supt.", for Superintendent. Exactly where this Chief's Buggy served is not indicated.

The roomy Five-Passenger Touring Car continued as one of Cadillac's most popular models. For 1914, the Cadillac tourer retained its standard price of $1,975. This price was identical to that of the companion Roadster and Phaeton. Large, open cars still accounted for the majority of Cadillac's sales, although closed bodies were steadily gaining in popularity. Bumpers were still considered a big-city frill.

For 1914, Cadillac advertised this big tourer as the "Cadillac Seven-Passenger Car." The previous year, the company had inserted a similar type of car, designated as simply the "Six-Passenger Car" into its touring car offerings between the four-passenger Cadillac Phaeton and the standard Touring Car for Five Passengers. The Seven-Passenger Car was equipped with two folding rear auxiliary seats, the tops of which are visible in this photo. The new Cadillac Seven-Passenger Car was priced at $2,075, or $100 more than the popular five-passenger tourer.

The Cadillac Motor Car Company offered its customers open-style touring cars in a choice of four, five and seven-passenger configurations. This is the 1914 Cadillac Phaeton, a four-passenger open car that carried a base price of $1,975. This price was identical to that of the standard five-passenger touring car. The special mouldings at the upper edge of the body were unique to the Phaeton. Standard equipment included a two-piece folding windshield and a pair of electric cowl lamps.

1914

At the upper end of its model range, Cadillac added a new, fully-enclosed limousine for 1914. A heavy, five-passenger car, this model was designated the Cadillac Inside-Drive Limousine and was priced at $2,800. The driver and his passengers entered the car through a center door. This handsome new model featured the same softly-rounded body lines embodied in the new Cadillac Coupe. Note the small assist handle to the right of the door handle. Oddly, this strange and rather awkward body style became very popular during this era, and was extensively used on such chassis as Ford, Dodge, Buick, Chevrolet, and Oldsmobile. In fact, Ford carried this centerdoor style as its prestige model up until 1922.

The most expensive standard model in Cadillac's 1914 model offerings was the formal Seven-Passenger Limousine. This semi-enclosed, chauffer-driven car was priced at $3,250 for the third consecutive year. For 1914, the stately limousine body featured simpler, straighter body lines. Standard equipment on this premium model was a speaking tube for communication between front and rear compartments, a pair of small carriage lamps and pull-down privacy shades in the luxuriously-appointed rear compartment. Dropped on this model was the side glass in the driver's compartment. Once again, the chauffer was exposed to the elements.

The Cadillac sales department encouraged its dealers to seek new, specialized business. A good example of this quest for new automotive applications is this interesting railroad inspection car, a modified 1914 Cadillac Five-Passenger Inside-Drive Limousine. No doubt the railway brass used this luxurious transportation as well as track inspection crews. Car 99 rolled smoothly along on flanged wheels. The front brakes were operated from the steering column. Rear-wheel brakes were controlled by the conventional brake pedal.

The highly-regarded Cadillac four-cylinder chassis was becoming more and more popular with custom body builders. This is a custom-built seven-passenger touring car with wire wheels and a Victoria-type canopy for the rear seat passengers. The builder of this handsome open car is not known. It is finished in a special light color rather than in Cadillac's usual Dark Blue with black running gear.

The hand-written caption accompanying this photograph states that this was the "first closed two-door body . . . made especially for me by Charles and Fred Fisher and L. Mendelssohn and mounted on a Cadillac four-cylinder chassis." The car is an open-front limousine fitted with wire wheels instead of the heavy wood artillery type wheels which were standard on all 1914 Cadillacs. The writer was not identified in the underline. It could have been Henry M. or Wilfred C. Leland. The chauffer's compartment has roll-down side curtains for use in bad weather.

1915

Introduction of the sophisticated, new eight-cylinder Cadillac marked another significant advance in American automobile design and further enhanced the Cadillac Motor Car Co.'s reputation for engineering leadership. Although V-type, eight-cylinder cars had been built more than a decade earlier, announcement of the 1915 Cadillac marked the V-8's first application in a standard production, mass-produced car. The V-8 engine has remained Cadillac's standard powerplant ever since, for an unbroken span of more than 65 years.

In 1903, Frenchman Clement Ader drove a racing car powered by a V-8 engine in the Paris-to-Madrid race. Two years later, Darracq built a car powered by a 200-horsepower eight-cylinder engine. Marmon exhibited a car with a 60-horsepower V-8 engine at the Indianapolis and New York automobile shows in 1906, and Hewitt in 1907 showed a $5,000 limousine that had a V-8 engine under its hood. But until the arrival of the new Type 51 Cadillac, the V-8 was a relatively exotic, premium powerplant of dubious practicality and reliability.

The Type 51 Cadillac V-8 engine was a 90-degree design of 314 cubic inch displacement. In its standard production version, output was officially rated at 70 brake horsepower at 2,400 RPM. Bore was 3-1/8 inches and stroke 5-1/8 inches. The Leland-engineered V-8 was a remarkably smooth and quiet passenger car powerplant capable of effortlessly delivering road speeds of 55 to 65 miles an hour. The man behind the design and development of the first Cadillac V-8 engine was D. McCall White, a Scottish-born engineer who had worked for Daimler and Napier before joining the Cadillac engineering staff. White was Cadillac's Chief Engineer until May, 1917.

The public announcement of the Cadillac V-8 in September, 1914 took the industry by surprise. Predictably, competitive detractors were both plentiful and vocal. To counter the critics, Cadillac early in 1915 had its advertising agency create an ad that is still regarded as a classic in the advertising business today.

Appearing in the January 2, 1915 issue of The Saturday Evening Post, the ad was entitled "The Penalty of Leadership" and focused on the plight of leaders in all endeavors who are thus natural targets for the inferior and the envious. Over the now-familiar Cadillac script, the ad concluded . . . "That which deserves to live . . . lives!" This classic advertisement was created by a former newspaperman, Theodore F. MacManus. The latter's Detroit advertising agency, MacManus, John and Adams, Inc. handled the Cadillac Motor Car Division account for many years.

Another interesting innovation on the 1915 Cadillacs was a tilt-beam headlight system. By flicking a switch mounted on the steering column, the driver could aim and adjust the car's headlight reflectors for night driving under varying weather and road conditions.

Cadillac's extensively re-engineered Type 51 was offered in a choice of 10 standard body styles — four open and six closed models. In a single model year, Cadillac's marketing emphasis shifted from open to enclosed-body cars, although open bodies would continue to be a staple item in Cadillac's lineup for many years to come. The wheelbase was increased another two inches, to 122 inches. Prices ranged from $1,975 up to $3,600 F.O.B. the factory in Detroit.

The revolutionary new eight-cylinder Type 51 Cadillac was an overnight success. The company built just over 13,000 of them during the 1915 model year, a noteable achievement for such an advanced automobile in its very first year on the market.

Standard Type 51 Cadillac models included the Cadillac Roadster, which carried a list price of $1,975; the four-passenger Cadillac Salon, a touring car also priced at $1,975; a standard five-passenger touring car and the Cadillac Seven-Passenger Car, both also priced at $1,975; the elegant Cadillac Coupe, priced at $2,500; the Cadillac Sedan for Five Passengers at $2,800; the standard Cadillac Limousine which listed at $3,450, and a new top-line car, the formal "Berlin Limousine", which was the company's most expensive offering at $3,600.

While Cadillac took the auto industry spotlight with its boldly-engineered new eight-cylinder cars, there were other eights: Remington and Briggs-Detroiter both exhibited cars powered by eight-cylinder engines at the 1915 National Automobile Show. Within a few years there were more than 20 V-8 engined cars on the U.S. market.

Also in 1915, the still-young General Motors Company paid out its first dividend. Things were looking up.

This is the powerplant that again catapulted Cadillac to the fore in American automotive engineering. The 90-degree, V-type eight-cylinder engine was the first of its kind used in a mass-production car. It was a superb design noted for its smoothness and quiet flow of power. Designed by D. McCall White, the Type 51 engine had a piston displacement of 314 cubic inches and was rated at 70 brake horsepower. Cadillac has used V-8 engines in its standard models ever since, for a continuous period of more than 65 years.

1915

CADILLAC "TYPE 51" COUPE

Cadillac's bold, new Type 51 chassis, powered by the company's pioneering V-type eight-cylinder engine, found immediate favor with custom body builders. The Seaman Body Co. of Milwaukee executed this handsome, yet entirely practical, coupe for two passengers on a standard 1915 Cadillac V-8 chassis. This little convertible coupe is fitted with what Seaman described as a "Jonas Top". The custom body complements this chassis quite nicely. The external landau irons are functional.

In general appearance, the new Cadillac Type 51 Coupe closely resembled Cadillac coupes produced in the previous two model years. Despite the major switch from a four-cylinder engine to the industry's first mass-produced V-8, exterior styling was almost unchanged. The 1915 Cadillac Coupe had a factory list price of $2,500. Cadillac's marketing emphasis was now on closed rather than open cars. Note the electric horn mounted at the lower edge of the cowl. This was a standard feature on all Type 51 Cadillacs.

The Cadillac Motor Car Co. continued to enjoy a fine reputation abroad, as well as at home. This Cadillac coupe was exported all the way to China. It was finished in a rather striking two-tone paint scheme. Standard closed Type 51 Cadillac models came from the factory finished in Calumet Green with black trim. This body style was mounted on a 122-inch wheelbase. The wood-spoked artillery type wheels carried 36 by 4-½ inch tires.

All set for a sunny day spin, this gentleman poses proudly with a shiny, new Cadillac Type 51 Touring Car. The top has been let down and is covered with a protective boot. Cadillac's open cars for 1915 were upholstered in hand-buffed, plaited long-grain English leather. This well-appointed touring car is equipped with an accessory front bumper and a rear-mounted spare tire carrier.

Although the company's clientele was beginning to show a clear preference for fully-enclosed cars, the big, open touring-type car remained a staple product in Cadillac's model lineup for some time. For 1915, Cadillac customers could choose from three open body styles — the standard Touring Car for Five Passengers illustrated here; a "Seven-Passenger Car", and the Cadillac Phaeton for Four Passengers. The 1915 Type 51 touring car was priced at $1,975 and was built on a new 122-inch wheelbase. Cadillac's open cars were finished in a deep Cadillac Blue with complementing black trim and running gear.

1915

A cold, pelting rain couldn't deter this hardy group from participating in a club run. Their car is a Type 51 Cadillac tourer equipped with Westinghouse air shock absorbers. Car No. 2 also wears side curtains, one of the more practical accessories available to touring car devotees.

This is a Cadillac Type 51 seven-passenger touring car in unrestored condition. It was photographed on the distinctive parqueted wood floor of the famed Henry Ford Museum in Dearborn, Michigan. Two folding auxiliary seats are visible in the rear passenger compartment. For some reason, Cadillac never referred to this model as anything other than the "Seven-Passenger Car". The big tourer sports an accessory spotlight attached to the left side of the windshield frame.

This phantom view reveals construction details of the 1915 Cadillac Type 51 seven-passenger touring car body. Note the complex arrangement of top bows and supports, all of which had to fold conveniently into a neat pile that would not intrude into the rear passenger compartment. Raising and lowering one of these tops was no job for a weakling. The folding auxiliary seats for two passengers are visible through the open rear door.

The open center door provides a good view of the interior layout of the new Type 51 Cadillac Sedan for Five Passengers. The modestly rounded, gable style hood sported embossed, integral trim moldings, as did the louvred hood side panels. These were stamped into the hood sheet-metal.

A new model in Cadillac's 1915 product lineup was this moderately-priced Sedan for Five Passengers. The eight-cylinder, Type 51 Sedan carried a list price of $2,800. The front side windows are gently curved to conform with the swelled body contours. Entry for driver and passengers was through a center door. Large window openings and softly curved body lines gave this car an elegant, semi-formal appearnace.

1915

Cadillac's carriage trade customers could choose between two luxuriously-appointed formal limousines. This is the standard open-front Type 51 Limousine for Seven Passengers. Formal automobiles of this style were invariably driven by appropriately-attired professional chauffeurs. A completely-enclosed Cadillac Limousine was also available for customers who preferred a closed car over the traditional open-front design.

Complementing the standard open-front limousine, Cadillac offered its well-heeled customers a fully-enclosed Type 51 Limousine. This is the 1915 Type 51 Inside-Drive Limousine for Five Passengers. Powered by the quiet, new Cadillac V-8 engine, this car and the companion Limousine for Seven Passengers elevated Cadillac to a new plateau in the very competitive American fine car market. Headroom was sufficient to accommodate gentlemen wearing top hats.

Lawbreakers in the Motor City certainly went to the hoosegow in style, at least when this Cadillac Type 51 paddy wagon responded to the call. The new Cadillac V-8 chassis proved immediately suitable for the mounting of all sorts of special purpose bodies. This long-wheelbase Police Patrol wagon was assigned to the Detroit Police Department's Service Division. Not too luxurious for the involuntary rear compartment passengers, perhaps, but definitely fast!

The powerful, ruggedly-constructed Cadillac Type 51 chassis formed the basis for the first completely armored car built and tested for the United States Army. The Davidson-Cadillac Armored Car was designed by Colonel Royal P. Davidson and underwent extensive testing at the Northwestern Military Academy in August, 1915. With a war raging in Europe, American industry was being called upon to provide an ever-increasing amount of armament and war materiels. The Davidson-Cadillac Armored Car carried Colt machine guns. The words "Armor Battery" and the distinctive Cadillac script were emblazoned on the battlecars's flanks.

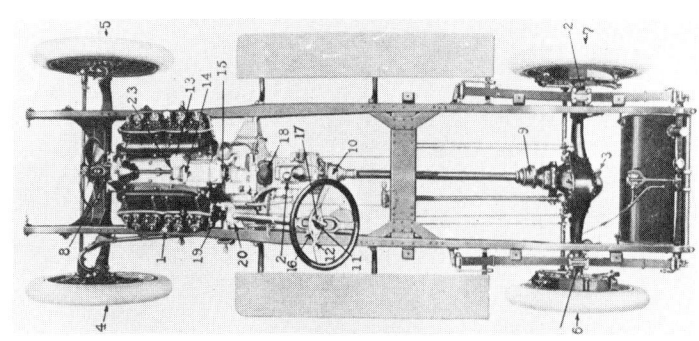

The new Cadillac Type 51 chassis proved ideal for all kinds of commercial applications, ranging from buses and police patrol wagons to ambulances and hearses. For some years, good, used Cadillac Type 51's were in considerable demand for conversion to funeral cars. This big eight-columned hearse with "rayed" bevelled glass side windows was built by A. Geissel and Sons of Philadelphia. Cadillac went on to eventually dominate the hearse and ambulance field in North America.

This overhead view shows to advantage the mechanical details of the new Type 51 Cadillac chassis for 1915. The new V-8 engine occupied about the same room under the Cadillac hood as did the four-cylinder engine previously used. The chassis frame was six inches deep at its center, 30 inches wide at the front and 33 inches wide at the rear. Transmission was of the sliding gear selective type with three forward gears and reverse.

1916

Cadillac's revolutionary new eight-cylinder cars were spectacular successes in their first year on the market. The V-8 Cadillac entered its second model year with a new series designation — Type 53 — and incorporated only minor engineering and styling refinements.

Hoods and body lines on the new 1916 Cadillacs were slightly higher than those on the Type 51 cars they replaced. The most significant mechanical changes were under the hood. Performance of the V-type-eight cylinder Cadillac engine was increased 10%, from 70 horsepower in 1915 to 77. This boost was accomplished primarily through the use of an enlarged intake manifold and modifications to the carburetor. Wheelbase on all standard models remained at 122 inches.

Cadillac's standard Type 53 model lineup included eight models. This array started with the dashing Cadillac Roadster for two passengers. Next came the still-popular five-passenger touring car which was now designated the "Cadillac Salon". The companion seven-passenger tourer was still officially referred to only as the "Cadillac Seven-Passenger Car".

The Type 53 Cadillac lineup also included a handsome, new Victoria for three passengers with a formal leather top. Next came the four-passenger coupe. Cadillac's standard sedan, a big five-passenger car, was now called the Cadillac Brougham. There was the traditional open-front Cadillac Limousine, and an enclosed-drive limousine called the Cadillac Berlin, or sometimes the Berline.

Base prices of the 1916 Cadillacs ranged from $2,080 F.O.B. the factory on up to $3,600. The Type 53 Roadster, the five-passenger Salon touring and the Seven-Passenger Car were identically priced at $2,080. The formal Seven-Passenger Berline was the company's most expensive offering for 1916.

The Type 53 Cadillacs also boasted some interior refinements and new standard equipment. A quality Waltham clock was standard, as was a handy "inspection light".

In May, 1916 a Cadillac established a new transcontinental speed record. E. G. "Cannonball" Baker, accompanied by W. F. Sturm, drove an eight-cylinder Cadillac Roadster from Los Angeles to New York in seven-and-a-half days, clipping nearly four days off their previous coast-to-coast record set in a Stutz Bearcat.

The booming Cadillac Motor Car Co. Plant in Detroit produced just over 18,000 cars during the 1916 model year. The plant was hard-pressed to keep up with orders in a period of extraordinary prewar prosperity. Potential customers were exhorted to place their orders for new Cadillacs early to avoid possible disappointment.

While Cadillac drew much industry and public attention with its new eight-cylinder cars, the competition wasn't standing still. The Packard Motor Company upstaged Cadillac with the introduction of the industry's first production twelve-cylinder engine. The new Packard Twin Six was announced in the latter part of 1915 as a 1916 model and did much to polish Packard's image as an innovative builder of fine cars. Cadillac would not offer its customers a V-12 engine for another 17 years.

Eight-cyliner engines were now very much in vouge. There were by this time no less than 18 makes on the U.S. market powered by V-8 engines. Besides Cadillac, they included such quality makes as Cunningham, Cole, Scripps-Booth and Stearns-Knight. Styling advances this year included a trend toward sloped windshields. The bold, new dual-cowl was the last word in contemporary automobile body design.

In industry news, Alfred P. Sloan formed his United Motors Corp. with himself as president, and William Crapo Durant made a dramatic return to again take control of General Motors.

Despite the protests of some Americans who preferred to remain neutral, the United States was gradually being drawn into the Great War that had been raging in Europe for two years. Led by Packard, the U.S. automotive industry collaborated in the design of a powerful new aircraft engine. Plans were being made to assign volume production of the twelve-cylinder "Liberty" aero engine to several manufacturers, including Cadillac, which had a proven record of precision V-type engine production already behind it.

The military was already taking a good, hard look at the eight-cylinder Cadillac chassis. Several foreign governments, including France and Canada, had ordered Cadillacs for war service overseas. Closer to home, generally unmodified Cadillacs had given an excellent account of themselves in the Pancho Villa uprising on the U.S.-Mexican Border.

Cadillac was about to go to war.

Now designated the Type 53, Cadillac's extremely successful eight-cylinder cars entered their second model year. The new 1916 Cadillac model lineup was essentially the same as that offered the previous year. Pacing the line was the 1916 Cadillac Type 53 Roadster which, at $2,080 F.O.B. the factory in Detroit, was priced identically with the Type 53 "Salon" five-passenger touring car and the Cadillac Seven-Passenger Car. The sporty little Type 53 Roadster was a two-seater. With a top speed of 65 miles an hour, it was the fastest car in Cadillac's stable.

1916

Construction details of the 1916 Cadillac Type 53 Roadster's convertible top are visible in this phantom view. The two-bow top could be easily raised and lowered by one person without leaving the seat. Top bows were of wood construction to keep weight down, with metal sockets and braces.

Specially painted and appointed cars have always been effective advertising mediums. But this Cadillac roadster had real "Moxie" — in the form of a gold-leaf coating applied over the entire body. This could have been the original Solid Gold (looking) Cadillac. Note the oversized rear tire carrier. Moxie was a popular American beverage at this time. It's interesting to note that this car sports the accessory spare tires that were fitted to virtually all Cadillacs. Yet, official Cadillac photos this year never show a spare tire in place.

A truly elegant addition to Cadillac's 1916 model line was the new Type 53 Victoria for Three Passengers. This smart automobile had a decidedly formal look about it, with its erect profile and fashionable carriage bows. The light colored insert in the upper door also added to the custom-built effect. Note the tiny windows in the upper rear roof area above the center of the landau irons. The Type 53 Victoria was priced at $2,800.

While more and more Cadillac customers were switching to closed cars, the old, reliable five-passenger Touring Car continued to sell well. For 1916, this model was formally designated the Cadillac Type 53 Salon for Five Passengers. The eight-cylinder Type 53 Salon carried a price of $2,080 — the same as the Type 53 Roadster and the companion Cadillac Seven-Passenger Car.

An open door affords a glimpse into the cozy interior of the 1916 Cadillac Type 53 Victoria, a new three-passenger model added to the line this year. All Cadillacs still sported two-piece ventilator type windshields. Bumpers were still considered unnecessary. Wheelbase remained at 122 inches for all standard 1916 models. Note that on this car, the passengers sat behind and to the right of the driver.

Side curtains of one type or another were vital necessities for anyone who owned a touring car in all but the driest climes. Most of these were of the folding canvas type. But those who desired something a little more permanent for use during the winter months could choose from a wide range of optional, aftermarket side curtains and enclosures. This is a Type 53 Cadillac five-passenger touring car equipped with what was called the "Bishop Enclosure" for touring cars with standard folding tops. Of semi-permanent construction, the Bishop Top offered the comfort and warmth of an enclosed car in winter or rainy seasons.

1916

For some obscure reason, the Cadillac Motor Car Company still refused to call this model anything other than the "Cadillac Seven-Passenger Car." The big tourer was equipped with two folding auxiliary seats, but was often utilized as a roomy five-passenger touring car. The Type 53 Seven-Passenger Car was priced at $2,080 — precisely the same as the standard five-passenger touring car and the sporty Cadillac Roadster.

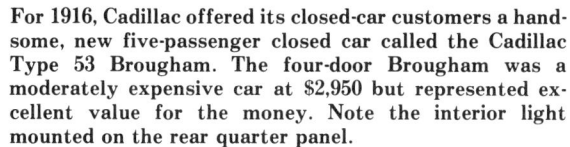

For 1916, Cadillac offered its closed-car customers a handsome, new five-passenger closed car called the Cadillac Type 53 Brougham. The four-door Brougham was a moderately expensive car at $2,950 but represented excellent value for the money. Note the interior light mounted on the rear quarter panel.

The seating arrangement of the Model 53 Limousine can be seen in this open-door view. A comparison of this car with the new Berline model indicates that the bodies had the same base, with the only real difference being in the treatment of the upper front compartment, between the A and B pillars. An interesting Cadillac feature was the large electric motor driven horn, mounted at the left base of the cowl. Bumpers and spare tires, considered accessories, were never illustrated in this year's official Cadillac photos. Apparently management felt these items spoiled the lines of the car.

The formal, open-front limousine still set the style for many fashion-conscious Cadillac customers. This is the 1916 Cadillac Type 53 Limousine for Seven Passengers, a correctly-formal conveyance that bore a standard price of $3,450. Customers who preferred a fully-enclosed limousine could opt for the even more expensive Cadillac Type 53 Berlin.

At the very top of Cadillac's standard 1916 model offerings was the regal Type 53 Berlin for Seven Passengers. This heavy, fully-enclosed limousine commanded a price of $3,600 — only $150 more than the open-front Type 53 Cadillac Limousine. A Waltham clock and an inspection light were made standard equipment on the 1916 Cadillacs, and the hood and body lines were slightly higher than on Cadillac's 1915 models. The Berline was fitted with a permanent non-removable windowed partition between the front and rear compartments.

1916

Open front and rear doors provide a peek into the posh interior of the 1916 Cadillac Type 53 Berlin, which was also sometimes advertised as the Cadillac Berline. The chauffeur's compartment was upholstered in the finest quality hand-buffed black leather while the rear passenger compartment, complete with a pair of folding auxiliary seats, was trimmed in fine cloth and mohair. The Type 53's V-8 engine was uprated to 77 horsepower for 1916.

The quality Cadillac eight-cylinder motor car chassis quickly became popular with custom body builders. The Seaman Body Co. of Milwaukee, Wisc. executed this crisply-formal open-front Town Car on the Type 53 Cadillac chassis without altering the standard 122-inch wheelbase. Formal cars of this type were almost invariably finished in dark colors — black, dark blue or deep Brewster green.

Although the underline indicates that it is a 1917, the style of hood on this car reveals that this chassis is of an earlier year. Perhaps construction of this custom-bodied limousine was not completed until 1917. The custom body was designed and built by the Seaman Body Co. of Milwaukee. Note how neatly the curved valance panel fills in the area between the running board and body sill. All Cadillacs had utilized such panels for some years.

Many years before the Detroit factory began to supply specially engineered chassis for this purpose, Cadillac hearses and ambulances were widely used all over the United States. This impressive eight-columned, carved-panel hearse was built on a Cadillac V-8 chassis by Owen Bros., of Lima, Ohio. Owen offered a complete line of hearses and ambulances built on the Cadillac eight-cylinder chassis or the slightly less expensive Buick Six.

The speedy Cadillac eight-cylinder chassis quickly acquired the respect of law enforcement agencies, and sometimes that of the criminal element they were frequently called upon to pursue. The Jersey City, N.J. Police Department once operated this fleet of Cadillac Police Patrols. The special patrol wagons have screened sides and a wide rear step. A heavily screened partition separates the driver from his usually involuntary rear seat (or bench) passengers.

Here is how the 77-horsepower V-type, eight-cylinder engine looked as installed in the 1916 Cadillac Type 53 chassis. The boost in horsepower, from 70 in 1915 to 77 this year, was accomplished primarily by refinements in the carburetor. Intake manifolds were also enlarged and the distributor was relocated from its former position at the rear of the engine to the front of the block.

1917

This year was destined to go down as one of the most eventful in Cadillac's 15-year history. Before the year was out, the United States would be drawn into the Great War in Europe, and the Lelands would be gone from Cadillac.

For 1917, the tremendously successful eight-cylinder Cadillac entered its third model year with relatively little change. There were several new body styles, however, and refinements in both exterior styling and engineering. Cadillac's 1917 cars carried yet another new model designation — Type 55. As the Type 55 went into Cadillac showrooms in the fall of 1916, the company proudly boasted that there were now more than 31,000 eight-cylinder Cadillacs on the road.

The Type 55 Cadillac lineup included 11 body styles. Eight were built on a new 125-inch wheelbase and three premium models were available on a special 132-inch wheelbase. The new, standard wheelbase was three inches longer than that of the Type 53 which the Type 55 succeeded. Cadillac's revised 1917 appearance included new, crowned fenders, redesigned nickel-trimmed, black enameled headlights and a 1½-inch wide molding around the upper edge of all open bodies. Cadillac's frames were increased to a hefty eight inches in depth and incorporated three tubular crossmembers. The Cadillac V-8 engine utilized new, lightweight pistons with an "hourglass" shape to reduce bearing surface.

The Type 55 lineup included two and four-passenger roadsters. The four-passenger ragtop was called the Cadillac Club Roadster.

The old, reliable standard five-passenger touring car had been replaced in the Cadillac lineup by the Phaeton for Four Passengers. There was still a companion "Cadillac Seven-Passenger Car", and a versatile new "Seven Passenger Convertible Touring Car" that offered the attributes of a closed car in colder weather and all of the pleasure of an open tourer in the summer.

Cadillac's closed body offerings included the elegant new Type 55 Coupe for Four Passengers, which bore a strong resemblance to electric cars of the period, and the Type 55 Five-Passenger Brougham. The "Convertible Victoria" also remained in the Cadillac line for 1917, but was now a four-passenger car. At the top of the range, on a new special 132-inch wheelbase, were three formal cars — the standard open-front Cadillac Limousine; an enclosed-drive limousine with division called the Imperial Limousine, and a stylish, new Cadillac Landaulet. The handsome Landaulet could be quickly converted from an enclosed limousine into a semi-open car by lowering the folding rear roof and sliding a hinged roof section forward. The rear side windows on this model also lowered adding to the open effect. Landaulets of this style were popular in Europe, and several U.S. luxury car builders and custom body builders offered them.

In April, 1917 the United States was finally drawn into the First World War. Cadillac was destined to play a significant role in this conflict. Early in the war, several foreign governments recognized the superiority of the eight-cylinder Cadillac chassis over anything else available. Hundreds of Cadillacs were shipped overseas for use as staff cars, ambulances and as other specialized military vehicles.

Thus when the United States was plunged into the war, the military did not have far to look for a suitable, all-purpose military automobile. In July, 1917 the U.S. Army conducted a series of punishing tests at Marfa, Texas to select an official car. Cadillacs underwent 7,000 miles of tests under the most appalling of conditions. The Type 55 touring car was ultimately selected as the official car of the U.S. Army. The U.S. Marine Corps chose the Cadillac, too. More than 2,300 were eventually shipped overseas.

As summer approached, there was increasing friction between the Lelands and W. C. Durant, who had recently returned to control General Motors. In his earlier tenure, Durant had left Cadillac management entirely to the Lelands. The highly patriotic Lelands wanted Cadillac to undertake the manufacture of Liberty aircraft engines for the U.S. Government, but Durant refused to allow this. In June, 1917 Henry M. Leland and his son, Wilfred, abruptly left the firm they had helped organize and bring to world-renowned prominence. Two months later Leland organized a new company to build Liberty aero engines. He named this company the Lincoln Company after the American president he had long idolized. The "Leland Era" at Cadillac had come to an end. Without the Lelands, the place just didn't seem the same. Richard H. Collins was named president and general manager of the Cadillac Motor Car Company. Benjamin H. Anibal was named chief engineer that year.

In typical Leland fashion, Lincoln soon occupied a huge, new plant in Detroit and had secured a contract to build 6,000 Liberty twelve-cylinder aircraft engines.

For the third time in as many years, Cadillacs for 1917 bore a new series designation — Type 55. The 1917 Cadillacs incorporated a number of improvements, and were built on two separate wheelbases. This is the Type 55 Cadillac Roadster for two Passengers, built on the new 125-inch wheelbase which was two inches longer than the Type 53 it succeeded. The base price, however, remained at $2,080, despite the fact that the car sported a brand new body with a greatly refined rear treatment.

1917

This is the 1917 Cadillac Type 55 Roadster as illustrated in a Cadillac Motor Car Co. salesman's book for that year. It shows interior and the new top construction details, and the new rumble seat in the open position. Note the rumble seat handrails, the spare tire carrier, and the new squared rear panel design.

For 1917, Cadillac offered sports-oriented buyers a choice of both two and four-passenger roadsters. The four-passenger ragtop was formally called the Type 55 Club Roadster, and at $2,080 was identically priced with the two-passenger version. Four models — the two and four-passenger roadsters, four-passenger phaeton and the seven-passenger touring car carried the same $2,080 price tag F.O.B. the plant in Detroit.

One of two "convertible" models in the 1917 Cadillac model lineup was this Type 55 Victoria for Four Passengers. Priced at $2,550, the "Vicky" could be quickly converted to a semi-open configuration by removing the "B" pillars and window glass. At this time the term "convertible" had not been applied to a car with a folding top. For 1917, Cadillac also offered a "convertible" seven-passenger touring car.

Cadillac's coupes were among the most elegant cars in the entire Cadillac line. Their styling was highly attractive, proportionately correct and not unlike the formal china cabinets of the era. These rather formal-looking coupes somehow looked less utilitarian than the company's other models. The Type 55 Coupe for Four Passengers, generally similar in appearance to Cadillac coupes sold for the previous few years bore a striking resemblance to the electric cars offered by other manufacturers. The gracefully-designed upper body area afforded exceptional visibility. The eight-cylinder Type 55 Coupe carried a base price of $2,800.

Among Cadillac's most popular models was the standard open tourer. For 1917, the standard touring car was this one, now called the Cadillac Four-Passenger Phaeton. The lithe-looking Phaeton replaced the five-passenger standard touring car. The Type 55 Phaeton, on the 125-inch wheelbase chassis, carried a base price of $2,080. A seven-passenger touring car was also offered in the Type 55 series for 1917.

1917

Again, we were unable to resist the temptation to throw in this photo of the Type 55 Phaeton with the top up and half-doors open. While wood-spoked, artillery-type wheels were standard on all models, wire wheels were optionally available at extra cost. These gave the Type 55 Cadillac a lighter, sportier look. While wire wheels were used on most cars being built in Europe, they were slow in gaining acceptance on this side of the Atlantic for a number of reasons. For one thing, wood-spoked wheels were stronger and they were definitely easier to maintain and to keep clean.

A companion car to the Type 55 Phaeton was the slightly larger Cadillac Seven-Passenger Car. This 1917 model was built on the same 125-inch wheelbase but had a longer rear body and carried as standard equipment two folding auxiliary seats. The Type 55 Cadillac Seven-Passenger Car was identically priced with the Phaeton at $2,080 and certainly represented top value for the money. Styling changes evident on the Type 55 Cadillacs included new, crowned fenders and a moulding around the top of the redesigned body.

In April, 1917 the United States was plunged into the First World War. Hundreds of Cadillacs had been shipped to Europe since the war began in 1914, and it came as no surprise when the Cadillac Seven-Passenger Car was selected as the standard car for the United States Army. Cadillac Type 55 touring cars were put through a gruelling series of competitive tests at Marfa, Texas in the summer of 1917 resulting in the U.S. Army's choice. Thousands of eight-cylinder Cadillacs were eventually shipped to the Front, where they served with distinction in a wide range of roles. Military modifications were minimal.

Several years before the United States entered the war, Cadillacs were to be found on the battlefields of Europe. English, French and Canadian governments ordered substantial numbers of Cadillacs which were used as staff cars, ambulances and other special military vehicles. Here, a Cadillac staff car crests a hill at Vimy Ridge carrying a load of British officers. Note the right-hand drive steering.

Closed bodies continued to gain in popularity, although they still accounted for only about 10 per cent of sales in this country. One of Cadillac's most popular closed cars was the roomy Five-Passenger Brougham, with a standard price of $2,950 making it one of the company's more expensive models. The four-door Brougham was built on the 125-inch wheelbase chassis and was equipped with what the catalog described as two "emergency seats" to accommodate two additional passengers.

1917

The Seven-Passenger Convertible Touring Car

An interesting new addition to the Cadillac model lineup this year was the Type 55 Cadillac Convertible Touring Car for Seven Passengers. This versatile new body style offered the best of two worlds — the comfort and practicality of a closed car with the fair-weather appeal of an open tourer. The new convertible tourer was constructed on the 125-inch wheelbase chassis and carried a standard price of $2,675. Although the solid top could not be folded down, the windows and pillars could be removed, thus creating one of the early "hardtop" styles.

The Seven-Passenger Limousine

For 1917, Cadillacs were built on two separate wheelbases. Eight Type 55 models were offered on the standard 125-inch wheelbase chassis. A new 132 inch wheelbase chassis was reserved for Cadillac's three most expensive formal limousine bodies. This is the 1917 Cadillac Type 55 Limousine for Seven Passengers, a traditional open-front limousine which listed for $3,600. Standard equipment included a pair of "disappearing" auxiliary seats. A fully-enclosed Type 55 Limousine was also available on the 132-inch wheelbase chassis, along with a new Cadillac Landaulet.

The majority of Cadillacs produced for military service were of the seven-passenger touring car type, but many modified limousines were provided for use as Army staff cars. The star in the lower left of the windshield calls attention to the high rank of the rear compartment occupant. Enlisted men habitually saluted any passing Cadillac, just to be on the safe side. Note the spare tires lashed to the roof-top luggage carrier and the extra gas tank on the running board.

The Seven-Passenger Landaulet

Another new addition to the eight-cylinder Cadillac line this year was the prestigious Type 55 Seven-Passenger Landaulet. This regal "convertible" could be quickly converted from a closed limousine into a semi-open car by lowering the folding rear roof and sliding the "fixed" rear roof section forward. The rear side windows could also be lowered. The Type 55 Landaulet was built on the 132-inch wheelbase chassis and carried the same $3,750 price as the Type 55 Imperial Limousine.

The Seven-Passenger Imperial

A new premium designation — Imperial — was added to Cadillac's vocabulary this year. The Type 55 Seven-Passenger Imperial was a richly-appointed, enclosed-drive limousine on the 132-inch wheelbase chassis. It carried a base price of $3,750. The "Imperial" designation would henceforth denote a formal Cadillac limousine with rear compartment partition well into the 1960's. The Cadillac Imperial Limousine is not to be confused with the top-line Chrysler which came along later.

1917

Exactly which member of the Henry Ford Family owned this pretty Cadillac Town Car isn't indicated, but the photo was headed "Cadillacs for the Henry Ford Family". The Ford Motor Company was still some time away from acquiring Lincoln, and outside of a few specially-bodied Model "T" 's, did not have a true luxury make to call its own, so it is entirely possible that the Fords kept a few fine cars in their family garages. This is a full custom-bodied car. The wire wheels seem to go well with the light, clean body styling.

This fascinating hybrid was simply too interesting to leave out. The special formal sedan was built by the Fleetwood Body Co. of Fleetwood, Pa. on a Packard chassis. But under the hood beats a Cadillac "heart", a Type 55 V-8 engine. Note the rakishly-angled windshield and dual spare tires. Custom body builders were busy creating all kinds of special bodies for customers who wanted something different. This Fleetwood "Packillac" certainly qualifies there.

Although more than a few professional car builders had already recognized the quality and endurance inherent in the Cadillac chassis, the marque still was not the primary figure on the American hearse and ambulance scene that it would become in later years. Part of the reason for this was the tremendous competition from dozens of other manufacturers for this business, plus the fact that most of the major professional car builders of this era were constructing their own chassis and many were even building their own engines and running gear. One company that utilized outside chassis, and even bought used chassis and reconditioned them, was the Hoover Wagon Body Co. of York, Pa. This firm turned out a wide variety of bodies, from the rather plain ambulance shown here to highly ornate carved-side hearses. In an effort to gain headroom for the attendants in this vehicle, the roof was built relatively high — in fact, much higher than the Cadillac windshield. To compensate for this mis-match, the builders simply attached a piece of canvas from the roof edge to the windshield top.

Although more than a few retired Cadillac hearses and ambulances were converted into rescue squads or other types of makeshift fire apparatus, factory-built Cadillac fire engines are exceedingly rare. This white-painted Type 55 Cadillac Combination Hose and Chemical Car was built by the Northern Pump Co. of Minneapolis, for the Redwood Falls, Calif., Fire Department. The soda-acid chemical tank is located under the seat. Cadillacs had long been popular as fire chief's runabouts.

One of the more significant American technical developments to come out of the First World War was the Liberty aircraft engine, which was developed by a consortium of U.S. automobile manufacturers spearheaded by Packard, and including Cadillac. Henry M. Leland wanted his Cadillac factory to build Liberty engines for the U.S. Government, but he encountered stiff opposition from William Crapo Durant, who had returned to control of GM. Consequently, H. M. and Wilfred Leland abruptly left Cadillac in July, 1917 to form an new company — Lincoln — for the purpose of producing Liberty engines. Ironically, Cadillac was soon building 12-cylinder Liberty aircraft engines, too.

1918

Despite the nation's involvement in a foreign war, the imposition of burdensome war taxes and the departure of the Lelands, the Cadillac Motor Car Company continued to prosper. The aging Cadillac plant, bursting at its seams, built more than 20,000 cars during the 1918 model year. The eight-cylinder Cadillac was now in its fourth year. Sales showed no sign of letting up.

For 1918, Cadillac redesignated its cars the Type 57. This series would be carried through virtually without change to the end of the 1919 model year. With its production capacity totally committed and limited, the company was making plans to build a new, much larger plant.

The new Type 57 Cadillacs had a higher radiator and a higher, longer hood. All models sported a new cowl. The proven Cadillac V-8 engine now had detachable cylinder heads. The transmission had been improved and a Kellogg tire pump was standard on 1918 models.

The Type 57 Cadillac was available in a choice of ten body styles on two wheelbases. There were two formal new body styles — a handsome Type 57 Town Limousine and a companion Town Landaulet. This limousine was built on the 132-inch wheelbase chassis and had a permanently-open roofless front and a narrower body than the standard seven-passenger Cadillac Limousine. The companion Town Landaulet had a collapsible rear roof.

The Type 57 model lineup included a two-passenger rumble seat roadster; the four-passenger Phaeton; the big Seven-Passenger Car; four-passenger Convertible Victoria and the popular Five-Passenger Brougham, all on the 125-inch wheelbase chassis.

At the top of the line, on the 132-inch wheelbase chassis, were the Limousine for Seven Passengers, the Seven-Passenger Imperial Limousine and the convertible Seven-Passenger Landaulet. Cadillac prices ranged from $2,805 on up to $4,345. At $2,085 the roadster, Phaeton and seven-passenger car were the lowest-priced Type 57's, while the big Imperial was the most expensive standard model.

Cadillac's chief rivals were still Packard, Pierce-Arrow and Peerless. Packard led the field at this point. Cadillac's finest cars were positioned a notch below Packard's. The Packard was sold as a gentleman's car — a cultured car with much emphasis on its correctness of appearance and appointments — while the Cadillac was sold on the basis of its excellent value for the money and the merits of its superb engineering. These two Detroit-based giants were not quite yet nose-to-nose rivals in the luxury car field. But the gap was closing.

Ironically, Cadillac eventually did manufacture Liberty aircraft engines despite pacifist W. C. Durant's original opposition. Cadillac shared this production with its sister GM car division, Buick. The Fisher brothers had been awarded a major contract for warplane production, and a huge, new plant was rushed to completion on West End Ave. on Detroit's near-west side. By 1918 this plant was turning out 40 planes a day. This plant was later acquired by the Fishers and became GM's big Plant 18. It produced bodies for a number of automobile manufacturers, primarily Buick. This plant is still a key element in Cadillac production today. It is known as the Fisher Fleetwood plant and produces most of Cadillac's standard bodies.

The man credited with creating Cadillac's world-renowned V-8 engine, D. McCall White, left the company this year to take a position as chief engineer of another new automaker, Lafayette. This was doomed to be a shortlived venture, out of business by 1926.

On November 11, 1918 an armistice was signed, ending the war that had over a period of four years left much of Europe in ruins. Production again returned to a peacetime market. Chevrolet in 1918 joined the General Motors family.

The nation was now preoccupied with the war in Europe. Most American automobile manufacturers had shifted over to war production. Civilian passenger cars were still rolling off the assembly lines in most cases, but were heavily taxed as luxuries. Cadillac's modestly changed 1918 models were designated the Type 57. This series was produced without significant change through the 1919 model year. This is the Type 57 Roadster for Four Passengers. A two-seat roadster with rumble seat was also available. Both were built on the 125-inch wheelbase.

The versatile Convertible Victoria remained in Cadillac's closed car offerings for 1918. This is how the pleasingly styled Type 57 Victoria appeared in "closed" form. The side window glass and roof pillars could be removed to convert this car into a semi-open car. One of 10 body styles, the Victoria on 125-inch wheelbase carried a list price of $3,205.

1918

This is a standard Cadillac touring car in civilian clothes. Cadillac for some reason was always juggling the descriptions of its larger, open cars. The standard five-passenger tourer came and went, as did the companion Phaeton. For the past several years the Phaeton was sold as a four-passenger car, but the Type 57 catalog listed it as the Phaeton for Five Passengers. In some 1918 advertising it's shown as a four. Whatever the seating, list price of the 125-inch wheelbase Phaeton is shown as $2,805.

Here is a standard Cadillac touring car in a real collegiate setting. This is obviously a posed publicity shot. The Type 57 tourer is finished in a two-tone color scheme. Mom seems proud of her all-star son. The 1918 Cadillacs feature detachable cylinder heads. Radiators and hoods were higher, and all Cadillacs had a new cowl.

As proved by the U.S. military, standard Cadillacs were easily adaptable to any number of specialized duties. This Cadillac touring car has been outfitted with large, flanged disc wheels for service as a railway track inspection car. This was not an uncommon chore for passenger cars at this time. Such cars were usually fitted with dual braking systems. Note the warning flag attached to the rear.

Military vehicles haven't changed all that much over the years. This big Cadillac staff car carries the no-nonsense warning "For Official Use Only" on its rear doors. The U.S. Army driver is chauffering a carload of civilian bigwigs, judging from the crop of silk toppers jutting out of the rear passenger compartment. The car is a Type 57 Seven-Passenger Car, as modified for military service.

When compared against the Healy-built town car in this chapter, the standard Cadillac Type 57 Town Limousine was a bit stodgy looking. Still, this was one of Cadillac's top prestige cars, sharing its spotlight only with the also-new Town Landaulet in style, and only with the Imperial Limousine in price. Built on the 132-inch wheelbase chassis, the 7-passenger car featured a chauffeur's compartment upholstered in fine-grain natural leather, while the rear compartment was done in quality broadcloth. The auxiliary jump seats seen through the open door could be folded against the forward bulkhead. It appears that Cadillac made no provision for weather protection for the driver on either this or the Town Landaulet. The "town car" designation was well deserved, as it was very rare indeed when one of these high-styled vehicles was seen outside of posh areas of major cities.

1918

Two versions of the Landaulet were offered in 1918 and 1919. One was the Landaulet shown here, with a permanent solid top over the chauffeur's compartment. The other was the Town Landaulet, in which the solid roof ended just forward of the rear quarter windows and a folding rear section of tonneau leather, fitted with functional landau irons. Both were priced at about $4,300.

War or no war, despite stiff war taxes, Americans still demanded style in their automobiles. Custom body builders had a wealth of quality chassis to choose from for mounting their custom creations. This inordinately handsome town car was executed on the eight-cylinder Cadillac Type 57 chassis by Healy & Co. of New York. A speaking tube juts up out of the narrow space between the front seats. This is a semi-collapsible version, meaning that the landau irons were functional, and the leather-covered quarter section aft of the door windows could be folded back to provide an open rear section. Other versions of this body, which was also installed on Packard chassis, were fully-collapsible and non-collapsible styles.

Several custom body shops were turning out some distinctive coachwork by this time, for discerning customers and families who wanted something really different and who had the money to pay for their whims. A good example of this is the special Victoria-type touring car body shown here. It was designed and built by the Fleetwood Metal Body Co., of Fleetwood, Pa. and is mounted on an eight-cylinder Cadillac Type 57 chassis. Its owner evidently did not like the standard Cadillac radiator, so he had it disguised. The sharply-raked windshield and wire wheels give this car a truly sporty look.

Another wartime Cadillac was this two-and-a-half ton crawler-type Artillery Tractor. Cadillac built 1,157 of these artillery tractors for overseas service. The V-type Cadillac eight-cylinder engine proved amazingly adaptable with a minimum of modification and fully lived up to its civilian reputation for unfailing dependability.

Cadillacs fulfilled a bewildering variety of duties on overseas battlefields during the First World War. This was one of them. The Cadillac Searchlight Car was built on a special 145-inch wheelbase chassis. The "mother" car transported the portable searchlight unit, which had a sky-piercing range of 15 miles. The giant searchlight had a 60-inch lens. This combination had a top speed of 50 miles an hour.

Yet another unique application of the Cadillac V-8 engine during the First World War was this special balloon winch which was employed to control the balloons of military observers to what were described as "dizzy heights to observe manuvers of the enemy, then lower them again to zones of safety when attacked by hostile aircraft." Thousands of Cadillacs were used as staff cars, ambulances and other specialized military purposes. Some 300 Cadillac limousines were supplied to the U.S. Army for use by senior officers.

1919

With the devastating war in Europe over, America's automobile factories shifted back to peacetime production. The economy was booming and there was a huge, pent-up demand for new automobiles.

The Cadillac Motor Car Co. carried its Type 57 models into the 1919 model year with little change. For the second year in a row, the company was able to build well over 20,000 cars.

The renowned Cadillac eight-cylinder V-type engine was now in its fifth year on the market. Relatively few changes or improvements had been made to this engine since its introduction. Displacement was 314 cubic inches. Bore and stroke were $3\frac{1}{8}$ inches by $5\frac{1}{8}$ inches. The Cadillac V-8 had a National Automobile Chamber of Commerce rating of 31.2 horsepower.

The Type 57 Cadillac line for 1919 included 10 standard body styles on two wheelbases. On the 125-inch wheelbase chassis were the Cadillac Roadster, four-passenger Phaeton and Seven-Passenger Car, the Victoria for Four Passengers and the Five-Passenger Brougham. On the 132-inch wheelbase chassis were the seven-passenger open-front Limousine, enclosed-drive Imperial and the convertible Landaulet. The Town Limousine and companion Town Landaulet introduced the previous year were continued for 1919, also on the long wheelbase chassis.

Late production Type 57 Cadillacs sported a new hood with finer, more numerous ventilation louvers.

Cadillacs still rode on wood-spoked artillery type wheels. Tires were 35 x 5 inches, except for coupes which had 34 x 4½-inch tires. Non-skid tires were standard on the rear wheels. Standard tread at this time was 56 inches, and all Cadillacs were equipped with 20-gallon fuel tanks.

Standard finish for Cadillac's open cars was a deep blue with black trim. Two-passenger cars sported cream-colored wheels. Closed-body Type 57 Cadillacs were finished in a dark Calumet Green with black trim. Running boards were linoleum-covered with metal bindings.

Despite the gradual trend toward closed bodies, some 90 per cent of all cars being built in the United States at this time were of the open type — roadsters and touring cars.

By 1919, the Fisher Bros. Body Corporation was building for General Motors exclusively. This year, GM acquired a controlling interest in Fisher Body.

In appreciation for his contribution to the United States' war effort, grateful Cadillac Motor Car Co. employees contributed from their own pockets to buy a special Cadillac closed car for General John J. Pershing, who had led the American Expeditionary Force overseas. This special car was finished in "Suburban Blue" and had the General's initials emblazoned on each rear door. All fittings were gold-plated.

Henry M. Leland's newly-formed Lincoln Motor Company had built more than 6,500 Liberty aero engines for the U. S. Government. With the war over, the Lelands were looking for new uses for their new Detroit plant. It came as no surprise to anyone in the industry when the Lelands announced that they would soon begin production of the finest motor cars ever built in America. These Leland-built cars would bear the Lincoln name.

Also virtually unchanged this year was the Type 57 Victoria. This four-passenger car was sold as the "Convertible Victoria", as the side window glass and center door posts could be removed to convert it into a semi-open car. The versatile Victoria was built on the 125-inch wheelbase chassis and carried a base price of $3,205 — the same as the previous year. This is an early model, with the heavy hood louvres.

The Type 57 Cadillac was carried through the 1919 with little change. Late production Type 57's however, had the type of hood shown on this car with more, and finer, ventilation louvres. This is the Type 57 two-seat roadster with rumble seat. Built on the 125-inch wheelbase, it was the companion car to the 4-Passenger Club Roadster. Base price for both roadsters, and the Type 57 Phaeton for four passengers, was $2,805.

1919

Large, open touring cars continued to hold an important place in Cadillac's product line. There were two in 1919 — the four-passenger Phaeton shown here and the big Cadillac Seven-Passenger Car. Both were on the 125-inch wheelbase. The Phaeton carried a base price of $2,805. Cadillac advertised this tourer as... "a snappy four-passenger sport type designed for the motorist who appreciates smart and racy lines."

The Seven-Passenger Car

The big Type 57 Seven-Passenger Car was priced the same as the smaller and lighter four-passenger tourer, at $2,805 F.O.B. Detroit. Both cars were built on the 125-inch wheelbase chassis. The seven-passenger touring car was equipped with a pair of auxiliary seats. Open cars were finished in Cadillac Blue with black trim, and upholstered in fine-grain high gloss black leather.

This is how the standard Type 57 Cadillac 7-Passenger touring car looked with the top up. All of Cadillac's open cars were equipped with what was described as the Cadillac "one-man" top, which could be easily raised or lowered by one person. Side curtains and a "dust envelope" cover for the top were also standard, but the twin spare tires on this example were an added cost accessory.

This is a Type 57 Cadillac seven-passenger touring car in fully-closed configuration. Standard equipment included the folding top, a "gypsy type" rear curtain with a rear window and side curtains, three-quarters of the area of which were transparent. Side windows swung open with the doors. The folding auxiliary seats can be seen in their folded position.

The list price of the Cadillac Brougham for five passengers remained at $3,650. Cadillac's standard closed models were finished in Calumet Green with black trim. Standard equipment included a Waltham eight-day clock, electric horn, power air compressor for inflating tires, a footrest, and robe hanger in the rear compartment. The whitewall tires shown in this section were definitely added-cost items, and were still not all that popular during this era.

1919

Even though the devastating war in Europe was over, Americans continued to lavish glory on their military heroes. As a token of their appreciation to General John J. Pershing, Commander-In-Chief of the U.S. Expeditionary Force in Europe, grateful Cadillac Motor Car Co. employees chipped in to buy this special Type 57 Cadillac Brougham for General Pershing. The car was finished in a rich "Suburban Blue", had gold-plated fittings throughout and rich, mohair antique upholstery. Note the General's initials on the rear door. Pershing was no stranger to Cadillac motoring — he had used Cadillacs as his standard staff and command cars throughout the war.

Priced just below the top-line Imperial was the Type 57 Town Landaulet. This elegant, formal car is shown here in the full-open position: the rear roof has been lowered, and the hinged roof over the passenger compartment has been folded forward, and the roof over the chauffeur's compartment has been removed. This must have been a truly dignified way to tour Central Park on a Sunday afternoon — whether to see or to be seen.

The postwar American economy was still booming, and all manufacturers were hard-pressed to meet pent-up demand for new civilian passenger cars. The luxury car bracket was still healthy, too, and Cadillac had no less than five formal models to offer its customers. This is the standard Type 57 Limousine, and open-front, chauffeur-driven car priced at $4,145. An enclosed-drive limousine was also available for those who preferred this style over the traditional open-front vehicle.

The Type 57 Town Limousine was now in its second year. This formal town car and the companion Town Landaulet were built on the 132-inch wheelbase chassis. The Town Limousine was priced at $4,160 and the Town Landaulet at $4,295. Both were sold as seven-passenger cars, as both featured folding auxiliary seats in the rear compartment.

The most expensive car in the Type 57 line was the Seven-Passenger Imperial, a formal, enclosed-drive limousine with divider partition. The stately Imperial was introduced in the previous series. For 1919, the standard price remained unchanged at $4,345. Cadillac's top five models were built on a special 132-inch wheelbase.

1919

The Seaman Body Co. of Milwaukee continued as a principal supplier of custom-built bodies for the eight-cylinder Cadillac chassis. This big seven-passenger closed car looks like an enlarged Brougham. It does not appear to have a divider. Note the slightly angled windshield. Seaman at this time was also supplying custom bodies to Packard, Lozier, and Locomobile on a regular basis, and was providing production bodies for Nash, Kissel, Velie, and Moline. Eventually, Seaman Body Co. became a division of Nash.

"Cadillac Proves Too Tough For The Germans" read the headline on this publicity photo. Even though the war was over, Cadillac's exploits on the battlefields of Europe were still being used as an effective sales tool by Cadillac salesmen. The standard Cadillac touring car had been the standard car of the U.S. Army. This battered Type 57 had obviously seen its share of action. Note the order papers stuffed between the cowl and windshield frame.

Certainly one of the most unusual of all Type 57 Cadillacs built was this two-passenger roadster, the entire body of which has been covered in DuPont fabric cord. Special creations like this one, of questionable durability, were often done for advertising and promotional purposes. The big Westinghouse air shocks and front bumper were also non-standard equipment. This conversion was obviously done after the car left the factory in Detroit.

Ever since its introduction five years earlier, the Cadillac eight-cylinder chassis proved irresistable to custom body builders. The Charles Schutte Body Co. of Lancaster, Pa. offered this special "Turtle Deck" sports body for mounting on the standard Cadillac Type 57 125-inch wheelbase chassis. Spare tires were carried in a well in the rear deck. The big disc wheels give this slick roadster a sturdy, speedy look. These, however, were not real disc wheels, but were the standard wood-spoke wheels enclosed between patent "Schutte Discs." For use in cold climates, the car could be provided with a top and side curtains. The strange looking unit just ahead of the rear fender is a folding extra seat. One can imagine the feelings of the extra passenger when precariously perched on this outboard seat, especially if the driver decided to put this speedster through its paces.

Healy & Co. of New York City was one of America's better known custom body houses. Healy did a fair number of Cadillacs, ranging from relatively staid styles that closely resembled production models to some thoroughly custom creations that bordered on the bizzare. This special touring car is undistinguishable as a Type 57 Cadillac. Noteworthy features are the high, gabled hood, special radiator shell, flat-planed fenders fore and aft and drum-style headlights. Through its ties with Uppercu-Cadillac, one of New York City's primary Cadillac agencies, Healy was to become one of the major eastern custom body companies building on Cadillac chassis in this era. Not exclusively locked in with Cadillac, Healy also added its touch to a multitude of Packard cars, and worked on other luxury chassis as well.

1920

Nineteen-Twenty was an eventful year for the American automobile industry, for General Motors, and for Cadillac. The nation was plunged into a depression following a prolonged economic boom that had continued through the First World War and the years immediately following. Sales of new passenger cars in the U.S. dropped substantially.

On the corporate front, the colorful William Crapo Durant lost control of General Motors for the second, and final, time. The flamboyant and ambitious Billy Durant had again overextended himself in his enthusiasm to expand General Motors, and he was finally ousted by a concerned board of directors. The corporation once again faced reorganization in order to survive and to untangle the financial morass left in the wake of its founder. Pierre S. duPont was named the new President, and Alfred P. Sloan implemented the concept of line and staff management that would eventually make General Motors Corporation the largest and most successful business enterprise in the world. In another important development at GM, Charles F. Kettering — who had brought Cadillac the self-starter and its second Dewar Trophy — organized the General Motors Research Corp. within the company.

As the third decade of the new century unfolded, General Motors had only 12 per cent of the U.S. automobile market. Still riding the spectacular success of a single, low-priced product, the Ford Motor Company had close to 60 per cent of the business. Dozens of smaller automakers, including Dodge Bros., Packard and Hudson to name a few, shared the rest.

The biggest industry news that year was the formation on January, 26 of the Lincoln Motor Company. Henry M. Leland, the man who had guided the formation of Cadillac, and his son, Wilfred, made good their promise to one day return to the automobile business. The Lelands had abruptly left Cadillac three years earlier in a disagreement with W.C. Durant over committing Cadillac's factories to the production of Liberty aircraft engines for the United States Government. The Lelands soon set up their own plant to make Liberty aero engines as part of America's war effort. Ironically, Durant was eventually forced to do the same using Cadillac facilities.

When the war ended, the Lelands wasted no time looking around for something else to do. It came as no surprise when the Master of Precision announced that he would soon place into production the finest automobile made in America. The Leland-designed and built Lincoln motor car would bear the name of the U.S. President old H.M. had long idolized. The first Lincolns came off the assembly line in the Lincoln Motor Company's Detroit plant in August, 1920. The Lincoln bore on its radiator a distinctive nameplate that carried the all-important statement . . . "Leland-Built".

Except for the radiator design, the new Lincoln car bore a strong resemblance to the Cadillac. Not surprisingly, the Lincoln was powered by a V-8 engine. The Leland-built Lincoln had a 60-degree, eight-cylinder engine of 357 cubic inch displacement and rated at 81 horsepower — two more than the 1920 Type 59 Cadillac V-8. With Packard and Pierce-Arrow already firmly entrenched in the luxury car field along with other makes like Peerless and Winton, and Cadillac gradually moving up into this niche, competition at the upper end of the market was finally becoming in-

Cadillac prices increased sharply when the new Type 59 Cadillacs were introduced for 1920. The lowest-priced cars in the line were the sporty Type 59 Roadster for two passengers, shown here, and the four-passenger Phaeton. Both of these models were priced at $3,590 compared with base prices of $2,805 the previous year. Even with its top up, the Type 59 Roadster, on 125-inch wheelbase, cut a dashing profile.

Here is a coming-at-you view of the Type 59 Cadillac two-passenger roadster. This was the first year that Cadillacs were equipped with a cowl ventilator. Other exterior changes included a new, flush-type hood hinge and generally flatter body lines. Note the Cadillac crest mounted in the center of the radiator and duplicated in miniature at the top of each headlight. While Cadillacs of this period were decidedly plain in appearance, these cars were handsome in a utilitarian way and did sport some nice finishing touches.

1920

tense. Unfortunately, the Leland Lincoln era was doomed to be short lived.

For the 1920 model year, Cadillac gave its modestly refined cars a new model designation — Type 59. This series would remain in production for the next two years. Mechanical improvements were relatively minor. All Cadillacs were now equipped with a cowl ventilator. The proven Cadillac V-8 engine was equipped with an exhaust-heated intake header. Exterior changes included flatter, straighter roof and body lines, a new flush-type hood hinge, and exterior door handles on the big Type 59 Seven-passenger Touring Car.

Cadillac prices increased sharply this year, ranging from $3,590 to $5,190. The 1920 model year marked the first time since the very limited production Model "L" Limousine of 1906 that Cadillac prices passed the $5,000 mark. As with the Type 57 of 1918 and 1919, the Type 59 Cadillac was built on two separate wheelbases — 125 and 132 inches depending on the body selected. There were 10 standard Type 59 Cadillac models including a two-passenger Roadster; the Phaeton for four passengers; the seven-passenger Touring Car; a four-passenger Victoria; the five-passenger Cadillac Sedan; a big, seven-passenger sedan called the Cadillac Suburban; a Cadillac Coupe; the standard Limousine, a Town Brougham and the top-line Imperial Limousine.

Three adventurous motorists drove a Cadillac through South Africa. On their travels they gave the 300-pound Queen of Swaziland a ride. Reports reaching these shores indicate that she was most impressed.

Despite the recession, the Cadillac Motor Car Company built 19,628 cars during the 1920 model year.

At this year's National Automobile Show, sponsored by the National Automobile Chamber of Commerce, manufacturers exhibited 119 closed cars compared with 79 a year earlier.

The U.S. automobile business had not heard the last of Billy Durant. The man who had conceived General Motors 12 years earlier was soon back in business building cars under his own name. Besides the Durant, he also later built cars named the Dort and the Star.

One of three open body styles left in Cadillac's model lineup, the Cadillac Seven-Passenger Touring Car was a companion to the popular four-passenger Type 59 Phaeton. The Type 59 seven-passenger tourer differed in appearance from the Phaeton in that it now sported exterior door handles of the stirrup type. A robe rail encircled the back of the front seat, and this car carried a pair of folding auxiliary seats. The big seven-passenger car was priced at $3,740, complete with top, side curtains, and black all-leather upholstery.

Cadillac's big touring cars rolled into the 1920's looking very much as they had for several years. This is the 1920 Cadillac Type 59 Phaeton for four passengers, which carried a standard price of $3,590. Wheelbase remained at 125 inches. A companion seven-passenger car was also available. The popular Cadillac Phaeton still did not have exterior door handles.

1920

Cadillac's modestly changed 1920 models were designated the Type 59. This is the dignified Cadillac Type 59 Victoria, a four-passenger car built on the 125-inch wheelbase. In addition to seating four passengers in somewhat roomy comfort, the Victoria offered storage space behind the driver's seat and in the gracefully curved rear deck. Standard price of this model was $4,340. Seating was of the modified cloverleaf design, with two passengers sharing the rear seat, the driver having his own position, and a fourth passenger perched on a fold-away front seat.

The American automobile industry was gradually shifting from open to closed body styles, and this trend was certainly evident in the body choices now being offered by the Cadillac Motor Car Co. The five-passenger sedan shown here was an attractive, comfortable and highly practical car. The boxy, four-door Type 59 sedan for five passengers was built on the 125-inch wheelbase and carried a starting price of $4,750. Interior appointments included armrests, a foot hassock, deep upholstery, and roll-up shades for each window in the rear section.

Cadillac's modestly refined 1920 models featured slightly altered profiles with flatter roof and body lines. Cadillac Type 59 sedans were quite boxy in appearance, but what they lacked in style they more than made up for in performance and fine appointments throughout. Note the standard location of the license plate, on the spacer bar between the headlights. This lofty perch kept it clear of road mud and readable. The upper windshield glass was hinged for ventilation. Ironically, the 1920-21 Lincolns were fitted with the same type of "bellflower" headlamps.

Among Cadillac's most popular models was the boxy four-door sedan. The wide doors permitted comfortable entry and egress. The 125-inch wheelbase Type 59 sedan was marketed as a five-passenger car. Note the standard spare tire mounting on the rear of the car. Bumpers were still optional. This car carries its owner's family crest on a tiny medallion mounted just below the beltline on the rear door.

For those with larger families or who frequently entertained guests, the Type 59 Suburban offered additional passenger capacity. The big Suburban was built on Cadillac's 132-inch wheelbase. This model was sold as a seven-passenger car. Price was $4,990 F.O.B. the factory in Detroit. The Type 59 Cadillac was powered by a V-type eight-cylinder engine uprated to 79 horsepower for 1920.

1920

For the first time since 1906, Cadillac prices for 1920 passed the $5,000 mark. The very limited production Model "L" Limousine built 14 years earlier had commanded a price of $5,000. Three Type 59 Cadillacs — the coupe, limousine and town brougham — were priced at $5,090. This is the stately Standard Cadillac Type 59 Limousine for 1920. Open-front limousines were still formally correct in the larger cities. Weather protection for the driver was by side curtains. The large glass tonneau partition was non-removable.

Cadillac's enviable reputation for quality and reliability was greatly enhanced by the marque's outstanding performance in a variety of roles during the First World War. More than a few military officers who had been assigned Cadillac staff cars purchased similar cars for personal use when they returned home after the war. General John J. Pershing was rarely seen in anything but a Cadillac. Patriotic and grateful Cadillac Motor Car Co. employees had presented General Pershing with a special closed Cadillac after the war. Here, General Pershing alights from an open front Cadillac Limousine modified for military service. Note the accessory Bi-Flex bumper and four-star license plate.

Topping the Type 59 Cadillac line was the regal Imperial Limousine, a lavishly-appointed, enclosed-drive limousine constructed on Cadillac's 132-inch wheelbase chasssis. This fine car carried a base price of $5,190. Privacy curtains flanked each side of the center divider partition between front and rear compartments. This premium model was intended for families who could afford a chauffeur, but who sometimes wanted a sedan type car. Thus, the glass partition could be lowered when the owner did his own driving. Several other makers referred to this style as a Berline.

The "Roaring Twenties" heralded the American age of style. Custom body design reached a peak during the next two decades that has not been seen since. The eight-cylinder Cadillac chassis had been popular among custom body builders since its introduction in 1915. The highly respected Derham Body Co., of Rosemont, Pa., built this elegant town Cabriolet with wicker trim and a collapsible rear quarter. Note the accessory bumper and unusual hood ornament. The side-mounted spare tire and demountable rim was also a custom touch. Sidemounts had still not become a styling innovation in this era. Weather protection for the chauffeur was by a removable snap-on canopy and side curtains.

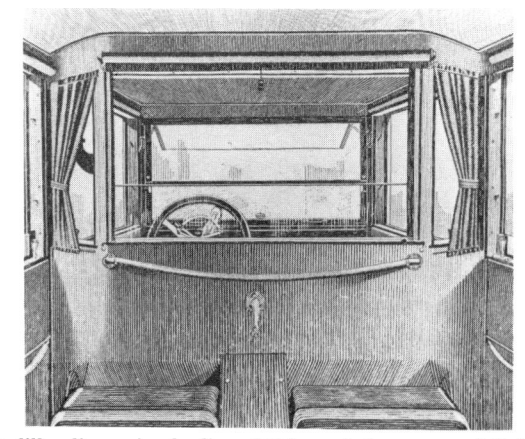

Cadillac limousine bodies of this period were exquisitely-appointed extensions of the family living room or parlor. The glass in the divider partition of the Imperial Limousine could be cranked down, Cadillac advertised . . . "for long tours or family parties when the owner drives." This rear passenger view of the 120 Cadillac Type 59 Imperial Limousine shows the folded auxiliary seats which could be raised for use as a seven passenger car. Above the glass is a large roll-up shade for privacy when desired. Similar shades were fitted to the side windows as well.

1921

This was to be another milestone year for the Cadillac Motor Car Co. It was in 1921 that the company moved into a completely new plant, one of the largest in the industry and the most modern in the world.

Steadily increasing Cadillac sales had severely taxed the company's production facilities, to the point where the division's operations were scattered among more than 70 different buildings around Detroit. The transfer of all operations over to the huge, new plant on Clark Avenue on Detroit's west side was a complex and time consuming one. As a result of this move, only 11,130 cars were built in the 1921 calendar year.

Cadillac's impressive new manufacturing, assembly and administrative complex covered some 49 acres and 2.5 million square feet. There were eight main buildings. Although Cadillacs have been assembled in various other General Motors assembly plants since 1968, the division's home address is still 2860 Clark Avenue in Detroit. After the big move, the old Cadillac Motor Car Co. plant at Cass and Amsterdam streets was sold to Fisher Body. It was later torn down.

For the 1921 model year, Cadillac carried its Type 59 models over from 1920 virtually without change. Two models — the Cadillac Coupe and the Town Brougham — were discontinued. The 1921 model lineup included only eight body styles, three open and five closed. Again, there were two Type 59 wheelbases, 125 and 132 inches. Prices remained unchanged from the previous year.

Models included the two-passenger roadster ($3,590); the old, reliable Cadillac Phaeton for four passengers ($3,590); the seven-passenger touring car ($3,740); the four-passenger Victoria ($4,340); the five-passenger Sedan ($4,750); the seven-passenger Suburban, ($4,990); the Standard Limousine, ($5,090) and the Imperial Limousine, ($5,190).

Standard finish on the Touring car and Phaeton was Brewster Green with Black trim. The sporty Roadster was finished in Cadillac Blue. The Victoria and Sedan also came in Brewster Green. The big Suburban could be had in either Brewster Green or formal black. The top-line Limousine and Imperial were finished in black. Other colors were available on special order and at extra cost.

All Cadillacs came with five lamps, a pair of headlamps with tilting reflectors and two small cowl

An open door affords an inviting look into the interior of the 1921 Cadillac Type 59 Victoria for four passengers. The right front seat folded forward under the dash when not in use or to allow rear seat passengers to enter and leave with ease. The staid-looking Cadillac Victoria had a small storage compartment behind the driver's seat and another in the curved rear deck. Interior appointments included taffeta roller sunshades on side and rear windows, a foot rail, outside key locks on both doors and armrests on the folding right front seat.

1921

1921

CADILLAC "TYPE 59" TOURING CAR
1921

For the 1921 model year, Cadillac carried over its Type 59 with virtually no changes. Part of the reason for this was the tremendous logistical problem of transferring all manufacturing and assembly operations into the new Cadillac plant on Detroit's west side. This is the 1921 Cadillac Type 59 Seven-Passenger Touring Car, which was built on the 132-inch wheelbase chassis. Standard finish was a rich Brewster Green with black running gear and trim. Price was unchanged at $3,940.

The Type 59 Victoria for four passengers continued as one of Cadillac's best selling closed cars. The Victoria or Coupe, was built on the 125-inch wheelbase chasis. Tires were 34 x 4-½-inch Goodyears or U.S. Fabric with plain tread on the front wheels and all-weather tread on the rear. This one is equipped with two spare tires, one of each tread for either front or rear use. Bumpers were still not considered necessary. Price of this model continued at $4,340 F.O.B. Detroit.

This would appear to be a custom-bodied Cadillac. The chassis is the eight-cylinder Cadillac Type 59, which was produced for the 1920 and 1921 model years. The style is crisply formal with sharp-edged body lines, four doors, enclosed rear quarters panels and decorative landau irons. Note the small trunk carried at the rear of the car between the body and the spare tire. The body builder is not known, but the lines hint very strongly of a design that the Judkins Co. of Merrimac, Mass., was turning out on a semi-custom basis, primarily for Lincoln, but also for other chassis.

lamps as well as a taillight with license holder. Open cars were equipped with a heavy weatherproof top and khaki whipcord lining that concealed the roof bows. The dash was equipped with a combination clock and speedometer, ammeter, oil pressure and gasoline gauges and a control for the auxiliary air pump. A tilting steering wheel was also standard on the Type 59 Cadillacs.

Cadillac was not the only component of General Motors that moved into new quarters this year. It was in 1921 that the imposing General Motors Building was opened in Detroit. This huge office building, which still serves as GM's world headquarters today, was originally to have been named the Durant Building, after the corporation's colorful founder, W. C. Durant who had left under less than pleasant circumstances the previous year.

In fact, the initial "D" is still clearly visible in some of the GM Building's ornate exterior stonework. The magnificent, 15-story General Motors headquarters was built on a site north of downtown Detroit that had been personally paced off by Bill Durant himself several years earlier. Several incredulous colleagues, including Alfred P. Sloan, were ordered to begin buying up the properties on the spot. The massive General Motors Building still dominates its West Grand Boulevard neighborhood today. For many years, several large automobile showrooms on the ground floor served as public showcases for the products of all GM divisions.

There were some important personnel changes at Cadillac this year. Herbert H. Rice was appointed President and General Manager of the Cadillac Motor Car Division in May, 1921. Ernest W. Seaholm was named the division's new Chief Engineer, and R. H. Collins left the Cadillac organization to head up rival Peerless in Cleveland.

In other industry news, Henry M. Leland's new Lincoln Motor Company had encountered serious difficulties. Production startup problems and delays kept production of the new Lincoln cars to a relative trickle. The company went into receivership in October, 1921. And the postwar depression continued to cut deeply into industry sales.

1921

Although this car has the appearance of a custom, it is not. This is a more or less standard Type 59 Cadillac touring car that has been fitted with a "California" top. This fairly popular accessory converted an open touring car into a semi-closed car. The enclosure could be removed with some difficulty in good weather. The small landau bows, running board scuff plates and door handles are not original. This car was photographed at the antique auto flea market at Hershey, Pa., many years ago.

Two-tone color schemes were applied to many Cadillac models at this time. Typically, the fenders, splash aprons and radiator shell were finished in black. The standard body color for this Type 59 five-passenger sedan was a dark Brewster Green. The hood and lower body were in this color while the "greenhouse" area above the body beltline was black. The new drum-type lamps and hubcaps were also black-finished. Special colors, however, were available on extra-cost special order. This sedan is on display at this year's Automobile Show in New York. Note the dust covers draped over the other cars in the Cadillac Motor Car Co. exhibit.

Cadillacs of the early 1920's were anything but stylish. Their lines were angular and severe. Cadillac's big closed cars were boxy and imposing. But what the Type 59 Cadillac lacked in dashing good looks, it more than made up for in dependable performance and comfort. This frontal view of a Type 59 shows to advantage the finely-detailed headlights, the headlight tiebar with license plate mounting clamps, the diminutive cowl lamps and functional windshield visor. Bumpers were still considered unnecessary.

Closed bodies were rapidly replacing open types throughout the U.S. auto industry. One of Cadillac's most practical models was the standard Type 59 sedan for five passengers. This plain-looking, but very roomy, four-door sedan was priced at $4,750 and was built on the 125-inch wheel base chassis. Standard finish on this model was Brewster Green. Note the softer roofline and the visor over the windshield. The spare tire was still carried on a holder attached to the rear of the car. This is a late model, fitted with the new drum headlights, which was a running change made about mid-year.

The big Cadillac Type 59 Suburban was carried forward without change. The Suburban rode on a 132-inch wheelbase and was designed to carry seven passengers in maximum comfort. Prices started at $4,990. This model was intended to be owner-driven, and did not have a partition between the front and rear seats. Standard color was Brewster Green with black trim. The auxiliary seats folded into the front seat-back when not in use.

1921

Automobile shows were still an important element in marketing stategy. All principal auto makers eagerly awaited the big shows to introduce new models and features and to show off their entire model range to the public. The biggest show was still the "National," held early in the year in New York City. This is a typical Cadillac Motor Car Co. exhibit in a classical setting with a stripped-down Type 59 chassis in the foreground. The chassis appears to have been fitted with a glass or plexiglass hood, cowl, radiator shell, and windshield.

Automobile styling as an art form was still several years off, but some exciting things in this area were going on out on the West Coast. Cadillac's California distributor, Don Lee of Los Angeles, had set up a special custom body department which restyled and modified standard cars into highly individual creations for affluent owners who desired something different. The chief designer at Don Lee Studios was a highly talented young man named Harley Earl. Cadillac eventually talked Earl into coming to Detroit to style Cadillacs, and the rest is history. This Victoria-style touring car was created for Mr. Lee's own personal use. Modifications include sweeping front fenders, wire wheels, side-mounted spares, a wide upper body molding and a sharply-raked windshield. The small carriage lamps mounted just behind the front doors were a Don Lee trademark. Except for the folding tonnuau top, no weather protection was provided on this car.

Depending on the whims of the customer, Don Lee creations ranged from the elegant to the bizzare. The Don Lee Studio designed and built this one-off, center-door limousine for Sussman Mitchell, of Visalia, Calif. Modifications include a padded top, roof luggage rack, flat fenders and coach-type splash guards, wire wheels and a wicker band around the belt line. The chassis is a Type 59 Cadillac. This type of fender styling was sometimes referred to as "6-fender style," and was considered the apex of formal-sport design.

This is yet another product of the Don Lee Studios in Los Angeles. This highly styled open front formal town car or cabriolet was commissioned for Jack Pickford. The chassis is a Type 59 Cadillac. Noteworthy features include wire wheels, a one-piece slanted windshield, leatherette covered spare and permanently-open chauffeur's compartment. A small oval opera window in the upper rear body quarter panels is framed by an ornamental landau bow. Note the very unusual tonneau windshield and the wide wicker belt molding around the tonneau.

1921

Here is yet another Don Lee creation. This is a severly formal Type 59 Cadillac Town Car of 6-fender design. Note the intentional isolation of the rear body from the rest of the car. The razor-edged body lines contrast with the softly rounded contours of the rest of the car. The liveried chauffeur's posture is as vertical as the windshield. No weather protection at all was provided for the driver. This type of body was often referred to as a panel brougham.

Designed by Harley J. Earl, this Don Lee transformation of a Type 59 Cadillac is quite elegant. The coach-type fenders, big Disteel disc wheels and running board scuff plates give this car a particularly dignified look. Contributing to the formal effect is the sloped windshield and closed upper rear body quarter panels with small opera windows and decorative carriage bows. The top was in fine-grain landau leather. The wide belt molding is in contrasting color. Most Earl/Lee creations favored wire wheels. Note the strange little splash guard between the sidemount and the running boards.

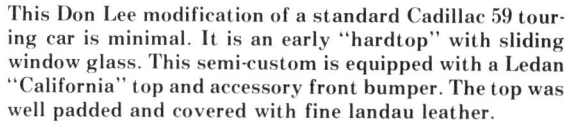

This Don Lee modification of a standard Cadillac 59 touring car is minimal. It is an early "hardtop" with sliding window glass. This semi-custom is equipped with a Ledan "California" top and accessory front bumper. The top was well padded and covered with fine landau leather.

Fatty Arbuckle, one of the most famous movie comedians in the early 1920s, shows off with one of his Don Lee Cadillacs. This particular model was a Type 59 Town Car, fitted with special wire wheels and very unusual body side molding running completely around the car from radiator shell to radiator shell. Arbuckle owned a series of Don Lee cars, including a famous Pierce-Arrow disguised to look like a Hispano-Suiza. Although Don Lee was Cadillac's California distributor, his custom body shop would work on any chassis that a customer desired. The custom body operation was closed when Cadillac lured Harley Earl, Lee's top designer, to Detroit in 1926.

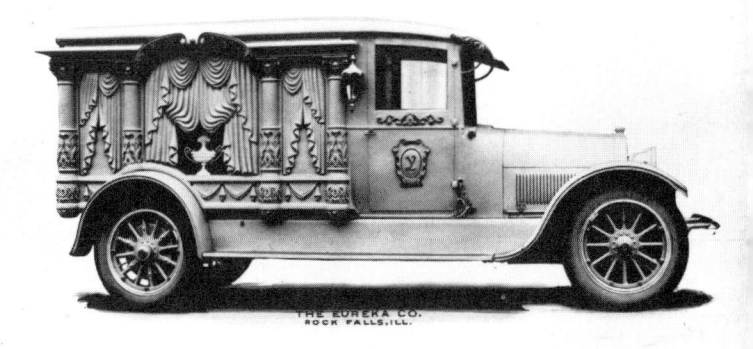

American funeral car and ambulance manufacturers and image-conscious funeral directors held the dependable Cadillac V-8 chassis in high esteem. The Eureka Co. of Rock Falls, Ill., designed and built this eight-columned, carved panel hearse with arched center windows on a used 1921 Cadillac Passenger car chassis. It is finished in light gray, a popular color for funeral livery equipment at this time. Note the owner's escutcheon on the front door. Eventually, Cadillac would dominate the funeral car and ambulance field, becoming the most popular chassis in America for installation of these professional car bodies.

1922

After a production run of two years, the Type 59 Cadillac was succeeded in September, 1921 by the refined Type 61. The Type 61 Cadillac was marketed in model years 1922 and 1923.

The new Type 61 Cadillac featured a number of mechanical and styling improvements. For one thing, the new Type 61 rode on smaller 33-by-5 inch tires which gave all models a much lower appearance. Type 61 styling refinements included a new, higher hood with softly-rounded "shoulders". The Type 61 Phaeton for Four Passengers and the Cadillac Sedan for Five Passengers featured a series of six aluminum rub strips mounted vertically on the rear body. These protective strips prevented the demountable trunk from marring the body finish. Inside, the Type 61 also sported a new walnut steering wheel and a revised instrument panel design.

Cadillac's renowned eight-cylinder, V-type engine entered its eighth model year. Displacement was 314 cubic inches. The bore was 3-$\frac{1}{8}$ inches with a 5-$\frac{1}{8}$ inch stroke. The most significant engineering improvement under the hood was a new, thermostatically-controlled carburetor.

All Type 61 Cadillacs were built on a common 132-inch wheelbase. Two new body styles were added to the line this year. They included a good-looking Coupe for Two Passengers and a new Four-Passenger Cadillac Coupe. Closed bodies continued to gain in favor with Cadillac's customers.

Other Type 61 Cadillac body styles offered for 1922 included the sporty two-passenger Cadillac Roadster; the perpetually-popular Phaeton for Four Passengers; the companion Seven-Passenger Touring Car; a Victoria for Four Passengers; the Five-Passenger Coupe; a Five-Passenger Sedan and a companion Seven-Passenger Suburban; the standard open-front Limousine and the most luxurious and expensive Type 61 Cadillac of them all, the Imperial Limousine.

Cadillac prices for 1922 ranged from $3,100 for the Type 61 Roadster for two passengers on up to $4,600 for the luxurious Imperial. Cadillac continued to sell some stripped-down, bare chassis to custom body builders and builders of special purpose commercial bodies. This bare-bones chassis was the standard 132-inch wheelbase unit. Most commercial body builders preferred to stretch the wheelbase to 145 inches. On request, Cadillac would lengthen the chassis at its Detroit branch.

Expansion and construction continued at Cadillac's new plant complex on Detroit's east side. There were now more than 8,000 people on the Cadillac payroll, and the huge, new plant was turning out more than 100 new cars a day. In June, 1922 the plant produced a record 150 Type 61 Cadillacs in one 24-hour period. Significantly, half of these were closed-body cars.

With the recession fading, custom automobile body builders across the country were prospering. There was strong demand for custom, highly individualized bodies on a wide range of premium car chassis. Price was literally no object. This relatively small but very prestigious segment of the U.S. automobile market would continue to grow in both size and stature. It was the true dawning of the golden custom-body era in America, and the real beginnings of the Cadillac Motor Car Company's influence on the U.S. luxury car market. At this time, Packard was the undisputed leader in the fine car field in the United States, but Cadillac — which had earned an enviable reputation for engineering leadership, innovation and realiability — was more and more taking aim at the top end of the market. Cadillac would soon begin to encroach menacingly on Packard's domain. The struggle for supremacy had barely begun.

The United States at this time boasted a large number of well-established luxury car builders. Besides Packard there was Pierce-Arrow and Peerless, Winton and Locomobile, an American offshoot of England's famous Rolls-Royce, and the new Lincoln Motor Car Co.

Unquestionably, the automotive industry story of the year was the sale of Lincoln to the Ford Motor Company. Despite the lengendary reputation and resources of the Lelands and the acknowledged high quality of the new eight-cylinder Lincoln car, the young Lincoln Motor Car Co. had not fared well. There were problems with suppliers, development and production launch delays. The company had quickly found itself in serious financial difficulties.

All this did not go unnoticed by Henry Ford. Mr. Ford, who had become rich and world-famous by building a simple, low-priced car for everyman, secretly longed for a product with which to cover the other end of the spectrum. Lincoln afforded him just such an opportunity. On February 4, 1922 the Lincoln Motor Car Co. was put up for sale by a receiver. Henry Ford bought it for $8 million. This was an especially satisfying deal for Mr. Ford, who must have recalled the humiliating early days of the old Henry Ford Company which eventually resulted in the creation of the Cadillac. Henry M. Leland and Wilfred C., however, were of considerably different temperament than Mr. Ford. Within four months the Lelands had left and Mr. Ford's son, Edsel B. Ford, was installed as President of Lincoln.

Once the production snags were straightened out, the Leland-designed Lincoln proved to be an exceptionally fine automobile in the true Leland tradition. The Ford-built Model "L" Lincoln remained in production from 1922 through 1931.

Priced identically at $3,150 with the two-passenger Roadster, the Type 61 Cadillac Phaeton was a versatile four-passenger car. The Phaeton was now equipped wa trunk rack between the rear body and the spare tire carrier shown here. Barely visible in this photo are a series of vertical aluminum ribs which prevented the removable trunk from scratching the rear body finish. The Type 61 model lineup also included a big seven-passenger touring car which was similar in appearance to the popular Cadillac Phaeton. This car also wears the accessory scuff plates on the apron, to prevent passenger's shoes from kicking the finish as they entered the car. These plates were made of polished aluminum.

1922

The new Type 61 Cadillac was introduced in September of 1921. This modestly refined series was the successor to the Type 59 of 1920 and 1921, and was produced for model years 1922 and 1923. One of the sportier models in the new Type 61 series was the two-passenger roadster, which was priced at $3,150. The rounded rear deck could be ordered with a disappearing auxiliary seat, or used as a conventional trunk. Even with the folding top up, this model looked racy and ready for action. Note the two spare tires mounted on a carrier behind the rear deck — one has plain tread for use on the rear wheels, while the other has a diamond pattern traction tread for use on the rear wheels. It was not uncommon in this era for cars to carry spares of each type of tread.

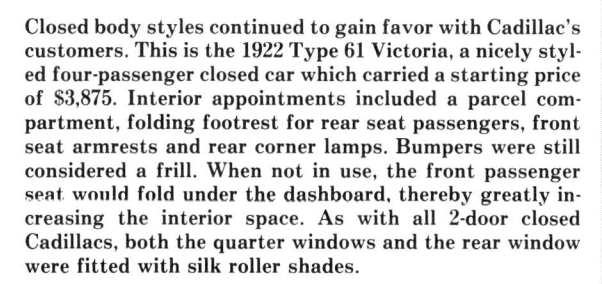

A new addition to the Cadillac model lineup for 1922 was this handsome Type 61 Two-Passenger Coupe, a companion car to the Roadster. Priced at $3,875, the new Cadillac Coupe was designed primarily for the business or professional man. The Coupe offered ample storage space in a compartment behind the seat and in the rear deck. Unlike other Cadillacs, the doors of this model were forward opening. This was the only Cadillac to have its doors hinged on the B-pillar.

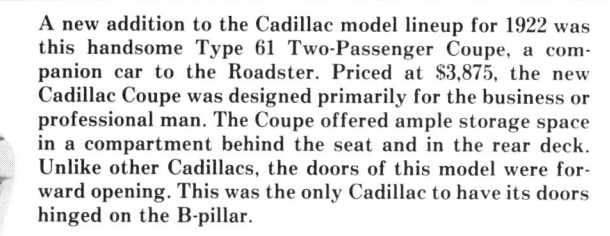

Closed body styles continued to gain favor with Cadillac's customers. This is the 1922 Type 61 Victoria, a nicely styled four-passenger closed car which carried a starting price of $3,875. Interior appointments included a parcel compartment, folding footrest for rear seat passengers, front seat armrests and rear corner lamps. Bumpers were still considered a frill. When not in use, the front passenger seat would fold under the dashboard, thereby greatly increasing the interior space. As with all 2-door closed Cadillacs, both the quarter windows and the rear window were fitted with silk roller shades.

The Cadillac Type 61 Five-Passenger Coupe was quite similar in appearance to the four-passenger Victoria and the new two-passenger Type 61 Coupe. The right front seat tilted forward to provide ample access to the rear seat but did not fold up completely as did the seat on the 4-passenger Victoria. The five-passenger coupe was generally intended to be owner-driven. Standard price for this model was $3,925.

For those customers who preferred the convenience of four doors, Cadillac for 1922 offered the Type 61 Sedan for Five Passengers. This roomy sedan offered excellent ventilation for summer driving. All windows, the ventilating-type windshield and cowl ventilator could be opened to permit an effective flow of fresh air. The six vertical aluminum rub strips on the rear of the body were designed to protect the finish when the trunk was placed on the trunk rack or taken off. The Type 61 Phaeton also had these strips. As was the case with all Cadillacs this year, map pockets were provided on all doors. These were handy for the storage of gloves and small parcels. Rear seat passengers were provided with an inclined foot rest and armrests at the outer edges of the seat.

1922

The big, seven-passenger Touring Car remained a popular model in Cadillac's offerings for 1922. Like all other Type 61 Cadillacs, the heavy seven-passenger tourer was built on a 132-inch wheelbase. The interior of this massive automobile featured deeply contoured seats upholstered in dull-finish black leather. Interior conveniences included a pair of folding auxiliary seats, an improved robe rail and a three-position adjustable foot rest. Price was $3,150 — the same as the companion Type 61 Cadillac Phaeton for four passengers. Because of the folding top, interior or tonneau lights on this car had to be installed on the outer ends of the front seat back, and thus could not be used for reading lights.

Near the upper end of Cadillac's closed body offerings for 1922 was the Type 61 Seven-Passenger Suburban. The imposing Cadillac Suburban was designed to be owner-driven. Note the large windshield visor and the diminutive cowl lights. The new Type 61 Cadillacs rode on 33 x 5 inch tires which contributed to the car's lower appearance. Silk roller shades were provided for all windows except those in the front doors and windshield. The folding auxiliary seats would store flush against the front seat back.

In a market already crowded with high quality cars — Packard, Peerless, Locomobile, Winton and the new Leland-built Lincoln, Cadillac's premium models represented excellent value for the money. Cadillac's top models were very competitively priced when compared with other luxury cars of the day. This is the standard Cadillac Type 61 Limousine for Seven Passengers, a stately open-front, chauffeur-driven car that carried a base price of $4,550. Storm curtains protected the driver in bad weather. The chauffeur's compartment was upholstered in bright enameled leather, while the tonneau was done in luxurious broadcloth. An intercompartment telephone allowed passengers to give orders to the chauffeur.

Cadillac offered a special, stripped-down Type 61 chassis to custom and specialty body builders. This is the standard Type 61 Cadillac chassis built on a 132-inch wheelbase. The engine was an eight-cylinder V-type of 314 cubic inch displacement with 3⅛ bore and 5⅛ inch stroke. Mechanical improvements for 1922 included a thermostatic carburetor. In addition to serving the carriage trade, Cadillac chassis were widely used for various commercial applications including ambulances and hearses, police patrol wagons, buses and even armored cars. Most body builders lengthened this chassis to 145 inches.

At the very top of the Cadillac model line was the formal Type 61 Seven-Passenger Imperial Limousine. This impressive closed car could be either owner or chauffeur driven and was equipped with a glass partition between the front and rear compartments. The glass could be dropped for informality. Prices started at $4,600. Like all other Type 61 Cadillacs, the regal Imperial Limousine rode on a 132-inch wheelbase. Other car makers often referred to this style as a "Berline" when the chauffeur's compartment was finished in the same material as the tonneau.

The eight-cylinder Cadillac chassis remained popular with custom body builders. This custom Station Wagon was built on a Type 61 Cadillac chassis by Healey & Co. of New York. This special car was designed by J. R. McLauchlen of Cadillac's Custom Body Department in New York City and was used mainly in the fashionable summer colonies in New Jersey. The body featured mahogany panels with ash slats. It is interesting to note that this style was referred to as a station wagon at this time. The resemblance to the crude depot hacks and utilitarian "shooting brakes" of the period was strictly superficial: this was a real quality creation. The seats were upholstered in fine grain leather, and side curtains were stored under the rear seat ready for use in inclement weather.

1923

The custom body era had dawned. Numerous custom body builders, many of them proud descendents of long-established carriage makers who had successfully made the difficult transition from horse-drawn to motor-propelled vehicles, catered to the needs of a large and wealthy clientele that demanded elegance and true distinction in motor cars. This beautiful four-passenger sedan on the Type 61 Cadillac chassis was built for the personal use of Mr. Judkins of the John B. Judkins Co. of Merrimac, Mass. This car was finished in Dibble's Japanese Blue with black superstructure and gold striping. The rear compartment was trimmed in Chase's blue oriental striped mohair and the front in blue Spanish leather. Other interior touches included inlaid mahogany below the window molding on the doors and also on the armrests, and special hardware designed by the Art Work Shops. Note the plated radiator shell. The head liner was in plain blue mohair, set off by two dome and two quarter panel lights also designed by Art Work Shops.

Judkins called this custom an "imitation collapsible sedan for four passengers". This car was commissioned for W.B. McSkimmon of Mansfield, Mass. and was very low in appearance for a closed car of its day. Note the large disc wheels and oval opera windows. The car was finished in Ditzler's Kenilworth Gray and black. The interior was finished in Wiese's Graniteweave gray cloth. Heavy Disteel disc wheels and dual sidemounts add an impressive touch and provide the only brightwork, as all parts normally nickel plated were finished in black enamel.

Cadillac's Boston distributor, A.L. Danforth, had this special two-passenger Type 61 Cadillac Coupe designed and built by the John B. Judkins Co. of Merrimac, Mass. Special features included parking lights set into the front pillars, folding armrests dividing the two seats and a rear deck that was divided into separate compartments for the trunk, golf bag and tools and chains. This car was finished in Willey's Celestial Blue with the upper body and fenders in Coach Body Blue. Striping was in gold. The interior was done in Wiese's Sedanweave cloth. The enameled landau irons were strictly decorative.

Here is another custom by Judkins. This is a special seven-passenger limousine-landaulet on the 1922 Cadillac Type 61 chassis. It was built for Mrs. Sumner Wallace of Rochester, N.H. Main body color was Valentine's Dust-proof Gray with black superstructure and running gear with striping in English Vermillion. The passenger compartment was upholstered in Laidlaw Bedford Cord, while the chauffeur's compartment was in semi-bright black leather. Disteel disc wheels and dual sidemounts give the car a European touch. Note the steep rearward rake of the windshield and curve of the connecting roof. The rear tonneau roof, with its extra-large blind quarter, was collapsible, and could be lowered. Thus, the enameled landau irons were functional.

Not all Don Lee creations were shocking or bizzare. This special Coupe for Two Passengers was designed and built by the Don Lee Coach and Body Works in Los Angeles. Note the sweeping fender lines, light wire wheels and vertical opera windows framed with decorative landau bows. The windshield contained unusual little triangular corner windows, and the entire superstructure was set off by the wide sill trim which flowed completely around the car in contrast to the very thin beltline trim at hood-panel level.

1923

Although anything but homely, Cadillacs of this period were more practical than they were pretty. The company's emphasis was decidedly on engineering excellence rather than styling. That was still a year or two down the road. The Type 61 Cadillac Roadster for two passengers, however, had a somewhat dashing look about it with its long hood, jaunty top and smooth, rounded rear deck. Even with those standard wood-spoked artillery wheels, the Type 61 Roadster looked sporty. It was an able performer, too. At $3,100 this was Cadillac's lowest-priced car.

Here is a three-quarter rear view of the Type 61 Roadster for Two Passengers. The rounded, sloping rear deck could be ordered with a rumble seat or used as a conventional trunk. Note the standard spare tire carrier and the massive walnut steering wheel. In addition to carrying the extra-cost accessory spare rim and tire, the carrier also provided a convenient mounting space for the stop light and rear license plate hanger. When mounted, the spare tire also doubled as a rear bumper of sorts. Leather upholstery and door panels were standard in Cadillac open models.

The open left-hand door affords a glimpse into the interior of the Type 61 Coupe for Two Passengers. This model had been introduced for 1922. Price was $3,875, the same as the Type 61 Victoria. Note the handy map pocket in the door panel. All closed Cadillacs this year were provided with these pockets in all doors.

The Type 61 Cadillac was continued through the 1923 model year without change. Cadillac proudly boasted in its advertising that it had produced more than 150,000 ninety-degree V-type eight-cylinder engines, not one of which was ever replaced by the factory for any reason. The company was also able to claim that more Type 61 Cadillacs had been sold than all other makes in its price class.

The new Cadillac Motor Car Co. plant on Detroit's west side had turned out 26,296 Type 61 Cadillacs during the 1922 model year. Another 14,707 Type 61's were built during the 1923 model year for a total of just over 41,000 in all.

The renowned Cadillac V-8 engine which had been introduced in 1915 was now in its ninth and final year. The second-generation Cadillac V-8 was being prepared for fall introduction in a new series of Cadillacs for 1924. Production would begin in the summer of 1923. During 1923, Cadillac announced the standardization of its parts prices all over the United States. This important step eliminated the great regional disparity that had existed in prices charged for factory parts around the country.

There were also several significant developments in the U.S. automobile industry this year. Cadillac's biggest rival, Packard, took the industry and public by surprise with its introduction of a straight-eight engine.

This was the first eight-in-line, L-head "Single Eight" to enter mass production, and it brought additional prestige to Packard which had clearly emerged as America's leading producer of luxury cars.

In 1923, the Ford Motor Company sold 7,875 Ford-built Lincolns, putting its recently-acquired Lincoln motor car division solidly into the black — something the Lelands had not been able to do. The Ford-built Lincolns were built on a 136 inch wheelbase, four inches longer than the Type 61 Cadillac of 1922-1923.

Also in 1923, Alfred P. Sloan became President of General Motors. Mr. Sloan brought to GM the staff-and-line concept of management that is credited with making General Motors the success that it is today. Under Sloan's new corporation management plan, each of the numerous "companies" within the corporation were reorganized as separate operating divisions, each relatively autonomous within GM but responsible for its own management. Important staff functions like finance, marketing and research were organized separately but were available to each of the divisions, of which Cadillac was one. Sloan's extremely successful management concept remains the cornerstone of General Motors to this day.

Cadillac's next product series — the Type V-63 — was being prepared for introduction in the fall of 1923 for the 1924 model year.

But the division's product planners were looking much farther down the road. Development work was already under way on a new type of car, one that would fit into the widening gap in the U.S. market that existed between GM's Buick and the Cadillac. Four years would pass before Cadillac's first "junior edition" would see the light of day.

1923

Going through a major style change was the Victoria, which now had its doors hinged at the rear, in the same manner as the two-passenger Coupe. As before, the dignified Victoria remained a steady seller in Cadillac's closed body offerings. A four-passenger car, the Type 61 Victoria was priced at $3,875. With the new door style, this model was very similar in appearance to the companion Coupe for two passengers, except of course that the Victoria had a larger body interior and a more pronounced superstructure to accommodate the split seating. The fourth seat continued to be of the fold-away type.

In later years, the term "suburban" would be reserved for station wagons designed to haul families from fringe residential areas to and from town. But from the midteens and into the 1920's, the Suburban was a large, semi-formal Cadillac sedan. The big Suburban was a seven-passenger car that could be either owner or chauffeur driven. This young lady looks right at home behind the wheel of a Type 61 Cadillac Surburban. Except for the extended rear body, this car was very similar in appearance to the companion five-passenger Cadillac Sedan. The car was equipped with a disappearing glass partition between compartments, but the forward compartment was upholstered like the rest of the car, and not done in formal leather.

Cadillac's largest, heaviest and most expensive car was the imposing Imperial Limousine for Seven Passengers. Standard equipment included a glass partition between the leather-upholstered chauffeur's compartment and the upholstered passenger compartment. The two folding auxiliary seats are visible in this view. This Type 61 Imperial Limousine is equipped with optional Westinghouse air shock absorbers at each corner. The speaking tube is visible at the top of the "B" pillar. All of the 4-door models have been fitted with a small courtesy light on the apron, just under the rear doors. Oddly, the 2-passenger car illustrated in this section also has this courtesy light — for what reason we don't know.

While closed bodies continued to attract more and more customers, there was still a sizeable market for large, open cars. The Cadillac Motor Car Co. met this demand with two models, the Type 61 Phaeton for four passengers and the standard Cadillac Seven-Passenger Touring Car, shown here. All Type 61 Cadillacs for 1923 were built on the same 132 inch wheelbase. Price was $3,150 for either the Touring Car or the Phaeton. This version has been fitted with dual spare tires, one with cross tread for the rear wheels and one with lateral or smooth tread for the front wheels.

1923

A new semi-custom addition to the Type 61 Cadillac series was this special sedan which was called the Cadillac Landau. Special styling touches included rounded rear window edges, leather-padded upper rear quarter panels and ornamental landau irons. This was a five-passenger car. Note the wide whitewall tires and dual spare tires. A late-year addition to the line, this car is sometimes classed as a 1924 vehicle, even though Cadillac's archives list it as a 1923 offering.

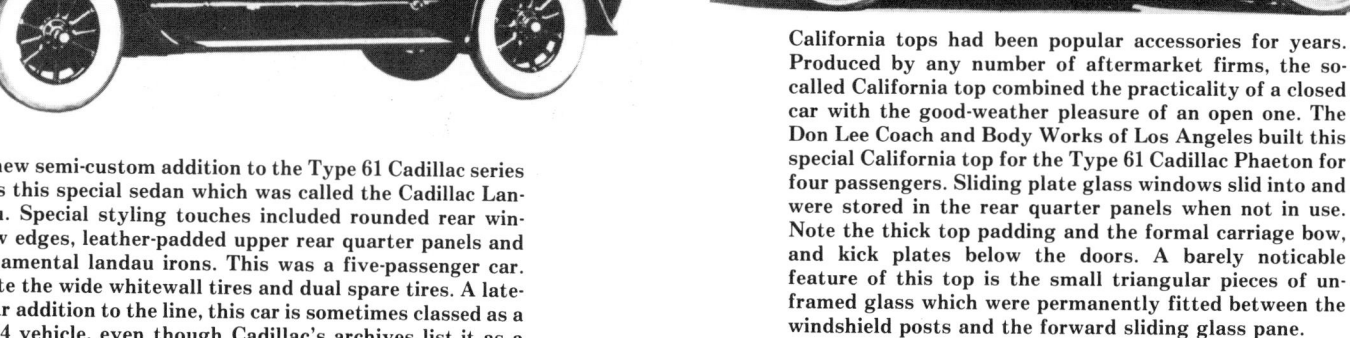

California tops had been popular accessories for years. Produced by any number of aftermarket firms, the so-called California top combined the practicality of a closed car with the good-weather pleasure of an open one. The Don Lee Coach and Body Works of Los Angeles built this special California top for the Type 61 Cadillac Phaeton for four passengers. Sliding plate glass windows slid into and were stored in the rear quarter panels when not in use. Note the thick top padding and the formal carriage bow, and kick plates below the doors. A barely noticeable feature of this top is the small triangular pieces of unframed glass which were permanently fitted between the windshield posts and the forward sliding glass pane.

The Don Lee Studios out in California continued to do some highly personalized conversions of standard Type 61 Cadillacs. This Victoria-type tourer is equipped with six disc wheels and a metal rear tonneau cover complete with windshield for the rear seat passengers. This car was built for a Miss Anne May. The design is in true 6-fender style, with the full extra fenders beginning just under the coach lights on the B-panels. Step-plates rather than running boards were often used with this fender design.

Here is another distinctive creation from the Don Lee Studio of Los Angeles. This is a custom touring sedan built on the Type 61 Cadillac chassis. Note the sweeping belt molding, wire wheels, angled windshield and dual roof vents. The design is of modified six-fender style which Lee favored. This involved having the front fenders end at the sidemount, and a small auxiliary fender begin just before the running boards, with these boards leading into a normal type of rear fender.

Since its introduction in 1915, the eight-cylinder Cadillac had won wide acceptance by builders of special-purpose commercial bodies. Not surprisingly, undertakers often chose the V-8 Cadillac chassis for mounting hearse and ambulance bodies. The funeral car business would soon become an important part of Cadillac's marketing plans. Owen Bros., of Lima, O., built this towering limousine style hearse body on the Type 61 Cadillac chassis, the wheelbase of which had been stretched from 132 to 145 inches. The unit showing through the rear side windows is the flower tray, which allowed the flowers to accompany the casket to the cemetery. When constructing vehicles of this type, some builders relied heavily on existing sheet metal of standard sedan bodies. However, it appears that the Owens brothers totally utilized their own designs on this particular model.

1924

Cadillac's next engineering triumphs were incorporated into yet another new series — the Cadillac Type V-63. The extensively redesigned Type V-63 was introduced to the public in September, 1923 and was produced, with various refinements and expanded model offerings, through the 1925 model year.

The most significant changes in the new Type V-63 were in the engine. Now in its 10th model year, the Cadillac V-8 was thoroughly redesigned for 1924. The new engine was equipped with what Cadillac described as the industry's first "inherently-balanced" crankshaft. The two-plane, 90-degree crankshaft was fitted with a series of counterweights which all but eliminated engine vibration and resulted in a remarkably smooth and quiet car. This was Cadillac's most important technological achievement since the company's introduction of the V-type, eight-cylinder engine in 1914. The new Type V-63 Cadillacs also had four-wheel mechanical brakes.

The new Type V-63 was initially offered in a choice of 11 body styles, all of them new. The new Fisher-built bodies were roomier, and standard equipment included a two-piece "VV" windshield, for ventilation and vision. Three additional body styles were added later in the model year, and five more semi-custom bodies were offered as the Type V-63 entered its second model year. There were two wheelbases — the standard 132 incher, and a special 138-inch chassis utilized for some of the custom and larger bodies. Horsepower had been increased to 83. The Type 61 which preceded the V-63 had been rated at 79 horsepower.

The Type V-63 model lineup, with list prices, was as follows: Two-Passenger Roadster, Four-Passenger Phaeton and Touring Car for Seven Passengers all priced at $3,085; Fisher Two-Passenger Coupe, $3,875; Five-Passenger Coupe, $3,950; Sedan for Five Passengers, $4,150; Five-Passenger Sedan with Imperial Division, $4,400; Seven Passenger Suburban, $4,250; Imperial Suburban, $4,500; Seven-Passenger Cadillac Limousine, $4,600; Town Brougham, also $4,500; the standard Sedan for Seven Passengers carried a price of $3,585. Also offered were a four-passenger Victoria at $3,275 and a semi-formal Landau Sedan for Four Passengers, at $3,835.

The Cadillac Motor Car Company built 18,827 Type V-63's during the 1924 model year. For reference purposes, the 1924 model was sometimes referred to as the Type V-63A, and the very similar 1925 V-63 as the V-63B. This was Cadillac's only passenger car series identified with a "V" prefix.

With the return of prosperity, the U.S. automobile industry had truly entered the custom-body era. The boom was on, and as fortunes piled up, so did demand

For those customers who sought the racy appearance of a ragtop roadster but the comfort and all-weather convenience of a closed car, Cadillac for 1924 offered an all-new Type V-63 Two-Passenger Coupe. Like the companion V-63 Roadster, this model had a special rear compartment for carrying golf clubs or small packages. Cadillac hailed this new model as ... "the season's smartest two-passenger closed car". The body was by Fisher. Wheelbase was 132 inches. Note the large windshield visor, plated headlights and formal, padded roof treatment complete with stately carriage bows. The Type V-63 Coupe for Two shown here rides on a set of optional wire wheels instead of the standard wood-spoked artillery wheels. Price was $3,875.

Introduced to the public in September, 1923 the new Type V-63 Cadillac was produced for 1924, and through the 1925 model year. The new Type V-63 Cadillac, successor of the Type 61 of 1922-23, was powered by a thoroughly redesigned V-8 engine with an inherently-balanced crankshaft, and boasted four-wheel brakes. This is the 132 inch wheelbase 1924 Type V-63 Roadster, a sporty two-seater which carried a base price of $3,085. The rear deck featured a golf bag compartment and a disappearing rumble seat for carrying two extra passengers.

With a proliferating choice of body styles, closed cars continued to win favor in all segments of the American automobile industry by the mid-1920's. Cadillac, of course, moved quickly to accommodate this trend. For 1924, the division marketed this practical Type V-63 Coupe for Five Passengers, which carried a list price of $3,950. Seating arrangement consisted of two individual seats up front and an extra-wide rear seat for three. The Cadillac V-type, eight-cylinder engine was now into its tenth model year. The Five-Passenger Coupe also featured a storage compartment in the rear deck.

1924

The venerable Cadillac Phaeton for Four Passengers soldiered on into the Series V-63. Standard V-63 wheelbase was 132 inches. The V-63 Phaeton carried a starting price of $3,085 list — the same as the price leader Roadster and companion seven-passenger tourer. A two-position, folding foot rest was provided for the comfort of rear seat passengers.

Cadillac's "other" big, open car was the ever-popular Touring Car for Seven Passengers. Despite the additional sheet metal and hardware required for the longer rear body, the Type V-63 Seven-Passenger Tourer was priced identically with the Four-Passenger Phaeton and Two-Passenger Roadster as one of the division's lowest-priced cars. Wheelbase was also identical, at 132 inches. Even with the top up, Cadillac touring cars had a sporty look about them.

Even though Henry M. and Wilfred had abruptly left the division some seven years earlier, Cadillac's products continued to reflect a definite "Leland look" in their styling. Except for minor differences in their radiators, Cadillac and "L" Series Lincoln touring cars of this era were almost indistinguishable at a glance. This Type V-63 Cadillac Tourer presents something of a problem in identity, inasmuch as it has no door handles. This would appear to be a Type V-63 Phaeton. Note the trunk carried between the body and the spare tire, and the protective boot buttoned into place over the folded-down top.

for special, premium automobiles. The custom body shops were busier than ever before catering to the whims of wealthy customers who demanded something beyond the very best offered by the leading fine car manufacturers of the day. Exclusivity became very important, and cost was no object. Typically, the customer selected the chassis of his or her choice and had a custom body builder design and build a special body for it. Some of these bodies were produced in small series — a dozen or 50 or so. Others were one-off creations.

Cadillac, at this point in its history, exercised little control over who mounted custom bodies on its V-8 chassis. The division's own custom body requirements were generally met by Fisher or Fleetwood, but a number were supplied by various leading custom body concerns like Willoughby, Brunn and others. Cadillac's big New York City distributor, Uppercu Cadillac Corp., even had its own custom-body department. Some of Uppercu's custom body designs were executed by Hollander and Morrill, of Amesbury, Mass. The luxury car business in America was on its way to unprecedented prosperity, a state of affairs which would last for another six or seven years.

This year's National Automobile Show was the first "all-gasoline" exhibition since the show had started at the turn of the century. There wasn't a single steamer or electric to be seen. Despite a booming U.S. economy, there were still casualties. One of the oldest and most esteemed names, Winton, had discontinued motor car production. But there was a strong newcomer in town, too. Walter Percival Chrysler, who had saved Willys and who had been brought in to work the same magic for Maxwell and Chalmers, announced a new car that bore his own name. Chrysler's new Light Six was powered by an advanced high-compression, six-cylinder engine and even boasted four-wheel hydraulic brakes. Chrysler went on to become one of the U.S. auto industry's leaders and giants.

In another milestone move, Alfred P. Sloan requested Cadillac to undertake a study concerning the possibility of building a smaller, less expensive companion car to the Cadillac. Three years would pass before this program would bear fruit.

Here is another view of the same Type V-63 Touring Car. The seating arrangement would indicate that it is a Phaeton, a four-passenger model. But the car has no exterior door handles — which were standard on both the V-63 Phaeton and Seven-Passenger Touring Car. Note the robe rail across the back of the front seat and the folding footrests. That open rear door sure looks inviting.

1924

Another distinctive, new body style available on the Type V-63 Cadillac chassis was the semi-formal Landau for Four Passengers. This smart closed car featured blind upper rear quarter panels covered in polished landau leather and spanned by graceful carriage bows, and a mildly angled windshield. Note the rear hinging of all four doors, and the nicely detailed window reveals. The Cadillac Landau carried a list price of $3,835.

For the 1924 model year, Cadillac's Type V-63 Sedan for Five Passengers was stretched seven inches to provide additional interior room and passenger comfort. This closed model was priced at $4,150 list. It had a redesigned trunk rack, six-window styling and ornamental landau irons. The new Type V-63 Sedan was offered in two versions — standard, or with an optional glass division between the front and rear compartments. When ordered with the partition, this model was designated the "Imperial Sedan for Five". The result was a close-coupled limousine. Two other Type V-63 models — the big seven-passenger Suburban and the top-line Limousine were also available with the Imperial division.

Considerably more expensive than the standard seven-passenger sedan was the Type V-63 Cadillac Suburban for Seven Passengers. The big Suburban was intended for both city and touring use and was priced at $4,250 list. Like the new Cadillac Sedan for Five Passengers, the Suburban could be ordered with an Imperial division between the front and rear compartments, for combined owner/chauffeur operation. When equipped with the disappearing divider glass, this model was called the Imperial Suburban. Other builders in this era would refer to this latter style as a "Berline."

Cadillac's standard big sedan was the Type V-63 Sedan for Seven Passengers, which had a list price of $3,585. These were by no means pretty cars, but their staid, imposing exterior looks were more than offset by interior comfort and top performance and dependability. By the end of the 1924 model year, Cadillac offered 14 body styles — three more than there were when the new series was announced the previous fall. Bumpers were still an extra-cost option.

At the top of the Type V-63 model lineup for 1924 was the correctly formal Cadillac Limousine. This stately conveyance was intended solely for operation by a professional chauffeur and had a semi-open front compartment. Base price was a hefty $4,600. The V-63 Limousine was designated a seven-passenger car. It was mostly at home in the big cities, plying the grand boulevards or the tree-lined streets of the most exclusive residential neighborhoods in town. Rear seat passengers were provided with a clock and a one-way telephone to the chauffeur.

1924

The very formal-looking Cadillac Town Brougham shared the top rung of the new Type V-63 product line with the Cadillac Limousine. The open-front town car was priced identically with the limousine, at $4,600. Cadillac advertised the elegant Brougham as a "fashionable town car, combining the desireable features of the limousine with the smartness of the landau style ... It is favored by women for shopping, theatre and social calls. The leather back and the landaulet joints add an aristocratic appearance," according to the Type V-63 catalog.

This is a custom-body proposal rendering by Brunn, based on the 1924 Cadillac V-63 chassis. Brunn and Co. of Buffalo, N.Y., described this design as a "Sporting Sedan Landaulet". By this time, the ritzy Automobile Salon — an exclusive, annual exhibition of custom body work restricted to the rich and very rich, was nearing its peak. Custom automotive creations like this one were specially prepared for the prestigious salons, and set the tone for automobile taste and style for the coming year. This design called for a pearl gray and black exterior color scheme, with a three-eighths-inch bright green stripe along the lower body edge, coupe pillar and window reveals. Upholstery was in Wiese gray broadcloth with broadlace trimming.

Described as a "Cabriolet Suburban", this full custom was designed by J.R. McLauchlen, who was custom body engineer for the Uppercu Cadillac Corporation, Cadillac's New York City distributor. Built for H. Bayard Swope, Editor-In-Chief of the New York World, it was constructed by Hollander and Morrill, of Amesbury, Mass., on the 138-inch wheelbase V-63 chassis. Really a berline-landaulet style, it sports the more rounded lines of a cabriolet. The roof was extended forward to form a visor over the sharply-angled windshield. This custom creation was finished in Cobalt Blue and black with nickelled trim accents. The lavish interior included Laidlaw broadcloth upholstery and Linden cabinetwork.

Willoughby and Co. of Utica, N.Y., was the builder of this custom creation on a stretched Cadillac chassis. The wheelbase was lengthened to 145 inches to accommodate an exceptionally roomy passenger compartment. Willoughby defined it as an Imperial Touring Sedan, or Seven-Passenger Berline. Note that only one spare tire has been fitted, even though the rack is designed to hold two rims.

The custom-body era had arrived. Dozens of custom bodybuilders were catering to a wealthy and increasingly style-conscious clientele, who insisted on something special above and beyond what was available from America's fine car manufacturers. The new Cadillac V-63 chassis, with its astonishingly quiet, inherently-balanced V-8 engine, was an ideal platform for mounting quality custom bodies. Willoughby crafted this six-wheel, cabriolet-type Brougham with wire wheels at its plant in Utica, N.Y. according to an automotive press write-up of the day, it was blessed with "a severely plain interior."

1924

This heavily-disguised Type V-63 Cadillac is another creation by J.R. McLauchlen, who headed the custom body department of Uppercu Cadillac in New York City. The custom bodywork, including the special radiator shell, was done by Hollander and Merrill of Amesbury, Mass. This job was described simply as a "Special Cadillac Brougham", and was built on a 138-inch wheelbase V-63 chassis. The front bumper is the new U.S.E. "panel" bumper, with a Cadillac nameplate mounted at the center. Note the large window openings and huge brake drums. Upholstery was in Wiese French bedford cord. Finish was in light blue with black trim and nickeled hardware.

Shotguns cradled in the arms of his two vigilant co-workers, a security guard totes a satchel of cash from a Cadillac armored car to a Detroit bank. With its rugged chassis and powerful V-8 engine, the new Type V-63 Cadillac was ideal for such special purpose conversions. Cadillac actively promoted commercial vehicle sales and encouraged its local dealers to do the same. This appears to be a standard 138-inch wheelbase chassis.

The funeral car and ambulance market was becoming increasingly important to Cadillac. Funeral directors had long recognized the suitability of the Cadillac V-8 chassis for hearse and ambulance work. The Cadillac's prestige and low rate of depreciation also made it a sound business investment. The E. M. Miller Co. of Quincy, Ill. was the builder of this Model 743-A Limousine Funeral Car. The longer, lower-looking limousine type hearse was now rapidly replacing the old carved-panel type, which had made the difficult transition from horse to motor power.

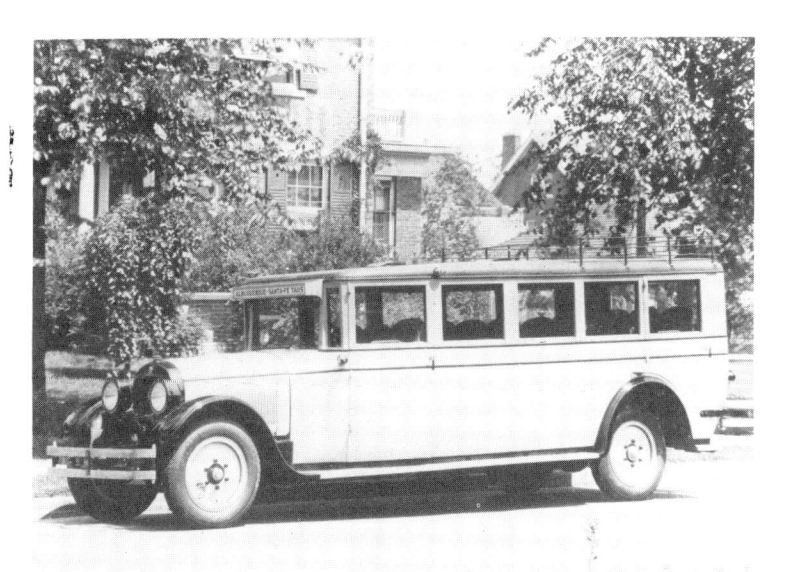

Like previous Cadillac eights, the new Type V-63 chassis with its new, inherently-balanced engine, was quickly adapted for numerous special commercial body applications. This V-63 has been greatly stretched to carry a bus body. Note the sturdy disc wheels, double-beam bumpers front and rear and rear roof luggage rack. The body builder is not known, but this motor coach saw service on the Albuquerque-Santa Fe-Taos route in New Mexico.

The most important change in the new 1924 Cadillacs were to be found under the Cadillac's round-shouldered hood. The new Type V-63 Cadillac was powered by an extensively redesigned Cadillac V-8 engine with an inherently-balanced, two-plane, 90-degree crankshaft. This new counter-weighted crankshaft virtually eliminated vibration resulting in astonishingly smooth and quiet engine performance. The Type V-63 Cadillac was also equipped with four-wheel mechanical brakes. Horsepower was increased to 83.

1925

This was to be another historic year for Cadillac. The most important development on Clark Avenue this year was the election in April, 1925 of 37-year-old Lawrence P. Fisher, Jr., one of the seven Fisher brothers of Fisher Body Co. fame, as President of the Cadillac Motor Car Division of General Motors.

The capable and colorful L. P. "Larry" Fisher would preside at Cadillac over the next nine very eventful years. Fisher's strong leadership and his quest for Cadillac supremacy in the luxury car field carried the division to new heights in both technical innovation and artistic achievement. The L. P. Fisher era at Cadillac was probably the division's finest. The art of custom body-building was nearing its peak, and the Golden Age of the classic American automobile was about to begin. Under Fisher's leadership, Cadillac successfully took on archrival Packard, setting new standards for the industry.

By the mid-1920's, Cadillac had found it necessary to contract for the entire annual output of the Fleetwood Metal Body Co. of Fleetwood, Pa. The U.S. luxury car market was continuing to grow at an unprecedented pace, and there was a seemingly-insatiable demand for special custom bodies for the Cadillac Type V-63 chassis. Finally, in 1925, Cadillac was forced to purchase Fleetwood outright, to ensure sufficient custom body-building capacity. The acquisition of Fleetwood also gave Cadillac close control over the design and style of special bodies fitted to its chassis.

Cadillac and Fleetwood would soon became synonymous in the public mind. To this day, Cadillacs still carry the Fleetwood name on their door sills.

Introduced only the previous year, the Type V-63 Cadillac was continued through the 1925 model year with only modest changes. During the series' second and final year, however, no less than six new models were added to the line.

Appearance changes on the new Type V-63 Cadillac included a new, higher radiator shell. Nickel-plating was standard, which did much to enhance the car's appearance. Early in 1925 the Type V-63 got a new scrolled-design radiator. This radiator was originally introduced on five new Fisher Custom models the previous October, but was made standard across the entire line in January, 1925. The Type V-63 also had a longer hood and new, deeply-crowned fenders. There were no significant mechanical changes under the hood.

The five new "Custom-Built" models included a Coupe for Two Passengers; a Five-Passenger Coupe; a Five-Passenger Sedan; a big Seven-Passenger Suburban and a top-line Cadillac Imperial for Seven Passengers. The new Custom Coupe was built on the 132-inch wheelbase chassis. The other four Customs utilized the 138-inch V-63 chassis.

Early in 1925, Cadillac announced a new, closed model. It was a two-door sedan called the Cadillac Coach. This practical car was designed to appeal to a wide segment of the market and was attractively priced down with Cadillac's lowest-priced standard models. The body was built by Fisher.

Cadillac produced 16,673 Type V-63's during the 1925 model year. In the summer of 1925, the production lines were cleared for the launch of another new generation of Cadillacs — the Series 314. The classic era at Cadillac was about to begin.

In other auto industry news that year, Walter P. Chrysler founded the Chrysler Corporation. Fully aware of what Packard and Cadillac were up to, Chrysler, too, had his eye on the luxury car market.

The Type V-63 Cadillac entered its second, and final, model year. The lightest and least expensive model in the 1926 lineup was again the sporty Cadillac Roadster. This fast two-seater was built on the 132-inch wheelbase chassis and in the standard form shown here carried a base price of $3,185 — the same price as the 1925 V-63 standard phaeton, touring car and the new Cadillac Coach.

Five new "custom-built" models were added to the V-63 model range for 1925. One of these was the handsome Cadillac Custom Coupe for Two illustrated here. This car was built on the 132-inch wheelbase chassis and was priced at $3,975. Styling features included double, half-round moldings and ornamental landau irons on the upper rear roof quarter panels. Note the golf club compartment door. Despite the name, these were mass produced cars, and certainly not "custom cars" in the true sense of the word.

Included in the 1925 Cadillac V-63 range of closed bodies was this practical Victoria. Designed for normal use by one or two, it had seating for four passengers on an "occasional" basis. The Cadillac Victoria was built on the 132-inch wheelbase chassis and in standard form commanded a price of $3,485. Not advertised in Cadillac's promotional material was the fact that the fourth passenger had to perch on a folding seat that normally lived under the dashboard.

1925

Despite a pronounced market shift toward fully-enclosed cars, the appealing Phaeton continued to command a loyal following among Cadillac's customers year after year. The Phaeton and the larger companion seven-passenger Cadillac Touring Car shared the 132-inch wheelbase chassis. Because it was fitted with a folding rear armrest, the Phaeton was sometimes advertised as a five-passenger car. Standard price for 1925 was $3,185. The Phaeton was first introduced away back in 1912.

Big touring cars of this era presented a massive, yet racy, appearance. This is the Cadillac Type V-63 Touring Car for Seven Passengers in standard road trim. Although it was somewhat larger and heavier, the seven-passenger touring car shared the same 132-inch wheelbase as the four-passenger Phaeton and was even priced identically, at $3,185. Bumpers were still an extra-cost accessory.

In January of 1925, Cadillac expanded its range of enclosed bodies with the addition of a new two-door model called the Cadillac Coach. The handsome coach was designed to transport five passengers and was aimed at a large market. The new Coach was mounted on the 132-inch wheelbase chassis and was by Fisher. It was priced down with Cadillac's lowest-priced V-63 models, at $3,185. This style, which was still sort of a novelty in 1925, was destined to become the most popular overall style on American roads, and in effect is the grand-daddy of all of today's full-size two-door models.

One of Cadillac's bread and butter models was the plain, boxy standard Cadillac Sedan. A closed, seven-passenger car, it utilized the 132-inch wheelbase and carried a base price of $3,885. No real attempt was yet being made — at Cadillac anyways — to "style" the product.

Early in the 1925 model year, Cadillac added five new "custom-built" models to the Type V-63 line. Four of these, including the Cadillac Custom Coupe for Five shown here, were built on the 138-inch chassis. The Fisher-bodied Custom models sported a slightly angled windshield. Standard models had a vertical windshield. The two-door Coupe for Five was priced at $4,350 and weighed 4,260 pounds.

When does a Cadillac become an Imperial? When you add a divider partition between the front and rear compartments. This is the 1925 Cadillac V-63 Imperial Sedan which was designed for use as either a "personal" sedan or as a chauffeur-driven vehicle. The division glass could be lowered for those informal occasions when the car was owner-driven. This deluxe model carried a $4,010 price tag. Unlike formal chauffeur-driven cars, in which the front compartment would usually be upholstered in leather, the Imperial Sedan carried its cloth upholstery throughout the car.

1925

Another one of Cadillac's new "custom-built" model offerings was the Custom Sedan for Five Passengers. Note the slightly raked windshield, which set the new Customs apart from the standard Type V-63 models. This one is finished in an attractive two-tone paint and rides on wire wheels rather than the standard wooden artillery wheels. All four doors open from the front. The Custom Sedan had a 138-inch wheelbase and weighed 4,490 pounds.

Another sooty-black, boxlike big Cadillac was this Type V-63 Suburban, which was offered in both standard and Imperial configurations. This is the Imperial Suburban for Seven Passengers, which means that it is equipped with a division glass between the front and rear compartments. The big Cadillac Suburban was mounted on the 138-inch wheelbase Type V-63 chassis and weighed 4,590 pounds. These cars were understandably quite popular with livery and limousine services all over the world.

In its standard model range, Cadillac offered its customers a semi-formal closed car called the Cadillac Landau. For 1925, this V-63 model became even more formal-looking with the addition of a stately oval opera window framed by a decorative carriage bow on the leather-covered upper rear quarter roof panels. The four-door Cadillac Landau was built on the 132-inch wheelbase chassis and was priced at $3,855.

During the 1925 model year, the Type V-63 Cadillac underwent some minor styling changes. These changes included a new "scrolled" radiator design, which is clearly visible here; a longer hoodline and new, deep-crowned fenders. This is a Custom Suburban for Seven Passengers, one of five new custom-built models added to the V-63 family for 1925. Note the slightly angled windshield and double belt moldings, which were exclusive to the premium Custom series. All Suburbans — standard or Custom — utilized a 138-inch wheelbase. This model was priced at $4,650.

This three-quarter rear view shows the standard trunk carrying arrangement on a 1925 Type V-63 Cadillac, in this instance the Custom Suburban. Note the triple taillights and the big, vertically-mounted rear shocks. The trunk is covered with a weatherproof cover and is lashed to the an extra-large accessory luggage rack. The spare tire is mounted in the left front fender instead of in its customary carrier at the rear of the car. This car has also been fitted with an unusual roof-top luggage carrier, of the type far more popular in Europe than in the U.S.

1925

Cadillac correctly advertised this massive four-door as a "dignified" car. That, it certainly was. This is the Type V-63 Cadillac Custom Imperial, a chauffeur-driven limousine on Cadillac's 138-inch wheelbase. The name "Imperial" indicated that the car was equipped with a division between the chauffeur and rear passenger compartments.

While Cadillac's standard models were decidedly plain and utilitarian in appearance, some of the "custom-bodied" creations mounted on the Type V-63 chassis were anything but. This beautiful Cadillac Town Brougham for Five Passengers is a good example of what we mean. Although this was a big, heavy car, its graceful proportions give it a light, artistically formal look. The Buffalo wire wheels, white sidewall tires and window reveals finished in the same color as the lower body contribute to the custom appearance. Note the absence of exterior door handles on the front doors and the slightly dropped upper door edges and moldings.

The Cadillac V-63 chassis, with its smooth, quiet V-8 engine, continued to appeal to funeral directors. Many hearse and ambulance builders preferred the Cadillac commercial chassis, which was spliced and stretched to carry funeral coach and ambulance bodies. The Eureka Company of Rock Falls, Ill. designed and built this Limousine Burial Coach on Cadillac V-63 chassis for George H. Sourbier of Harrisburg, Pa. The body was identified as Eureka Style #194.

Cadillac's Detroit Factory Branch from time to time got into the commercial vehicle business on its own. The company built this "bandit-proof" armored car for Michigan's Wayne County and Home Savings Bank. Built on a special 145-inch wheelbase chassis, the body features 14-gauge aluminum panels covering an interior lining of chrome-nickel steel. The space between the panelling was lined with steel wool, and all glazing was bulletproof. The door on the driver's side could be opened only from the inside.

As cars grew more complex and sophisticated, so did the garages that sold and serviced them. Cadillac had built up a superb dealer organization and insisted on top-notch servicing and maintenance for its products in every area where they were sold. This is an early automobile wash rack. Positioned on it is a 1925 Cadillac Coach, a new Fisher-bodied two-door model introduced early in the year. Even the underside gets a thorough hosing down.

1926

This was another momentous year for Cadillac, and for the American automobile industry. General Motors acquired the Fisher Brothers Body Corporation, creating a new Fisher Body Division, and Cadillac President Lawrence P. Fisher brought a bright, young automobile stylist named Harley J. Earl to Detroit. He had been the top stylist with the famed custom body firm of Don Lee or Los Angeles, and had been responsible for a number of fine custom designs on Cadillac and other fine chassis.

In July, 1925, for the 1926 model year, Cadillac announced its new Series 314 model line. The new Series 314 was the successor to the Series V-63 of 1924 and 1925, and was the first in a long line of classic Cadillacs that derived their series name from the cubic inch displacement of their powerplants. The new Series 314 was an extensively re-engineered car. Chassis weight was reduced a substantial 250 pounds. The redesigned V-8 engine alone was 150 pounds lighter than the V-63 eight-cylinder engine, but horsepower was increased from 80 to 87 horsepower. There were improvements in the cooling and lubrication systems, too, and all 1926 Cadillacs boasted an electric gasoline gauge. Parking lights were incorporated into the headlamps, and standard equipment on the new Series 314 Cadillacs included low-pressure 33 X 6.75 tires.

There were two distinct series within the Series 314. Cadillac offered a choice of six "Standard" models on a 132-inch wheelbase. In the higher price range there were seven "Custom" models — six on a 138-inch wheelbase and one, a Cadillac Custom Roadster, on the shorter wheelbase. Later in the model year these 13 models were augmented by a new series of custom bodies available on special order from such prestigeous shops as Judkins, Fleetwood, Le Baron and Brunn.

In January, 1926 Fisher introduced a new "VV" (vision and ventilation) slanted windshield on some of its custom bodies. This design incorporated small triangular window panes on each side of the car just forward of the windshield pillars. On models equipped with this feature, the instrument panel was moved three-and-a-half inches forward.

The handsome new Series 314 Cadillacs also sported a bright, nickel-plated radiator shell.

Cadillac's new general manager, Larry Fisher, was determined to revitalize Cadillac and make its products more exciting. On a tour of West Coast dealerships, Fisher met young Harley Earl who was then designer for the Don Lee Studio in Los Angeles, the custom car division of Cadillac's biggest distributor in California. Lee had been doing some interesting things for several years, primarily creating highly individual (if sometimes bizzare) custom cars for wealthy and famous people. Fisher talked Earl into coming to Detroit, and Cadillac, on a consulting basis. Fisher wanted to create a specialized vehicle design and styling activity to be called the "Art and Color Section", and he wanted Earl to head it up. Fisher put his new designer to work on a new, smaller car Cadillac was planning to introduce the following year. The car would be called the LaSalle.

During the 1926 model year, Cadillac built its 200,000th car powered by a V-type engine. This milestone car was delivered to aviation pioneer Glenn H. Curtis. Also that year, the Cadillac Motor Car Division finally moved into its new Administration Building on Clark Avenue in Detroit, construction of which had begun in 1919.

Cadillac produced 14,249 Series 314 passenger cars during the 1926 model year. Prices for 1926 were lowered as much as $500.

In another development that would have far-reaching consequences for Cadillac, Owen M. Nacker was placed in charge of an extremely ambitious

The new Series 314 Cadillacs for 1926 were offered in two series — Standard and Custom. Standard models were built on a 132-inch wheelbase, while all but one of the Custom-Built bodies were mounted on a longer, 138-inch wheelbase chassis. The only 314 Custom on the shorter wheelbase was the Cadillac Custom Roadster for Two Passengers, shown here. This model carried a basic price of $3,250. The car illustrated is equipped with optional disc wheels and standard wind wings.

All 1926 Series 314 closed-body Cadillacs featured a new Fisher one-piece ventilating-type windshield. This is the Standard Series 314 Two-Passenger Coupe. Built on the 132-inch wheelbase chassis, it commanded a price of $3,045. The new Series 314 lent itself very nicely to two-tone color treatments. This car is equipped with optional wire wheels. Note the padded leather top and the elegant carriage bow that framed the small side window.

1926

program to develop a new family of high-performance engines — a V-16 and a V-12. This program was initiated and successfully kept under extreme, strict secrecy.

For some years, Cadillac's sales department had observed the popularity of its eight-cylinder commercial chassis with funeral car and ambulance builders. After a careful study, the company decided to become directly involved in this relatively small but very prestigious, and lucrative, market. For 1926, Cadillac introduced a line of quality funeral cars and ambulances built on a special 150-inch wheelbase commercial chassis. The handsome limousine-style hearses and ambulances were designated "Custom Imperials" and were well-received by the American funeral directing profession. Hearses and ambulances would remain a small but important part of Cadillac's business for many, many years.

In industry news, Cadillac's sister division, Oakland, came out with a new car called the Pontiac. This car would eventually replace the Oakland.

Good times had returned and the U.S. economy was strong. There was furious activity in the stock market, and the luxury car market was growing by leaps and bounds. Fashion was the rage, and custom body builders were swamped with orders for special custom bodies.

An industry upstart — Chrysler — had introduced a high-performance luxury car which Walter P. Chrysler named the Chrysler Imperial 80. The numeral was no gimmick: the well-engineered Imperial was guaranteed to do an honest 80 miles an hour. The Imperial would remain Chrysler's flagship car and its top luxury model for the next 49 years.

Also in 1926, Erratt Lobban Cord bought control of Duesenberg. Cord formed the Duesenberg Motor Company in Indianapolis and confidently announced that the new firm would soon introduce the finest and most powerful luxury car the world had ever seen.

Cadillac now had its own, high-toned dealer publication — "The Crest". In August of 1926, the company announced its refined Series 314-A for 1927.

Both Standard and Custom Series 314 Cadillacs featured new, improved Fisher-built bodies. This is the handsome Series 314 Custom Coupe for Five Passengers, which was priced at an even $4,000. Front and rear bumpers were standard equipment on the premium Custom Series. This was one of six Custom models built on the 138-inch wheelbase Series 314 chassis. Equipped with wire wheels, this Series 314 Custom Coupe is on display at an auto salon.

The lowest-priced car in the 1926 Series 314 was the Cadillac Standard Brougham for Five Passengers. Built on the 132-inch wheelbase chassis, it was priced at $2,995. The popular Brougham was equipped with extra-wide 37-inch doors. This one rides on the standard wood-spoked artillery wheels. Cadillac's use of the name brougham for this relatively plain coach has always been misleading, as the word "brougham" in this era usually referred to a very formal chauffeur-driven car.

The practical Standard Series Five-Passenger Sedan rode on a 132-inch wheelbase, and carried a price of $3,195. Interior upholstery in Standard Series models was double-tufted mohair on Marshall seat springs. Three Duco exterior color choices were offered on Standard Series 314 cars. This model is shown with its standard wood-spoke wheels and demountable rims.

One of six Standard Series 314 Cadillacs for 1926, the Four-Passenger Victoria was built on the 132-inch wheelbase chassis and carried a price of $3,095. A nicely-proportioned car, this Series 314 Victoria is equipped with optional wire wheels. Bumpers were optional, extra-cost accessories on the Standard series but were included as standard equipment on the more expensive Custom Series. The full spare tire cover was a dealer-installed accessory. Access to the rather ample trunk was made difficult by the rear mounted spare.

1926

The largest car in the 1926 Series 314 Standard line was the big Seven-Passenger Sedan. Like all other Standard models, this one utilized the 132-inch wheelbase chassis. Price was $3,295. A companion Standard Seven-Passenger Imperial, with a partition between front and rear compartments, was offered at $3,435 and was the most expensive model in the Standard line. Outwardly, the Imperial and the Seven-Passenger Sedan were identical.

The tilt-beam headlamps previously used on all Cadillacs were replaced for 1926 with new drum-type headlamps which gave the Series 314 Cadillac a more massive frontal appearance. This is the Series 314 Standard Seven-Passenger Sedan. All closed bodies now featured two-inch wide wood roof slats, covered with padding and leather fabric, to reduce "drumming".

Almost overnight, Cadillac's big, open cars markedly changed in appearance from staid to sporty. The new Series 314 Phaeton and the companion Seven-Passenger Touring Car were extremely attractive cars with performance that matched their racy good looks. This is the 1926 Cadillac Custom Phaeton on the 138-inch wheelbase. Price was $3,250. This four-passenger car is equipped with wire wheels and a trunk.

A really fine looking car was the Series 314 Cadillac Custom Phaeton with its top folded down. These sporty cars carried handsomely-styled Fisher bodies. General Motors Corporation acquired Fisher Body this year. This car was exhibited at one of the exclusive Automobile Salons, prestigious exhibitions of fine cars staged solely for the benefit of the affluent who appreciated and could afford the luxury cars exhibited.

The Cadillac Custom Series 314 Touring Car for Seven Passengers, at $3,250, was priced identically with the smaller, four-passenger Custom Phaeton. Both of these cars were built on the 138-inch wheelbase chassis. This profile shows to advantage the Touring Car's longer rear body. Note the small wind wings attached to the windshield frame.

1926

This is the Series 314 Cadillac Custom Touring Car in full road trim, with its top up. Despite its massive proportions and strong lines, this car was a real performer. The car shown was finished in a light blue with pale yellow insert in the upper rear door, and black running gear, moldings and trim. The wood-spoke wheels were standard — all other types (wire or disc) were extra-cost options. This car has been fitted with an accessory tonneau windshield.

The Custom Series 314 Sedan for Five Passengers was simply a lengthened version of the Standard. This premium model was priced at $4,150 or $955 more than the Standard Series Five-Passenger Sedan. Custom models offered a choice of six standard Duco finishes, while unlimited special colors were available on sixty days' notice to the factory. This Custom, on 138-inch wheelbase chassis, is equipped with extra-cost wire wheels.

A new, slanted "VV" windshield with small, triangular windows positioned just ahead of the "A" pillar was introduced on closed Custom Series 314 Cadillacs in January, 1926. This is the imposing Series 314 Custom Seven-Passenger Suburban equipped with this new Fisher windshield design. The Custom Suburban, on 138-inch wheelbase chassis, was priced at $4,285. The most expensive car in the line was the Series 314 Custom Seven-Passenger Imperial, at $4,485.

This Custom Sport Landau Sedan on the 132-inch wheelbase chassis was introduced on Cadillac's 1927 models. But the battery box visible above the lower end of the front fender on this car indicates that this is a 1926 chassis. The battery box and tool box on the other side of the car were eliminated on Series 314-A Cadillacs for 1927. This may have been a late-production 1926 or a pilot model. This five-passenger body was by Fisher.

In addition to six Standard and seven Custom bodies — 13 in all — the new Cadillac Series 314 was available with a wide choice of special custom bodies designed and built by a number of well-known custom body houses. This elegant Berline was executed by the John B. Judkins Co. of Merrimac, Mass. The car is seen as exhibited in that year's Salon at the Hotel Commodore in New York City. The Seven-Passenger vehicle contained a disappearing division window behind the front seat which could be raised when the car was to be used for formal engagements.

1926

Resplendent in its luxurious Salon setting is this elegant Coupe for Two Passengers. The body builder is not known but Judkins or Locke are likely candidates. These Salons were lavish affairs, with invitations strictly limited to a select clientele. Tire-kickers were definitely not welcome. Note the upholstered footstool positioned to allow Salon guests to inspect the interior of this car. Evening gowns, tuxedos and top hats were considered proper attire at the Salon shows, and salesmanship was definitely frowned upon. Of course, orders would be discreetely taken.

In later years, the Cadillac Motor Car Division would exercise strict control over what companies outside General Motors would be permitted to mount custom-built bodies on the Cadillac chassis. During the 1920's, however, a number of well-known custom body builders created special bodies for the highly-regarded Cadillac eight-cylinder chassis. Brunn & Co. of Buffalo, N.Y. designed and built this special Collapsible Touring Coupe for Two Passengers on the Series 314 chassis. This was a true convertible, with wind-up windows and a weather-tight passenger compartment.

The Fleetwood Body Corporation of Fleetwood, Pa. offered a wide range of custom body creations on the new Cadillac Series 314 chassis for 1926. One of these was this rather heavy-looking two-passenger coupe, which the builder described as an "Inside-Drive Cabriolet with Rumble Seat". Note the additional ventilation louvers in the cowl, and the forward-angled windshield of the Brewster-type. This car had a fixed top, with decorative landau irons.

This Series 314 Phaeton was styled by Roberts-LeBaron and is very similar in appearance to the standard production Cadillac Custom Phaeton for Four Passengers. Special touches include custom hardware and windshield and the curved, half-round moldings forward and aft of the doors. The striping on the hood and engine compartment side panels is also unique to this special Salon creation. LeBaron, Inc. was started in 1920 by Thomas L. Hibbard and Raymond H. Dietrich as LeBaron Carrossiers, Inc. of New York City and Bridgeport, Conn.

This is another custom Brunn & Co. body for the Series 314 Cadillac. A particularly graceful design, this four-passenger sedan had extremely narrow windshield pillars and a landau-type upper rear roof area. The car was finished in maroon with black trim and an interior done in "Maroon Drab" Boyriven cloth.

1926

Custom body builders' nomenclature — the names used to describe a particular body style — was becoming more and more complex. For instance, the Fleetwood Body Corporation officially described this formal four-door model as an "Inside-Drive" Stationary Cabriolet". Special touches include drum-type cowl lamps and landau irons on the upper rear body quarter panels, and a division window between compartments.

Another creation from the Fleetwood Body Corporation was this big car, described by its builder as an Enclosed Limousine for Seven Passengers. This car has the new "VV" windshield, and the body and window moldings are of the half-round type. This limousine utilized the Series 314 Custom 138-inch wheelbase chassis. A formal vehicle, it was fitted with divider partition between compartments, and a leather-covered front seat.

The Fleetwood Body Corporation, of Fleetwood, Pa. was doing more and more custom body work for Cadillac. Fleetwood created this formal-looking Series 314 Town Car with special six-wire-wheel equipment. Note how the upper front door molding sweeps up over the cowl. This car was built on the Custom Series 138-inch wheelbase chassis. Known as a Semi-Collapsible Cabriolet, the model featured a leather rear quarter which could be folded down from the C-pillar rearward.

Here is another Fleetwood-bodied Series 314 Town Car, again on the 138-inch Custom chassis. Fleetwood described this as a "Cabriolet-type" town car. The windshield side wings, a feature popularized by Fleetwood, operated with the doors. This elegant car was finished in Sagebrush Green above, striped with Ivory, with Moss Gray lower body panels. The leather-covered top was non-collapsible, and the landau irons were strictly decorative.

Fleetwood also designed this open-front Town Car, formally designated as a Cabriolet for Six Passengers. This model was offered in both stationary and collapsible variations. The vertically-positioned windshield frame was nickeled, as were the windwing frames. A collapsible fold-away top provided weather protection for the chauffeur.

1926

The builder of this severely-formal Series 314 Town Car is unknown, though Willoughby or Brewster are likely candidates. The razor-edged upper body lines contrast with the softly-rounded contours of the Series 314 Cadillac's radiator, hood and fenders. This stately car, complete with carriage lamps, rides on Cadillac's standard wood-spoked artillery wheels. This particular style, with large undecorated blind quarters, was usually referred to as a Panel Brougham.

Across the Atlantic Ocean, British and European coach-builders were also producing custom bodies for the new Cadillac Series 314 chassis. Van den Plas of Belgium produced this special semi-collapsible Cabriolet for Colonel V. van Strydonck, of the famed Belgian Regiment des Guides. The top of this town car dropped down, but the pillars remained in position. Note that because of the side-mounts, the battery box had to be moved part way down the running board.

A Belgian coach builder built this long-wheelbase "Tourist Coach" on a Series V-63 Cadillac chassis, completing it in 1926. The open-front motor coach was designed to accommodate 10 passengers and their luggage. Note the open-driver's compartment and roof luggage rack.

Shown here is the business end of the 1926 Cadillac Series 314 chassis. The boxes faired into the trailing edges of the fenders contained the battery and a tool kit and disappeared when the Type 314-A Cadillac was introduced for 1927. The refined V-type, eight-cylinder engine displaced 314 inches and was 150 pounds lighter than the V-63 engine of 1925. The 314 was rated at 85.5 horsepower at 3000 rpm. Compression ratio was 4.7 to 1.

The V-8 Cadillac chassis had become fairly popular among builders of buses. The chassis frame was cut and often extended to considerable lengths. The Keystone Vehicle Co. of Columbus, O. was the builder of this six-door sedan motor coach conversion of a Cadillac Series V-63 chassis, that was completed and delivered to an operator in New Mexico in 1926. Such conversions often took several months to complete and extended into the following model year.

1926

The sturdy Cadillac eight-cylinder chassis was successfully employed in a wide range of commercial service roles. This is a 1926 Series 314 chassis that has been fitted with an armored car body. The considerable extra weight imposed by the installation of this heavy bulletproof body necessitated a switch to steel disc wheels.

For years, the Cadillac commercial chassis had enjoyed wide acceptance among American funeral car and ambulance manufacturers. Cadillac's sales department had been closely watching this situation, and in 1926 the company entered this relatively small but profitable and prestigious market on its own. Cadillac offered limousine-style hearses and ambulances on a special 150-inch wheelbase "Custom Imperial" chassis. This Series 314 Funeral Coach is equipped with rear window draperies and a full-length wooden flower tray above the casket compartment.

This is the Cadillac Custom-Built Series 314 Limousine Ambulance on the Custom Imperial 150-inch commercial chassis. Standard equipment included frosted rear compartment window glass, a Lorraine spotlight, choice of wood-spoked artillery or steel disc wheels, spring covers, a Moto-Meter, and two heaters, one for the front and one for the rear compartment. Tires were 33 x 6.75 inchers.

Included in Cadillac's new funeral car and ambulance offerings was this special Combination Livery and Pallbearers' Coach. This heavy car, designed for use by undertakers and funeral car livery services, featured a padded leather top with "landau back" styling and a formal oval opera window, as well as the new Fisher "VV" slanted windshield. This one was sold to C. Kampp & Sons of Chicago.

This Model 945 Cadillac Funeral Coach was built by Owen Bros. of Lima, Ohio. Like Cadillac's own hearse and ambulance models, this car was marketed as the "Owen Imperial". Note the half-length flower tray and the ornamental carvings around the upper window openings. This body was known as Owen's Model 945, and could be installed on a number of compatible chassis in addition to Cadillac's.

1927

The introduction of the new LaSalle overshadowed all else at Cadillac this year. Introduced to officially commemorate Cadillac's Silver Anniversary in March 1927 — toward the end of the 1927 model year — Cadillac's handsomely styled, lower-priced companion car was enthusiastically received by the public.

For several years, Cadillac's market planners had been carefully studying the price gap that existed between General Motors' highest-priced Buicks and the lowest-priced Cadillac. There was clearly room for a new, moderately-priced car in this price class, and Cadillac moved quickly to fill this need. The "small" Cadillac development program was already well along when Cadillac chief Lawrence P. Fisher brought in Harley Earl from the former Don Lee firm in California. Earl was assigned to contribute his touch for colors and aesthetics to the project. The new LaSalle was consequently the first production car anywhere to reflect the artistic influence of a stylist.

The highly-successful Cadillac car bore the name of the founder of Detroit, Antoine de la Mothe Cadillac. It was particularly appropriate, then, that Cadillac's new junior edition would also carry the name of a noted French nobleman and explorer — Rene Robert Cavelier de la Salle, who in 1682 had claimed Louisiana for his King, Louis XIV.

Several of GM's other passenger car divisions would also eventually come out with lower-priced "companion" cars: Buick, the Marquette; Oldsmobile, the Viking, and Oakland, the Pontiac.

The Pontiac, of course, ultimately replaced the Oakland. The LaSalle remained an important element of the Cadillac Motor Car Division's product plans for the next 14 model years, while ironically, both the Viking and the Marquette vanished after little more than a year following their introductions.

While still fondly recalled by many, the LaSalle must be considered Cadillac's only real failure. While sales were robust for its first three years on the market, the LaSalle suffered badly through the Depression, and when rivals Packard and Lincoln introduced new lower-priced cars in the mid-1930's, LaSalle was never really able to catch up. Only when the LaSalle name was replaced with a Cadillac nameplate on a new series of lower-priced Cadillacs for 1941 did Cadillac really begin to close in on its rivals. It hasn't looked back since.

Officially designated the Series 303, rather unimaginatively for the new car's engine displacement, the LaSalle was initially offered in a choice of 11 body styles, all by Fisher. Eight of these were on a 125-inch wheelbase that was seven inches shorter than the standard 1927 Cadillac's. Three rode on a longer 134-inch chassis. A wide selection of semi-custom Fleetwood bodies were also available.

The all-new LaSalle was essentially a smaller, lighter Cadillac. It was intended to be owner-driven. Cadillac went out of its way to point out that this fine, new car was a real "blood brother to the Cadillac", built by Cadillac in the Cadillac factory in Detroit. Introductory prices ranged from $2,495 to $2,685.

The LaSalle was a totally-new car. It was powered by its very own 75-horsepower, 90-degree, L-head, V-type eight-cylinder engine of 303 cubic inch displacement and $3\frac{1}{8}$-inch bore and $4\frac{15}{16}$-inch stroke. In appearance, the LaSalle bore a strong resemblance to the Hispano-Suiza, a European luxury car which Harley Earl had openly admired. Earl had carefully studied Europe's finest cars at shows on the Continent and had made numerous sketches of features he liked. The "Hisso look" was especially evident in the strong, peaked radiator shape with its winged emblem, the badge tie bar between big, stanchion-mounted bowl headlights, and long, graceful "tablespoon" front fenders. Earl also made highly effective use of colors and two-tone exterior paint treatments.

For 1927, Cadillac offered an incredible 50 body styles in no less than 500 color combinations. In outward appearance, the new Series 314-A Cadillacs were little changed from those of the previous year. This is the Series 314-A Roadster, on 132-inch wheelbase which carried a base price of $3,350. Cadillac at this time offered its products in two distinct price classes. These included standard-production models in a "Standard" series, and a choice of limited-production, special-order bodies in a premium "Custom" series. This Roadster was listed as a "Custom". In true roadster fashion of the era, its windshield could be folded flat across the cowl. Weather protection was via side curtains. Upholstery was of fine-grain leather.

This was Cadillac's first, true production convertible, with a weather-tight body, roll-up side windows and functional landau irons. One of five new Fisher-built bodies, the Series 314-A "Custom Convertible Coupe" sold for $3,450 and was built on the 132-inch wheelbase. Despite the name, this was actually a semi-custom body style. Cadillac this year offered a choice of 18 Fisher bodies in standard production and special-order "Custom" series.

1927

Three months after the LaSalle made its splashy public debut, Willard "Big Bill" Rader drove a stripped-down Series 303 LaSalle Roadster chassis more than 950 miles in a blistering, 10-hour durability run. Accompanied by well-known dirt track racer Gus Bell, Rader — who headed up Cadillac's experimental staff at the General Motors Proving Grounds near Milford, Mich. — averaged an astonishing 95.3 miles an hour on the run.

A total of 26,807 LaSalles were sold in the 1927 and 1928 model years. Cadillac's new companion car was off to a spectacular start. Cadillac Motor Car Division sales soared to a new record 47,136.

While the new LaSalle was in the forefront, the refined Series 314-A Cadillac for 1927 was doing well, too. For 1927, Cadillac offered its customers an unprecedented 50 different styles and types of bodies in no less than 500 color combinations.

The Series 314-A Cadillac was available with a choice of 18 Fisher-built bodies, five of them new, and 32 limited production, semi-custom bodies supplied by outside coachbuilders. Cadillac continued to market its cars in two distinct price classes. The "Standard" series included the ordinary regular production bodies. The premium "Custom" series included numerous semi-custom bodies. Beyond these, Cadillac's customers could choose from yet another series of special-order bodies designed and built by outside firms including Fleetwood, Brunn and Willoughby. With few exceptions, "Custom" models sold by Cadillac were actually semi-custom limited production bodies. Cadillac listed, advertised and sold these as "Customs", however.

Exterior changes to the Series 314-A Cadillacs were few, and minor. All '27 Cadillac hoods sported 12 louvers on each side. The boxes built into the lower front fenders and aprons on the Type 314 of 1926, which housed the battery on the left and a tool kit on the right, were eliminated and incorporated into compartments directly below the front doors on all 1927 models. Front fenders were new one-piece stampings instead of the three-piece assembly used previously. The Series 314-A Cadillac sported a new Cadillac medallion, and the big drum headlights previously used only on premium "Custom" models were extended across the line. Inside, there was an attractive new instrument panel.

Prices ranged from $2,995 for the standard Five-Passenger Brougham (Coach) on up to $4,350 for the semi-custom Seven-Passenger Custom Imperial.

For the third year, Cadillac continued to market its own line of "Custom-built" funeral cars and ambulances on a special 150-inch wheelbase commercial chassis. These very expensive quality hearses and ambulances were called Cadillac Imperials. The special bodies for these cars were built for Cadillac by the Meteor Motor Car Company of Piqua, Ohio.

In other industry news that year, General Motors stockholders shared more than $134 million in dividends, and Henry Ford built his last Model "T".

We just couldn't resist publishing this one. Exactly what that eagle-eyed gunner is doing standing in the rumble seat of this Series 314-A Fisher-bodied convertible coupe — a "Custom Series" offering — is unclear. The fellow in the passenger seat is ready for action, too. Maybe they were out gunning for Packards. Note that when the convertible top was lowered, the windshield visor remained in position, providing a slightly unusual appearance. The top, when folded, bulked much more than did the top on the Roadster.

The lowest-priced model in Cadillac's 1927 lineup was the standard production, "Standard Series" Five-Passenger Brougham. This practical and roomy closed car, on 132-inch wheelbase, carried a base price of only $2,995. Its styling closely followed the popular "Coach" design, which was quickly becoming General Motors' best-selling body style. Note the extra-wide doors and generous glass area. The space between the rear panel and the spare tire was designed to hold an accessory trunk, and more often than not these Broughams sported such a unit, either dealer supplied or purchased as an after-market item.

Also at the low end of the Series 314-A Cadillac price list was the Five-Passenger Victoria. This attractive five-window closed car, on 132-inch wheelbase, was priced at $3,195. The Victoria's trunk was in the rounded rear deck. The companion Series 314-A Brougham had a folding, external trunk rack. The '27 Cadillac hood had 12 louvres on each side.

1927

Cadillac's "Standard Series" Two-Passenger Coupe for 1927 was priced at $3,100. It, too, utilized the 132-inch wheelbase chassis. Note the continuous molding which ran from the radiator shell back to the base of the rear deck, giving the car a longer appearance. The battery was now housed in a compartment built into the left apron beneath the front door on all models. The landau bars on this model were strictly decorative, but gave the car an appearance quite similar to the Convertible Coupe, which utilized functional landau irons.

For those customers who wanted a little more flair in their closed coupes, Cadillac for 1927 offered this Series 314-A Sport Coupe for Two Passengers. At $3,500, this car was priced $400 higher than the Standard Series Two-Passenger Coupe. The car illustrated is equipped with six disc wheels, an extra-cost option. Note the cover on the sidemounted spare and the color insert around the window which complemented the fenders and splash aprons. The dummy landau irons on this model were nickeled, but were painted black on the plain version. The bumpers shown on some of these cars were still considered extra-cost accessories, installed by the local dealers.

Above and beyond the 18 standard and limited-production Fisher-built bodies for 1927, Cadillac offered its customers a wide choice of semi-custom, special-order bodies by Fisher and several outside firms. One of these was this "Custom Series" Cabriolet Coupe for Five Passengers, which was mounted on the 132-inch wheelbase Series 314-A chassis. Note how the ornamental landau iron is integrated into the upper corner of the rear window, an interesting styling touch. The use of accessory sidemounts rather than a rear spare made entrance to the integral trunk much easier and also allowed for the mounting of a trunk rack so that even more luggage could be carried. Wire wheels were also as extra-cost item, above the $4,095 basic price.

Included in the limited-production Series 314-A "Custom" Series for 1927 was this extremely handsome "Custom Coupe for Five Passengers". The car shown features the new Fisher "VV" slanting windshield and optional disc wheels. Note the fine detailing and reveals around the window openings. This model was priced at $3,855. The disc wheels were extra-cost accessories.

Cadillac's Series 314-A for 1927 included two four-passenger Phaetons. This is the "Custom Series" Phaeton for Four Passengers, a limited production Fisher body on the 138-inch wheelbase chassis. This big open car had a starting price of $3,450. A more expensive companion Cadillac Sport Phaeton was also available. Although wood-spoke wheels were standard fare for this car, they certainly didn't enhance the overall appearance as did the wire spokes shown on the Sport Phaeton.

One of five new Fisher bodies available on the Cadillac Series 314-A chassis for 1927 was this beautiful Cadillac Sport Phaeton, with folding rear tonneau deck and windshield. The rakish "Custom Sport Phaeton" was priced at $3,975, or $525 more than the standard Cadillac Phaeton that year. This one is equipped with six wire wheels and a trunk. Eventually, this style would come to be known as a dual-cowl Phaeton. The rear cowl and windshield would swing upward when either rear door was opened, to allow easier entry for rear seat passengers.

1927

The limited production "Cadillac Custom Sport Phaeton" was an extremely impressive car and it drew admiring glances wherever it was seen. Here is one in full regalia, complete with nickeled disc wheels and a big, stanchion-mounted swivel spotlight on the left running board. The use of the latter accessory gave this very masculine-looking sportster the appearance of a fire engine.

Cadillac's 1927 model line included three big, open cars. These included the Phaeton for Four Passengers, the premium "Custom Sport Phaeton" and this "Custom Touring Car for Seven Passengers". All were built on the 138-inch Series 314-A chassis. This Touring Car and the base Phaeton were identically priced at $3,450. For 1927, the color insert panel was extended to include both front and rear doors: on 1926 models it was used on the rear doors only. Cadillac "Customs" were actually limited-production, semi-custom cars available only on special order, and often sporting refinements specified by the buyer.

One of Cadillac's most popular closed cars was the practical, if plain-looking, Five-Passenger Sedan. This is the Series 314-A Standard Series Sedan for Five Passengers, a 132-inch wheelbase car priced at $3,250. Several other more luxurious versions of this body style were also available this year. On this model, silk roller shades were provided on the rear and rear quarter windows. When this car wore its standard wood spoke wheels it had a rather stodgy appearance.

Fisher Body introduced five new bodies for the 1927 Cadillac Series 314-A. One of these was this sporty "Sport Sedan for Five Passengers" on the 132-inch wheelbase chassis. This model was priced at $3,650. Although the lower body was identical to that of the standard five-passenger sedan, the rear quarter area above the beltline was significantly different, having much smaller quarter windows, a covering of fine-grain landau leather, and sporting decorative landau irons.

For 1927, Cadillac continued to offer two distinctly different product series, a regular production "Standard" Series and a premium "Custom" Series, the latter consisting of a selection of limited-production, semi-custom bodies furnished by several outside coachbuilding houses. This is a Series 314-A "Custom Sedan for Five Passengers", with the distinctive Fisher "VV" (Ventilation and vision) windshield, and six-wire-wheel equipment, the latter being above the $3,595 base price.

At the top of the semi-custom "Cadillac Custom" Series was this imposing Custom Cabriolet Sedan for Five Passengers. Features include the Fisher "VV" slanted windshield and ornamental landau bows integrated into the upper rear corner of the side windows. This model was available only on special order. Six wire wheels further contribute to the custom look. This vehicle's prices began at $4,095 and ranged upward, depending on specifics ordered.

1927

The largest and heaviest standard production Series 314-A Cadillacs were the massive Seven-Passenger Sedans. This is the "base" seven-passenger closed car, the "Standard Series" model which carried a starting price of $3,350. A companion "Custom Suburban for Seven Passengers" was also offered, as well as the top-line Seven-Passenger Imperial.

Up near the top of Cadillac's "Standard Series" closed car offerings was the big "Custom Suburban for Seven Passengers". In stock form, this semi-custom, 138-inch wheelbase car sold for $4,125. This Seven-Passenger Series 314-A Suburban rides on the standard wood-spoked artillery wheels. Wire or disc wheels were available at extra cost.

The most expensive car in the standard production Series 314-A model lineup for 1927 was the impressive Seven-Passenger Imperial. All Cadillac "Imperials" featured an "Imperial Division" between the front and rear compartments, and were intended to be chauffeur-driven. Photographed outside Cadillac's Clark Avenue plant in Detroit, this 138-inch wheelbase car is painted formal black. This model was base priced at approximately $3,500.

Beyond the "Standard" Seven-Passenger Imperial and the premium "Custom Imperial" was this special Series 314-A "Custom Cabriolet Imperial" base priced at $4,445. Note the very formal upper quarter treatment with stylish "leather back" and decorative carriage bows. Fine, bright window moldings heighten the custom look, as do the extra-cost wire wheels and sidemounted spares. Note that all of these models were fitted with small courtesy lights on the aprons, to light the running boards when the doors were opened at night.

Even at the pinnacle of Cadillac's 1927 model offerings, there were standard-production and premium series. This is the "Custom Imperial for Seven Passengers". Actually a limited-production, semi-custom, this model was priced at $4,350, or $800 higher than the "Standard" Seven-Passenger Imperial. Wheelbase was a regal 138 inches. The Fisher-built body incorporates a "VV" sloped windshield. Even on these cars, the bumpers were considered an accessory. If any sort of long distance traveling was contemplated, these cars would have had to be ordered in six-wheel style, so that a luggage rack could be mounted in place of the standard rear spare.

In addition to 18 standard production and semi-custom bodies by Fisher, Cadillac for 1927 offered a choice of no less than 32 limited-production special bodies built for Cadillac by outside coachbuilders. Eighteen of these premium semi-customs were supplied by Fleetwood. This is the Series 314-A Cadillac Fleetwood Town Cabriolet, an open-front town car of graceful proportions, with prices starting at $5,000. The side-mounted spare tires are recessed into the front fenders, and the side window glass opened with the front doors. Wire wheels and landau upper quarter panels accentuate the stern, formal look. A wide variety of color options — ranging right into the wild side — was available for this car.

1927

This beautiful open-front town car was called the Series 314-A Cadillac Fleetwood Limousine Brougham, a conveyance, Cadillac observed..."which is in great demand for business and social occasions." The appropriately lavish interior was upholstered in Weise broadcloth. The exterior of this car was finished in Deep French Gray and Light Fleetwood Gray. As far as can be ascertained, Fleetwood offered no other color combinations on this model, though undoubtedly customer pressure could have resulted in any hues that the buyer demanded. It appears that this was the highest priced semi-custom Cadillac offered by Fleetwood, with its base price starting at $5,500, exclusive of transportation, taxes, or dealer handling.

Another Fleetwood Body creation on the 1927 Cadillac Series 314-A chassis was this Fleetwood Transformable Town Cabriolet, which sold for $5,000. As the name implied, the front compartment could be left open as shown or closed in. The front door window glass lowered into the doors and a weatherproof roof could be snapped into place. On the other open-front Fleetwood bodied cars, front compartment weather protection was by a folding top and sidecurtains.

This is one of four special, limited production bodies for the Cadillac Series 314-A chassis designed and built by Brunn & Co. of Buffalo, N.Y. This elegant style was called the Cadillac Brunn Town-Cabriolet. The leather-covered rear quarter was collapsible, permitting the car to be used as a semi-open vehicle in fair weather. Note, too, the graceful sweep of the body molding extending from the cowl to the B-pillar. Upholstery was in Laidlaw broadcloth, while Willey's lacquer was used on the exterior. Available colors were Geneva Blue; Orriford Lake (maroon); and Channel Green, all with black fenders, aprons, and trim.

Here is yet another Fleetwood-bodied Cadillac town car. This one was called the Cadillac Fleetwood Town-Brougham. It was advertised as a car..."of austere dignity with beautifully-balanced body lines and a smartly narrow rear quarter window." Color choices included Fleetwood Blue and black; Persidio Gray and black; Larchmont Blue and black; Orriford Lake and black, and Renault Green and black. Tonneau upholstery was in Weise broadcloth, while the chauffeur's compartment was done in natural leather — usually in black.

Cadillac's comprehensive "special custom body line" for 1927 included 18 Fleetwood bodies, four by Brunn and 10 others supplied by other well-known coach builders. Willoughby and Company, of Utica, N.Y. designed and built this Town Cabriolet body for the Cadillac Series 314-A chassis. It was available in Black, Cornelian Red and Black; Willey's Claret Wine Light and Black, and Roman's Normandy Blue and Black, with Weise broadcloth interior upholstery. Other Willoughby touches include the stirrup-type door handles for the rear compartment, and interior handles only for the front doors. Note that the speaking tube is almost concealed on this style, visible only at the center of the driver's seat, whereas the Fleetwood bodies mounted this tube very conspicuously on the left B-pillar. The Willoughby company, incidentally, had the reputation for providing the most comfortable seating and finest interiors of any of the custom body firms of the era.

Cadillac continued to offer the funeral directing and medical professions a series of special, "custom-built" hearses and ambulances. These were built on a special "Custom" 150-inch wheelbase chassis. This is the 1927 Cadillac "Custom Imperial Limousine Funeral Coach". Note the Fisher "VV" windshield, full-length sliding flower tray above the casket compartment and disc wheels. Bodies for these vehicles were built by the Meteor Motor Car Co. of Piqua, Ohio.

1927

This is the imposing Cadillac "Custom Imperial Limousine Ambulance" for 1927. Embellishments included big, nickeled drum headlamps, matching cowl lamps, nickeled cowl band and Fisher "VV" angled windshield. The leaded glass in the rear windows was an extra-cost option. The interior was upholstered in two-tone gray leather and the floor was covered in ship linoleum. Equipment in the patient's compartment included a dome light, electric fan, and exhaust heater. This compartment measured 93½ inches long, 42 inches wide between the wheels, and 46 inches wide at the rear access doors. Two headroom styles were available, the 53-inch model shown here, or a lower 50-inch version.

What could be more representative of the halcyon days of 1927 than Colonel Charles A. Lindberg's "Spirit of St. Louis" and the all-new LaSalle? Lindberg made his epic solo flight across the Atlantic in May, 1927—two months after Cadillac announced its new, lower-priced companion car, the LaSalle. Its windshield and top folded down, the $2,525 LaSalle Roadster looks right at home snuggled up under the wing of Lindberg's world-famous Ryan Monoplane. For license plate buffs, that is a 1927 Missouri tag on the front.

Cadillac announced its new LaSalle with much fanfare in March, 1927. Introduction of this new, lower-priced companion to the Cadillac marked Cadillac's Silver Anniversary. The LaSalle was officially known at Cadillac as the Series 303. The 1927 Cadillac was called the Series 314-A. This is the light and lithe-looking 1927 LaSalle Roadster, available in rumbleseat form only. The LaSalle was the first production car designed by a stylist. The stylist was Harley J. Earl, whom Cadillac chief Larry Fisher had enticed to Detroit from the West Coast the previous year. In California, Earl had been the chief stylist for the famed Don Lee firm, which had produced numerous custom cars on Cadillac chassis.

The LaSalle Coupe for Two Passengers was one of 11 body styles offered on the new 1927 LaSalle. All were built by Fisher. Several semi-custom Fleetwood bodies were added later. The new LaSalle was built on a 125-inch wheelbase, seven inches shorter than the standard Series 314-A Cadillac for 1927. LaSalle made effective use of two-tone color combinations, painting the body one color and the hood another complementing color. The effect was visually quite pleasing. As with the similar bodied Cadillac, the LaSalle Coupe used a leather-covered top and decorative landau irons. A roll-down rear window allowed front seat passengers to converse with those in the rumble seat when that facility was in use.

The dashing new LaSalle bore more than a slight resemblance to the Hispano-Suiza. Harley Earl admittedly admired the "Hisso", and incorporated some of that European luxury car's styling features into the new Cadillac companion car. This influence is most noticeable in the radiator and winged emblem, the badge tiebar between the big bowl-shaped headlights, and the long "tablespoon" front fenders. This is the 1927 LaSalle Series 303 Convertible Coupe, which was priced at $2,635.

1927

This is one of the special Fleetwood semi-custom bodies offered for the new Series 303 LaSalle chassis for 1927. This is a rather formal-looking coupe with rounded rear deck and two-tone paint treatment that emphasizes the angular body. Six-wire-wheel equipment, a thickly padded roof and decorative landau irons contribute to this coupe's distinctive appearance. The unique paint job gives the impression of a curved coach sill. Fleetwood listed this body as its Style 3110, built on special order only, and priced at $3,600 in basic form, which probably did not include the chromed wire wheels or dual sidemounts.

The all-new LaSalle was offered in a complete range of body styles, including this beautiful four-passenger Series 303 Phaeton. A companion dual-cowl Sport Phaeton was also available. This 1927 LaSalle Phaeton has the striking two-tone paint combination used extensively on this new series. The upper body molding panel is painted in a complementing lighter shade. The body was by Fisher. Surprisingly, this was LaSalle's lowest priced car, having a base tag of $2,495, which was $30 less than the Roadster.

The sportiest of all the new LaSalles was this racy-looking Series 303 Dual-Cowl Phaeton for Four Passengers. The striking two-tone paint, wire wheels, folded-down windshields and stanchion-mounted spotlight give this car a powerful, speedy look. Available as equipped here for less than $3,000 this car was a truly remarkable bargain for the money. The basic body was the same as used on the 4-passenger Phaeton, but the extra $500 bought a ton of class.

The 1927 LaSalle was introduced as a lower-priced, lighter and smaller companion car to the Cadillac, bridging the price gap between the Buick and the Cadillac. The Series 303 LaSalle was powered by a new L-head, 90-degree V-8 of 303 cubic inches displacement, and rated at 75 horsepower. This is the elegant LaSalle Victoria, shown in two-tone paint with solid disc wheels, which were probably a non-cost option included in the $2,635 price. The interesting rear quarter treatment and decorative landau bars gives the body a custom look.

Cadillac produced 10,767 LaSalles during the marque's first model year. All LaSalles were strikingly-styled cars which stood out wherever they went. Public interest in this new, lower-priced companion car to the Cadillac was very high. Even the standard Series 303 Five-Passenger Sedan imparted a semi-custom look when "dressed" with wire wheels and LaSalle's distinctive two-tone exterior color choices. LaSalle prices began at $2,495 for the Four-Passenger Phaeton and ranged up to $4,700 for the Transformable Town Cabriolet. The Five-Passenger Sedan was at the lower end of the price scale, with a base tag of $2,685.

1927

At the top of the new LaSalle model lineup were three big closed cars with special limited-production bodies. These models were built on a special 134-inch wheelbase and included the big Series 303 Seven-Passenger Sedan at $2,795 shown here, a Five-Passenger Imperial Sedan with divider at $2,795 also, and a Seven-Passenger Imperial Sedan at $2,895. Note that the Seven-Passenger car used the same rear quarter treatment as did the Victoria.

This is one of several special semi-custom bodies later offered on the new Series 303 LaSalle for 1927. This is the Imperial Sedan for Five Passengers, a correctly-formal model not to be confused with the 134-inch wheelbase sedans. The straight, angular body lines contrast with the softly-rounded contours of the LaSalle hood, radiator and sweeping front fenders. Like the bigger Cadillacs, LaSalles with an "Imperial" designation had divider windows between the front and rear compartments and were usually chauffeur-driven. Built by Fleetwood on the 125-inch wheelbase chassis, this special order car was designated Style No. 3120, with prices starting at $3,800.

In its extremely successful first year on the market, the new LaSalle was available in a complete range of body styles from sporty roadsters and coupes on up to stately limousines and formal town cars. At the top end of the new Series 303 model range was this beautiful LaSalle Town Cabriolet by Fleetwood with open front and six wire wheels. Note the wind wings which open with the front doors, a feature pioneered by Fleetwood. This model, on the 125-inch wheelbase, was designated Style 3130, and had a base price of $4,500. The top was covered in black landau leather, and on this model featured a set of decorative landau irons painted black. A folding top provided weather protection for the driver. As could be expected, this vehicle was available on special order only.

During the Classic Era, luxury car builders tempted prospects with lavish, beautifully illustrated deluxe "Salon" sales brochures that are highly-prized collector's items today. The commercial artists who illustrated these colorful brochures often took extreme artistic liberties with the proportions of the vehicles they were illustrating. Hoods, fenders and bodies were stretched to impossible lengths, but the result was always pleasing. This is a Salon catalog illustration of the 1927 LaSalle Transformable Town Cabriolet by Fleetwood, Style 3051. Contrast this illustration with the photo of the 1927 LaSalle Fleetwood Town Cabriolet, both of which shared the same body, and basic appearance. The Transformable, priced at $4,700, was the most expensive body style on a LaSalle chassis this year. It featured a rear quarter which could be lowered, the top dropping just aft of the C-pillar.

In a convincing demonstration of the new LaSalle's durability, Willard "Big Bill" Rader, who headed Cadillac's experimental staff at the General Motors Proving Grounds at Milford, Mich., drove a stock, stripped-down Series 303 Roadster chassis for 951 miles in 10 hours, at an average speed of 95.3 miles an hour. This was an astounding performance at the time. The test was run three months after the LaSalle's public introduction. Accompanying Rader on this blistering run was famous dirt-track racer Gus Bell. The run began at 6:30 a.m.

1928

With the announcement of its 1928 models in September, 1927, Cadillac replaced the Series 314 Cadillac of 1926 and 1927 with another new series of cars — the Series 341. Again, the new numerical designation indicated the cubic inch displacement of the Cadillac V-8 engine.

The highly successful new LaSalle went into its second model year as the Series 303.

The 1928 Cadillacs were the first styled by Harley J. Earl. Earl's artistic touch was evident in the strong, new Cadillac front-end ensemble, which was closely patterned after that used on the Earl-designated 1927 LaSalle. The new Series 341 Cadillac was equipped with huge, bowl-shaped headlights mounted on stanchions and separated by a Hispano-Suiza-type tie bar with the Cadillac script in a rectangular emblem at its center. Hood louvers were more numerous, and finer. The 1928 Cadillac was a longer, more streamlined car. Wheelbase was increased to 140 inches from the 132 and 138-inch wheelbases offered previously. The refined engine was now rated at a robust 90 horsepower. Cadillac continued to offer its customers more than 50 body styles by Fisher and Fleetwood, in an array of more than 500 interior and exterior color choices. Mechanical changes included a larger rear axle, improved cooling and a new double-disc dry clutch.

The 1928 Cadillac had a redesigned instrument panel. For the first time, instruments were laid out in linear fashion along the dash, rather than in a confusing cluster, or group.

Cadillac's new, lower-priced companion car, the LaSalle, entered its second model year with few changes. The most visible exterior change was in the hood, which now had 28 fine vertical louvers instead of the 12 heavy louvers used in 1927. Three new standard-production bodies were added to the LaSalle model lineup for 1928. These included a new four-passenger coupe and five and seven-passenger "Family Sedans". At the upper end of the range, Fleetwood offered five new LaSalle bodies including a four-passenger Victoria, a two-passenger Business Coupe, a five-passenger Fleetwood Sedan and five and seven-passenger Imperial Sedans.

Cadillac also continued to cater to the funeral directing and medical professions with its own line of "Custom-Built" funeral cars and ambulances. These were built on a redesigned 152-inch wheelbase commercial chassis with special bodies again built for Cadillac by the Meteor Motor Car Co. of Piqua O.

The Cadillac Motor Car Co. built an astonishing 56,038 cars in the 1928 model year, including 40,000 Series 341 Cadillacs and 16,038 LaSalles, a record which would stand until 1941.

In one of the more interesting developments on Clark Avenue that year, famous World War I flying ace Captain Edward V. Rickenbacker joined the sales department of Cadillac-LaSalle as a sales troubleshooter.

The American automobile industry was being swept up in a new wave of prosperity. The stock market was booming, and demand for luxury cars was reaching new highs.

The prestigious Automobile Salons, which had been held annually since 1904, provided custom body builders and luxury car manufacturers with an exclusive setting in which to display their most artistic wares, and to show them off to well-heeled prospects who were eager to buy style at any price.

Each new model year, the opening Salon took place in the Hotel Commodore in New York City. It then moved on to the Congress Hotel in Chicago, and later to the Drake in Chicago. Automobile Salons were also staged at the Biltmore in Los Angeles and the Palace Hotel in San Francisco. The lavishly-illustrated Fisher-Fleetwood salon catalog for this year featured "Color Creations from Nature's Studios".

In industry news of the year, Studebaker in August, 1928 acquired control of Pierce-Arrow. Packard dropped out of the six-cylinder car market, and Buick observed its Silver Anniversary. Walter P. Chrysler's rapidly growing enterprise astounded the auto world when it bought Dodge Brothers, then introduced two completely new car lines — Plymouth and DeSoto.

On December 1, 1928 Fred and August Duesenberg announced the fantastic new, limited production Model "J" Duesenberg chassis. Powered by a 420 cubic inch eight-in-line engine, rated at an astounding 265 horsepower, the chassis alone cost $8,500. Introduced at the New York Salon, the Model "J" Duesenberg chassis was built in two sizes — 142 and 153-inch wheelbases.

Duesenberg sold only the chassis. Customers could choose an appropriate body for it from any one of several custom body builders. Duesenberg in a single stroke had sharply escalated the American luxury car race. But Cadillac had something in the works, too.

The restyled 1928 Cadillacs carried another new series designation—Series 341, the numeral again indicating the cubic inch displacement of the V-8 engine. This is the 1928 Cadillac Series 341 Roadster. Basically a two-passenger car, the Roadster could accommodate two additional passengers in a folding rumble seat built into the car's rear deck. The snappy Roadster listed at $3,350. This one, with disc wheels and folded-down windshield, carries a full complement of three flapper passengers. Cadillac's open car weighed 4,590 pounds, and was available in rumble seat form only.

1928

All Series 341 Cadillacs for 1928 were built on a common 140-inch wheelbase that was two inches longer than the longest-wheelbase models previously offered. This is the 1928 Series 341 Convertible Coupe, a handsomely proportioned two-passenger car that carried a starting price of $3,495. Note the heavy-looking standard wood-spoked artillery wheels. Cadillac's 1928 model offerings also included a fixed-roof coupe for two passengers, which was priced at $3,295. Both models came in rumble seat form only. The fixed-roof model weighed 4,820 pounds, while the Convertible Coupe weighed 4,665 pounds.

The Series 341 Cadillac Five-Passenger Coupe was identically priced with the Convertible Coupe, at $3,495. Cadillac boasted that this practical closed car offered an exceptionally roomy passenger compartment with the added convenience of a large, fully enclosed luggage compartment in the rear deck. This Coupe is equipped with six wire wheel equipment, including two sidemounted spares, these being an added-cost accessory. In plain form, the car weighed 4,760 pounds.

One of the smartest-looking of Cadillac's numerous closed body styles for 1928 was the Series 341 Town Sedan, an attractive semi-formal model for five passengers. This one carries an outside-mounted trunk. The large stanchion-mounted bowl headlights, rectangular Cadillac emblem in the center of the headlamp tie bar and strong radiator shape were inspired by the new LaSalle, which had been introduced late in the previous model year. Cadillac often referred to this model as a four-passenger car, in that seating was basically designed for four, with a fold-up armrest in the rear seat providing space for an extra passenger when necessary. Weight was 4,875 pounds.

For those who desired the sporty looks of a roadster or convertible but the comfort and practicality of a closed car, Cadillac offered this Series 341 Coupe for Two Passengers. The lowest-priced car in the 1928 line, it had a base price of $3,295. Note the graceful landau irons, the baggage compartment door and disc wheels. Features also included a rumble seat and an adjustable rear window so rumble seat passengers could communicate with passengers in the front seat. The graceful lines of the new fenders are evident in this illustration. This was the first year that Cadillac's fenders were stamped in one piece.

The 1928 Cadillacs were the first Cadillacs to reflect the distinctive styling touches of Harley Earl, who had strongly influenced the appearance of the new LaSalle. Even the lowline, standard Fisher-bodied Series 341 Cadillac Five-Passenger Sedan clearly showed Earl's artistic influence. This is evident in the strong front end ensemble, delightful window and molding detail and the wide choice of available exterior colors, including multiple tones. This model sold for $3,595, and weighed 4,880 pounds in its standard form, which included wood-spoke wheels with demountable rims and a single rear spare rim.

1928

Again, the new Series 341 Cadillacs were marketed in two distinct price classes—a standard production series with Fisher bodies and a premium "Custom" series with a choice of special, semi-custom bodies designed and built by Fleetwood. This is the 1928 Fleetwood Sedan for Five Passengers, Style 8025. There was also a Five-Passenger Imperial Cabriolet, Style 8035, with glass partition and two folding "opera" seats. The upper rear quarter panels were leather-covered. Interior amenities included a smoking set, vanity case and imported eight-day clock. The Fleetwood Sedan wore a price tag of $4,095 and weighed 4,995 pounds in basic form.

Here is the Series 341 Cadillac Five-Passenger Sedan Cabriolet by Fleetwood. This model had fully-enclosed rear quarter panels, leather-covered, with ornamental carriage bows. This model was listed as Style 8045. A companion Imperial Cabriolet, Style 8055, was equipped with a glass divider partition and two folding "opera" seats. The Sedan Cabriolets were priced the same as the Fleetwood Sedans, while the Imperial versions cost $4,245. Also, the weights on the two Sedans were identical, while the Imperial versions weighed 5,035 pounds.

The Fisher-bodied Series 341 Cadillac Five-Passenger Imperial was designed to be either owner or chauffer-driven. The "emergency" seats, which faced to the rear, folded into the rear of the front seat when not in use. The Series 341 Cadillac line for 1928 also included a premium Fleetwood Five-Passenger Imperial. The Fisher-bodied Imperial was priced at $3,745 and weighed 4,925 pounds in plain form, which included wooden spoked wheels, single rear-mounted spare rim and no luggage rack.

A higher-priced companion to the Fisher-bodied Cadillac Five-Passenger Imperial was the "Custom Fleetwood Imperial for Five Passengers". Externally, this car was very similar in appearance to the less expensive Fisher-bodied version. The Fleetwood-built variant, however, has a different molding design and had a more opulently appointed interior. The rear compartment of this luxurious car was equipped with two "occasional seats" and two-toned, gold, bright-edged, satin-inlaid Ternstedt hardware. All this went to a buyer for a base price of $4,245, which was exactly $50 more than the Fisher-bodied version.

Cadillac's new Series 341 included a bewildering variety of big sedans, all quite similar in appearance at a glance. This is the low-end, standard-production Fisher-bodied Seven-Passenger Sedan on 140-inch wheelbase, which sold for $3,695. Buyers could also specify a more expensive companion Fleetwood Seven-Passenger Sedan. The Fisher version weighed 4,965 pounds as shown.

1928

The most expensive model in the standard production Fisher-bodied Series 341 Cadillac line was the big Seven-Passenger Imperial Sedan. This stately car, with Imperial Division, sold for $3,895. This one was photographed in front of the Scott Fountain on Detroit's picturesque Belle Isle, a popular photo background often used by Detroit auto manufacturers. A bit unusual is the fact that this car has been fitted for dual sidemounts, but retains the wood-spoke wheels and demountable rims rather than wearing the more expensive wire wheels that graced a majority of sidemounted cars. Probably the owner wanted the convenience of a luggage rack, hence the sidemounts, but did not want to go the extra cost for the wire wheels. In base form, this car tipped the scales at 5,025 pounds.

Cadillac called its more expensive standard models "Custom-Built" jobs, although these were actually semi-customs built in batches by various body builders. True one-off full customs were rare. This Seven-Passenger Sedan by Fleetwood was listed as a "Custom". A stylish cowl molding curves gracefully forward from the base of the windshield to the sill. This model was designed Style 8000. A companion Seven-Passenger Fleetwood Imperial with glass divider was sold as the Style 8010. The 8000 model sold for $4,195 and weighed 5,080 pounds, while the Imperial version cost $4,445 and weighed 5,135 pounds.

At the top of Cadillac's lineup of big, closed cars was the imposing Series 341 Seven-Passenger Imperial Cabriolet by Fleetwood, Style 8015. For many years, the name "Imperial" as applied to Cadillacs referred to a large, closed-body car with a glass divider partition between the front and rear compartments. Most Cadillac Imperials were chauffeur-driven. The 7-passenger Imperial Cabriolet was priced at $4,445 and weighed 5,135, the same as the Imperial Sedan. The top was of Burbank cloth. On the Salon version, the color scheme was Arbor Lake (beige) and Woodsmoke Brown, with all interior and exterior brightwork plated in brass. Interior hardware was by Ternstedt, and consisted of units plated in gold and inlaid in wine color cloisonne. Copies of this car were available from Fleetwood on special order.

Fleetwood showed this model, Style 8005, as a Seven-Passenger Sedan Cabriolet. At the upper end of this range was the Style 8015, a Seven-Passenger Imperial Cabriolet with glass partition between the front and rear compartments. It will be noted that Cadillac's own promotional material for the year used the same illustrations for the Styles 8000, 8010, and 8005. The Style 8010 cost $4,195, and weighed 5,080 pounds, the same as the Sedan.

Looking very much at home in its posh country club setting, the 1928 Cadillac Series 341 Phaeton for Four Passengers cuts a dashing figure indeed with its top down and windshield folded forward. For those who liked big, open cars, Cadillac also offered a more expensive companion Sport Phaeton with folding tonneau cover and second windshield. The standard production Cadillac Phaeton shown here was priced at $3,450 and was identically priced with the Series 341 Touring Car. It weighed 4,640 pounds. Note the attractive contrasting color wicker insert on the rear door molding. Wire wheels on this model were an accessory, but were standard on the Sport version.

1928

This is the Fisher-bodied Series 341 Four-Passenger Sport Phaeton. Equipment included a folding tonneau cover with its own, separate windshield for the comfort of rear seat passengers. This Phaeton has a spotlight mounted atop a nickeled running board stanchion. Some Cadillac Phaetons were equipped with two spotlights, one on each side of the cowl. Note the return to a color insert panel in the rear door only. This racy-looking car has a removeable trunk, wire wheels and canvas covers for the sidemounted spares, all of which were considered standard fare included in the $3,950 price tag. The car weighed 5,125 pounds.

The Series 341 Cadillac Touring Car for Seven Passengers was priced at $3,450, the same as the companion standard Four-Passenger Phaeton. Both were built on a 140-inch wheelbase chassis and had high-styled Fisher-built bodies. This tourer is equipped with Cadillac's standard wood artillery wheels. Even so, and with the top up, it is a handsome, sporty-looking car. Surprisingly, it weighed only 4,630 pounds, which was ten pounds lighter than the Phaeton.

This is the 1928 Cadillac Series 341 Touring Car for Seven Passengers in profile. Although outwardly similar in appearance to the Four-Passenger Phaeton, the big Touring Car sported a slightly longer rear body and carried two folding auxiliary seats. There was also a folding armrest in the wide rear seat. The driver's seat was fully adjustable. It is interesting to note that Fleetwood did not offer companions to the Fisher open cars.

It was 1928, and the Golden Age of the truly classic American car had begun. The new Series 341 Cadillacs were the first to bear the mark of stylist Harley Earl, who had been brought in to spruce up the appearance of the all-new LaSalle for 1927. In addition to a wide choice of standard and semi-custom bodies by Fisher and Fleetwood, the 140-inch wheelbase Series 341 chassis could be ordered with an equally wide selection of special order bodies by Fleetwood. One of these was the Transformable Town Cabriolet, Style 3525, shown here. This very beautiful formal town car has a vertically-mounted "V" type windshield with nickeled frame. Note also the bold color panel that sweeps majestically from the base of the cowl over the hood to a point at the rear of the radiator. The elegant Grosse Pointe setting is entirely appropriate here.

This is another 1928 Series 341 Cadillac Transformable Town Cabriolet by Fleetwood, with the removable chauffeur's compartment cover buttoned into place and the front door window glass raised. This car was photographed outside the factory in Detroit. Although this car has the bold sweep panel on its cowl and hood, it is finished in a monotone exterior color scheme rather than the striking two-tone paint treatment sometimes seen on this model. These cars could be ordered with either a fixed rear quarter, or a collapsible unit with functional landau irons. In fixed-top style, this car listed for an even $5,000, while the collapsible version added an extra $500 to the tab. These vehicles weighed 5,135 pounds with six wire wheels and sidemounts, which seem to be standard equipment.

1928

This is another Cadillac Transformable Town Cabriolet by Fleetwood, Style 3520. This style features small quarter windows, the very stylish "V" type vertical windshield and two-tone cowl "sweep" color panel. The front compartment roof cover has been snapped into place for operation in less-than-ideal weather. It appears that both the 2-window and 3-window versions bore the same $5,000 price tag. All windows in the tonneau section were equipped with silk roller shades for privacy. The chauffeur now had roll-up windows in his doors, and no longer had to rely on sidecurtains for protection in cold or wet weather.

This semi-custom, limited-production Fleetwood body, Style 3591, was designated a "Cadillac Transformable Limousine Brougham". This handsome model featured full window glass in the rear compartment. It is shown here with the chauffeur's compartment glass raised, but without its folding roof. This Limousine Brougham has the trend-setting Fleetwood "V"-type vertical windshield with nickeled frame, and distinctive cowl sweep panel. This car listed at $5,005, and weighed 5,135 pounds. All of Fleetwood's open-front town cars shared the same basic lower body, with the major variations appearing above the belt line. All were 7-passenger cars, fitted with twin auxiliary seats.

This is the Salon version of the 1928 Cadillac Series 341 with semi-custom Transformable Town Cabriolet body designed and built for Cadillac by Fleetwood. The car illustrated here was finished in Midnight Blue and Sable, with double molding treatment on the hood and cowl. The opulent interior was upholstered in Weise broadlace—blue with stars in silver needlepoint. Hardware was silver inlaid with deep blue. The fold-away auxiliary seats were equipped with unusually high backs for added comfort. Yet, these seats folded well out of the way when not in use.

Even when viewed from the rear, these classic cars presented an imposing formal appearance. This is the Salon version of the Series 341 Cadillac Transformable Town Cabriolet with Body by Fleetwood. The hood and cowl have Fleetwood's bold double sweep molding, which effectively complements this car's otherwise straight and angular body lines. This model was finished in Arden Green throughout, with a special leather top material in "forest-like" pattern. The rear seat cushions were done in an exclusive Wiese fabric with forest needlepoint design. Tonneau carpeting was in natural sheepskin, while the hardware was in green gold.

Here is a glimpse into the interior of a typical Fleetwood semi-custom body. The car is a Cadillac Limousine Brougham. The richly-appointed rear passenger compartment is upholstered in the best Wiese broadcloth embroidered in a medallion effect in needlepoint. The crank handle visible at the right is for raising and lowering the Imperial division glass between the front and rear compartments. Note the window roller shades and the folding auxiliary seats. The smoking set just below the C-pillar assist cord has been fitted with its own clock—probably a Waltham. Buyers of these cars had a wide range of selections as to colors, types and designs of fabrics, hardware plating, and even seat fitting.

1928

The Fleetwood custom body catalogs prepared for the high-toned Automobile Salons were richly-illustrated works of art and are much in demand by collectors today. One issued in 1928 was entitled "Color Creations From Nature's Studios" and highlighted custom-bodied Cadillacs finished in natural tones inside and out. The Fleetwood custom body catalog used by dealers and their salesmen contained highly-detailed renderings of each body style offered. This delightful line drawing is typical of dozens in one of these bodybuilder's books. This is the Style No. 3591 Transformable Limousine Brougham for the Series 341 Cadillac chassis. It includes complete interior dimensions. The backs of these books sometimes contained packets of such diagrams on tissue.

Some European coachbuilders also created special bodies for the Series 341 Cadillac chassis. This attractive town car body was executed by Hibbard and Darrin, of Paris. The wheel discs used on this car with conventional hubcaps were popular on large cars in Europe. Howard A. "Dutch" Darrin eventually returned to the United States and set up his own custom design operation in California. He was later known for his work on Packards, Kaisers and some special bodies for West Coast Rolls-Royces.

This very nice-looking Series 341 Cadillac was exhibited at this year's important Brussels Salon, in Belgium. It was described as a "Glass-Quarter Brougham with All-Weather Front". The car was finished in a pleasing Straw Yellow with black chassis, moldings and upper body. Interior cabinetwork was in curly ash. It cannot be ascertained who built this body, but the coachwork looks very much like that turned out by Van den Plaas, a well-known Belgian firm.

Here is another European Series 341 Cadillac as exhibited at the prestigious Brussels Salon. The car shown was formally described as a "Cabriolet-Type Town Car with All-Weather Front". The doors extended down to the running boards with a pronounced "turn-under". Note also the absence of any sort of belt molding. A note accompanying this photo described the auxiliary seats folding into a convex division cabinet of burl amboyna, a Molluccan Island wood.

Cadillac's new Series 341 again included a special line of "Custom-Built" funeral cars and ambulances. These were built on a new 152-inch wheelbase commercial chassis. Bodies were supplied by the Meteor Motor Car Company of Piqua, Ohio, an established hearse and ambulance manufacturer. This is the 1928 Cadillac Series 341 Limousine Funeral Coach. The customer could specify either wood-spoked or disc-type wheels. One spare was standard, carried in the left fenderwell. The body was finished in Duco, in any desired color combination, while the front compartment was in bright finish black leather. The rear compartment could be done in taupe mohair or two-tone gray leather, with either silk drapes or roller shades.

This is the Cadillac Series 341 Limousine Ambulance on the new 152-inch wheelbase hearse and ambulance chassis. Funeral car and ambulance bodies were built by Meteor. Suspension components included four Lovejoy hydraulic shock absorbers. Duco finish was available in any combination, but black fenders were standard. In the rear compartment, roller shades of rubber composition were provided on the windows. The floor was done in battleship gray linoleum. A Bomgardner cot with padding and side-arms and two attendant's seats were provided for rear passengers. The compartment also had an electric fan and a Kelch heater.

1928

After some years of dependable service as a passenger car, this Series 341 Cadillac embarked on a second career —as a fire engine. The V-8 chassis has been equipped with a small front-mount fire pump and a hose body probably transferred from an older truck. Three lengths of hard suction hose are slung across the left side of the car. Where this classy rig served is not known. More than a few classic Cadillacs were converted into squad cars by fire departments and rescue squad outfits.

The new LaSalle was a resounding success in its first year on the market. More than 10,000 LaSalles were sold in 1927. The Series 303 LaSalle was carried over into the 1928 model year with only minor changes. This is the 1928 LaSalle Series 303 Roadster, a fleet-looking two-seater with body by Fisher. Two extra passengers could be carried in the rumble seat in the rear deck. This sporty car weighed 3,755 pounds and had a factory price of $2,485.

For 1928, LaSalle continued to offer two Phaetons. These included the standard production Five-Passenger Phaeton shown here, and a premium Sport Phaeton. The lowline Phaeton weighed 3,770 pounds and was priced at $2,485—the same as the 1928 LaSalle Roadster. The Sport Phaeton, at $2,975, cost $310 more and weighed 4,190 pounds. Note the curved cowl molding and the wide belt at the top of the body, which was carried up over the cowl.

The most visible exterior change in the 1928 LaSalle was in the hood, which now had a series of 28 fine vertical louvers instead of the 12 much coarser louvers used on the 1927 model. All of the 1928 LaSalles were high-styled, good-looking cars but this Two-Passenger Coupe was a standout. The Fisher-bodied Coupe was built on the 125-inch wheelbase, weighed 3,770 pounds and had a starting price of $2,450. Wire wheels, and a leather "landau back" roof with landau irons contribute to this car's pleasing looks. The cowl and hood are painted a darker color than the lower body.

A new model in the 1928 LaSalle line was this formal-looking Fleetwood Business Coupe for Two Passengers. The rounded rear deck contained a folding rumble seat. Other new Fleetwood bodies for the 1928 LaSalle included a four-passenger Victoria, a five-passenger Fleetwood Sedan and five and seven-passenger Imperial Sedans. The new Business Coupe was priced at an even $3,000 and weighed 3,935 pounds. Wheelbase was 125 inches.

1928

Companion to the 1928 LaSalle Five-Passenger Coupe was this Fisher-bodied Series 303 Coupe for Four Passengers The double belt molding, curved cowl molding and a small, painted landau iron on the rear roof panel give this 125-inch wheelbase coupe real style. The rear quarter retained the unusual landau design that was used on the 1927 Victoria, with the panels covered in black landau leather and decorative carriage bows (landau irons) cutting into the upper rear window corners.

As in 1927, most of LaSalle's 1928 models were built on a 125-inch wheelbase. Several models, however, utilized the longer 134-inch wheelbase chassis. One of these was the Series 303 Coupe for Five Passengers shown here. This model carried a price of $2,625 and weighed in at 4,050 pounds. The car shown rides on heavy-looking wood-spoked artillery wheels.

Reflecting the increasingly-popular coach body style, this is the 1928 LaSalle Town Sedan, a pleasingly-styled five-passenger car by Fisher. The Series 303 Town Sedan was priced at $2,650 and weighed 3,975 pounds. The small cowl lamps were miniature versions of the big, bowl-shaped headlamps. This Town Sedan is equipped with six wire wheels, with the sidemounts allowing space for a rear mounted luggage rack. All of these items were extra-cost accessories.

Cadillac listed this model as the 1928 LaSalle Series 303 "Standard Sedan for Five Passengers". Note the three-tone color scheme—darker hood and cowl, lighter body with even lighter belt molding. Six wire wheels, a "landau back" roof and carriage bows worked into the upper rear corner of the window give real flair to what would otherwise be a fairly plain-looking four-door sedan. Armrests and a rear window roller shade added touches of luxury to this $2,495 car. The vehicles weighed an even two tons.

Three new models were added to the standard-production Series 303 LaSalle lineup for 1928. Two of these were five and seven-passenger "Family Sedans". The other was a four-passenger Coupe. This is the new Family Sedan for Five Passengers, on 125-inch wheelbase. Note the wide belt molding which provided a natural color break between the upper and lower portions of the body. Quick identification could be made between this and the Standard Sedan by noting the difference in the rear quarter treatment. Priced at $2,350, this model ranked with the Business Coupe as being LaSalle's lowest priced styles for the year. Its weight was 4,060 pounds.

1928

The largest car in LaSalle's regular production series was the roomy Seven-Passenger Sedan. Two versions of this big car were offered for 1928. One was the "standard" version, shown here, which was priced at $2,775 and weighed 4,345 pounds. A new companion Seven-Passenger Family Sedan was priced $200 lower and weighed 45 pounds less. The sedan shown carries a spare tire on the rear instead of the increasingly popular sidemounts.

All LaSalles of this vintage were beautiful cars. This 1928 Series 303 Transformable Town Cabriolet is especially graceful. The wire wheels, side-mounted spares and thickly-padded roof give this Fleetwood creation a pleasing, correctly-formal look. A speaking tube is visible on the door post, approximately at the level of the chauffeur's left ear. The Fleetwood Transformable Town Cabriolet was priced at a hefty $4,700 and was LaSalle's most expensive offering this year. The main difference between the Transformable and the Town Cabriolet was that the Transformable version had roll-up windows in the front doors and a front compartment that could be made completely weather-tight, while the Town Cabriolet was fitted with wind-wings, and relied on sidecurtains for weather protection.

This is another 1928 LaSalle Transformable Town Cabriolet by Fleetwood. The car illustrated is finished in a dark monotone, which works well with the padded, formal roof. Wire wheels gave these luxurious cars an efficient, lighter look. A canvas top could be quickly buttoned into place over the chauffeur's compartment when the weather changed for the worse. At $4,700 this was LaSalle's most expensive 1928 model. All Fleetwood-bodied LaSalle town cars had a basic weight of 4,100 pounds.

LaSalle continued to offer a wide selection of semi-custom Fleetwood bodies for those who desired a correctly-formal car, but at a price lower than the Cadillac. This is the 1928 LaSalle Town Cabriolet by Fleetwood, an elegant open-front seven-passenger town car which had been introduced in 1927. Note the absence of front door handles. This five-passenger car carried a price of $4,500. A Transformable Town Cabriolet was also available at an additional $200. The Salon version of this car was finished in Nassau Sand with Tidewater Blue striping. The top was in tan leather, and the hardware by Ternstedt was in satin nickel inlaid in light blue. Upholstery was in tan broadlace highlighted with blue accents.

Since its acquisition by Fisher Body three years earlier, virtually the entire output of the Fleetwood Body Co. of Fleetwood, Pa. was contracted to Cadillac. A few other coachbuilders were building small batches of semi-custom bodies for mounting on Cadillac and LaSalle chassis. Willoughby and Co. of Utica, N.Y. designed this "Silhouette Sedan" for the Series 303 LaSalle chassis for this year's ritzy Salons. This car reportedly drew much attention with its "perfection of detail and unobtrusive two-tone gray exterior." The interior was done in Copra Lusterweave. The Willoughby Co., which designed many bodies for Lincoln, Wills St. Claire, and Packard, in addition to other prestige chassis, was noted for its high quality workmanship and the comfort of its deeply padded seats.

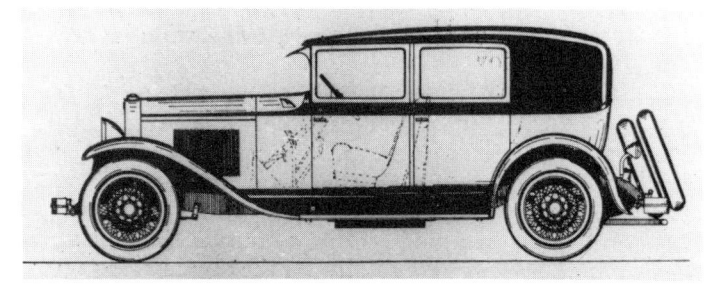

1929

This was a year of significant product refinement for Cadillac. The new Series 341-B Cadillacs for 1929 went into production on Clark Avenue in August of 1928.

Exterior changes were minor. The most visible change in appearance from the 1928 models was the relocation of the parking lamps from the cowl to the tops of the front fenders—a subtle modification which somehow gave the car a sportier look.

But beneath the generally unchanged sheet metal lay a host of important mechanical improvements. The most important of these was a new "clashless" Synchro-Mesh Silent-Shift transmission which did away with double-clutching and marked a notable advance in smoother, easier gear selection.

The Series 341-B Cadillac also employed a new mechanical four-wheel braking system that required considerably less pedal pressure than earlier models. As a safety feature, all Series 341-B Cadillacs boasted shatterproof Security Plate glass in all window openings. And that wasn't all. The car's improved suspension system was equipped with new double-acting Delco hydraulic shock absorbers. Interior improvements included fully-adjustable seats on all models except for the top-line Imperial, which always had a fixed-position front seat. Apparently the chauffeur's comfort wasn't considered all that important.

Cadillac's 1929-model prices ranged from $3,295 to $7,000.

The 140-inch wheelbase chassis was continued through the 1929 model year. In addition to 11 Fisher bodies, the Series 341-B Cadillac was offered with a choice of 12 special Fleetwood bodies in what was marketed as the "Fleetwood Custom" Line.

The Series 303 LaSalle of 1927 and 1928 was succeeded by the modestly-changed Series 328. Again, the numerical designation reflected the cubic inch displacement of the car's V-type eight-cylinder engine. As on the Cadillac, parking lamps on the 1929 LaSalle were relocated from the cowl to the top of the car's long tablespoon front fenders.

The Series 328 Roadster and the two 1929 Phaetons retained the 125-inch wheelbase chassis previously used. All closed models, however, rode on a 134-inch wheelbase that was only six inches shorter than that of the 1929 Cadillac. Again, LaSalle offered a dazzling choice of body styles ranging from a sporty Roadster on up to regal Fleetwood-bodied Transformable Cabriolets. Prices ranged from $2,345 to $4,900.

During the 1929 model year, the Detroit plant produced 18,004 Series 341-B Cadillacs. Production of 22,961 Series 328 LaSalles actually exceeded Cadillac's, for a total of 40,965 cars.

Prosperity had reached new and dizzying heights. Then came the Stock Market Crash of October, 1929. No one knew it at the time, but the bubble had burst. Things would never again be the same.

Conversely, this year was truly an epochal one for the American automobile industry. At General Motors, Oldsmobile and Buick fielded new junior editions named the Viking and Marquette. Cord introduced its revolutionary front-wheel-drive L-29, and Pierce-Arrow announced a new 125-horsepower, L-head Straight Eight. The Model J Duesenberg was the first of the true supercars, and a proud, old name in the luxury car field—Locomobile—was no more.

The Automobile Salon, Inc. was still the correct coming-out place for all that was new and fashionable in luxury cars.

In mid-December of 1929, Lawrence P. Fisher sent out to all Cadillac dealers a letter announcing the most fabulous Cadillac of them all—a totally new sixteen-cylinder series to be introduced early in 1930. The first showing of this up-until-then top-secret ultra-luxury car would be at the factory in Detroit on December 27. Fisher promised three V-16's for the public announcement at the New York Automobile Show in January.

Packard's supremacy in the fine-car field was about to be seriously challenged.

Designated the Model 341-B, Cadillac entered the 1929 model year with relatively minor, yet technologically significant, changes. All 1929 Cadillacs were equipped with safety plate glass in all windows and an innovative, clashless "Synchro-Mesh" transmission. The most visible exterior change was the relocation of parking lamps from the cowl to the tops of the front fenders. This is the 1929 Cadillac Series 341-B Coupe for Two Passengers, a 4,800-pound car which listed for $3,295. Nickeled decorative landau irons and a leather covered top added a luxury-sport touch.

The modestly restyled 1929 Cadillac featured a finely-detailed radiator with a series of fluted, vertical shutters. Big bowl-type headlamps were mounted on stanchions and separated by a tie bar with the Cadillac script in a rectangular badge at its center. This is the Series 341-B five-passenger Town Sedan, which sold for $3,495 and weighed in at 4,750 pounds. Heavy belt moldings were pinstriped in complementary colors. Standard equipment included the removable rear trunk, which could be fitted with custom luggage.

1929

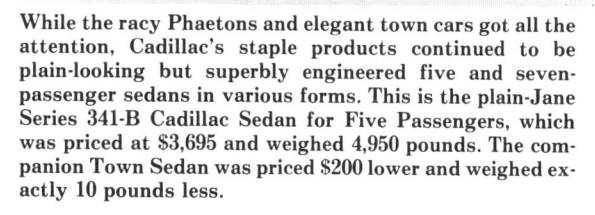

While the racy Phaetons and elegant town cars got all the attention, Cadillac's staple products continued to be plain-looking but superbly engineered five and seven-passenger sedans in various forms. This is the plain-Jane Series 341-B Cadillac Sedan for Five Passengers, which was priced at $3,695 and weighed 4,950 pounds. The companion Town Sedan was priced $200 lower and weighed exactly 10 pounds less.

Wide whitewall tires on wire wheels, a pair of large, bumper-mounted driving lamps and a pleasingly-contrasting two-tone color scheme give this otherwise plain-looking Series 341-B Seven-Passenger Sedan a truly handsome appearance. Louvers on the 1929 Cadillac hood extended only two-thirds of the length of the side panels.

The big Series 341-B Cadillac Sedan for Seven Passengers was a 5,050-pound car priced at $3,795. For another $200, customers could specify the more formal Seven Passenger Imperial, which weighed an additional 40 pounds. This 1929 model rides on the standard wood-spoked artillery wheels. Note the visor border and bright cowl band, both of which were finished in nickel plate.

At the top of Cadillac's standard sedan line was the stately Seven-Passenger Imperial. This formal model featured a retractable glass partition between the front and rear compartments. The 1929 Model 341-B Seven-Passenger Cadillac Imperial was priced at $3,995 and weighed 5,090 pounds in standard form. This one is equipped with optional wire wheels and dual sidemounts, both of which greatly enhanced the car's appearance.

Closed bodies now dominated the Cadillac line. But there was still a place for sporty, open models. This is the 1929 Series 341-B Phaeton for Four Passengers, which weighed 4,635 pounds and carried a base price of $3,459. Also offered was a Sport Phaeton, at $3,950 and a Seven-Passenger Touring Car at $3,450. The Phaeton illustrated is equipped with optional disc wheels and contrasting two-tone exterior paint of rather unusual design.

1929

In addition to 11 standard Fisher body styles, the Series 341-B Cadillac for 1929 was available with an equal number of special Fleetwood bodies. These premium bodies were listed as the "Fleetwood Custom" Line, even though some were obviously produced in quantity. Here is the Cadillac Five-Passenger Sedan, Style 3830-S. A companion model with glass partition and two folding auxiliary seats, Style 3830, was also offered. The car shown was priced at $4,195 and weighed 4,950 pounds. Style 3830 carried a price of $4,345 and weighed 100 pounds more. The dash was done in French walnut and carried an 8-day clock.

Designated Fleetwood Style 3861-S, this is the Series 341-B Cadillac Club Cabriolet for Five Passengers which weighed 4,850 pounds and was priced at $4,395. Cadillac noted in this catalog that an Imperial version of this Club Cabriolet was not available, but could be obtained on a special order basis. Both the top and the trunk were finished in black landau leather, while the blind quarters were decorated with black landau bars with nickel trim.

Yet another Fleetwood special body offering was this Cadillac Sedan Cabriolet for Five Passengers, Style 3855-S. The suffix "S" indicated that this car was not equipped with a divider partition or folding opera seats. When so equipped the "S" was dropped from the description. Thus the Five-Passenger Imperial Cabriolet, Style 3855, had the Imperial divider and folding auxiliary seating. This model had full leather rear quarters, weighed 4,920 pounds and was priced at $4,195.

Cadillac now described its big, open cars as "All-Weather Phaetons". This is the 1929 Series 341-B Five-Passenger All-Weather Phaeton by Fleetwood, Style 3180. Like the closed cars in the premium Fleetwood series, this one could also be ordered with a movable glass partition between the front and rear seats. In this configuration it was listed as the All-Weather Imperial Phaeton, Style 3880. The standard job shown was priced at $5,750 and weighed 4,880 pounds. The Imperial version carried a price of $5,995 and weighed only 10 pounds more. The trunk was covered in the same material as the top, but it could not be ascertained if this was a standard item or an extra cost accessory.

The Fleetwood-bodied Cadillac Seven-Passenger Sedan for 1929, Series 341-B, was offered in two versions—the Style 3875-S shown here, and a more expensive Style 3875 Imperial. Like all Cadillac Imperials, the Style 3875 had a divider partition between the front and rear compartments. Note the stylish cowl sweep panel used on all Fleetwood-bodied models. The Style 3875-S was priced at $4,295. The Imperial commanded a price of $4,545. At 5,090 pounds, the top-line Imperial weighed only 40 pounds more than the standard version.

At the top of the Series 341-B Cadillac model line was a series of special Fleetwood-bodied town cars. This is the 1929 Cadillac Transformable Town Cabriolet, Style 3525. Like all Pennsylvania-built Fleetwoods, it sports a vertically-mounted, V-type windshield. This stately car was priced at $5,250 and weighed 4,980 pounds.

1929

Cadillac's 1929 model offerings included no less than three Fleetwood-bodied Transformable Town Cabriolets. This is the Style 3520, which looks like a limousine brougham with landau quarter panels. The most expensive of all Series 341-B Cadillacs, it commanded a price of $5,500 and weighed 5,070 pounds.

Priced identically with a Transformable Town Cabriolet at the very top of Cadillac's price list this year was the imposing Cadillac Transformable Limousine Brougham, Style 3591. This big Fleetwood-bodied car weighed a whopping 5,030 pounds. Its starting price of $5,000 was equally impressive. The car shown here is finished in a striking three-tone paint combination with cowl sweep panel and is equipped with six wire wheels. The distinctive vertical V-windshield frame was exclusive to Fleetwood bodies built in the old Fleetwood, Pa. plant. Those bodies built in Detroit used a flat windshield.

A variation of the Club Cabriolet for Five Passengers was this model, fitted with a collapsible rear quarter and functional landau irons. This model was not included in Fleetwood's line of semi-production models, but was available on special order. When so equipped, the car was designated the Model 3861-SC, with the "C" suffix standing for "Collapsible". This treatment was not exclusive to the Model 3861—for a fee, any Fleetwood closed body this year could be special ordered with collapsible rear quarters. In each case, the "C" would be added to model number on the bill of sale.

A very interesting car was the Series 341-B Transformable Cabriolet, Style 3550, which was another Pennsylvania-built special order body by Fleetwood. Priced at $5,500, the car was the divison's heaviest, weighing 5,070 pounds. It is shown here totally buttoned up for inclement weather.

When an owner wished to use his Style 3550 Transformable Cabriolet for pretentious town car purposes, it would be up to the chauffeur to remove the folding top over the forward compartment. With that done, the car had this ultra-formal appearance. Both the tonneau and the forward top were done in the same canvas-type material, as were the sidemount covers. Probably this material was either Burbank or laidlaw cloth.

1929

For summer touring around one's country estate or local club, the Style 3550 Transformable Cabriolet was hard to beat. For then, the chauffeur could be instructed to completely lower the tonneau top and the division window, with the result being that the car was now transformed into a very attractive open touring model. Note that on this model, the landau irons were functional and the B-pillars were removable above the sill-line. The rear of the body was scalloped in order to allow the folded top to lie flat in the same horizontal plane as the sill-line. Although these cars were actually limited-production vehicles, few were built unless the customer's order was in hand.

In addition to its so-called "Custom Line"—actually a series of 13 semi-custom bodies produced in varying limited quantities, Fleetwood designed and built a very small number of true, one-off full customs for Cadillac. This five-passenger Inside-Drive Cabriolet with collapsible rear quarters was commissioned for an unnamed "prominent motor car official". It was finished in a deep maroon and Paris gray. The wire wheels were chrome-plated. Styling touches included a sweeping "mailcoach" sill and Burbank-grain light leather top and trunk. The extremely thin, raked windshield pillars were a precursor to later Madam "X" Cadillacs. Note the roof ventilator. The interior was done in maroon Aero leather with dark maroon snakewood on the doors and division friezes.

Another Fleetwood full custom was this Inside Drive Imperial Cabriolet created for this year's prestigious Salons. The trade publication, "Autobody", described this car as actually a seven-passenger berline with opera-type seating for "emergency" passengers. The gently sloped windshield, long hood and cowl and gracefully curved coach sill give this car a suggestion of both fleetness and power, Autobody opined. It was finished in deep maroon and tan with a tan leather top. Note that both doors are rear-hinged.

By this time, the Cadillac Motor Car Division was closely controlling which firms were permitted to mount custom bodies on the Cadillac V-8 chassis. Very few Cadillacs were equipped with anything but Fisher or Fleetwood bodies. Small numbers of chassis, however, were shipped to Europe. Belgian custom body builder Van den Plas executed this very nicely styled open-front town car with collapsible rear bodywork on the Series 341-B chassis of 1929.

Here is another view of the fully convertible Cabriolet by Van den Plas on a Series 341-B Cadillac chassis. Even with its driver's compartment top in place, the car imparts the speedy, low-slung look of a Phaeton or touring car. Wind wings are attached to the thin, gently sloped windshield pillars.

1929

The largest custom body builder in Belgium, and one of the most prominent houses in Europe was Van den Plas. That firm designed and built this impressive Landau Limousine with collapsible rear quarters on the 1929 Series 341-B Cadillac chassis. When lowered, the bulky top certainly presented an untidy appearance. Note the roof luggage rack, which was commonly found on European luxury cars.

Here is the same Series 341-B Landau Limousine by Van den Plas with the rear quarter in the raised position. Those massive landau irons were entirely functional, if somewhat untidy in appearance. This massive car is equipped with six disc wheels and front bumper-mounted driving, or fog lamps. The disc wheels, sidemounts and long body overhang give this car a huge, almost tank-like appearance.

A prominent German coachbuilder, Papler Karosseriewrk G. m. b. H., designed and built this extremely long "Pullman Cabriolet" on the 1929 Series 341-B Cadillac chassis. The six-window car must qualify as among the largest convertibles ever built. The entire upperstructure could be removed and stored to create a completely open car. The upper portion of the door hinge pillar folded into a compartment in the division. All windows were nickel-framed. In a manner similar to the Fleetwood Transformable Cabriolet, the top on this car could be lowered completely, or be left up with only the chauffeur's portion folded.

The highly successful LaSalle swept into its third model year with a new model designation—Series 328—succeeding the Series 303 of 1927 and 1928. The 1929 LaSalle model lineup was juggled, with the original 125-inch wheelbase limited to the Series 328 Roadster and two Phaetons, and all other bodies on the big 134-inch wheelbase chassis. This is the Series 328 Fisher-bodied Convertible Coupe for Two Passengers, which carried a base price of $2,595. Lowest-priced car in the LaSalle line for 1929 was the Roadster which listed at $2,345.

LaSalle for 1929 retained its dashing good looks. As on the companion Cadillac, parking lamps were relocated from the cowl to the tops of the front fenders. This is the LaSalle Series 328 Coupe for Five Passengers, a 134-inch wheelbase car which weighed 4,275 pounds. Prices started at $2,625.

1929

For 1929, LaSalle offered two five-passenger sedans. The standard Series 328 Sedan for Five Passengers shown here was priced at $2,595 and weighed 4,430 pounds. A slightly less expensive Series 328 Family Sedan cost $2,450.

LaSalle's 1929 model lineup included a choice of 12 Fisher-built bodies and two semi-customs by Fleetwood. This is the Fisher-built Seven-Passenger Sedan on 134-inch wheelbase, which weighed 4,555 pounds. Price was $2,775. For an extra $100, a Seven-Passenger Imperial Sedan was available with Imperial divison between the front and rear seats. The big Imperial weighed 4,700 pounds.

The Series 328 LaSalle catalog showed two four-door open cars for 1929—the Four-Passenger Phaeton illustrated here and a more luxuriously equipped and appointed Sport Phaeton. Both were built on the 125-inch wheelbase which was also shared by the Roadster. The standard Phaeton commanded a price of only $2,295 and weighed 4,080 pounds. The Sport Phaeton cost a rather substantial $580 more.

"Replete with smart trappings, handsomely embellished with chromium plate, the Sport Phaeton is a striking LaSalle creation", the 1929 ads read. The Series 328 LaSalle Four-Passenger Sport Phaeton rode on a 125-inch wheelbase and carried a $2,875 price tag. The tonneau cowl was equipped with a folding Security Plate Glass windshield. This model weighed in at 4,405 pounds. Note the dual belt molding and stanchion-mounted spotlight on the left running board.

At the top of the 1929 LaSalle price list were two Transformable Town Cabriolets. These elegant cars were every bit as impressive as their Cadillac counterparts, with the finest Fleetwood coachwork. Styling touches included a flat, plated windshield frame, rear hinged doors and a rear-mounted spare tire. Sidemounted spares and/or wire wheels could be specified at extra cost. This formal car is fitted with the standard wood-spoked artillery wheels. This model weighed a hefty 4,916 pounds and carried a price of $4,900. The black enameled landau irons were strictly decorative.

1930

This year is still remembered as probably the most important for new product introductions in Cadillac's entire history.

Before the end of calendar year 1930, Cadillac had launched a bold bid for supremacy in the luxury car field and was offering the public no less than four separate passenger car lines—a LaSalle V-8; the Cadillac V-8; the monumental new sixteen-cylinder Series 452 Cadillac and a companion Cadillac V-12. But General Motors was fated to make these moves after the stock market crash of 1929, and just as the national economy began its long slide into a disastrous depression.

The promising 1930 model year began quietly enough, with the announcement of the modestly-changed Series 353 Cadillacs and the refined Series 340 LaSalle. The sensational new Cadillac V-16 was not introduced until early in 1930, and the equally impressive Series 370 Cadillac V-12 came along the following fall, for 1931. These new Super Cars are dealt with later in this book. We will confine ourselves in this chapter to the bread-and-butter Cadillac V-8's and LaSalle.

The 1930 Cadillacs carried a new series designation, Series 353, again taken from the displacement of its engine. The second-generation Cadillac V-8 used in the Series 341 of 1928 and 1929 was bored out another 1/16th of an inch, to 3-3/8 inches, and horsepower was increased to 95. For the third consecutive year, Cadillac's wheelbase remained at 140 inches.

The most visible exterior styling change was the extension of the vertical ventilator louvers to the full length of the hood. Windshields were slightly angled, and the division's closed models were pre-wired for the installation of Delco-Remy radios. The radio antenna was concealed in the car's top.

Cadillac customers could choose from a total of 21 body styles, seven by Fisher and 14 semi-customs from Fleetwood. The Fisher line included a two-passenger Coupe; two-passenger Convertible Coupe; five-passenger Coupe and Town Sedan; five and seven-passenger sedans and a seven-passenger Imperial Sedan with glass division.

Fleetwood offered 14 bodies in what it advertised as the "Fleetwood Special Custom Line". All of these bodies carried colorful catalog identification names. These were as follows: Fleetdowns Two-Passenger Roadster; Fleetway Four-Passenger All-Weather Phaeton; Fleetwing Four-Passenger Sedanette Cabriolet; Fleetwind Four-Passenger Sedanette; Fleetdene Five-Passenger Sedan and Five-Passenger Imperial Sedan; Fleetmere Five-Passenger Sedan Cabriolet and Imperial Cabriolet; Fleetdale Seven-Passenger Sedan and Seven-Passenger Imperial Sedan; Fleetwick Town Cabriolet with Opera Seats; Fleetmont Town Cabriolet with Quarter Windows; Fleetcrest Town Cabriolet with Full Rear Quarters, and a Fleetbourne Limousine Brougham.

The 1930 LaSalle also had a new series designation —Series 340. All LaSalles now utilized a 134-inch wheelbase, only six inches shorter than the more expensive Cadillacs.

For 1930, the LaSalle was powered by essentially the same V-8 engine that had been used in the 1928 and 1929 Cadillacs. This powerplant was now rated at 90 horsepower. In its styling, the LaSalle was beginning to break away from the strong Hispano-Suiza look and was being styled more and more like a Cadillac. There were seven Fisher bodies and six semi-custom body choices by Fleetwood, for a total of 13 in all. Fisher Body choices for the Series 340 LaSalle included a two-passenger Coupe and Convertible Coupe; five-passenger Coupe, Sedan and Town Sedan and a seven-passenger Sedan and Imperial Sedan.

Fleetwood "Special Custom" body choices included a two-place Roadster; five-passenger Phaeton; seven-passenger Touring Car; five-passenger All-Weather Phaeton; Sedanette Cabriolet and a five-passenger Sedanette. Some semi-custom bodies over and above these, including open-front Town Cabriolets, were also available on a special order basis.

The 1930 LaSalle's radiator was 2-1/2 inches higher than on previous models.

The declining U.S. economy was already being reflected in much lower new car sales. Cadillac was no exception. Only 11,005 Series 353 Cadillacs were built in the 1930 model year, considerably less than the 18,000 Series 341's sold the previous year. Production of 14,986 LaSalles also fell far below the 22,961 Series 328 LaSalles built for 1929. Cadillac also produced 3,250 Series 452 V-16's in 1930.

The new Series 353 Cadillacs for 1930 went into production in September, 1929. Wheelbase remained at 140 inches, and customers could choose from seven Fisher-built bodies and no less than 14 semi-custom creations by Fleetwood. One of the least expensive models in the V-8 Series 353 model line was the Fisher-bodied Convertible Coupe for Two Passengers, which had a base price of $3,595 and weighed 4,845 pounds. Three passengers could be accommodated on the front seat if necessary, and two more in the folding rumble seat, which was equipped with such refinements as foot rail and arm rests. The chromed landau irons were fully functional.

The least-expensive model in the semi-custom Fleetwood body series for the Type 353 Cadillac was the Roadster for Two Passengers, such as this sidemount-equipped model owned by John Pfleider of Minneapolis. Fleetwood called this style the "Cadillac Fleetdowns." The sidemounts hide the attractive cowl louvers, which matched the hood louvers, and were a special styling touch on all 1930 Fleetwood-bodied open cars. The roadster weighed 4,510 pounds and was base priced at $3,450 with rear-mounted spare. The standard rumbleseat was equipped with arm-rests and a foot rail. The top was fitted to chrome-plated slats with ebony roof bows.

1930

The lowest-priced model in the V-8 Series 353 Cadillac line was this two-passenger Coupe by Fisher. Three passengers could ride on the front seat in a pinch, and two more could be carried in the folding rear deck seat. The Series 353 Coupe weighed 4,940 pounds and carried a $3,295 price tag. This model bore Fisher Body Style 30-158. Easy luggage or golf bag access was provided by the door in the right rear quarter panel.

The Series 353 Cadillac Five-Passenger Coupe carried a base price of $3,595. Body was by Fisher and was designated Style No. 30-172. Weight was 4,930 pounds. Interior appointments included a smoking set with detachable catalin-cased lighter and two ash receivers, and a vanity case with mirror, ash tray and even a memo book with silver pencil! The car shown is equipped with steel disc-type wheels, which were an option. Both dome and corner reading lamps were provided, with the dome light being automatically activated when either door was opened.

Many of America's finest cars of the golden Classic Era rolled on premium-quality Vogue Cord tires. Movie actor William Boyd obviously preferred Vogues, as attested by this autographed photo which bore the message... "Success to Vogue Cords". Boyd's car is a Series 353 Cadillac "Fleetway" All-Weather Phaeton by Fleetwood, which has been customized by the addition of a pair of distinctive Woodlite headlights. This car also has the curved "sweep" molding which extends from the base of the cowl to the front of the hood.

Similar in appearance to Fleetwood's Style 4380 All-Weather Phaeton on the sensational new V-16 chassis was the "Cadillac Fleetway" All-Weather Phaeton for Four Passengers by Fleetwood. This 4,975-pound open car had a starting price of $4,700. A glass partition between the front and rear seats could be used as an Imperial division, as a tonneau windshield or lowered out of sight. Note the V-type windshield. The top was of Burbank cloth. The folding trunk rack could be fitted with an accessory trunk, styled and painted to match the car, and containing a set of specially designed luggage.

The most prolific of the numerous body styles offered on the Series 353 Cadillac chassis for 1930 were a series of five and seven-passenger sedans by both Fisher and Fleetwood. Here is the standard Fisher-bodied Cadillac Five-Passenger Sedan, Style 30-159, which was priced at $3,695. Front-seat legroom was improved this year by recessing the instrument panel into the cowl. The rear seat was provided with side armrests and a center pull-down armrest. When fitted with the extra-cost sidemounted spares, a fold-down luggage rack could be carried.

The Fisher-bodied Five-Passenger Town Sedan, Style 30-152, was one of the most attractive of all body styles available on the eight-cylinder Cadillac Series 353 chassis for 1930. The close-coupled body featured closed rear quarters, and rear seat leg room was increased by a cutaway under the lower back of the front seat. This four-door closed car carried a factory list price of $3,495 and weighed just over 5,000 pounds.

1930

The Fleetwood-bodied Sedan for Five Passengers on the 1930 Series 353 chassis was listed as the "Cadillac Fleetdene". Price was $4,195 and it weighed in at 5,135 pounds. These semi-custom Fleetwood closed bodies had different moldings and window openings than the less expensive Fisher-bodied versions. The car shown also has the distinctive Fleetwood cowl sweep molding which tapered to a point just behind the radiator cap.

At the upper end of the standard Series 353 model line were a pair of big seven-passenger sedans, one by Fisher and a more expensive offering by Fleetwood. This is the Fisher-bodied Series 353 Seven-Passenger Sedan for 1930, which carried Fisher Body Style No. 30-X. Price was $3,795. A companion Fisher Seven-Passenger Imperial with glass division was available for an extra $200. Weight was 5,155 pounds.

The most expensive of the seven Fisher bodies available on the 1930 Cadillac Series 353 chassis was the Seven-Passenger Imperial, which listed for $3,995. This model weighed only 40 pounds more than the standard Fisher seven-passenger sedan. The 1930 Cadillac catalog boasted . . ."Three passengers in opera cloaks find an abundance of room in the rear seats with generous clearance between them and the auxiliary seats. There is an inbuilt telephone for conversation with the driver." An adjustable glass division window separated the chauffeur from his charges in the rear compartment. The front compartment was upholstered in leather.

The Fleetwood "Special Custom Line" of bodies for the Series 353 Cadillac included this ponderous-looking formal car, the "Fleetdale Seven-Passenger Imperial". This model sold for $4,595 and weighed 5,305 pounds. The companion "Fleetdale" Seven-Passenger Sedan was priced at $300 less. The wire wheels were optional. Note the large windshield visor and cowl sweep panel up and over the long hood.

At the very top of Cadillac's eight-cylinder model offerings for 1930 were a series of formal semi-custom bodies by Fleetwood. This is the "Cadillac Fleetcrest", an open-front Town Cabriolet with V-windshield, cowl sweep panel and enclosed upper rear quarter panels ornamented with stately carriage bows. This model listed for $5,145 and weighed 5,135 pounds.

1930

The Series 353 Cadillac "Fleetmont" was a Fleetwood-bodied, formal open-front Town Cabriolet with small rear quarter windows and decorative landau irons. At $5,145, this model was priced identically with the Town Cabriolet with closed quarter panels, and the Fleetwood Limousine Brougham.

The Cadillac Series 353 Limousine Brougham by Fleetwood combined the open look of a conventional limousine with the formality of an open-front town car. This luxurious model commanded a price of $4,145 and weighed 5,305 pounds. It had a slightly raked "V" windshield and cowl sweep panel. Fleetwood listed this impressive model as the "Cadillac Fleetbourne" in its 1930 Book of Fleetwood Styles.

Cadillac's well-heeled customers—at least those who had thus far not been wiped out by the stock market crash and the ensuing Depression—could select the 1930 Cadillac of their choice from dozens of enticing illustrations in the Fleetwood Body Books supplied to all Cadillac Motor Car Co. dealers for this purpose. Not all of the body styles illustrated were actually produced. This is a Fleetwood rendering of a Sport Phaeton for the V-8 Series 353 chassis. An identical body was also available on the monumental new Cadillac V-16 chassis as Fleetwood Style 4260.

This is a very early pre-production Series 353 five-passenger sedan by Fisher Body which was photographed in the Cadillac garage. Styling was now a recognized and important design function at General Motors, and the Art and Colour Section headed up by Harley J. Earl was exerting a strong influence among other GM divisions as well. Art and Colour later evolved into the GM Styling Staff, with responsibility for the appearance and detailing of all of the corporation's products.

Now designated the Series 340, the modestly-changed 1930 LaSalle was available in a choice of 13 body styles—seven by Fisher and six semi-custom body offerings by Fleetwood. All were on a 134-inch wheelbase and were powered by the same 340 cubic inch, 90-horsepower engine that had been used in the 1928 and 1929 Cadillac. This is the fleet-looking Fisher-bodied two-passenger Roadster. A Fleetwood two-passenger Roadster was also available this year.

1930

Companion car to the Series 340 Roadster was the Fisher-bodied Convertible Coupe for Two Passengers. This car has an interesting two-tone paint treatment on the upper body moldings and windshield frame, and functional, chromium-plated landau irons. Price was $2,590. The Convertible Coupe weighed 4,435 pounds in standard trim. Both this car and the Roadsters were rumble seat equipped.

For those who preferred a sporty, open car, LaSalle for 1930 offered two-passenger Roadsters in two separate price classes. This is the premium model with Fleetwood coachwork. The lower-priced Roadster had a Fisher body. This dashing model, Style 4002, was also shown in the Fleetwood Body catalog as the "LaSalle Fleetcliff". Price was $2,450. The louvered cowl was a distinctive styling feature of Fleetwood-bodied Cadillac and LaSalle open models this year. Standard wood spoke wheels and rear spare take away much of the sporty look provided by the optional extra-cost wire wheels and sidemounts.

The lowest-priced car in LaSalle's 1930 model line was this solid-looking two-passenger Coupe. Factory price was $2,490. The car shown is equipped with the standard spoked artillery wheels, but wire or disc wheels were available as extra-cost options. The popular Series 340 Coupe weighed 4,465 pounds. A rumble seat was provided for extra passengers, while the "golf-bag door" on the right rear quarter panel made it easy to load luggage or parcels into the rear compartment.

The very practical five-passenger coupe remained a popular model in LaSalle's offerings for 1930. This is the Series 340 Coupe for Five Passengers, a 4,485-pound car which was priced identically with the two-passenger Series 340 Convertible Coupe and five-passenger Town Sedan at $2,590. Body was by Fisher. This was the only LaSalle to be equipped with its own integral trunk compartment.

The Series 340 LaSalle model lineup also included this handsome five-passenger Phaeton, Fleetwood Style 4060. This open model was also listed as the "LaSalle Fleetshire". A similar appearing seven-passenger Touring Car was also available. The Fleetshire Phaeton had a $2,385 price tag and the big seven-passenger Tourer $2,525. Like the LaSalle "Fleetcliff" Roadster and Series 353 Cadillac Roadster, this model has a louvered cowl.

1930

The largest open car in the Series 340 LaSalle model line for 1930 was the Seven-Passenger Touring Car with body by Fleetwood. This model was also listed as the "LaSalle Fleetlands". The heavy-looking artillery wheels were standard. Wire wheels were a popular option, as were the sidemounts shown here. The big Fleetlands Touring Car was priced at $2,525. It weighed 4,435 pounds. Four of the six models in the Fleetwood "Special Custom" line this year were open cars.

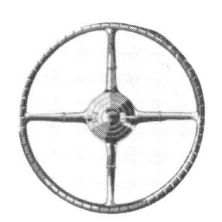

Cadillac called its convertible sedans All-Weather Phaetons. This is the Series 340 All-Weather Phaeton, a five-passenger model by Fleetwood. This style was listed in the Fleetwood Body Book for this year as the "LaSalle Fleetway". The "Fleetwood Custom Line" for the 1930 LaSalle chassis also included a standard five-passenger Phaeton and a seven-passenger Touring Car. The Fleetway All-Weather Phaeton listed for $3,995 and weighed 4,670 pounds. The tops on these cars were of Burbank cloth.

A true forerunner of the hardtop body style of later years was this special Fleetwood-bodied Series 340 LaSalle. Called the "Fleetwing Sedanette Cabriolet", it was a five-passenger car that effectively combined the open look of a Phaeton or Touring Car with the all-weather practicality of a closed car. Price was $3,725. Fleetwood Body Style 4081, this model weighed 4,600 pounds in stock form.

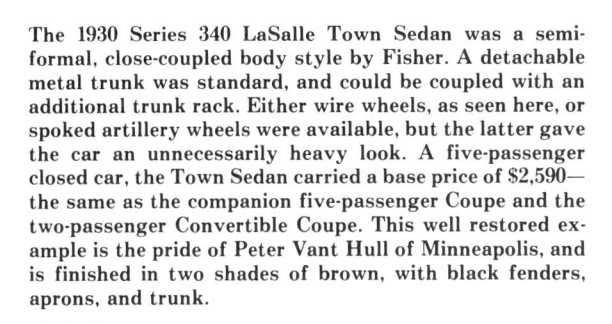

The 1930 Series 340 LaSalle Town Sedan was a semi-formal, close-coupled body style by Fisher. A detachable metal trunk was standard, and could be coupled with an additional trunk rack. Either wire wheels, as seen here, or spoked artillery wheels were available, but the latter gave the car an unnecessarily heavy look. A five-passenger closed car, the Town Sedan carried a base price of $2,590—the same as the companion five-passenger Coupe and the two-passenger Convertible Coupe. This well restored example is the pride of Peter Vant Hull of Minneapolis, and is finished in two shades of brown, with black fenders, aprons, and trunk.

Another interesting Fleetwood body style available on the 1930 LaSalle Series 340 chassis was this "Fleetwind Sedanette", Style 4082. A five-passenger car, this model was quite similar to the companion Style 4081 Sedanette Cabriolet, but had full windows in the upper rear quarters which gave it an open, airy look. The Sedanette weighed 4,600 pounds in standard road trim and had a $3,825 price tag—$100 higher than the Fleetwing Sedanette Cabriolet. All windows could be lowered flush with the sill, and the window posts could be removed, making the car a true open-air vehicle with a solid top. The top itself was covered with Burbank cloth, and was fitted with a roof ventilator.

1930

All of LaSalle's four-door sedans for 1930 had Fisher bodies, while Series 353 Cadillac buyers could choose from both Fisher and Fleetwood five and seven-passenger sedans. The Five-Passenger Sedan was the most popular model in the LaSalle Series 340 line this year. Starting price was $2,565 and the car weighed 4,645 pounds. This one has six wire wheel equipment and the extra-cost folding luggage rack.

The only visible difference between the LaSalle Series 340 five and seven-passenger sedans was the length of the rear body and window. It takes a careful eye to spot the dimensional difference. This is the 1930 LaSalle Seven-Passenger Sedan, which sold for $2,775 and weighed 4,745 pounds. For another $150, an Imperial Sedan with adjustable glass division was available.

The biggest and most expensive model in the Fisher-bodied Series 340 LaSalle line for 1930 was this formal Imperial Sedan for Seven Passengers. Intended for chauffeur operation, it had an Imperial glass division between the front and rear compartments. A standard seven-passenger sedan without division was also available. The LaSalle Imperial Sedan carried a list price of $2,925 and weighed 4,820 pounds. The wood-spoke wheels were standard, but the sidemounts and folding trunk rack were extra-cost items.

This formal open-front town car does not appear on LaSalle's 1930 price list, but the chassis is very definitely a Series 340 LaSalle. The coachwork is unmistakably Fleetwood-Detroit. Fleetwood did build special-order bodies over and above the so-called "Fleetwood Special Custom Line", and this is evidently one of them. Cadillac referred to this style of car as a Transformable Town Cabriolet. A weatherproof cover buttoned into place over the chauffeur's compartment. Styling touches include a one-piece flat windshield set at a slight angle, six wire wheels and a padded rear roof area with black decorative landau irons.

Under the guidance of Harley Earl, Cadillac's styling department was now making extensive use of full-scale vehicle mockups in wood and clay in developing future product designs. This is an early styling model of a big LaSalle seven-passenger sedan. The front end is a 1930 Series 340, but the Fleetwood four-door sedan body with wide double belt molding, rounded rear window shape and gracefully curved windshield pillars did not show up until the 1931 models were introduced. Various details were tried out—and approved or shot down—on these full-size models before production decisions were made.

V-16

In December, 1929, Cadillac Motor Car Company President Lawrence P. Fisher sent a letter to all of the division's dealers in which he formally announced the most fabulous Cadillac of them all—the sixteen-cylinder Series 452.

Cadillac's highly successful V-8 passenger car line, Fisher said, would be supplemented early in 1930 by a totally new, ultra-luxury car series which had been quietly under development for more than three years. With the prestigious new V-16, Cadillac was taking dead aim at the very top of the U.S. luxury car market which had long been dominated by Packard.

In one masterful stroke, Cadillac had outmanuvered its rivals and set new standards in automobile design and performance.

The new sixteen-cylinder Cadillac made its public debut with great fanfare at the 1930 New York Automobile Show, which opened on January 4th of that year. Only one of the mighty Series 452 Cadillacs was ready for exhibition when the curtain went up on the important New York show. It was a majestic Imperial Landau Sedan with Fleetwood coachwork and a collapsible rear quarter.

The announcement and introduction of the new Cadillac V-16 created a sensation in the motoring world—just as Cadillac's pioneering V-8 had some 15 years earlier—even though the U.S. economy was already sliding into a Depression. The factory was besieged with requests for pictures and information about the company's new Super Car, and orders poured into Detroit from eager would-be owners who had not even seen one.

In an era of fours, sixes and straight-eights, the public was dazzled by the very notion of a sixteen-cylinder powerplant. The Owen Nacker-designed Cadillac V-16 engine was a mechanical and artistic masterpiece. It was the first automobile engine anywhere to bear the mark of a stylist. The new V-16 was a truly beautiful creation of bright chrome, polished aluminum, porcelain and gleaming enamel. There wasn't a trace of unsightly wiring or hoses: everything was tidily out of sight or under appropriately detailed covers.

The new engine was essentially two in-line, eight cylinder engines that shared a common crankshaft. The cylinder banks were angled at a very narrow 45 degrees, and each had its own fuel distribution and exhaust system. Bore was three inches and stroke four inches. Displacement was a massive 452 cubic inches, from which the series derived its numerical designation. The overhead-valve engine featured hydraulic valve silencers and was conservatively rated at 165 horsepower. The powerful V-16 engine was capable of propelling these very heavy cars at speeds ranging from 80 to 100 miles an hour. Only the brawny Duesenberg was faster.

The Series 452 Cadillac of 1930 and the virtually unchanged Series 452-A of 1931 utilized a 148-inch wheelbase chassis. Customers could choose from a

Despite the nation's slide into a depression, the stunning new sixteen-cylinder Cadillac burst upon the automotive scene like a bombshell. The only Series 452 Cadillac ready for public exhibition at the 1930 New York Automobile Show, held in January of that year, was this regal Imperial Landau Sedan for Five Passengers by Fleetwood. This model was also referred to as an Imperial Landau Sedan, or Landaulette. Officially designated Style 4108-C, this beautifully restored example was photographed with its rear quarter let down. Only four were built. The 4100-series designation placed this model in the legendary "Madam X" body series, even though the Pennsylvania vertical V-type windshield differed from the flat, slightly sloped style used on most other true "Madam X" V-16 Cadillacs. This photo, through the courtesy of Automobile Quarterly, is of the car owned by David Towell of Ohio.

Probably the raciest of all Series 452 Cadillacs was this body style, the hairy-looking Roadster for Two Passengers, which was listed as Style 4302. This sporty job was also shown in the Fleetwood body style books under the name "Fleetdowns". Even with its top up, this car projected an image of power and speed — which it certainly lived up to. With its windshield folded down over the cowl and with two passengers in the rumble seat, a ride down an open road at speed in this brute must have been a memorable motoring experience. Factory price was $5,350 and the Roadster weighed 5,310 pounds — nearly a dollar a pound.

V-16

bewildering array of semi-custom Fleetwood body styles—54 in all—ranging from sporty roadsters, convertibles and coupes to big sedans, phaetons and stiffly formal limousines and open-front town cars. Prices started at a hefty $5,350 and ranged on up to a truly stratospheric (for the time) $9,700, at a time when a brand-new Chevrolet could be had for $495.

The exquisitely styled Fleetwood bodies specially created for the big Series 452 chassis came from two plants—the old, original factory in Fleetwood, Pa. and the new Detroit Fleetwood plant which had just commenced operations. Pennsylvania-built Fleetwood V-16 bodies utilized a vertical, V-type windshield while the new Detroit-built bodies usually had a flat, slightly angled, one-piece windshield.

The Series 452 and 452-A Cadillacs included an exclusive, limited production series of special four-door sedans which bore the intriguing designation of "Madam X", after a popular stage play of the day. This series was limited to Fleetwood's premium 4100 body series, although flat, one-piece "Madam X" windshields were sometimes found on other V-16 bodies. Pennsylvania-built versions had the vertical V-type windshield. Distinctive "Madam X" styling touches included very thin, slightly raked windshield pillars which flowed gracefully into a tapered belt molding at the cowl and thin, chrome edgings around all side windows.

The appropriately long Cadillac V-16 hood had five ventilator doors on each side instead of the usual louvers. Some V-16 hoods had straight upper moldings while others had the bold curved cowl sweep panel introduced in 1928. Customers could specify a conventional straight sill with a series of bright chrome streamers between the body and running board, or a more formal curved coach sill that dipped below the frame. The tall Series 452 grille was protected by a chromed screen. A pair of bright horn trumpets were mounted below huge bowl headlights. A circular medallion at the center of a concave tie bar between the headlights bore the imposing "V-16" emblem.

Following the major auto shows, a trio of the mighty new V-16's were dispatched on a triumphal tour of fashion-conscious Europe, where they were ecstatically received. The Series 452 Cadillacs of 1930 and 1931 marked the peak of Cadillac's Golden Age. The stunning Style 4264-B Transformable Town Brougham, with its razor-edged rear body lines and richly detailed simulated canework, was probably the most ostentatious Cadillac ever placed into production. To many automotive historians, the 452 Super Series was Cadillac's Finest Hour.

But despite its technological innovations coupled with high quality semi-custom coachwork, the big V-16 was not a success. Only 4,403 were built over a period of 11 model years, from 1930 through 1940. Nearly two-thirds of these were built in the car's first year on the market. The introduction of the companion Cadillac V-12 for 1931 obviously cut into V-16 sales. But the deepening Depression forever changed the complexion of the American luxury car market, bringing to an end the era of conspicuous luxury and the custom-body business as it had long been known. Dozens of automobile manufacturers and custom body houses fell by the wayside as the Depression took its economic toll.

Cadillac sold 3,251 sixteen-cylinder cars in 1930 and 1931. Of these, 2,887 were Series 452 Cadillacs sold in the 1930 model year. Only 364 Series 452-A's were sold in model year 1931. But the legendary V-16 overnight established Cadillac as a serious contender for a nervous Packard's throne.

The Series 452 Roadster was probably the most exciting — and fastest — of all of the 1930-1931 sixteen-cylinder Cadillacs, and this style is still eagerly sought by collectors. The superbly restored example shown here is owned by the Craven Foundation and is on display in the Foundation's Museum and Restoration Shop in Toronto, Ontario. This V-16 Roadster is equipped with a huge bumper-mounted driving lamp, wind wings and an externally-carried trunk. Despite its 5,310-pound weight, the big V-16 engine could push this car to blinding highway speeds.

The single-piece flat windshield on this handsome Cadillac Series 452 Convertible Coupe reveals that it came from Fleetwood's Detroit plant. Pennsylvania-built Fleetwood-bodied V-16's utilized the distinctive vertical, V-type windshield frame. This model was called the Style 4235, of which 94 were produced — a relatively high production for this style. The Pennsylvania variant, Style 4335, sported the V-type windshield and the curved cowl sweep panel. Factory records show 100 of the latter were produced. The Detroit-built convertible shown weighed 5,655 pounds and carried a base price of $6,900. The Pennsylvania model was priced at $1,000 less. Two additional passengers could be carried in the folding rumble seat. Note the golf club compartment door.

V-16

Cadillac's Series 452 of 1930-1931 included three two-door soft-tops — a Roadster, a Convertible Coupe and this pleasingly proportioned Victoria. Style 4285, it featured the Pennsylvania Fleetwood V-type windshield and a dropped coach-type sill. Cadillac formally designated this model an All-Weather Sport Cabriolet. Only two examples were produced on the 1931 Series 452-A chassis. This was a five-passenger car, and in effect was one of the forerunners of the modern 5-passenger convertible, which would not reach popularity until almost a decade later. The car contained a true integral trunk, with rather substantial luggage space for the era.

One of the handsomest closed bodies available on the new Series 452 Cadillac V-16 chassis was this gracefully-proportioned Coupe for Two Passengers, Style 4276. This coupe was available with the curved coach sill shown or could be ordered with a conventional straight sill as Style 4476. Although not included in the exclusive 4100 "Madam X" body series, this model had a "Madam X" windshield. Seventy Style 4276 Coupes were delivered, along with another 11 straight sill models. Prices started at $5,800. Both models were listed at 5,750 pounds.

Fleetwood Style 4376 was a fixed-roof Coupe for Two Passengers. This model had a V-type Pennsylvania windshield, cowl sweep molding and straight sill. Note the "Madam X"-type bright chrome moldings around the side window and the large, externally-carried trunk. Ninety-eight of these coupes were built. Price was $5,800, weight was 5,750 pounds.

The new Series 452 sixteen-cylinder Cadillac chassis was available with a bewildering choice of open and closed body styles by Fleetwood. This is the Series 452 Town Coupe for Five Passengers, Style 4381. This model had a right front seat that folded forward to permit three passengers to comfortably enter and leave the rear seat area. The Town Coupe was priced at $5,950 F.O.B. Detroit. A spacious luggage compartment was a major feature of this style.

This is probably the most-published of all photographs of sixteen-cylinder Cadillacs. Posed on Michigan's Belle Isle, against a truly appropriate background of Detroit affluence of the early 1930's, including a huge steam yacht, the car is an early production Style 4260 Phaeton. This sporty car featured dual cockpits, a second windshield for rear-seat passengers which could be lowered into the divider, and a gently curved rear body. A five-passenger car, the 4260 listed at $6,500. A total of 85 were sold. Wheelbase was 148 inches.

V-16

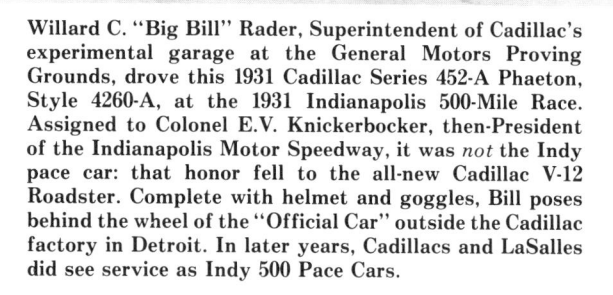

Willard C. "Big Bill" Rader, Superintendent of Cadillac's experimental garage at the General Motors Proving Grounds, drove this 1931 Cadillac Series 452-A Phaeton, Style 4260-A, at the 1931 Indianapolis 500-Mile Race. Assigned to Colonel E.V. Knickerbocker, then-President of the Indianapolis Motor Speedway, it was *not* the Indy pace car: that honor fell to the all-new Cadillac V-12 Roadster. Complete with helmet and goggles, Bill poses behind the wheel of the "Official Car" outside the Cadillac factory in Detroit. In later years, Cadillacs and LaSalles did see service as Indy 500 Pace Cars.

Cadillac factory records indicate that only one of these huge Series 452 Convertible Touring Cars was built. Like most of the big Cadillac touring cars produced over the years, the prototype V-16 Touring Car was an impressive, exceptionally roomy seven-passenger car. Two models were planned — the Style 4257-H shown here and a companion Style 4257-A, a five-passenger Touring Car with tonneau cowl and folding rear windshield. The seven-passenger Tourer shown does have a folding rear windshield with large wind wings. Weather protection on this car was by side curtains, which lived in their own special compartment when not in use.

Two distinctively different All-Weather Phaetons were available on the Series 452 Cadillac V-16 chassis. This very beautiful convertible sedan, Style 4280, was a product of the new Fleetwood Body plant in Detroit. Introduced late in the 1930 model year, and intended for full production in the Series 452-A for 1931, this model had the dropped coach-type sill and a gently-raked, one-piece flat windshield. The deepening depression, however, intervened and only three of these gorgeous cars were ever built.

This big convertible sedan was a companion car to the sportier-looking Series 452 Phaeton. Officially listed as the Style 4380 All-Weather Phaeton, this model featured a straight sill, the distinctive Cadillac cowl sweep panel and the vertical V-type windshield assembly common to Pennsylvania-built Fleetwood bodies for the new Cadillac V-16 chassis. This example has canvas spare tire covers bearing the Cadillac crest and its radiator is adorned with the classic Cadillac "flying goddess" hood ornament.

V-16

The Style 4380 All-Weather Phaeton was just as good-looking with its top up as it was in full-open road trim. Prices began at $6,650. Even at that price, this was an amazingly popular Series 452 model, with 250 built. This big four-passenger convertible sedan weighed 5,675 pounds. Note the fore-and-aft door hinging and the very wide belt molding which is dipped at the rear of the body to make room for the bulky folding top.

The Series 452 All-Weather Phaeton, Style 4380, was offered in two versions. The standard fully-open model carried a $6,650 price tag. A premium version with glass division between the front and rear seats was priced at $7,350. Both models were shown as weighing 5,675 pounds. In this photo the side window and partition glass have been lowered to the halfway position, which meant that the window posts had to remain in place. This imposing convertible sedan was a product of Fleetwood's Pennsylvania plant.

A close-coupled sedan, Style 4161 was a true "Madam X" Cadillac in the exclusive 4100 series Fleetwood body grouping. This elegantly-proportioned five-passenger closed car was available in two versions — the Style 4161-S Club Sedan shown here, or Style 4161, an Imperial Club Sedan with glass partition. The Club Sedan was listed at $6,950. Only 43 were built.

Most Fleetwood closed bodies for the Cadillac V-16 chassis were offered in two basic forms — standard, and in an Imperial limousine configuration with a glass partition between the front and rear compartments. Some models were available with the distinctive "Madam X" angled windshield. This is a Series 452 Cadillac V-16 Imperial Cabriolet for Five Passengers, Style 4355, of which 52 were built. Without the partition, this model was called the Style 4355-S Sedan Cabriolet, of which 81 were sold. The Sedan Cabriolet carried a price of $6,125. The Imperial Cabriolet listed at $6,350.

Another five-passenger closed body style offered on the Series 452 Cadillac V-16 chassis was this Club Sedan for Five Passengers, which was listed as the Style 4361-S. When specified with the Imperial glass division, usually for chauffeur operation, it was simply designated the Style 4361 without the "S" suffix, and became the Imperial Club Sedan. This Style 4361-S Club Sedan has the cowl sweep panel, a straight sill and externally-carried trunk. Base price was $5,950. A fairly popular model, 258 were built. But, only two Imperial Club Sedans were ordered.

V-16

Externally, the Style 4355-S Sedan Cabriolet and the more expensive Style 4355 were identical. The only real difference was inside, in the form of a glass partition between the front and rear compartments for use when the car was being driven by the family chauffeur. This is the standard Style 4355-S Five-Passenger Sedan Cabriolet, sporting a very attractive 3-tone paint job and a top covered in fine leather.

Of the 11 body styles included in the exclusive 4100 "Madam X" Series, four were various configurations of the exquisitely detailed five-passenger formal sedan shown here. This is Fleetwood's Style 4155, which was listed as an Imperial Cabriolet. Variations included the Style 4155-S Sedan Cabriolet which was identical except for the absence of a glass divider between the front and rear seats; Style 4155-C, an Imperial Landau Cabriolet, and Style 4155-SC, a Landau Sedan Cabriolet. The Style 4155 was the most prolific, 10 examples having been produced. The 4155 listed at $7,350 and the 4155-S at $7,125. This car weighed 5,925 pounds. The thin, gently angled windshield pillars and fine chrome window moldings exclusive to the true "Madam X" models are shown to advantage in this photo. The top and rear quarters were covered in fine landau leather. The top was fitted with a ventilator, while the quarters were decorated with painted landau irons.

While the dashing Roadsters, Convertible Coupes and Phaetons are the most sought-after V-16 Cadillacs among collectors today, the fact remains that the bread-and-butter cars of the Series 452 line were the rather plain and heavy-looking five and seven-passenger sedans. One of the most popular of these was the Style 4330-S, a five-passenger sedan of which 394 were built. A companion Style 4330 Imperial Sedan for Five Passengers was also available. In stock form, the Style 4330-S was priced at $5,950 and weighed 5,835 pounds.

Cadillac's Series 452 sedans were heavy-looking, extremely solid cars that were constructed like bank vaults. But despite their staid looks, these were fast and powerful cars. The cowl sweep panel extending to the tip of the hood of this Style 4330-S Five-Passenger Sedan somehow breaks up the severe, angular appearance of the body with its thick window pillars. The three chrome streamers on the tool compartment panels below each door are a distinguishing feature of straight sill Series 452 Cadillacs built in 1930 and 1931. Most Cadillac V-16's rode on premium wide whitewall tires.

The suffix "S" as applied to Fleetwood body style numbers on Series 452 Cadillacs indicated that the car in question did not have an Imperial Division or a pair of folding opera seats. Thus this car, Style 4330, is a Five-Passenger Imperial Sedan. Fifty of these inside-drive limousines were built, compared with 394 Style 4330-S Five-Passenger sedans. The Style 4330 Imperial Sedan carried a list price of $6,300 and weighed 5,905 pounds.

V-16

This is a genuine "Madam X" Cadillac. True "Madam X" Cadillacs carried a Fleetwood Series 4100 body designation. The most distinctive styling feature of a true "Madam X" is a slightly sloped, one-piece flat windshield with extremely thin side pillars and fine chrome moldings around all side windows. The "Madam X" style windshield, however, was occasionally found on other Fleetwood V-16 bodies. The car shown is a Style 4130 Imperial Sedan for Five Passengers, only 17 of which were produced at $7,300 a copy. Fleetwood also built 49 generally similar Style 4130-S Five-Passenger Sedans which sold for $6,950.

The big seven-passenger, four-door sedan was the most popular single body style produced by the Fleetwood plants for the Series 452 Cadillac V-16 chassis through model years 1930 and 1931. This model was offered in three configurations: Style 4375-S, the standard Seven-Passenger Sedan; Style 4375, the Imperial Sedan for Seven Passengers shown here, and Style 4375-C, an Imperial Landau Sedan with collapsible rear quarter, only two of which were built. Style 4375 was a formal inside-drive limousine with Imperial division and two folding, forward-facing opera seats. A total of 438 of these were built. List price was $6,525 and the heavy-looking 4375 weighed just over three tons. Despite its massive proportions, it was a fast car.

With its mysterious-sounding name taken from a popular play of the day, the "Madam X" remains the most legendary of all sixteen-cylinder Cadillacs in the mighty Series 452 of 1930 and 1931. The "Madam X" name actually applies only to a group of four-door sedans in Fleetwood's "4100" body series. Most true "Madam X" Cadillacs had the flat, angled windshield seen on this Style 4175 Imperial Sedan for Seven Passengers — although the Pennsylvania V-style windshield was used on some. The side windows on true "Madam X" sedans were edged with a thin chrome molding. Fleetwood built 110 of these big cars, with Imperial division and two forward-facing folding auxiliary seats. Prices started at $7,525 and the 4175 weighed a substantial 6,005 pounds.

The most popular model in the classic Cadillac Series 452 line of 1930 and 1931 was this one, the Style 4375-S Seven-Passenger Sedan. Just over 500 were built. The massively-proportioned four-door sedan had a list price of $6,225 and weighed 5,965 pounds. Except for the length of the rear body and a slightly longer third side window, Series 452 five and seven-passenger sedans were almost indistinguishable in appearance. A companion Style 4375 Seven-Passenger Imperial Sedan had a glass division and two forward-facing auxiliary seats.

The Seven-Passenger Sedan, Style 4175-S, was built in considerably fewer numbers than its more expensive companion, the Style 4175 Imperial Sedan. Only 47 of these elegantly styled V-16 sedans were built, compared with 110 inside-drive Imperials in the exclusive 4100 "Madam X" Fleetwood body series. List price for this model was $7,225 and the car weighed 5,965 pounds. The distinctive "Madam X" windshield was modestly angled and supported by extremely thin windshield pillars which flowed harmoniously into the wide belt molding at the cowl. This molding then tapered into a thin, straight molding which extended to the front of the long V-16 hood.

V-16

Certainly the most elegant of all of the Series 452 Cadillacs were the gracious open-front town cars. These were built in five and seven-passenger versions, with straight or coach-type sills and with flat or V-style windshields. This is the Style 4212, which was officially listed as a Transformable Town Cabriolet for Five Passengers. Note the closed rear quarters, swooping coach sill and flat windshield. Only six of these were built. List price was $8,750.

Another open-front town car on the ultra-luxury Cadillac Series 452 chassis was Fleetwood's Style 4312, a Transformable Town Cabriolet for Five Passengers. This model has a slightly angled V-type windshield and straight sill. Fleetwood built 24 of these bodies for mounting on the Cadillac V-16 chassis. Price was an even $7,000. Although it was considered a 5-passenger car, the forward tonneau partition contained two folding opera seats for transforming the car into a 7-passenger model when necessary.

Generally similar to the five-passenger Style 4212, this is the Style 4225 Town Cabriolet for Seven Passengers. Fleetwood built only six of these Transformable Town Cabriolets, which listed for a lofty $8,750. One Style 4225-C, with a collapsible rear quarter, was also built. This model tipped the scales at 5,835 pounds. The Town Cabriolet shown sports the one-piece flat windshield and a dropped coach sill. The plush interior was fitted with two forward facing auxiliary seats. The chauffeur's compartments on all of these town car types were upholstered and panelled in fine grain leather.

Cadillac designated its formal, open-front town cars as "Town Cabriolets". Most had closed, leather-covered upper rear quarter panels and were officially listed in the Fleetwood sales catalogs as "Transformable Town Cabriolets". This is the Style 4325, a seven-passenger town car of which 35 were built. Prices started at $7,150. The one shown has a V-type windshield, front-opening forward doors and a straight sill.

When built with quarter windows, this V-16 style was known as a Limousine Brougham. Fleetwood called town cars with closed rear quarters Town Cabriolets. This is the Style 4291 Transformable Limousine Brougham for Seven Passengers, an especially elegant model which was priced at $8,750. The example shown has a flat windshield, dropped coach sill and an attractive two-tone paint treatment. Fourteen of these were produced.

V-16

For those who desired the correctly formal appearance of a Town Cabriolet but the more open look of a limousine, Fleetwood offered the elegantly-styled Transformable Town Cabriolet for Seven Passengers, Style 4220. Only nine of these stately cars were built. They sold for a very expensive $8,750. This model had smaller rear quarter windows than the companion Style 4391 but carried small ornamental landau irons on their upper rear quarters. Style 4320 was similar but had a straight sill and V-type windshield. Twenty-five of the latter were delivered.

The Style 4391 town car was similar to Style 4291, but had a V-type windshield and a straight sill. Like the Style 4291, this model was officially listed as a Transformable Limousine Brougham for Seven Passengers. Fleetwood built 30 of these cars, each with a base factory price of $7,150. This particular example, shown with the top raised over the chauffeur's compartment, is owned by Richard Gold of Minneapolis. It is shown at the 1981 Midwestern Classic Concourse at Fort Snelling, Minn.

Fleetwood Style 4320 was a Transformable Town Cabriolet for Seven Passengers. This model was similar in appearance to the companion Style 4220 but had the sweep cowl panel that tapered to a point at the front of the hood, a straight sill and the stylish V-type Pennsylvania windshield. Fleetwood built 25 of these formal bodies for the Series 452 Cadillac V-16 chassis. List price was $7,150. A waterproof cover could be quickly buttoned into place over the chauffeur's compartment in inclement weather. This cover was stored in a compartment above the division behind the front seat. Forward facing folding auxiliary seats were standard on these models.

Factory records reveal that Fleetwood built only 10 of these exotic, razor-edged panel brougham-type bodies for the sixteen-cylinder Series 452 Cadillac chassis. Four were of this style with a full coach-type sill. Cadillac described this severely formal town car as a Transformable Town Brougham, Style 4264. The other six had a unique sill treatment and were listed as Style 4264-B. Three of the latter were delivered in a monotone lacquer finish and the other three had simulated canework on the lower rear body. The imposing Style 4264 was among the most expensive V-16's offered, at $9,200. The simulated cane-work was an incredibly time consuming project. The "cane" strips were actually strips of thick paint composition, meticulously applied by hand, using a small squeeze tube. Despite the immense amount of time needed to produce a cane-work effect this method resulted in a far better job than did cementing real cane-work to the body, as the real cane had a tendency to break loose from the metal in a relatively short time.

With the announcement and subsequent public display of the fabulous new sixteen-cylinder Cadillac, the factory in Detroit was deluged with frantic requests for pictures and detailed information on the new Series 452. Cadillac was caught short: literature and illustrated catalogs were not ready. The company hastily compiled packets of Fleetwood body drawings to show potential V-16 customers the wide range of bodies available on the new super chassis. Here is a typical Fleetwood body builder's drawing of the Style 4264-B Transformable Town Brougham in standard monotone paint trim and with the special dual-contour sill. A companion Style 4264 had a full coach sill. The Style 4264-B could also be ordered with elaborate simulated canework on the lower rear body panels.

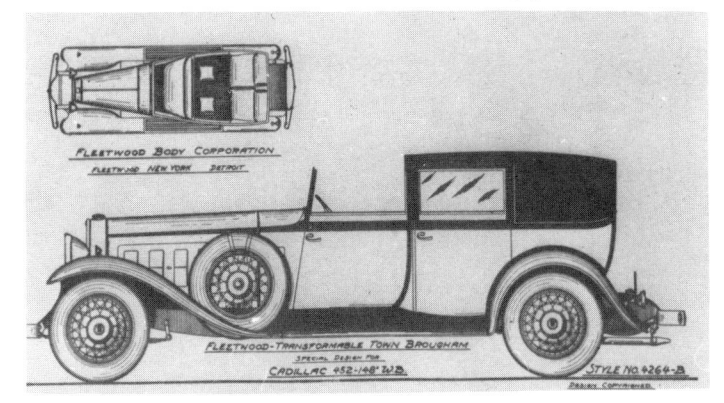

V-16

This was probably the most ostentatious production Cadillac ever built. Designated Fleetwood Style 4264-B and listed as a Transformable Town Brougham, it was the most expensive model in the super-luxury Cadillac V-16 line. List price was a stratospheric (for the time) $9,700. Only six of these spectacular creations were built, three with simulated canework and three without. The entire lower rear body of this car was finished in delicate simulated canework which was painstakingly applied with a special paint fixture. The upper rear quarters of the metal body were lacquer-finished rather than leather-covered. The special sill is of the straight style below the front doors and is of the dropped full-coach type in the rear. The lavish interior was equal to this stately car's stunning exterior. This magnificent example, which marked the peak of Cadillac's styling individuality, was photographed on Detroit's fashionable Belle Isle.

In an epic case of unfortunate timing, the mighty sixteen-cylinder Cadillac was launched only months after the stock market crash of 1929, and as the economy was sliding into a deep depression. But at the time the mighty V-16 made its public debut early in 1930, there was still a large and eager market for this new ultra-luxury car. This photograph was taken through the showroom window of Cadillac's big Chicago branch. Prominently displayed is the most lavish V-16 of them all, the Fleetwood-bodied Style 4264-B Transformable Town Brougham with simulated rear body canework and a razor-edge upper rear body. Cadillac was making a determined — and eventually successful — bid to break Packard's grip on the American luxury car market.

All Fleetwood bodies designed and built for the fabulous new Cadillac V-16 chassis were special, although they were not true customs in the full sense, because most were produced in batches of from two or more to several hundred. This, however, is a full-fledged Fleetwood custom, a one-off four-window Berline created for the new Series 452 chassis. Fleetwood called this a Special Imperial Cabriolet for Seven Passengers. Styling features included an oversized windshield visor of amber celluloid, padded upper rear quarters and a dropped, coach-style sill. The car was finished in deep maroon, with a tan superstructure and tan leather top finished to resemble Burbank cloth.

Cadillac rigidly controlled distribution of its new V-16 chassis, with only a handful being released to body builders other than Fleetwood. Very few were exported. As far as is known, no Series 452 chassis were built for commercial use. But at least one Cadillac V-16 was converted into a hearse. This one-off limousine style funeral coach was built by The Eureka Co. of Rock Falls, Ill., probably using a late-model Series 452 for the conversion. The original V-type windshield has been retained. The chassis has been lengthened just ahead of the rear fenders.

The awesome, new sixteen-cylinder Cadillac was as impressive under the hood as it was outside. The Series 452's engine was probably the first automobile powerplant that was actually "styled". Much attention was paid to the V-16 engine's exterior appearance. The result was an unusually attractive, tidy appearance of polished aluminum and bright enamel, with no unsightly clusters of wire or hoses. Cadillac went so far as to proclaim its new mechanical masterpiece a work of art. The Owen-Nacker-designed V-16 was of 45-degree design with 452 cubic inches of displacement, three-inch bore and four-inch stroke.

1931

This year marked the peak of Cadillac's model proliferation and styling individuality. For 1931, Cadillac blanketed the American luxury car market with no less than four separate passenger car lines powered by eight, twelve and sixteen-cylinder engines.

In a single year, Cadillac had dramatically outgunned its rivals and was solidly on its way to eventual domination of the fine car field. But the worsening Depression was taking its toll, and sales of all luxury cars were already down sharply.

The sixteen-cylinder Series 452, introduced early in 1930, had been a conspicuous success in its first six months on the market. More than 1,000 had been shipped within four months of the V-16's introduction, and this figure had doubled by June.

But in April, Cadillac Motor Car Co. President Lawrence P. Fisher informed his dealers that the fabulous V-16 would soon be joined by yet another multi-cylindered supercar — a Cadillac V-12. Thus the new Series 370 arrived in Cadillac-LaSalle showrooms in October, 1930. Cadillac now saturated the luxury car market with its moderately-priced LaSalle V-8; a full line of eight-cylinder Cadillacs; the new V-12 and the sixteen-cylinder Series 452-A, which went into its second model year without change.

The new Cadillac V-12 engine was essentially the Owen Nacker-designed V-16 minus four cylinders. It was a shorter, 45-degree V-type engine of 368-cubic inch displacement.

Bore was an eighth of an inch larger than the V-16 but its four-inch stroke was the same. Like the V-16, the new twelve with its hydraulic valve silencers was noted for its smooth, extremely quiet delivery of useable power. The V-12 was officially rated at 135 brake horsepower — 30 less than the sixteen — but more than a few automotive historians consider the V-12 the better of the two powerplants.

> In October, 1930, Cadillac again astounded the automotive world with the introduction of yet another entirely new series of multi-cylinder supercars. Supplementing the ultra-luxury sixteen-cylinder Cadillac introduced early the same year was the new Series 370 powered by a V-type twelve-cylinder engine. The new Cadillac V-12's were available with a complete range of Fisher and semi-custom bodies by Fleetwood. This is the 1931 Series 370-A Roadster for Two Passengers which commanded a price of $3,945. It weighed 4,910 pounds. Series 370/370-A models are easily identified by the single chrome streamer between the running board and body sill, at the center of which is a courtesy light that comes on when the door is opened.

The new V-12 Cadillac was introduced as the Series 370 in the latter part of 1930 and became the Series 370-A in 1931. A complete range of 11 body styles by both Fisher and Fleetwood were available on the new twelve-cylinder chassis. These included sporty roadsters and convertible coupes, phaetons, convertible sedans and touring cars, two and five-passenger closed coupes, cabriolets and town sedans, five and seven-passenger sedans and various formal limousines and town cars. Series 370/370-A prices ranged from $3,795 to $4,995. Serial Numbers ran upward from 1000001.

The semi-custom bodies created for the new V-12 chassis by GM styling chief Harley Earl and Ernest Schebera, who headed up the Fleetwood Body division, were equal in every respect to those designed for the prestigious V-16. By now, Fleetwood had completed the transfer of all production from the old, original Fleetwood Body Co. plant in Pennsylvania to the new Fleetwood Body plant in GM's former Plant 18 on Detroit's near-west side.

The Cadillac V-12 sold well — but at the expense of the V-16. From the time of the Series 370's introduction, Series 452 sales plummeted.

This was due partially to the effects of the deepening Depression, but there is no doubt that some potential V-16 customers opted for the equally impressive, but less expensive, V-12 instead. More than 2,800 Series 452 Cadillacs were built in 1930. Only 364 were sold in 1931. A total of 5,725 Series 370 and 370-A Cadillac V-12's were produced in 1930-31. A gleaming white Cadillac V-12 Roadster with white leather upholstery and silver striping was the pace car for the 1931 Indianapolis 500-Mile Race. Cadillac test chief Willard Rader drove the Series 370-A Roadster to a 100 MPH-plus record at Indy.

While the spotlight was on the spectacular V-16s and V-12s, changes were being made in Cadillac's

> Twelve-cylinder Cadillacs produced during the 1931 model year were listed either as Series 370 or 370-A. Production of the new V-12 series began in September, 1930 with public introduction the following month. Cars built in 1930 carried "370" identification. Those built in 1931 were listed as Series 370-A. This is a Series 370 Convertible Coupe for Two Passengers. The same body style was available on the 1931 Cadillac V-8, the lower-priced LaSalle and on the top-line Series 452 sixteen-cylinder chassis. The Series 370/370-A Convertible Coupe had a factory price of $4,045, and weighed 5,005 pounds.

1931

The new V-12 Cadillacs were available with a selection of semi-custom bodies by Fleetwood. The 1931 Series 370-A catalog showed no less than three Phaetons on the new V-12 chassis. These included four and five-passenger Cadillac Series 370-A Phaetons of this style, and a four-door convertible sedan listed as an All-Weather Phaeton. The Phaeton for Five Passengers was priced at $4,045 and weighed 4,950 pounds. The Series 370 also included a big Seven-Passenger Touring Car.

As had been Cadillac's practice for some years, this very beautiful four-door convertible sedan was listed in the 1931 Fleetwood Body book as an All-Weather Phaeton. The Series 370-A All-Weather Phaeton was among the most expensive offerings in the line, with a base price of $4,895. "This car", Cadillac stated in its advertising this year, "has California appeal." A five-passenger car, the All-Weather Phaeton was equally handsome with the top up or down. It weighed and impressive 5,290 pounds.

Series 370/370-A twelve-cylinder Cadillacs for 1931 were built on two wheelbases, 140 and 143 inches. Most models, like this Series 370-A Coupe for Two Passengers, were on the shorter 140-inch wheelbase. This handsome Coupe was the lowest-priced model in the 1931 V-12 line, with a factory price of $3,795, and a weight of 5,035 pounds.

bread-and-butter cars. The 1931 Cadillac V-8s and lower-priced LaSalle were given numerous engineering and styling changes. The Cadillac V-8 and LaSalle were now almost identical mechanically. Both shared the same 134-inch wheelbase. The LaSalle's engine was the same V-8 that had been used in the 1930 Cadillac and would continue for the next three years. Even though its 353-cubic inch engine was basically unchanged, Cadillac got a new series designation — Series 355 — which it would carry with an alphabet suffix through the 1934 model year. Horsepower was increased to 95.

The 1931 Series 355 Cadillac and Series 345-A LaSalle were given some significant styling changes. Ventilator doors replaced hood louvers on both, and a chrome-plated screen covered their radiators — just like the V-12 and V-16. Most models now had pleasingly curved sills which dipped below the new, lower chassis frame rails.

Only two semi-custom LaSalle models still had V-type windshields. All others had sloped, one-piece flat windshields. There were 12 body styles in all. A total of 10,700 Series 355 Cadillacs were produced during the 1931 model year, along with just over 10,000 LaSalles. Prices had been reduced by $900 to stimulate sagging sales.

Also in 1931, the Marmon Motor Co. of Indianapolis had announced the Marmon Sixteen — the first real competition for the Series 452 Cadillac. Ford's Lincoln Motor Division introduced the new Model "K" Lincoln, on a 145-inch wheelbase and powered by a 120-horsepower V-8. One of the prestigious "three P's" — Peerless — went out of business, an early victim of the deepening Depression. The entire industry was being strangled by the failing economy, and the outlook for luxury car manufacturers was not good.

The largest of the six open models in the new Series 370 Cadillac V-12 model lineup was the big Seven-Passenger Touring Car. Most of Cadillac's V-8 and V-12 models for 1931, including the LaSalle, had the curved sill visible on this car which was suggestive of the dipped "coach sill" used on the most expensive formal Cadillac V-16's. The double upper body molding on the Series 370 Touring Car is reminiscent of the one used on the very first Cadillac Phaetons many years earlier. This car weighed 5,005 pounds, the same as the Convertible Coupe.

1931

The Series 370-A Coupe for Five Passengers was a 140-inch wheelbase car that sold for $3,895. Cadillac's twelve and sixteen-cylinder cars carried exquisitely detailed semi-custom Fleetwood bodies. This model featured a full-length body molding which followed the softly-rounded contour of the rear deck. The belt molding swept up to the base of the windshield pillars and around the upper body greenhouse. The "A" pillars, center post and window reveals were finished in a darker color that complemented the principal body color, and the fenders and running gear were black. The over-all effect was quite attractive. This model tipped the scales to 5,055 pounds.

Fleetwood called this formal five-passenger car a Sedan Cabriolet. It has the rather heavy lines of the Fleetwood bodies created for the big V-12 and V-16 sedans, characterized by a very thick roofline and wide windshield and door posts. A companion Imperial Cabriolet with glass division was also available. The Series 370-A Sedan Cabriolet has no bright exterior moldings or trim, contributing to the severely-formal appearance. This car has a leather-covered roof and rear quarters, painted landau irons and a Cadillac "Heron" mascot atop its radiator cap. A 5,380-pound car, it sold for $4,095.

Two distinctly different styles of five and seven-passenger sedan bodies were available on the 1931 Cadillac Series 370/370-A V-12 chassis. This is the Fisher Five-Passenger Sedan. The Fisher-built bodies had wide belt moldings, elegantly-curved windshield pillars and a nicely rounded third side window opening. The more expensive Fleetwood five-passenger sedan was more angular and heavier-looking. The Fisher V-12 Five-Passenger Sedan was priced at $3,895 and was not available in an Imperial configuration. At 5,215 pounds, it was the lightest of the 4-door closed cars.

The special Fleetwood-built bodies designed by Harley Earl and Ernest Schebera for the V-12 and V-16 Cadillac chassis were probably the most artistic and distinctive ever to bear the Cadillac name. This beautifully proportioned, close-coupled Town Sedan for Five Passengers on the V-12 Series 370-A chassis of 1931 is a standout in every respect. The Town Sedan, on 140-inch wheelbase chassis, had a factory price of $3,945, and weighed 5,230 pounds.

Fleetwood Style 4830 for the new Cadillac V-12 Series 370 chassis was a massively-constructed but heavy-looking five-passenger sedan. Compare this style with the lighter, more graceful body lines of the Fisher-bodied Series 370 five-passenger sedan. Like its Fisher counterpart, this model was not available with an Imperial division. Priced at $4,095, this was the least expensive Fleetwood V-12. It weighed 5,345 pounds.

1931

The Series 370 Cadillac V-12 chassis was available with big seven-passenger sedan bodies by Fisher or Fleetwood. This is the Fisher Body version, which could be ordered as either a standard seven-passenger sedan at $4,195 or as an enclosed-drive limousine at $4,345. The enclosed-drive limousine configuration was shown as the Fisher Seven-Passenger Imperial. The Imperial option included an adjustable division glass between the front and rear compartments and a pair of folding opera seats. For a relatively plain-Jane four-door sedan, this was an unusually handsome motor car. The sedan version weighed 5,345 pounds, while the Imperial model was the heavyweight of the Fisher line at 5,420 pounds.

This is the Fleetwood-bodied Series 370-A Sedan for Seven Passengers at $4,245, which was also offered with an Imperial division at $4,445 if it was to be chauffeur-driven. The formal Fleetwood bodies for the V-12 chassis were almost devoid of bright moldings or trim, except for the chrome streamer between the sill and running board. Most other V-12 body styles had modestly-curved sills. Only the senior Fleetwood five and seven-passenger sedans and town cars retained the straight sill. The Sedan style weighed 5,460 pounds, while the limousine was 40 pounds heavier.

As with the flagship V-16, the most expensive body styles available on the V-12 Series 370/370-A chassis were a series of formal, open-front town cars. These semi-custom bodies, produced in very small quantities, were by Fleetwood. This is a Series 370-A Transformable Cabriolet for Seven Passengers. Note the padded roof and rear quarters, painted landau irons, straight sill and flat, one-piece windshield. This model and the Limousine Brougham shared the same $4,945 price tag, the highest V-12 semi-custom price for 1931.

This stately car is a Fleetwood Seven-Passenger Limousine Brougham on the 1931 Cadillac Series 370-A twelve-cylinder chassis. Open-front town cars were still in demand, although the Depression was beginning to severely depress the American luxury car market. This correctly-formal car has a gently sloped one-piece windshield and straight sill. It is almost totally devoid of any bright exterior trim. All Fleetwood town car styles weighed approximately 5,600 pounds.

Fleetwood identified this special-order body style as either a Transformable Cabriolet or a Limousine Brougham with quarter windows. The impressive example shown has a padded roof and rear quarters and small ornamental landau irons. A speaking tube is mounted on the "B" pillar just behind the chauffeur's left ear. The windshield frame is plated. The 1931 Cadillacs sported a pair of bright trumpet horns mounted below the huge bowl headlights.

1931

The relatively short, erect five-passenger rear body of this Series 370-A Fleetwood Town Cabriolet emphasizes the length of the restyled Cadillac hood. With a rear-mounted spare instead of the sidemounts, it would look even longer. Like other formal V-12's, this open-front town car has almost no bright exterior trim. Even the landau irons are painted black to match the leather-covered roof and upper rear quarters. This style cost $4,795 and was also available in a seven-passenger version at $4,995. The 5-passenger car weighed 5,380 pounds.

Very few V-12 or V-16 Cadillac chassis were made available to outside body builders, or adapted for commercial purposes. Here is a notable exception. This is a 1931 Cadillac V-12 Limousine Funeral Coach built on the Series 370-A chassis by The Eureka Company of Rock Falls, Ill. This firm also mounted at least one similar hearse body on a V-16 chassis. This side-loading hearse was built for a Chicago undertaking firm and was finished in light gray with black upper body and chassis gear.

For 1931, the eight-cylinder Cadillac was given a number of engineering and styling changes. All Series 355 Cadillacs rode on a new 134-inch wheelbase — a full six inches shorter than 1930 models and precisely the same as the lower-priced LaSalle. Ventilator doors replaced hood louvers, and a chromed screen covered the radiator. This Series 355 Cadillac Roadster for Two Passengers has a pleasingly curved sill. The dual upper body moldings flow nicely into the outer edges of the rear deck. The Roadster was priced at $2,845. It was the lightest Cadillac of the year, weighing only 4,340 pounds.

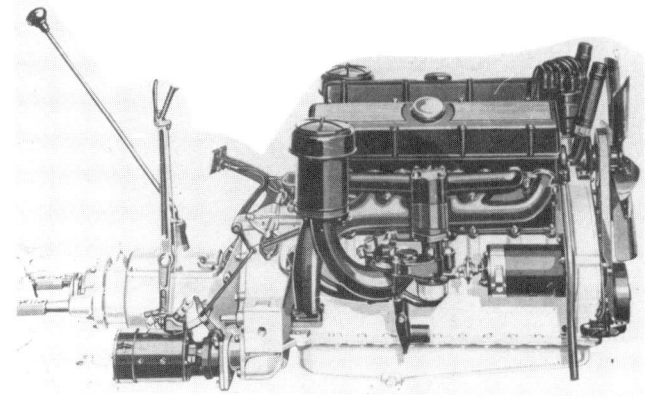

Cadillac's new V-12 engine was actually little more than the companion V-16 minus four cylinders. It was based on the monumental Owen Nacker-designed 1930 Cadillac V-16. Displacement was actually 368 cubic inches, but the car it was designed for was known as the Series 370. The highly efficient new V-12 was rated at 135 brake horsepower, 30 less than the sixteen. Bore was 3-1/8 inches — one-eighth larger than the V-16. Four-inch stroke was identical. Like the V-16, the V-12 was an aesthetic masterpiece. Its performance more than lived up to its efficient appearance.

The 1931 Cadillac Series 355 Convertible Coupe for Two Passengers was priced only $100 higher than the companion Roadster, at $2,945. All Series 355 Cadillacs rode on a new, lower chassis frame. All of the 1931 Cadillac V-8's had gracefully curved sills which extended down over the edges of the new chassis frame. All standard model Series 355 Cadillacs had sloped, one-piece windshields, but two premium 1931 LaSalles retained V-type windshields. The Convertible Coupe weighed 4,435 pounds.

1931

Cadillac's 1931 V-8 model lineup included four open cars, including this racy-looking Phaeton for Five Passengers which weighed 4,380 pounds. A five-passenger car, the Phaeton carried a factory price of $2,945 — the same as the Convertible Coupe and Seven-Passenger Sedan. This Series 355 Cadillac Phaeton has the flowing "goddess" hood ornament available as an accessory on both Cadillacs and LaSalles through the 1933 model year. All Cadillacs and LaSalles now had single-bar bumpers.

By far the most expensive model in Cadillac V-8 model offerings for 1931 was the convertible sedan, officially listed as an All-Weather Phaeton. This versatile open car carried a starting price of $3,795 — a full $700 higher than the most expensive closed model, the Seven-Passenger Imperial. The All-Weather Phaeton by Fleetwood was sold as a five-passenger car. The upper body molding complements the curved sill while the lower molding continues in a straight line to the front of the hood. The car weighed 4,670 pounds.

This imposing Seven-Passenger Touring Car is not shown on Cadillac's 1931 standard price list, but appeared on later lists at $3,195. It weighed 4,435 pounds. The elongated rear body belies its 134-inch wheelbase. The chromed wire wheels — six of them including dual sidemount spares — give this big open tourer a custom look. At the height of the golden age of classics, Cadillac appropriately photographed this car in front of a Grosse Pointe mansion.

Ventilator doors replaced hood louvers on the new Series 355 Cadillacs for 1931. Five were used on each side of the hood, and there was one vent door on each side of the cowl. This is the Series 355 Cadillac Coupe for Two Passengers, a popular closed model which had a factory list price of $2,695, and a weight of 4,465. Like all other 1931 eight-cylinder Cadillacs, this car has a gently-dipped sill. Use of a rear-mounted spare instead of sidemounts enhances the long look of the restyled hood.

Cadillac V-8 buyers could choose from a wide selection of closed and open body styles for 1931 by both Fisher and Fleetwood. This is the Series 355 Coupe for Five Passengers, a practical and good-looking offering by Fisher. The five-passenger Cadillac Coupe listed for $2,795 and weighed 4,485 pounds. The car shown has wire wheels and a rear-mounted spare tire.

1931

The semi-formal Town Sedan was available on both Cadillac V-8 and LaSalle chassis for 1931, as well as the new Series 370 Cadillac V-12. The Series 355 Cadillac Town Sedan was priced at $2,845 — the same as the Series 355 Roadster. The example shown has a padded roof and upper rear quarters. Note the graceful curvature of the windshield pillars, and the dual body moldings. At 4,660 pounds, it was one of the heaviest 5-passenger V-8s.

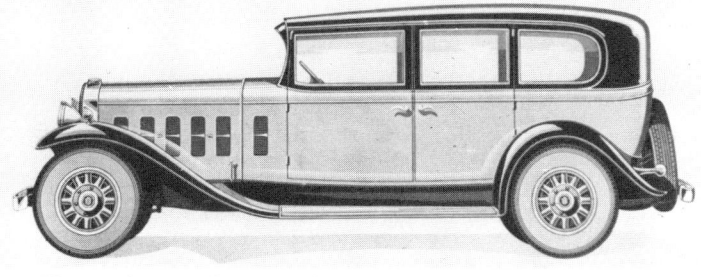

The most common of all 1931 eight-cylinder Cadillacs were the big five and seven-passenger sedans. This is a Fisher Series 355 Sedan for Five Passengers, which sold for $2,795 and weighed 4,645 pounds. The car shown has the "Goddess" radiator ornament. No "Imperial" version with glass divider was offered, but a Seven-Passenger Imperial was available. The single-bar front bumper is gently Veed. Bumperettes flank each side of the rear-mounted spare.

More and more, the spare tire was being relegated from the front fenderwell to the rear end of the car, to emphasize the length of the restyled Cadillac hood. All 1931 Cadillacs now sported stylish ventilator doors instead of hood louvers. This is the Cadillac Series 355 Sedan for Seven Passengers with body by Fisher. The Seven-Passenger Sedan was priced at $2,945, or only $100 more than the companion Series 355 Five-Passenger Sedan. The heavy-looking wood-spoke wheels make the car appear to weigh more than its 4,745 pounds.

The new Fleetwood Body plant in Detroit was extremely busy turning out a wide range of semi-custom body styles for the new Cadillac V-16 and V-12 super chassis. By this time, operations from the old, original Fleetwood Body Co. factory in Fleetwood, Pa. had been transferred to Detroit. The 1931 Cadillac Series 355 eight-cylinder chassis was available with bodies by both Fisher and Fleetwood. Beyond the standard range, Fleetwood did offer some limited-production, semi-custom bodies on the V-8 Cadillac chassis. This is a catalog illustration of a formal Transformable Town Cabriolet for Five Passengers. The town car shown has a sharply-raked, Madam "X"-ish windshield. Twelve-cylinder Cadillacs had a single chrome streamer below the body sill: V-16's had three chrome trim pieces. Priced at $4,995, this car weighed 5,135 pounds.

For an additional $150, a formal enclosed-drive limousine version of the Series 355 Cadillac was available based on the big Seven-Passenger Sedan. The Fisher-bodied Seven-Passenger Imperial, with glass division between the front and rear compartments, carried a factory price of $3,095. The ventilator doors in the hood and cowl were sometimes chrome-plated to make the V-8 look more like the prestigious new twelve and sixteen-cylinder Cadillacs. At 4,820 pounds, it was the heaviest production V-8 in Cadillac's 1931 book.

Even though the worsening Depression was taking its toll of well-heeled customers, Cadillac continued to cater to America's wealthy families. In a noteworthy effort to improve service to its customers, Cadillac dealers in larger cities instituted a "residential" pick-up and car delivery service. In response to a service call, the dealer dispatched a service man on a motorcycle who went to the owner's home to pick up the car. With the motorcycle attached to the rear bumper, the serviceman took the car back to the dealership for service. When the work was done, he returned the car and rode the motorcycle back. The car is a 1931 Series 355 Cadillac Town Sedan. Note the snappily uniformed service attendant.

1931

Very few Cadillac chassis of this era bore coachwork done by any other builder except Fisher or Fleetwood. Waterhouse and Murphy executed some one-off bodies on the Series 452 V-16. Full customs on the standard V-8 chassis were exceedingly rare. This nicely-done Victoria Cabriolet was by Saoutchik, a French coachbuilder later known for some very flambouyant creations, often on Cadillac chassis.

The vast General Motors Proving Grounds near Milford, Mich. has seen some strange vehicles over the years. One of them was this interesting test rig which was cobbled up to evaluate various experimental suspension systems. The car is a 1931 Series 355 five-passenger sedan. Bumpers and the front badge bar have been removed. Weights were added or removed to simulate various load and road conditions. Maurice Olley, who did much ride and suspension development work at Cadillac in the early 1930s which led to the independent front suspension adopted several years later, used a similar test-bed called the "K-2 rig".

The modestly-restyled Series 345-A LaSalle for 1931 had much in common with the 1931 Series 355 Cadillac, including the same 134-inch wheelbase and a bored-out version of the same V-8 engine that had been used in the eight-cylinder 1930 Cadillac. Like the 1931 Cadillac, the 1931 LaSalles sported new chromed screens over their radiators and had five ventilator doors on each side of the hood instead of louvers. The Series 345-A LaSalle Roadster for Two Passengers was among the lowest-priced cars in the line at $2,245, and at 4,330 pounds was the lightest LaSalle in the barn.

Priced only $50 higher than the two-passenger Roadster was the 1931 LaSalle Series 345-A Convertible Coupe. This model was also listed as a two-passenger car, although two extra passengers could be accommodated in the folding deck, or rumble seat. This model weighed 4,440 pounds, or 110 pounds more than the Roadster. List price was $2,295. The chromed landau irons were functional. The belt and hood molding and windshield frame are painted a lighter color than the rest of the body.

At a glance, this attractive body style looks like a convertible sedan, or what Cadillac and Fleetwood referred to as an All-Weather Phaeton. But close examination reveals that it is actually a Sedanette, which had a fixed roof, complete with ventilator, and removable center window posts. In a real sense, this was a precursor to the four-door hardtop body style which came along much later. Like the companion Series 345-A All-Weather Phaeton, the Sedanette had a wide upper body molding. Note the V-windshield and padded top. The Sedanette and companion Sedanette Cabriolet were identically priced at $3,245. A 4,650-pound weight was shown for the Sedanette.

1931

LaSalle

The six-window Sedanette Cabriolet was one of two special body styles carried over from 1930. The Sedanette Cabriolet was similar to the companion Sedanette, except for full windows in the rear quarters which gave it an open, phaeton look. Both models carried the same base price — $3,245 — but the Sedanette Cabriolet was 25 pounds heavier at 4,675 pounds. Like the Sedanette, it has a slightly-sloped "V" windshield and small roof ventilator.

Among the most attractive of all body styles offered on the 1931 Series 345-A LaSalle chassis was the popular Town Sedan, a beautifully-proportioned five-passenger closed car with a decidedly formal look about it. Wire wheels and a two-tone paint treatment further enhance the pleasing appearance of the example shown. The 1931 LaSalle Town Sedan was priced at $2,345 and weighed 4,665 pounds. The roof is painted rather than covered and the removable trunk complements the car's body lines.

The lowest-priced model in the 1931 LaSalle Series 345-A line was the Coupe for Two Passengers. Factory price was only $2,195. The car illustrated is equipped with a rear-mounted spare tire instead of sidemounts and rides on heavy-looking standard wood-spoked artillery wheels. The LaSalle Coupe weighed in at 4,470 pounds. Wheelbase was 134 inches.

The most expensive cars in the 1931 LaSalle line were the Sedanette, Sedanette Cabriolet and All-Weather Phaeton. Next came the big seven-passenger sedans. Except for the length of the rear body and longer rear side windows, LaSalle's five and seven-passenger sedans were almost indistinguishable. The Series 345-A Seven-Passenger Sedan weighed 4,750 pounds and was priced at $2,475. The companion Seven-Passenger Imperial with glass division, for chauffeur operation, listed for $2,595 and was 75 pounds heavier. LaSalle's new screened grille is shown to advantage in this view.

Again, the bread-and-butter cars in the 1931 LaSalle line were the boxy-looking but very practical five and seven-passenger sedans. This is the Series 345-A Sedan for Five Passengers, with six wire wheel equipment. The five-passenger LaSalle Sedan had a base price of $2,295, the same as the Five-Passenger Coupe. The Sedan weighed 4,650 pounds. How about that mansion in the background?

1932

Despite the worsening Depression, this was another important year of product development for Cadillac and the American automobile industry.

All of Cadillac's four passenger car lines were extensively restyled for 1932. Introduced early in January, the Series 355-B Cadillac V-8 line, the Series 370-B Cadillac V-12, the refined Series 452-B Cadillac Sixteen and the Series 345-B LaSalle all had softer, more rounded body lines, slightly smaller dimensions and a stronger family resemblance. Many historians feel that these resulted in the best looking classic Cadillacs ever to roll from Detroit.

Styling changes included sweeping new front fenders which flowed into gracefully curved running boards. Bodies now extended all the way down to the running boards, completely covering the frame and eliminating the sill. Up front, all of Cadillac's 1932 models had their first real grilles. Gone were radiator shutters or screens. The tall, narrow radiator shell instead housed a flush, die-cast aluminum honeycomb. The distinctive Cadillac "V" emblem, with "8", "12" or "16" numerals depending on the series, was now affixed directly to the upper center of the grille. Huge, bullet-shaped headlights flanked the attractive new grille design. Headlights, parking lights and taillamps had windsplits. The headlight tie bar used on all Cadillacs since 1928 was gone, but was retained for one more year on the LaSalle.

This would be the last year of classical Cadillac styling, of tombstone radiators and clamshell fenders. Of all of the many cars which he styled at GM over the years, these would remain Harley J. Earl's personal favorites.

The 1932 Cadillac V-8 and LaSalle were powered by the same "353" engine which had been uprated to produce 115 horsepower. The Series 345-B LaSalle model line included seven standard body styles, all by Fisher. Four rode on a 130-inch wheelbase while the other three senior models utilized a 136-inch wheel base chassis. The V-8 Series 355-B line offered 21 body styles on 134 and 140-inch wheelbases. There were 13 Fisher bodies and seven semi-customs by Fleetwood. The V-12 Series 370-B included an identical model choice. The only way one could tell a V-8 Cadillac from a twelve was by looking at the radiator badge or the hubcap inserts. LaSalle, Cadillac V-8 and V-12 models had six ventilator doors on each side of their hoods: the mighty V-16 had seven.

For the first time, eight Fisher bodies were available on the prestigious Series 452-B sixteen-cylinder chassis. A total of 16 semi-custom Fleetwood bodies were also offered. Series 452-B models rode on two wheelbases — 143 and 149 inches. The lighter Fisher-bodied V-16's were noted for their impressive top speeds, which easily exceeded 100 miles an hour. Mechanical changes to Cadillac's 12 and 16-cylinder engines were minimal.

Technical improvements in Cadillac's 1932 models included a vacuum-operated automatic clutch and "free-wheeling". Two-way hydraulic shock absorbers could be adjusted from the driver's seat. Silent helical gears were now used in all three forward speeds in Cadillac's synchromesh transmission.

The Depression continued to have a devastating effect on new car sales. The luxury car market was particularly hard-hit. Cadillac's sales dropped to 8,084 cars during the 1932 model year. Fewer than 300 V-16's were sold, and the LaSalle outsold the eight-cylinder Cadillac. The U.S. auto industry chalked up its worst production year since 1918.

Cadillac's competitors, however, were moving quickly to counter Cadillac's trend-setting V-12 and V-16 supercars. Both Packard and Lincoln introduced V-12 engines for the 1932 model year and were quickly followed by Pierce-Arrow. At the other end of the scale, Packard also chose 1932 to introduce a price-leading Packard Light Eight. This model was dropped after only one year when it proved unprofitable.

Packard's V-12 was a 160-horsepower, 455 cubic inch powerplant of $3\text{-}7/16$-inch bore and four-inch stroke. Lincoln's V-12, designed for the 145-inch wheelbase "KB" chassis, was of 447 cubic inch displacement and was rated at 150 horsepower. Bore was $3\text{-}1/4$ inches and stroke $4\text{-}1/2$ inches. Pierce-Arrow offered two V-12's, of 429 and 398 cubic inch displacement and Lincoln eventually came out with a second smaller twelve. Not to be outdone, Duesenberg announced its incredible supercharged SJ straight-eight which developed a fantastic 320 horsepower.

Henry Martyn Leland, the "Master of Precision" who was instrumental in the formation of the Cadillac Automobile Co. 30 years earlier, and who had guided Cadillac to greatness through its first 15 years, died in 1932 in his 90th year. Leland's cherished concept of parts interchangeability had made the modern automobile industry possible.

The 1932 Cadillac Series 355-B incorporated some important styling changes. These included long, sweeping fender lines and a new flush, die-cast aluminum grille. The headlight tie bar was dropped, and the Cadillac V-8 medallion was attached to the center of the honeycomb grille. This is the dashing Series 355-B Two-Passenger Roadster, which weighed 4,635 pounds and had a factory price of $2,895. Wheelbase was 134 inches. An identical Series 370-B V-12 Roadster was available for $3,595, also on the 134-inch wheelbase.

1932

The Series 355-B Cadillac V-8 for 1932 sported a pair of huge, bullet-shaped headlights that flanked the new, flush honeycomb grille. These were duplicated in miniature in the parking lights atop the car's graceful new front fenders. This is the Fisher-bodied Two-Passenger Convertible Coupe, a 134-inch wheelbase car which listed for $2,945. Note the functional plated landau irons and the golf club compartment door. The identical V-12 Convertible Coupe cost $3,645 and weighed 5,060 pounds

Shown on the roof of Cadillac's Detroit home is this Five-Passenger Convertible Coupe by Fleetwood on the 1932 Cadillac Series 370-B chassis. A V-12, it sports an attractive two-tone paint treatment and has a rather massive windshield fame. The seats are upholstered in cloth — not usual for a convertible. From this angle, the Convertible Victoria body style has a European look about it. It was priced at $4,195.

This Fleetwood-bodied 1932 Series 370-B Convertible Victoria has its top in the raised position. Note the extreme length of the chrome-plated, fully-functional landau irons. The integral trunk is an extension of the rear body, but this car is also equipped with a folding luggage rack for carrying extra baggage. This handsomely-proportioned five-passenger car, on the 134-inch wheelbase, weighed 5,225 pounds.

Cadillac offered no less than four Phaeton models on the 1932 Series 355-B chassis. All were by Fisher, and all four utilized the 140-inch wheelbase chassis. The line included a standard Five-Passenger Phaeton; a Special Five-Passenger Phaeton, a dual-cowl Sport Phaeton and a four-door convertible sedan which was listed as an All-Weather Phaeton. The standard Five-Passenger Phaeton had a price of $2,995; the Special Phaeton $3,095 and the Sport Phaeton $3,245. This Series 355-B Phaeton has chrome-plated hood ventilator doors. The Series 370-B twelve-cylinder lineup included four identical models. The V-8 weighed 4,700 pounds while the V-12 weighed 5,240 pounds.

Probably the sportiest of all body styles offered on the handsome 1932 Cadillac chassis was the racy Sport Phaeton, with rear tonneau and folding rear windshield. The Fisher-bodied Phaetons were built on the 140-inch wheelbase chassis. The Series 370-B Sport Phaeton had a factory price of $3,945. Also available was a Standard V-12 Phaeton at $3,695 and a Special Phaeton at $3,795. An identical model line was also available on the V-8 and V-16 chassis this year.

1932

Cadillac's 1932 model offerings included a vast selection of Phaetons powered by V-8, V-12 and V-16 engines on two separate wheelbases. All were nearly identical in appearance. Although more accurately described as a convertible sedan, this model was officially listed as an All-Weather Phaeton for Five Passengers. This was a Fisher body style. The Series 355-B and Series 370-B All-Weather Phaetons, with V-8 and V-12 engines respectively, were built on a 140-inch wheelbase. The V-16 Series 452-B All-Weather Phaeton was on the 149-inch chassis. The Series 355-B All-Weather Phaeton had a factory price of $3,495. The V-12 version was priced at $4,195 and the big Sixteen at $5,195.

This extremely impressive twelve-cylinder Cadillac Touring Car does not show on the Series 370-B price list. It is a classic seven-passenger tourer built on the 140-inch wheelbase. Possibly a one-off custom, it was photographed outside the Cadillac Motor Car Co. plant and office complex on Clark Avenue in Detroit. The big, open body style goes well with the car's clean front-end ensemble, but that huge, tent-like top must have been a bear to put up in a hurry.

Certainly one of the prettiest body styles offered on the 1932 Cadillac chassis was the beautifully contoured Two-Passenger Coupe by Fisher. The sporty-looking coupe body was available on the V-8, V-12 and even the long V-16 chassis. The Series 355-B Coupe shown here was the lowest-priced model in the 1932 line, at only $2,795. Weight was 4,705 pounds. The V-12 Series 370-B Coupe on identical wheelbase had a factory price of $3,595.

By 1932, body sheet metal on all Cadillacs was extended all the way down to the running board, completely eliminating the sill. This was an important step toward streamlining. The sweeping front fenders flowed into a curved running board for a coach sill-effect. Hubcap inserts told the viewer whether he was looking at an eight, twelve or sixteen-cylinder Cadillac, as did the badge affixed to the center of the grille. This Series 370-B Five-Passenger V-12 Coupe had a factory price of $3,695 and weighed 5,220 pounds. Wheelbase was 140 inches.

Cadillac's revised 1932 styling included this car's first real grille. Gone were shutters or a screen. The new flush grille had a die-cast aluminum honeycomb texture with the Cadillac "V" medallion mounted in its center. Big, bullet-shaped headlights had windsplits and were no longer separated by a tie bar. The effect was a fresh, tidy frontal appearance. This is the Style 355-B Five-Passenger Coupe, a 140-inch wheelbase model priced at $2,995. An identical model was offered in the V-12 line.

1932

Similar in appearance to the Five-Passenger Coupe, but with a squared-off instead of rounded rear deck, was the all-new Fleetwood-bodied Cadillac Town Coupe. Also a five-passenger car, the Town Coupe had an integral trunk faired into its rear body design. The total effect was quite pleasing, as the trunk complemented this model's softly rounded body contours. The V-8 Series 355-B Town Coupe, on 140-inch wheelbase, had a factory price of $3,395. It weighed 4,915 pounds. The companion V-12 model was priced at $4,095.

Cadillac's Five-Passenger Town Sedan was a popular four-door closed model with a semi-formal profile. The Series 370-B illustrated was a V-12 version which weighed 5,370 pounds and had a factory price of $3,795. Wheelbase was 140 inches. The Cadillac Heron radiator cap ornament was quite popular on the 1932 models. The flowing "goddess" hood ornament could also be specified.

The most prolific of all body styles available on the 1932 Cadillac chassis, regardless of engine choice, were a series of nearly identical five and seven-passenger sedans. All of these were four-door, six-window models with softly-rounded window openings. The 1932 Cadillac price list showed no less than eight five-passenger sedans and an equal number of seven-passenger sedans. The Fisher-bodied Series 355-B Five-Passenger Sedan on 134-inch wheelbase was priced at $2,895; a "Special Five-Passenger Sedan" on 140-inch wheelbase had a factory price of $3,045, and at the top of the line was a Fleetwood-bodied five-passenger sedan which listed for $3,395. This Fisher Five-Passenger Sedan was photographed on the roof of the General Motors Building in Detroit.

The Town Sedan for Five Passengers continued on into the 1932 model year. This body style was quite similar in appearance to the Town Coupe but had a longer body and four center-opening doors instead of two. The integral trunk is barely visible in this view. Built on the 140-inch wheelbase Series 355-B chassis, this car weighed 4,980 pounds. Factory price was $3,095. A V-12 version was also available.

Cadillac's 134 and 140-inch wheelbase five-passenger sedans were virtually indistinguishable. This is a 1932 Cadillac Series 355-B V-8 Sedan for Five Passengers on the 140-inch wheelbase. The car has an attractive two-tone paint treatment with black running gear and upper body and the hood and body finished in a lighter shade. The use of a rear-mounted spare tire instead of sidemounts emphasizes the length of the front end. Note the bright chrome cowl band and wire wheels.

1932

The 1932 Cadillac V-12 line included three five-passenger sedans. These were a 134-inch wheelbase model with a factory price of $3,595; a "Special Five-Passenger Sedan" priced at $3,745 on a 140-inch wheelbase, and a Fleetwood-bodied 140-incher with a price of $4,095. This is the V-12 Series 370-B Sedan for Five Passengers on the 140-inch wheelbase. The car sports white sidewall tires and has no bright exterior moldings or trim. The Detroit skyline looms to the south in this view taken from the roof of the General Motors Building.

For 1932, Cadillac offered its customers a choice of no less than eight big seven-passenger sedans. The V-8 engined Series 355-B line included two seven-passenger sedans, both on the 140-inch wheelbase. The standard Sedan for Seven Passengers had a factory price of $3,145. The Seven-Passenger Imperial with glass division between the front and rear compartments was priced at $3,295 and was sometimes listed as the Seven-Passenger Limousine. This is the Seven-Passenger Sedan, which weighed 5,110 pounds. The Limo was 40 pounds heavier.

The V-12 Series 370-B model lineup for 1932 included two seven-passenger sedans. This is the Seven-Passenger Imperial, an enclosed-drive limousine which sold for $3,995. The companion Seven-Passenger Sedan cost only $150 less — $3,945. Both were built on the 140-inch wheelbase chassis. The Series 370-B catalog also showed a Fleetwood Seven Passenger Sedan for $4,245 and a Fleetwood Seven-Passenger Limousine, at $4,445. Both of these cars were also on the 140-inch wheelbase.

The most expensive cars in the 1932 Series 370-B Cadillac line were three formal semi-customs by Fleetwood. All utilized the 140-inch wheelbase chassis. This is the V-12 Fleetwood Town Cabriolet for Five Passengers, an open-front town car which had a factory price of $4,795. The example shown has the "metal", or painted, back and no bright exterior trim. A Seven-Passenger Town Cabriolet was also available, with a price of $4,945. The other model in this Fleetwood body series was a Limousine Brougham for Seven Passengers which also had a factory price of $4,945. All of these models weighed approximately 5,500 pounds.

Prestige was for many years an important consideration in the funeral and ambulance business, which is why most hearses and ambulances were built on luxury car chassis. Luxury makes also depreciated slowly, making them better investments than more common makes. Patients were certainly transported to the hospital in style in the limousine-type ambulance on 1932 Cadillac V-12 chassis, built by the Silver-Knightstown Body Co. of Knightstown, Ind. Stylish touches include a "goddess" radiator mascot, small carriage lamps on the "B" pillars and leaded, colored glass in the rear window openings. A big coaster siren mounted in front of the grille cleared the way. This ambulance was built for the firm of A.S. Turner.

1932

The James Cunningham, Sons Co. of Rochester, N.Y. was one of America's finest builders of low volume, extremely high quality custom automobiles. Cunningham was also a major manufacturer of premium funeral cars and ambulances. In 1932, Cunningham began mounting its funeral coach and ambulance bodies on selected luxury makes, in addition to its own custom-built chassis. This is a Cunningham limousine-style hearse body on a 1932 Cadillac V-8 Series 355-B chassis. Wire wheels were not too common on hearses.

The Flxible Co., of Loudonville, Ohio also built some hearse and ambulance bodies on Cadillac V-8 chassis. This is a limousine-style funeral coach with full window draperies and half-length flower trays. Note the full-chrome wheel covers, a styling feature which would soon become more prominent on Cadillac passenger cars.

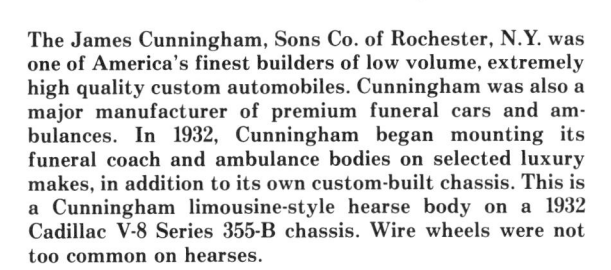

For the first time, a series of Fisher bodies were available on the sixteen-cylinder Cadillac chassis, along with a selection of Fleetwood semi-customs. This is the wood-and-clay styling mockup of the Fisher Style 155 Roadster for Two Passengers on the 143-inch wheelbase Series 452-B chassis. Only three of these huge roadsters were eventually built. Price was $4,595 and weight was 5,065 pounds. A Convertible Coupe for Two Passengers was also offered this year, with a factory price of $4,645 and a scale register of 5,530 pounds.

Just one of these classically-proportioned Five-Passenger Convertible Coupes was built on the 1932 Cadillac V-16 chassis. At least one was also produced on the V-12 chassis. A true Victoria with built-in trunk, this model was built on the 149-inch wheelbase Series 452-B chassis. Fleetwood listed this design as Style 5185. It was priced at $4,645 and weighed 5,505 pounds.

Only six true Phaetons were built on the 1932 Cadillac Series 452-B chassis. All had Fisher coachwork. These included three Style 280 "Special Phaetons", two Sport Phaetons, Style 279, and one "Standard Phaeton", Fisher Style 256. The Fisher Five-Passenger Phaeton was priced at $4,695. The example shown appears to have a half-tonneau, but no folding windshield.

1932

Among the more popular body styles for the refined 1932 Series 452-B sixteen-cylinder chassis was the practical, yet highly attractive Town Coupe for Five Passengers. Fleetwood built 24 of these cars on the 149-inch wheelbase V-16 chassis. Factory price was $5,095. This model was priced identically with the Series 452-B Five-Passenger Sedan, and was listed as Fleetwood Style 5181. It weighed 5,605 pounds.

In addition to three conventional open Phaetons, the 1932 Cadillac Series 452-B line of sixteens included an All-Weather Phaeton, also with Body by Fisher. Thirteen of these imposing convertible sedans were built on the 149-inch wheelbase chassis. Factory price was $5,195. The sidemounted spares on this example are painted a different color than the rest of the car. The Series 452-B All-Weather Phaeton weighed 5,195 pounds.

The 1932 Cadillac V-16 model lineup included five-passenger sedans by both Fisher and Fleetwood. The Series 452-B Fisher Five-Passenger Sedan, Style 159, was built on the 143-inch wheelbase chassis and weighed 5,625 pounds. Factory price was $4,595. The Fleetwood-bodied version was priced $500 higher and was mounted on the 149-inch wheelbase chassis. This view shows Cadillac's handsome new aluminum honeycomb grille with the V-16 emblem at its center. A "goddess" mascot tops the radiator cap.

Both Fisher and Fleetwood offered practical five-passenger sedan bodies for the 1932 Cadillac V-16 chassis. The Fleetwood-bodied version was mounted on the big 149-inch wheelbase Series 452-B chassis, while the less expensive Fisher version utilized the 143-inch chassis. Fleetwood Style 5130-S carried a factory price of $5,095. Note the huge bullet-shaped headlights which were duplicated in miniature in the parking lights atop the long, sweeping front fenders

The only seven-passenger sedans on the 1932 Cadillac V-16 chassis were a pair of 149-inch wheelbase cars by Fleetwood. This is the standard Fleetwood Style 5175-S Sedan for Seven Passengers, of which 47 were built. Price was $5,245. A Fleetwood Seven-Passenger Imperial with glass division was also available for $5,445 and was listed as the Fleetwood Limousine. A Style 5175-FL Limousine Cabriolet version was available in Fleetwood's semi-custom body offerings this year.

1932

This is a true Fleetwood Custom. Only one of these gigantic eight-passenger sedans was built on the 1932 Series 452-B sixteen-cylinder chassis. Positively hearse-like in its dimensions, it was constructed on a special 165-inch wheelbase chassis for the Annenburg family. It was shown as Style 5177, a Fleetwood Imperial Sedan. It has a full-length roof luggage rack for extended touring and pull-down shades in all rear windows.

Only five Fleetwood special-order bodies for 1932 were available with the very distinctive "Madam X" windshield. Twelve had been offered on the first-series V-16 chassis for 1930-31. The flat, one-piece "Madam X" windshield was raked at a bold 18 degrees and was framed by a pair of extremely slender, delicately curved windshield pillars which flowed into a tapered belt molding at the cowl. Fleetwood Style 5131, a "Special Imperial Sedan for Five Passengers", has the striking "Madam X" windshield treatment. This style was also available as Style 5131-S, a five-passenger sedan without Imperial division. The long, flowing fender lines and upward-swept rear window shape are delightfully compatible with the rakish windshield profile.

Among the rarest of all special-order Fleetwood bodies for the mighty Cadillac V-16 chassis was the stately 5-passenger Imperial Cabriolet with Collapsible Rear Quarter. Very few of these regal limousines were built between 1930 and 1936. Just one was executed on the 1932 Series 452-B chassis. Fleetwood listed this formal model as Style 5155-C. It has the distinctive "Madam X" windshield and was built on the 149-inch wheelbase chassis.

Fleetwood Style 5155 was also an Imperial Cabriolet for Five Passengers, but with solid rear quarters. Only four of these were built, all on the 149-inch wheelbase V-16 chassis. One Style 5155-C with collapsible rear quarter was also built by Fleetwood. All had the rakish "Madam X" windshield.

The 1932 Cadillac Series 452-B model line included several formal four-door sedans with closed upper rear quarters. This is Fleetwood Style 5175-FL, which was listed in the 1932 Fleetwood Body book as a Seven-Passenger Imperial Sedan. A five-passenger version, Style 5130-FL, was also available. These elegantly-proportioned cars were also referred to as Imperial Cabriolets.

1932

This Five-Passenger "Transformable Town Brougham" has a full metal back and no ornamental landau irons. Fleetwood listed this 1932 model as Style 5164. The roof over the chauffeur's compartment folded into the top of the solid tonneau roof.

Four of these open-front town cars, Fleetwood Style 5112, were built on the 1932 Cadillac V-16 chassis. This model was officially listed as a Fleetwood Town Cabriolet for Five Passengers. Factory price was $5,795. Two seven-passenger Town Cabriolets of this design, Style 5125, were also built. At $5,945 they were the most expensive cars in the line. The roof and rear quarters were leather-covered. Fleetwood often referred to this style of car as a "leather back".

Although a true Limousine Brougham, Fleetwood also listed this Series 452-B body as a "Transformable Town Cabriolet", Style 5191. Seven of these big seven-passenger, open-front limousine style town cars were built on the 149-inch wheelbase chassis. Factory price was a very expensive $5,945. The 1932 Fleetwood catalog listed it among the "metal back" models.

Actually a Limousine Brougham, this formal town car was also listed as a "Transformable Town Cabriolet", Fleetwood Body Style 5120. It has rear quarter windows and small, ornamental landau irons. All of these special-order Fleetwood bodies were available for both the 149-inch V-16 wheelbase or the 140-inch V-8 and V-12 wheelbase.

The fascinating Fleetwood Body catalog for 1932 included a number of interesting body designs for the Series 452-B sixteen-cylinder chassis, some of which were never built. Others were produced in extremely small quantities — sometimes only one, making them true customs. Style 5150 offered the truly discriminating buyer the best of three worlds. This fully-collapsible model is shown in its open-front Town Car configuration. It could be quickly converted into a big convertible sedan, or completely enclosed as a formal seven-passenger car.

Another Fleetwood body style for the 1932 V-16 chassis which was apparently never built was this oddly-proportioned formal sedan. It appears to be a seven-passenger car on a five-passenger wheelbase, with resultant rather awkward-looking rear overhang. This model was shown as Style 5183. The roof looks as if it was fabric-covered. The thin window edges are suggestive of first-edition "Madam X" bodies, although a standard windshield is shown. The window pillars appear to be removable when the windows were lowered, so that the car could be used as a 7-passenger touring, with solid roof.

1932

The 1932 LaSalle incorporated significant changes inside and out. Like its more expensive brother, the Cadillac V-8, the new Series 345-B LaSalle had a finely-detailed new grille design. The headlight tie bar with "LaS" emblem at its center, a holdover from the first LaSalle built five years earlier, was retained for the last year. The prim young lady in riding attire is about to step into her 1932 Series 345-B LaSalle Convertible Coupe, one of seven LaSalle body styles offered this year. The LaSalle Roadster had been dropped from the line. This pretty 130-inch wheelbase car had a factory price of $2,545.

The 1932 LaSalle and the 1932 Cadillac V-8 were almost identical. In fact, the LaSalle buyer was actually getting a Cadillac in everything but name, and at a lower price. Both cars were powered by the same "353" 115-horsepower engine. The LaSalle had started to shed its now-dated Hispano-Suiza look and was almost indistinguishable from the 1932 Cadillac. This is how the Series 345-B Convertible Coupe looked with its top up. The example shown has metal-covered side-mounted spare tires and is finished in an attractive light color.

The lowest-priced model in the Series 345-B LaSalle model lineup for 1932 was this beautiful Two-Passenger Coupe, which sold for only $2,395. This attractive coupe body was available on all other Cadillac Motor Car Co. chassis this year including the Series 370-B twelve and even the mighty Series 452-B sixteen! This V-8 Coupe has metal-covered sidemounted spares and the graceful "Heron" hood ornament. Wheelbase was 130 inches and weight was 4,660 pounds.

The five-passenger LaSalle Town Coupe was a practical and especially handsome car. One of four 1932 LaSalle body styles on the 130-inch wheelbase Series 345-B chassis, the Town Coupe was factory-priced at $2,545—the same as the Convertible Coupe. Note how the trunk is faired into the rear body. A single spare tire is mounted on a carrier behind the trunk, though sidemounts could be ordered when a folding trunk rack was desired. Weight was 4,695 pounds.

1932

This is a well-equipped 1932 LaSalle Series 345-B Five-Passenger Sedan. The body is finished in a light color with the belt molding and upper body in a complementing darker hue. Fenders, metal-covered sidemounts and wire wheels were painted black. This four-door sedan weighed 4,840 pounds. Wheelbase was 130 inches.

LaSalle's standard four-door sedan was this nicely-styled car, seen here without sidemounted spares and in more or less stock form. The Series 345-B Sedan for Five Passengers utilized the 130-inch wheelbase and had a factory price of $2,495. The five-passenger sedan was price-positioned between the Two-Passenger Coupe, which cost $100 less, and the Convertible Coupe which was priced $50 higher.

The 1932 Series 345-B LaSalle was available on two separate wheelbases. Four models utilized a 130-inch wheelbase. Three others were built on the longer 136-inch wheelbase. The stylish Five-Passenger Town Sedan was one of the three 136-inch wheelbase LaSalles. This model, with its rounded, integral trunk, resembled the Town Coupe but had four doors. The Series 345-B Town Sedan had a factory price of $2,645 and weighed 4,895 pounds.

In its styling, the 1932 LaSalle was a much-changed car. The new Series 345-B LaSalle had more rounded body lines and long, sweeping fenders. This frontal view of a 1932 LaSalle sedan shows the car's handsome new grille design to advantage. The same style of finely-textured grille, without shutters or a screen, was also used on all of the 1932 Cadillacs, including the V-12 and V-16. Only the LaSalle, however, retained the headlight tie bar. This car, with light-painted six wire wheel equipment and metal-covered sidemounts, has the popular "goddess" hood ornament. Prices ranged from $2,395 to $2,795.

The largest and heaviest cars in the 1932 LaSalle line were a pair of big seven-passenger sedans, both built on the 136-inch wheelbase. The Series 345-B Seven-Passenger Sedan weighed 5,025 pounds and had a factory price of $2,645. A LaSalle Seven-Passenger Imperial was also available with glass division for limousine duty. The Imperial was priced at $2,795.

1933

This was a grim year for the beleaguered American automobile industry and for the entire U.S. economy as the Great Depression hit rock-bottom. Automobile sales and production slowed to a trickle. The era of conspicuous automotive luxury had come to an abrupt end.

Cadillac sales also continued to fall, with only 6,655 cars sold in the entire 1933 model year in all four lines.

But 1933 was an important evolutionary year for Cadillac styling. All of Cadillac's 1933 models—LaSalle, the Cadillac V-8, V-12 and V-16—were introduced early in the new year showing startling styling changes which marked a sharp break with the past. Gone forever was the cobbled-up, rectangular look. The trend now was toward streamlined, integrated body lines which flowed smoothly from bumper to bumper.

There were some important functional improvements, too. An extremely important innovation on all closed-body 1933 Cadillacs and LaSalles was the introduction of Fisher no-draft ventilation. Small, pivoted window panes were fitted to the front and rear side windows for a controlled flow of fresh air through the car at the driver's or passenger's discretion. Air could also be introduced into the car through a cowl ventilator. Window vent panes of this type would be universally used on virtually all cars well into the 1960's.

The 1933 Cadillacs and LaSalles bore a strong family resemblance.

All models had new, fully-skirted fenders and a windsplit V-type radiator shell and grille. Hood ventilator doors were now horizontal, emphasizing the length of the tapered hood. The flagship Cadillac V-16 was given its own distinctive front-end styling to set it apart from the more plebian Cadillac V-8's and V-12's. The 1933 Series 452-C Sixteen sported a larger, restyled "goddess" hood ornament that was now exclusive to the V-16 line. The long V-16 hood was graced on each side by three functional horizontal ventilator "spears", which were duplicated in miniature on the lower front fender skirts. The 1933 Cadillac V-16's also had massive, new four-bar bumpers front and rear.

The 1933 Series 345-C LaSalle continued to use the Cadillac "353" V-8 engine for the third and final year. Seven Fisher body styles were again offered, four on a 130-inch wheelbase and three on a 136-inch chassis.

The V-8 Series 355-C Cadillac model lineup included a choice of 16 bodies, 10 by Fisher and six by Fleetwood on 134 and 140-inch wheelbases. An identical choice was available in the more expensive companion Series 370-C twelve-cylinder line, also on 134 and 140-inch wheelbases. The only way one could tell the 1933 Cadillac V-12 from the V-8 was too look at the radiator emblem or hubcap inserts, or peer under the hood.

Ten standard Fleetwood body styles were offered on the refined Series 452-C sixteen cylinder Cadillac chassis, all on a 149-inch wheelbase.

While no Fisher bodies were offered on the V-16 chassis this year, at least two were delivered as Series 452-C Sixteens, mounted on leftover 143-inch V-16 chassis. Fleetwood also offered a bewildering variety of semi-custom and special order bodies for the 1933 V-16 chassis, including five with "Madam X" windshields. Only 126 sixteen-cylinder Cadillacs were sold in the 1933 model year.

Despite the gloom of the Depression, all eyes were on American industry at the 1933 Chicago "Century of Progress" Exposition. For the huge General Motors exhibit, which even included an operating Chevrolet assembly line, Cadillac created a stunning experimental show car built on a production V-16 chassis. Called the V-16 Aero-Dynamic Coupe, it was a dramatically-streamlined, teardrop-shaped fastback coupe with a narrow V-type windshield and pontoon fenders. Gorden Beuhrig, who was working at Cadillac at this time, was deeply influenced by this design and incorporated its basic shape in his famed Cord 810 three years later. The Aero-Dynamic Coupe went into production the following year. Twenty were built between 1934 and 1937.

Three other special luxury cars were exhibited at the Chicago Exposition. These included a golden-bronze Packard V-12 Sport Sedan based on a Raymond Dietrich design and called the "Car of the Dome"; a radically-streamlined Pierce-Arrow V-12 fastback named the Pierce Silver Arrow, and Duesenberg's Rollston-bodied "Twenty Grand", a name indicative of the latter's staggering price tag. Ford showed a John Tjaarda-styled dream car called the Zephyr, which ultimately evolved into the Lincoln Zephyr of 1936.

For 1933, Lincoln introduced its new KA Series powered by a smaller 382 cubic-inch V-12 engine of 125 horsepower. This car was designed as a companion car to the larger, more expensive Lincoln KB.

All 1933 Cadillacs and LaSalles showed major styling changes. For the first time, the radiator was enclosed in a painted shell behind a V-type grille. Front and rear fenders were fully skirted, and new horizontal ventilator doors with chrome accents were used on the hood. Eight and twelve-cylinder Cadillacs were equipped with the split-style bumper shown here with three thin center bars. This is the 1933 Cadillac Series 355-C Convertible Coupe with Fisher body, a 134-inch wheelbase car which sold for $2,845. A companion Roadster was also offered, for $2,795. This sporty body style was also available in the V-12 line.

1933

Again, the twelve-cylinder Cadillacs were indistinguishable from the lower-priced V-8's except for the grille badge and hubcap inserts. The 1933 Cadillac Series 370-C Convertible Coupe for Two Passengers carried a factory price of $3,545 and weighed 5,125 pounds. Like the lower-priced V-8, it rode on a 134-inch wheelbase. A Series 370-C Roadster was also listed, with a factory price of $3,495.

The lowest-priced model in the 1933 Cadillac V-8 line was the Two-Passenger Coupe. Five wire wheels were standard. Body was by Fisher and wheelbase was 134 inches. The Series 355-C Coupe had a factory price of $2,695 and weighed 4,855 pounds. An identical Two-Passenger Coupe was also available on the Series 370-C twelve-cylinder chassis for $3,395. This would be the final year for the graceful "Heron" hood ornament which was available on Cadillac V-8's and twelves and the 1933 LaSalle.

The 1933 Cadillac Series 355-C Five-Passenger Coupe was built on the 140-inch wheelbase chassis and weighed 4,850 pounds. Factory price of this popular model was $2,895. Body was by Fisher. New, horizontal hood ventilator doors emphasized the length of the tapered hood. This practical body style was also available in the V-12 Series 370-C. The twelve-cylinder model had a base price of $3,595 and weighed 5,200 pounds.

This would be the last year that Cadillac would refer to this body style as an All-Weather Phaeton. From 1934 onward, this model would be listed as a convertible sedan. For 1933, the Fisher-bodied All-Weather Phaeton was available in both V-8 and V-12 versions on the 140-inch wheelbase chassis. There was also a Fleetwood All-Weather Phaeton on the big V-16 chassis. The Series 355-C eight-cylinder All-Weather Phaeton shown here was a five-passenger car which had a factory list price of $3,395. Weight was 5,110 pounds.

The five-passenger Cadillac Town Sedan was offered in both V-8 and V-12 versions, both on the 140-inch wheelbase chassis. Cadillac noted that this body style..."combines the smart lines of a sport car and the generous capacity of a five-passenger sedan". Five wire wheels were standard. This is the Series 370-C V-12, which sold for $3,695. The V-8 Series 355-C Town Sedan had a factory price of $2,995. Note the integral trunk.

Externally identical to the V-8 Fisher All-Weather Phaeton was the companion V-12. You had to look at the radiator emblem or hubcaps, or raise the hood, to be sure. The 1933 Cadillac Series 370-C All-Weather Phaeton listed for $3,595 and was built on the 140-inch wheelbase chassis. The V-12 convertible sedan weighed 5,200 pounds. This is how it looked with the top down.

1933

The raciest of all body styles offered on the 1933 Cadillac V-8 or V-12 chassis was this extremely fleet-looking Sport Phaeton by Fisher. Standard equipment included a separate cowl for the rear-seat passengers and a folding second windshield. This style is also accurately described as a dual-cowl Phaeton. The rear tonneau was hinged at the front for passenger entry and exit. Note the striking two-tone paint and full chrome wheel covers. This is a Series 370-C V-12. As a V-8, it sold for $3,245 and as a V-12, it jumped to $3,945.

The 1933 Cadillac Series 370-C All-Weather Phaeton shown here has a bustle-back with integral trunk. The window center posts could be removed and the side glass lowered into the doors to convert this car into a full convertible. The V-12 All-Weather Phaeton had a starting factory price of $4,095 and was marketed as a five-passenger car. Later in the season, prices on these cars dropped about $100 per model.

The most important technical innovation on Cadillac's 1933 models was the introduction of Fisher no-draft ventilation on Cadillac and LaSalle closed-body models. Small, pivoted window panes in the front and rear side windows permitted driver or passengers to direct a flow of fresh air into and through the car. Air flow could also be introduced through a cowl ventilator. This marked a significant advance in passenger comfort, and no-draft windows were eventually used on all American cars well into the 1960's. This profile view of a 1933 Cadillac Series 370-C Five-Passenger Sedan shows the no-draft vent panes in the front door and rear side window openings.

This is the 1933 Cadillac Series 355-C Five-Passenger Sedan, which was also offered in a V-12 version in the more expensive Series 370-C line. Body was by Fisher, and the four-door sedan utilized the 140-inch wheelbase chassis. The Series 355-C Five-Passenger Sedan tipped the scales at an even 5,000 pounds and had a factory price of $2,895—the same as the V-8 Five-Passenger Coupe and All-Weather Phaeton. Note the bright trumpet horns mounted below the large headlights and the three-section front bumper. This publicity photo was posed in one of Detroit's affluent neighborhoods.

The 1933 Cadillac Series 370-C Five-Passenger Sedan with V-12 engine was externally identical to the lower-priced V-8 version and was built on the same 140-inch wheelbase. It weighed 335 pounds more, however, and at $3,595 was priced $700 higher. This is a prototype V-12 sedan posed in the styling studio for GM management approval. The parking lights atop the new skirted front fenders are miniature duplicates of the headlights. Customers could specify sidemounted spare tires or a rear-mounted spare. This preproduction model has dual-faced white sidewall tires and the graceful "Heron" hood ornament.

1933

The most expensive models in the 1933 Cadillac V-8 and V-12 Fisher Body lines were the big Seven-Passenger Sedans. The Series 355-C Seven-Passenger Sedan by Fisher had a factory price of $3,045 and weighed 5,105 pounds. A Fisher Seven-Passenger Imperial with glass division was also offered for $3,195. In the V-12 Series 370-C line for 1933, the Fisher V-12 Seven-Passenger Sedan had a price of $3,745 and the Imperial sold for $3,895. All were on a 140-inch wheelbase.

Cadillac's comprehensive 1933 model lineup included five and seven-passenger sedans and seven-passenger Imperial limousines in both V-8 and V-12 lines with seven-passenger bodies by both Fisher and Fleetwood. This is a V-12 Series 370-C Seven-Passenger Sedan by Fleetwood, which sold for $4,145. The companion Seven-Passenger Fleetwood Limousine with Imperial division had a factory price of $4,345. The five and seven-Passenger sedans were generally identical in appearance except for the length of the rear body.

In addition to six standard body styles, Fleetwood offered a series of semi-custom, special-order bodies which could be mounted on the 1933 Cadillac V-12 and V-8 chassis. This is a formal seven-passenger sedan which was listed as the Fleetwood Seven-Passenger Limousine Cabriolet. It has closed rear quarters and a leather-covered top which was described in the catalog as a "full-leather back". This model was Fleetwood Body Style 5475-FL, which bore a base price of $4,795 with V-12 power and $4,095 with a V-8. The landau bars were strictly decorative.

The most expensive models in the 1933 Cadillac V-8 and V-12 model lines were the elegant Limousine Broughams —big, very formal open-front town cars available only with seven-passenger seating. Fleetwood Body Style 5491, the Limousine Broughams had full windows in the upper rear quarters. Both eight and twelve-cylinder models were built on the 140-inch wheelbase chassis. The Series 370-C Fleetwood Limousine Brougham had a factory price of $4,845 and weighed a hefty 5,575 pounds. The V-8 version was priced at $4,145. Again, in April, these prices went up $100 per car.

Among the most expensive V-8 and V-12 Cadillacs offered for 1933 were a series of formal, open-front town cars by Fleetwood, which were listed in the Fleetwood catalog as Transformable Town Cabriolets. These were available in five and seven-passenger configurations, both on 140-inch wheelbase. The Series 370-C Town Cabriolet for Seven Passengers, Fleetwood Series 5425, had a factory price of $4,695. The seven-passenger model cost $4,845. The V-8 Five-Passenger Town Cabriolet had a factory price of $3,995 and the seven-passenger Series 355-C version $4,145. Shown is the V-12 Fleetwood Five-Passenger Town Cabriolet. By April, all of these prices were adjusted upwards by $100 per model.

1933

The V-8 Cadillac chassis continued to enjoy wide popularity among funeral directors and funeral car and ambulance manufacturers. The Eureka Co. of Rock Falls, Ill. designed and built this "Chieftan" limousine funeral coach body for the 1933 Cadillac chassis. Arched flower trays are visible in the rear side windows, and this handsomely-styled hearse has six wire wheel equipment, including metal-covered spares.

The Depression was having a devastating effect on all custom-body builders. One of the hardest-hit was the prestigious old firm of James Cunningham, Son and Co. of Rochester, N.Y. Orders for the company's custom-built luxury cars had all but disappeared and the company was forced to rely on its hearse and ambulance business for survival. Cunningham was now mounting its premium funeral car and ambulance bodies on commercially-available chassis, primarily Cadillacs. Cunningham built this massive carved-panel town car "Cathedral" hearse with arched center panels for the St. Louis, Mo. firm of Hermann & Son.

Ten Fleetwood body styles were available for the 1933 Cadillac V-16 super chassis, all on a 149-inch wheelbase. Only four of these Five-Passenger Town Coupes, Style 5581, were built at $6,250 a copy. This one was custom-built for A. K. Striker, Jr. and was shown as a Fleetwood V-16 "No. 20" Town Coupe. On lesser makes, this two-door body style would be called a Coach. Note the two small vertical ventilator doors below the front of the top hood ventilator "spear" on this particular car.

For 1933, Cadillac's prestigious sixteen-cylinder models were given their own, distinctive styling identity, setting them apart from the V-8's and V-12's. The long V-16 hood had three horizontal ventilator "spears" which were duplicated on the lower edges of the deep-skirted front fenders. The Series 452-C Cadillacs also had massive four-bar bumpers and stylish new optional spinner-type wheel covers. While only Fleetwood bodies were offered on the 149-inch wheelbase V-16 chassis this year, two Fisher bodies were mounted on carryover 143-inch V-16 chassis. This is the Fisher Five-Passenger Series 452-C Coupe, priced at $4,595. The other 1933 Fisher-bodied V-16 was a Two-Passenger Convertible Coupe priced at $4,645.

This stunning four-door touring car was the last true dual-cowl phaeton built by Fleetwood. Style 5559, it was listed as a Series 452-C Dual-Cowl Sport Phaeton. The wind wings lowered into the doors and rear-seat passengers had their own windshield. This body style was also sometimes called a Tonneau Phaeton. Like most 1933 Cadillac V-16's, the headlights and parking lamps on this imposing car were painted rather than plated. The bright, full-length upper body molding is also worthy of note. Only one of these big five-passenger phaetons was built, making it a true Fleetwood Custom. Note that the rear wind wings met with both tonneau windshield and top when raised—making the tonneau totally weather-tight.

1933

Only two of these elegant Five-Passenger Convertible Victorias were built on the Series 452-C sixteen-cylinder Cadillac chassis for 1933, as Fleetwood Style 5585. Price was a steep $7,500. This three-quarter rear view shows the integral trunk, four-bar rear bumper and metal-covered sidemount spares. Note the very long landau bows.

This would be the last year for the Fleetwood All-Weather Phaeton. Henceforth, this versatile body style would be known at Cadillac as a convertible sedan. The Fleetwood Style 5579 All-Weather Phaeton for Five Passengers was by far the most expensive car in the 1933 sixteen-cylinder Cadillac line. Factory price was a substantial $8,000. The Series 452-C All-Weather Phaeton shown is the famed Al Jolson car, now in the Harrah Automobile Collection in Reno, Nev. This example has a straight back.

Just one of these impressive Fleetwood Imperial Cabriolets was built on the 1933 Cadillac V-16 chassis. A formal five-passenger sedan, it was Fleetwood Body Style 5530-FL and had a factory price of $5,540. The roof and upper rear quarters are fabric-covered and this car has a straight back. Plated decorative landau irons span the formal rear quarters. The Imperial division glass between the front and rear compartments qualifies this car as an inside-drive limousine. The Series 452-C Imperial Cabriolet weighed 6,100 pounds. Wheelbase was 149 inches.

By far the most famous of all sixteen-cylinder Cadillacs built was the daring Aero-Dynamic Coupe which was specially created for the General Motors exhibit at the 1933 Chicago Century of Progress Exposition. This stunning show car was based on the production Series 452-C Cadillac V-16 and utilized the same 149-inch wheelbase chassis. The Aero-Dynamic Coupe's most striking styling features included a true "fastback" body design, a wide V-type windshield and pontoon fenders. This car also had an all-metal roof and aluminum running boards. A four-passenger car, the Aero-Dynamic Coupe was shown as Fleetwood Body Style 5599.

The boldly-styled Cadillac Aero-Dynamic Coupe was the forerunner of the numerous GM fastback and torpedo body designs which appeared in the early 1940's. Although only one was built in 1933, the Aero-Dynamic Coupe did eventually go into limited production. A total of 20 were built between 1934 and 1937. This rear view shows the experimental V-16 Aero-Dynamic Coupe's sweeping fastback roofline. The license plate was set into the center of the deck lid. Gordon Buehrig, who at the time was working for GM, was very much impressed with this design and adopted its basic shape for his famed Cord 810 in 1936.

1933

Only 126 sixteen-cylinder Cadillacs were sold in 1933 as the Great Depression hit bottom. Even fewer would be sold in subsequent years. While the Book of Fleetwood offered a fascinating variety of special semi-custom bodies, many were never built and only one example of others were eventually delivered. Factory records indicate that just one of these Five-Passenger Close-Coupled Fleetwood Imperials was actually produced. This formal four-door was listed as Style 5561-S.

At the time of introduction, Cadillac strongly advised potential V-16 purchasers to place their orders early as only 400 sixteen-cylinder Cadillacs would be built for the entire 1933 model year. This advice proved wildly optimistic, as only 126 Cadillac V-16's were sold in the year the Great Depression hit rock bottom. This is the 1933 Cadillac Series 452-C Sedan for Five Passengers by Fleetwood, a 6,070-pound car that had a factory price of $6,250. It was listed as Fleetwood Style 5530-S. The blades on the optional spinner hubcaps were actually handles used to take the hubcaps off. This style of wheel cover proved irresistable for generations of hot-rodders and car customizers.

The 1933 Cadillac V-16 line included two big seven-passenger sedans of this style. The Fleetwood Seven-Passenger Sedan had a factory price of $6,400. For another $200 a Fleetwood Seven-Passenger Limousine with Imperial glass division was available. The car shown is an early styling model photographed in the GM styling studio where it was presented for senior management approval. The 10 small, vertical louver ornaments on the front fender skirts did not appear on production cars. Instead, three horizontal spears like those on the hood were used.

The 1933 "Book of Fleetwood" contained some intriguing individual body designs created for the Cadillac V-16 chassis. Many of them were never actually built. Here are three of them. Above is the Style 5586 Convertible Four-Passenger Coupe, which is very Teutonic in its styling. In the center is a big Seven-Passenger V-16 Touring Car, Style 5557. Below is a special Fleetwood Seven-Passenger Limousine, Style 5566. All of these catalog illustrations show the small vertical louvers on the front fender skirts which did not go on production cars. The Fleetwood catalog also listed five 1933 body styles with "Madam X" windshields, and even a razor-edged open-front Town Brougham with simulated canework, like the few built on the first-generation V-16 chassis in 1930-31.

1933

This big 1933 Cadillac Seven-Passenger Sedan by Fleetwood is equipped with white sidewall tires and metal side-mount wheel covers. This profile shot shows the 1933 sixteen-cylinder Cadillac's flowing body lines. No-draft ventilator windows were an important new feature this year and were used on both front and rear side windows. The new style "goddess" hood ornament was exclusive to the flagship V-16. This sedan was posed outside the Cadillac plant on Clark Avenue in Detroit. At the top of the 1933 Series 452-C model lineup were five and seven-passenger Town Cabriolets and a seven-passenger Limousine Brougham.

In the new age of high automotive fashion, America's funeral directors had become very style-conscious. Funeral car manufacturers did their best to keep up with styling trends in Detroit. This open-front, carved-panel town car hearse by Eureka utilizes a V-16 Cadillac hood and sharply-raked windshield. The bumpers and fenders and the type of hood ornament used would indicate that it is actually a V-8. Nevertheless, it represented a truly classic way to go.

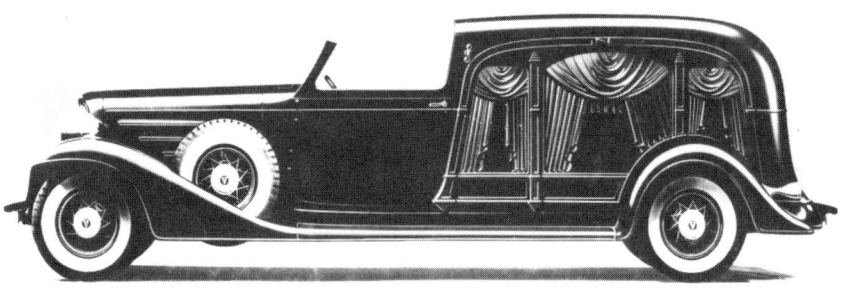

Like the 1933 Cadillacs, the 1933 LaSalle showed major styling changes. These included new skirted fenders, a V-shaped radiator shell and grille and horizontal hood ventilator doors. Again, there were seven Fisher body styles available on two wheelbases. This is the Series 345-C Convertible Coupe for Two Passengers, a sporty 130-inch wheelbase model that sold for $2,395. Although listed as a two-passenger car, two more passengers could be accommodated in the folding rumble seat in the rear deck. The rumble seat compartment was even equipped with a foot rail. This model weighed 4,675 pounds.

LaSalle's fresh 1933 styling marked a dramatic shift away from its traditional appearance and a bold step toward streamlining. The Series 345-C Town Coupe was a beautifully-styled car with an abundance of curves and clean, good looks. Five wire wheels and an integral, lockable and weathertight trunk were standard. The five-passenger LaSalle Town Coupe was built on the 130-inch wheelbase chassis and had a factory price of $2,395—the same as the Series 345-C Convertible Coupe.

The lowest-priced models in the 1933 LaSalle model line were the Two-Passenger Coupe and the identically-priced Five-Passenger Sedan. The very attractive Series 345-C Coupe had a factory price of only $2,245 and was one of four 130-inch wheelbase models. The Coupe illustrated has a rear-mounted spare tire and "Heron" radiator ornament. Note the golf club compartment door.

1933

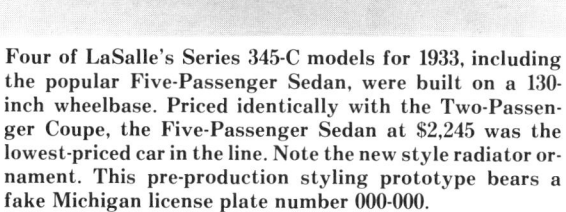

Four of LaSalle's Series 345-C models for 1933, including the popular Five-Passenger Sedan, were built on a 130-inch wheelbase. Priced identically with the Two-Passenger Coupe, the Five-Passenger Sedan at $2,245 was the lowest-priced car in the line. Note the new style radiator ornament. This pre-production styling prototype bears a fake Michigan license plate number 000-000.

An attractive two-tone paint treatment highlights the clean, flowing lines of the 1933 LaSalle Town Sedan. This was one of three Series 345-C LaSalles built on the 136-inch wheelbase chassis. A four-door, five-passenger car with integral trunk, the Town Sedan sold for $2,495 and weighed 4,915 pounds. This well-equipped example has six wire wheels.

Three 1933 LaSalle models were built on the longer 136-inch wheelbase. These included the Town Sedan, a Seven-Passenger Sedan and a Seven-Passenger Imperial with glass division. This Series 345-C sedan has a special upper rear quarter styling treatment with small quarter windows, ornamental carriage bows and a leather-covered roof and back. The Seven-Passenger Sedan was priced at $2,495. The most expensive model offered was the LaSalle Imperial, at $2,645.

With its thoroughly-restyled 1933 models, LaSalle broke away for good from its rather staid earlier appearance. This profile view of a Series 345-C sedan shows the stylish, new skirted fenders, rounded window openings with Fisher no-draft ventilation and horizontal hood ventilator doors which gave the car a longer frontal appearance. All 1933 LaSalles were powered by the same Cadillac "353" V-8 engine that had first been used in LaSalle's 1931 models. This would be its final year.

Up to this time, very few hearses or ambulances were built on the LaSalle chassis. Cadillac would eventually supply funeral car and ambulance builders with a special LaSalle commercial chassis designed for this purpose. The A. J. Miller Co. of Bellefontaine, Ohio built this handsome Limousine Ambulance on a stretched 1933 LaSalle chassis for the Stonington Ambulance Corps. The artistically-leaded rear side windows provided a degree of privacy for the patient and attendants.

1934

This would be another highly eventful year for Cadillac. All of the division's four car lines underwent thorough styling changes, from the totally-new 1934 LaSalle on up to the flagship Series 452-D Cadillac Sixteen. Under the sleek, new sheet metal, mechanical improvements were equally impressive.

The 1934 Cadillacs were the first in the industry to banish the spare tire out of sight and into the trunk, although side-mounted spares were still available for those who preferred them.

One of the most noticeable improvements was in riding comfort. All of Cadillac's '34 cars featured new "knee-action" independent coil spring suspension. Pioneered by former Rolls-Royce suspension and ride engineer Maurice Olley, this advanced vehicle suspension system had been under development at GM for several years. Other important mechanical changes included Hotchkiss steering and a new dual X-type chassis. Inside, the handbrake was now mounted on the dash.

All of Cadillac's 1934 cars had fresh, new streamlined styling. The Series 355-D Cadillac V-8's and Series 370-D twelves were identical in appearance with pontoon fenders, wide V-type grilles set back at a dashing angle and huge, bullet-shaped headlights mounted high up on each side of the radiator shell. Front fenders had a horizontal crease line at the center of their leading edges. Parking lights were incorporated into the headlight stanchions.

The most interesting exterior innovation was the use of unique "bi-plane" bumpers on all of Cadillac's 1934 cars, including the LaSalle. These art deco bumpers consisted of a pair of thin blades separated by two bullet-shaped spacers. On impact, the spring-mounted bumpers recoiled up to two inches. The biplane bumpers, however, were more stylish than they were practical. They were easily damaged, difficult and costly to repair and were discontinued after only one year.

Despite a trend within General Motors toward fewer parts and greater interchangeability, the 1934 Cadillac line included a bewildering variety of wheelbase lengths and body choices. The Series 355-D V-8 line offered a choice of 25 body styles on no less than three separate wheelbases. There were six Fisher bodies on a 128-inch chassis; seven Fisher bodies on a 136-inch chassis and 12 Fleetwood bodies with "modified V" windshields on a 146-inch wheelbase. The V-8 engine was now equipped with aluminum piston and was uprated to 130 horsepower.

The V-12 Series 370-D line offered a choice of 18 Fleetwood bodies, all on the same 146-inch wheelbase. Six had straight, or flat, windshields. The other 12 had V-type windshields. The Cadillac twelve-cylinder engine was rated at 150 horsepower.

The redesigned 1934 Cadillac Series 452-D Sixteen rode on a huge, new 154-inch wheelbase chassis—the longest ever used on a production car. The restyled V-16 featured a tall, narrow grille design not unlike the one used on the style-setting new LaSalle.

A total of 18 Fleetwood semi-custom body styles were available on this spectacular, new chassis, six with flat windshields and 12 with the V-type windshield. At introduction in January, Cadillac grandly announced that it would build only 400 sixteen-cylinder cars for the entire 1934 model year.

The 1934 Cadillac was an extensively restyled and re-engineered car. The V-8 Series 355-D Cadillac's new streamlined styling included a sloped V-type grille, big bullet-shaped headlights and unusual twin-bladed biplane bumpers. The V-12 Series 370-D Cadillacs had identical exterior styling. The Fisher-bodied Two-Passenger Series 355-D Coupe was available on two wheelbases, a 128-inch model priced at $2,545 and a 136-inch version at $2,745. A Fleetwood Coupe on 146-inch wheelbase was included in the V-12 line with a $4,795 price tag. The spare tire on most 1934 Cadillacs was stowed out of sight in the trunk, although sidemounted spares were still available. Fisher coupes had flat windshields, Fleetwoods a V-type windshield.

The restyled Series 355-D Cadillacs for 1934 were built on three different wheelbases—128, 136 and 146 inches. Thirty-one body styles by Fisher and Fleetwood were available with flat and V-type windshields. This is the Series 355-D Convertible Coupe for Two Passengers by Fisher. The 128-inch model had a factory price of $2,645. The 136-inch model was priced an even $200 higher. A Series 370-D V-12 Convertible Coupe by Fleetwood cost $4,945. Wheel discs painted the same color as the body covered the wire wheels and a bright, chrome molding ran the length of the body. Doors opened from the front. Most of these convertibles had flat windshields. Fleetwood-bodied versions had the V-windshield shown here.

1934

Prospective purchasers were sternly advised to place their orders early if they wished to join this elite group. That advice turned out to be embarrassingly optimistic: only 60 Series 452-D Cadillacs were sold in 1934. Next year would be even worse. The ultra-luxury automobile market was dying. Basically the same engine that had been introduced in 1930, the V-16 was by now putting out 185 horsepower.

The Cadillac Aero-Dynamic Coupe, which had created quite a stir at the 1933-34 Chicago Century of Progress Exposition, went into limited production in 1934. Fleetwood built only 20 of these sleek two-door coupes between 1934 and 1937, powered by V-8, V-12 and V-16 engines. The V-16 was listed as Fleetwood Body Style 5899; the V-12 Style 5699 and the eight, Style 5799.

But the most changed of all Cadillac's 1934 models was the dramatically restyled new Series 350 LaSalle. Legend has it that the LaSalle was about to be dropped in depression retrenchment, but when Harley Earl showed the GM brass what he had in store for 1934, this decision was quickly reversed.

Styled by Jules Agramonte under the tutelege of GM styling chief Harley Earl, the new LaSalle's most striking styling feature was a tall, extremely slender grille. Large, bullet headlights were attached to stubby airfoils that sprouted from the sides of the radiator shell. Hood ventilators set into a convex horizontal panel consisted of five portholes on each side of the hood. The hood itself was long and handsomely tapered. Three chrome chevrons ornamented the leading edge of the pontoon front fenders. And, of course, there were those biplane bumpers. The LaSalle would retain its narrow, vertical radiator styling theme until the marque came to the end of the line in 1940. Bill Rader drove a LaSalle Convertible Coupe as the pace car for the 1934 Indianapolis 500-Mile Race.

Only four LaSalle body styles were offered, all on a shorter 119-inch wheelbase. LaSalle's 1934 prices dropped well below $2,000 for the first time placing this car in a new, lower price class. It was no longer a junior Cadillac. The new Series 350 LaSalle had Cadillac's "knee-action" independent suspension and boasted hydraulic brakes a full two years before Cadillac got them.

Even the LaSalle's engine was brand-new. Gone was the venerable "353" Cadillac V-8. In its place under the 1934 LaSalle's slim hood was a 240 cubic-inch straight-eight built to Cadillac specifications by sister GM division, Oldsmobile. This 95-horsepower engine would be used in the LaSalle through the 1937 model year and was the only in-line eight ever used in a Cadillac Motor Car Co. product.

Cadillac built 13,014 cars in the 1934 model year, 7,195 of them LaSalles. Despite the lingering depresssion, sales were beginning to pick up.

An era came to an end on June 1, 1934 when Nicholas Dreystadt succeeded the colorful Lawrence P. Fisher as Cadillac's chief executive. The title of President was dropped. Henceforth, the top man at Cadillac would have the title of General Manager. Fisher had presided over Cadillac's golden, classic age.

Across town, Chrysler Corporation introduced its radically styled, highly controversial Chrysler Airflow —a breakthrough in automobile design, but years ahead of its time. Although a superbly engineered car, the spectacular Airflow never sold very well. Chrysler's customers admired it but opted for more conventional styling.

The handsome Series 355-D Five-Passenger Town Coupe was offered only on the 128-inch wheelbase. Factory list price was $2,695. Body was by Fisher, and this stylish two-door model weighed 4,630 pounds. All 1934 V-8 Cadillacs sported a graceful new "goddess" hood ornament. The spare tire on this Town Coupe is mounted outside the car, behind the trunk.

The pleasingly-proportioned Series 355-D Town Sedan by Fisher was available on two separate wheelbases. The 128-inch model had a factory list price of $2,695. A 136-inch version cost $2,895 or exactly $200 more. For the last year, Cadillacs would have composition roofs with a fabric-covered center section. The sloped grille consisted of a series of thin vertical bars divided by five horizontal bars. A rectangular Cadillac emblem, which also denoted the number of cylinders under the hood, was mounted at the upper right.

Starting in 1934, Cadillac no longer referred to this body style as an All-Weather Phaeton. Instead, it was more accurately listed as a Convertible Sedan. The V-8 Series 355-D Convertible Sedan by Fisher was built on two wheelbases. The 128-inch model had a factory list price of $2,845. A 136-inch wheelbase version cost $3,045. Although the spare tire on most 1934 Cadillacs was now concealed in the trunk, this Convertible Sedan is equipped with dual sidemounts.

1934

Cadillac's sixteen-cylinder Aero-Dynamic Coupe created a real sensation at the General Motors exhibit at the 1933-34 Chicago Century of Progress Exposition. This ultra-streamlined five passenger coupe went into limited production in 1934. Fleetwood built only 20 Aero-Dynamic Coupes between 1934 and 1937, powered by V-8, V-12 and V-16 engines. This is the 1934 Cadillac Series 355-D Fleetwood Aero-Dynamic Coupe on 146-inch wheelbase chassis and powered by the 130-horsepower V-8 engine. The V-12 Series 370-D Aero-Dynamic Coupe used the same chassis. The sixteen-cylinder version rode on that series' huge, new 154-inch chassis.

The 1934 Cadillac Series 355-D eight-cylinder line included three conservatively-styled five-passenger sedans, two by Fisher and one by Fleetwood. The Fisher-bodied Five-Passenger Sedan shown was available on two separate wheelbases. The 128-inch wheelbase version cost $2,645 and the 136-inch wheelbase model $2,845. The Fleetwood Five-Passenger Sedan utilized the 146-inch wheelbase chassis and was considerably more expensive, at $3,495. The 128-inch wheelbase four-door sedan shown has light-colored wheel discs.

The restyled Series 370-D twelve-cylinder Cadillacs for 1934 were identical in appearance to the V-8 Series 355-D, except for the radiator emblem and nameplates on the sides of the hood. A total of 18 body styles were offered, all by Fleetwood and all mounted on the same 146-inch wheelbase V-12 chassis. Six had flat windshields and 12 used the "modified V" windshield shown on this 1934 Series 370-D Convertible Imperial Sedan. This five-passenger car weighed 5,800 pounds and had a factory list price of $5,195. New pontoon-type front fenders had a horizontal crease line on the leading edges. This car rides on wire wheels and for some reason sports a 1932 Michigan license plate.

The restyled Series 355-D Cadillac model lineup for 1934 also included a series of big seven-passenger sedans. The Fisher Seven-Passenger Sedan and a companion Seven-Passenger Limousine were built on the 136-inch wheelbase chassis. The Seven-Passenger Sedan sold for $2,995. The Seven-Passenger Imperial with glass division between front and rear passenger compartments had a factory list price of $3,145. A Fleetwood Seven-Passenger Sedan and Limousine were also available, on the 146-inch chassis, with a choice of either flat or V-type windshields.

The big, open touring car was a vanishing body style. This very impressive Fleetwood Seven-Passenger Touring Car does not appear on the 1934 Cadillac price list, and was likely a special-order model. The V-type windshield is chrome-plated and this car rides on chromed wire wheels. Wheelbase was 146-inches. At a glance, this car looks like a giant 1934 Ford Phaeton.

Cadillac's 1934 twelve-cylinder car line included six closed Fleetwood body styles with flat windshields. One of these was this sleekly-styled Fleetwood Five-Passenger Town Sedan, Style 6133-S, which had a factory list price of $4,245. Its four doors are hinged at the center. This model was also available as a "Special Fleetwood Town Sedan", Style 5733-S, with a V-type windshield. Price was $450 higher, at $4,695. Note how the trunk complements this four-door sedan's graceful roofline.

1934

This was certainly one of the most beautiful bodies ever designed for the Cadillac V-12 chassis. It is an elegantly-proportioned four-door Formal Sedan by Fleetwood. This body style has the appearance of a convertible sedan, but has a fixed, leather-covered roof. Other notable styling features include a V-type windshield and bright-plated window frames, chrome-plated piano-type door hinges and metal-covered sidemounted spare tires. This design includes some of the elements later incorporated into the style-setting Cadillac Sixty Special of 1938 and is a true Fleetwood custom creation.

Twelve of Cadillac's 1934 Series 370-D V-12 models had Fleetwood bodies with what was described as a "modified V" windshield. Ten of these were closed cars. This is the impressive Series 370-D Special Seven-Passenger Fleetwood Limousine, Style 5775, which sold for $4,995. A companion Seven-Passenger Sedan, Style 5775-S, was also available for $200 less. These big seven-passenger cars were built on the 146-inch wheelbase chassis. The example shown has sidemounted spare tires without the optional metal covers.

This is a 1934 Series 370-D Fleetwood Seven-Passenger Limousine with the flat-style windshield. It was listed as Style 6175 and had a factory price of $4,545. The Seven-Passenger Sedan model was Style 6175-S and cost $200 less. The car shown is equipped with sidemounted spare tires with metal covers.

The 1934 Cadillac chalked up another industry first when it tucked the spare tire out of sight inside the trunk. Side-mounted spare tires were still available for those who preferred them, and some 1934 Cadillacs had a rear-mounted spare. This is a Series 370-D Five-Passenger Sedan by Fleetwood which was available with either a flat or V-type windshield. Note the twin-bladed biplane rear bumper. It cost $4,645 and had a base weight of 5,735 pounds.

The 1934 Cadillac V-12 Series 370-D model line included five and seven-passenger Imperial Cabriolets with both flat and V-type windshields. This formal Fleetwood body style featured closed upper rear quarters and a covered roof. The flat-windshield five-passenger Imperial Cabriolet, Style 6130-FL was priced at $4,595. The seven-passenger variation cost $150 more and was Style 6175-FL. The V-windshield models were called "Special Imperial Cabriolets". Style 5730-FL was the five-passenger model and had a price of $5,045. The seven-passenger model, Style 5775-FL, cost $5,195. All had an identical 146-inch wheelbase. This is the flat-windshield, seven-passenger Style 6175-FL.

1934

Cadillac's fresh, new 1934 styling included a wide, angled V-type grille design consisting of a series of fine, vertical bars with five horizontal dividers. The grille texture was carried down over the valance panel at the bottom. Parking lights were designed into the headlight stanchions. The unique biplane bumpers consisted of two flat blades with big, bullet-shaped spacers. These bumpers deflected two inches on impact. These stylish bumpers proved too fragile, however, and were used only on Cadillac's 1934 models.

In 1935, Cadillac introduced a special commercial chassis designed and built for funeral car and ambulance manufacturers. Up until then, hearse and ambulance builders had to cut and splice conventional passenger car chassis to achieve the desired wheelbase length. This is a handsomely-styled Limousine Funeral Coach body mounted on a 1934 Cadillac V-8 Series 355-D chassis by the A. J. Miller Co. of Bellefontaine, Ohio. Note the rear-hinged doors and plated hinges. Miller introduced this curvaceous "beavertail" rear styling for 1933.

Among the most impressive of all the 16-cylinder Cadillacs built were a series of sporty four-passenger bodies mounted on the gigantic 154-inch wheelbase chassis used between 1934 and 1937. These two-door bodies emphasized the extreme length of the V-16 hood and the extraordinarily long rear deck. Five of these sleek Fleetwood stationary Coupes, Style 5876, were sold in 1934 at $7,750 a copy. This racy-looking model was available only with the V-type windshield. Weight was 5,900 pounds. Rather than use a rumble seat, this car was equipped with two tiny auxiliary seats behind the main bench seat.

At the top of the 1934 Cadillac V-12 model line were a series of three formal, open-front town cars. The Series 370-D Fleetwood Five-Passenger Town Cabriolet, Style 5712, had a factory price of $6,395. The Seven-Passenger Town Cabriolet, Style 5725, cost only $100 more and was the most expensive model in the V-12 line. Both had the wide V-type windshield and were built on the 146-inch wheelbase chassis. A Seven-Passenger Fleetwood Limousine Brougham with quarter windows, Style 5791, rounded out the line and had a $6,395 price tag.

Only two of these four-passenger Convertible Coupes were built on the 1934 Series 452-D sixteen-cylinder Cadillac chassis. Like the companion Stationary Coupe, this model was available only with the V-type windshield. Factory price was $7,900. Weight was a ponderous 5,900 pounds. A set of small auxiliary seats, behind the main seat, occupied the interior, while a trunk deck graced the rear panel.

1934

Just one of these big five-passenger Convertible Coupes was built during the 1934 model year on the Series 452-D sixteen-cylinder Cadillac chassis. The extreme length of the rear body behind the front doors resulted in attractive, true Convertible Victoria proportions, even on this gigantic chassis. Style 5885, this model sold for a steep $8,150.

Cadillac's experimental Aero-Dynamic Coupe created a great deal of excitement at the 1933-34 Chicago Century of Progress Exposition. By popular demand, this visually-stimulating five-passenger coupe was placed into limited production in 1934. Fleetwood built only 20 on V-8, V-12 and V-16 chassis between 1934 and 1937. Eight of these were Sixteens, three on the 1934 Series 452-D chassis. Designated Fleetwood Style 5899, the 1934 sixteen-cylinder Aero-Dynamic Coupe commanded a price of $8,150.

Shown in fore and aft views is this beautiful Convertible Sedan, Style 5880, with body by Fleetwood. Owned by Andrew Darling of Minneapolis, the auto appeared at the 1981 Midwestern Concourse at Fort Snelling, Minn. Especially suited to the imposing dimensions of the very long V-16 chassis, this four-door convertible style was no longer called an All-Weather Phaeton as it had been in the past. Standard equipment included an Imperial glass division between the front and rear compartments. This model is one of only five such Fleetwood Imperial Convertible Sedans built in 1934. In addition, one Style 5880-S was built. This was an identical car, except it had no Imperial division. When new, the Imperial version had a factory price of $8,150, while the Style 5880-S cost $200 less.

Two Series 452-D Five-Passenger Town Sedans were available. These included a flat-windshield Style 6233, and the V-windshield Style 5833-S illustrated. The straight-windshield model cost an even $7,000 while the V-windshield car was priced $650 higher. These renderings from the 1934 Book of Fleetwood do not include the three ornamental "spears" on the lower edges of the front fenders used on all production cars.

Like the rest of the 1934 Cadillac line, the 16-cylinder Series 452-D featured all-new streamlined styling highlighted by a tall, vertical grille of distinct design. A total of 18 Fleetwood bodies were available on the huge, 154-inch wheelbase Series 452-D chassis—six with the flat style windshield shown here and 12 with what Fleetwood called a "modified V" windshield. Only five of these imposing Series 452-D Seven-Passenger Sedans, Style 6275-S were built in 1934. Nine Style 6275 Seven-Passenger Limousines with Imperial division were also sold this year. The big Sedan weighed 6,190 pounds and had a factory price of $7,100. The Limousine cost $200 more and weighed 6,210 pounds.

1934

The 1934 Series 452-D line of sixteen-cylinder Cadillacs included five and seven-passenger Imperial Cabriolets. Style 5875-FL was listed as the Fleetwood Special Seven-Passenger Imperial Cabriolet, only two of which were built in 1934 at $8,150 a copy. The five-passenger version was Style 5830-FL and had a price of an even $8,000. Flat-windshield versions were also listed—Style 6230-FL being the five-passenger car and the seven-passenger Style 6275-FL.

This is the "modified V" windshield version of the 1934 Cadillac Series 452-D Seven-Passenger Sedan by Fleetwood. This model was listed as the "Special Seven-Passenger Sedan", Style 5875-S and had a factory price of $7,750—$650 more than the flat-windshield version. The companion V-windshield Special Seven-Passenger Limousine, Style 5875, was priced at $7,950, also $650 higher than the straight windshield model. Five-passenger Series 452-D sedans of this body style, with a choice of flat or V-type windshields, were also included in the 1934 Cadillac V-16 line.

For those who could still afford such luxuries, Fleetwood continued to offer a wide variety of special-order custom and semi-custom bodies for the Cadillac V-16 chassis. This is a Fleetwood Five-Passenger Town Cabriolet with collapsible rear quarters, Style 5812-C. Factory records indicate that none of these were sold during the 1934 model year. Note the extreme length of the clean V-16 hood, which was extended rearward to all but cover the cowl.

In a grandiose bid for exclusivity, Cadillac solemnly announced at the beginning of the year that it would build only 400 sixteen-cylinder cars for the entire 1934 model year and urged prospective customers to get their orders in early to avoid disappointment. This objective fell an embarrassing 340 short of that number, when a mere 60 Series 452-D Cadillacs were sold. Next year would be even worse. The most expensive of the 18 models in the 1934 sixteen-cylinder Cadillac line was the aristocratic Fleetwood Seven-Passenger Town Cabriolet, Style 5825, which sold for a very substantial $9,250. This classic, open-front formal town car weighed an equally impressive 6,390 pounds. The price, incidentally, was about four times the average working man's annual salary in this era—if work could be found at all.

At the top of the 1934 sixteen-cylinder Cadillac line, along with the Seven-Passenger Town Cabriolet, was the Fleetwood Seven-Passenger Limousine Brougham, another formal open-front town car with full windows in its rear quarters. Style 5891, the impressive Limousine Brougham carried a factory list price of $9,150.

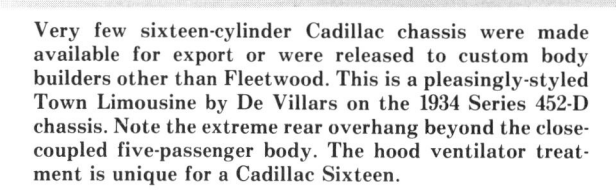

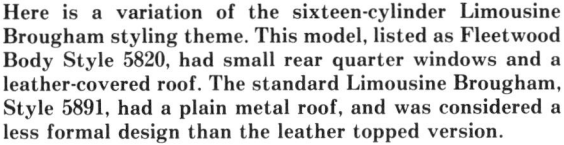

Very few sixteen-cylinder Cadillac chassis were made available for export or were released to custom body builders other than Fleetwood. This is a pleasingly-styled Town Limousine by De Villars on the 1934 Series 452-D chassis. Note the extreme rear overhang beyond the close-coupled five-passenger body. The hood ventilator treatment is unique for a Cadillac Sixteen.

Here is a variation of the sixteen-cylinder Limousine Brougham styling theme. This model, listed as Fleetwood Body Style 5820, had small rear quarter windows and a leather-covered roof. The standard Limousine Brougham, Style 5891, had a plain metal roof, and was considered a less formal design than the leather topped version.

Both Cadillac and Fleetwood stylists were given a free hand in designing proposed custom bodies for the 1934 16-cylinder chassis. One such design was this very clean Fleetwood Roadster, which would have been style number 5802, had any been built. The car features a radically swept-back windshield and a metal boot to totally conceal the top when folded. Stylized rear fender skirts and interestingly scalloped doors give the car an almost toy-like look. There is no record of this car ever having been built. Note the lack of running boards on this and the Phaeton.

Another Fleetwood proposal that was on the books but which drew no orders was the Phaeton or Sport Phaeton, style number 5859. This huge V-16 car utilized the same concealed top as found on the proposed Fleetwood roadster. It appears that the tonneau windshield was permanently fixed, and could not be lowered when the top was up. Both front and rear windshields appear identical. This car was on the drawing boards, and theoretically could have been ordered as a full-custom—had any buyers been found who wanted an ultra-expensive sport vehicle of this type.

Possibly designed as a parade car for official state functions was the Fleetwood 7-passenger Touring Car, style number 5857. This huge Series 452-D vehicle was of fairly conventional styling, except for the swept-back windshield. Obviously, no governmental bodies, either in the U.S. or overseas, desired to add this car to their fleets this year, as none were sold. And, the days of family touring in such open vehicles was gone forever, even if Fleetwood or Cadillac had found a family rich enough to afford such a car. Note that all three of these proposed but unbuilt cars utilized the same forward-opening front door design and the same windshield design without a horizontal upper bar.

1934

Everything about the 1934 LaSalle was new. From its tall, narrow radiator shape, portholed hood and thin-bladed biplane bumpers, the completely redesigned 1934 LaSalle was a styling sensation. Even the Cadillac V-8 engine was gone from under the new LaSalle's long, tapered hood. In its place was a straight-eight built for Cadillac by sister GM division, Oldsmobile. This is the 1934 LaSalle Series 350 Coupe, at $1,595 the lowest-priced car in the 1934 line. Designated Model 6376, it weighed 3,815 pounds and was advertised as a two-passenger car.

Only four body styles were offered in the totally-restyled and redesigned 1934 LaSalle line, all on a new, shorter 119-inch wheelbase chassis. All were by Fleetwood. This is the Series 350 Convertible Coupe, a two or four-passenger car depending whether or not the folding rumble seat was being used. Model 6335, it had a factory price of $1,695 and weighed 3,780 pounds.

Three of the four 1934 LaSalle models had identical price tags. The Convertible Coupe, Five-Passenger Sedan and the distinctive, new Club Sedan were all priced at $1,695. This is the formal-looking Series 350 Club Sedan, Model 6333-S, a four-door, five-passenger car which weighed 3,960 pounds. With its new pricing stance, LaSalle moved down into a lower, but broader, price class.

The narrow, vertical grille design of the boldly-restyled 1934 LaSalle would remain a distinctive LaSalle styling feature until the marque came to the end of the line in 1940. The new LaSalle's streamlined styling was pure Art Deco—from the twin-blade biplane bumpers, porthole hood vents and headlights mounted on airfoils extending from the sides of the radiator shell, to the three chrome chevrons on the leading edge of the front fenders. Even the wire wheels were covered by chromed discs. Cadillac Motor Car Co. General Manager Nicholas Dreystadt, right, who had just succeeded the legendary L. P. Fisher as Cadillac's top man, delivers a Series 350 Four-Door Sedan to an unidentified VIP customer.

The dramatically-restyled 1934 LaSalles were as advanced in their engineering as they were in appearance. Independent, knee-action front suspension was standard on all models and the Series 350 LaSalle boasted hydraulic brakes a full two years before Cadillac got them. This is the 1934 LaSalle Four-Door Sedan, Model 6330-S, which sold for $1,695—the same price as the companion Club Sedan and the Convertible Coupe. Weight was 3,960 pounds.

In an unprecedented departure for Cadillac, the all-new 1934 LaSalle was powered by a non-Cadillac engine—the only straight-eight ever used in one of the division's cars. The 240 cubic-inch, 95-horsepower in-line eight was built to Cadillac specifications by the Oldsmobile Division of General Motors and was used in the LaSalle through the 1936 model year. Bore was three inches and stroke four-and-a-quarter inches. This engine had 20 less horsepower and 113 fewer cubic inches of displacement than the Cadillac "353" V-8 previously used, but was a respectable performer nonetheless.

1935

Although Cadillac's 1935 models, from the LaSalle on up to the V-16, showed little outward change, all were given new series designations.

The Cadillac V-8 passenger car line for 1935 included the Series 10, 20 and 30 on three individual wheelbases. The Series 40 was the Cadillac V-12. The 1935 LaSalle was redesignated the Series 50, and the flagship Cadillac V-16 became the new Series 60. In some quarters, the V-8 was known as the Series 355-D and the Sixteen as the Series 452-D, the same designation the redesigned V-16 had carried the previous year. But henceforth, all Cadillac models would be identified by two-digit numerical series names, a practice that would persist into the mid-1960's.

Series 10 eight-cylinder Cadillacs for 1935 were built on a 128-inch wheelbase. The Series 20 utilized the 136-inch wheelbase chassis. Series 30 Cadillacs rode on a 146-inch wheelbase. All 1935 Cadillac V-12's — Series 40 cars — were built on the same 146-inch chassis. The mighty V-16 retained the giant 154-inch wheelbase chassis introduced the previous year.

Eight-cylinder Cadillacs for 1935 were offered with a wide choice of body styles by both Fisher and Fleetwood, with flat and V-type windshields. Fleetwood crafted 18 bodies — 12 with V-windshields and six with flat glass — for the V-12 and an identical model range for the Sixteen.

The big news at General Motors this year was the introduction of the one-piece, all-steel "turret top", a major breakthrough in body construction that did away with wooden roof slats and the fabric-covered center section.

Roofs for GM's new turret-top models were stamped out on huge, 500-ton presses. The all-steel, one-piece top was also considered an important advance in vehicle safety. Cadillac's V-12 and V-16 models did not offer this feature in 1935 but got it later.

Mechanical improvements in the 1935 Cadillacs included a new road stabilizer, which kept the body level when the car was making sharp turns or rounding highway curves. The 1935 LaSalle was equipped with a new variable-output generator that kept a constant, adequate charge on the car's battery.

All 1935 Cadillacs, including the LaSalle, had wide, new single-piece bumpers front and rear. Gone were the dainty two-bladed biplane bumpers that had been used only in 1934. The massive new bumpers weren't as stylish but they were a good deal more practical.

The 1935 LaSalle retained its 1934 styling, except for a switch to a V-type windshield on all models and new bumpers front and rear. The LaSalle's wheelbase was stretched one inch, to 120 inches, and the Series 50 LaSalles for 1935 were about an inch lower. The Oldsmobile-built straight-eight was uprated to 105 horsepower. The four-door Club Sedan was dropped, but a handsome Two-Door Touring Sedan was added

This attractive two or four-passenger Convertible Coupe was offered in four versions for 1935. The Series 10 Convertible Coupe by Fisher on 128-inch wheelbase was priced at $2,445; the Series 20 Fisher Convertible Coupe on 136-inch chassis listed for $2,645; there was a Fleetwood Convertible V-8 Coupe with V-windshield on 146-inch chassis, Series 30, for $4,045, and a Series 40 Fleetwood V-12 Convertible Coupe with V-type windshield at the top of the line for $4,745. Wire wheels were covered by discs and had eight-inch chrome hubcaps.

Except for new single-piece bumpers, Cadillac's V-8 and V-12 models for 1935 showed no exterior changes. All carried new series designations, however. The V-8 line continued on three separate wheelbases. This two or four-passenger Coupe was offered in four versions; Series 10 Fisher Coupe on 128-inch wheelbase at $2,345; Series 20 Fisher Coupe on 136-inch chassis at $2,545; Series 30 by Fleetwood on 136-inch chassis with V-windshield at $3,895, and the Series 40 Fleetwood V-12 on 146-inch wheelbase with "V" windshield at $4,595. Series 10 and 20 Coupes had the flat, or "straight" windshield shown here.

The 1935 Cadillac Five-Passenger Convertible Sedan was available in three V-8 versions and one V-12: The Series 10 Fisher Convertible Sedan on 128-inch wheelbase at $2,755; Series 20 Fisher on 136-inch chassis at $2,955; Series 30 Fleetwood Imperial V-8 Convertible Sedan on a 146-inch wheelbase at $4,295, and the V-12 Series 40 Fleetwood Imperial Convertible Sedan with its 146-inch wheelbase at $4,995. Series 30 and 40 models had a V-type windshield.

1935

The four-door Convertible Sedan was a sporty body style offered for 1935 with a choice of both flat and V-type windshields. Here is how this model looked with its top down. Cadillac supplied this striking white car with optional white sidewall tires and metal-covered sidemount spares for the use of Imperial Potentate Dana S. Williams at the 61st Annual Shrine Conclave in Washington, D.C. Cadillac's new General Manager, Nicholas Dreystadt, sits behind the wheel. With one foot on the running board is the division's General Sales Manager, J.C. Chick.

The 1935 Cadillac V-8 Town Coupe was offered in only one model, a Series 10 five-passenger two-door with Body by Fisher. The Town Coupe rode on a 128-inch wheelbase and utilized the flat, or "straight", windshield This two-door coach carried a $2,495 price tag and weighed 4,630 pounds.

Cadillac's 1935 model line included six five-passenger sedans, four with V-8 engines and two V-12's. In the eight-cylinder line there was the Series 10 Fisher Five-Passenger Sedan on 128-inch wheelbase at $2,445; Series 20 Fisher Sedan on the 136-inch chassis, $2,645; Series 30 Fleetwood with flat windshield on the 146-inch wheelbase at $3,295 and a "Special" Fleetwood Five-Passenger Sedan, also on 146-inch chassis with V-type windshield for $3,745. The Series 40 twelve-cylinder line included a Fleetwood Five-Passenger Sedan with flat windshield at $3,995 and a V-windshield version with a price of $4,445. Both V-12's were built on the 146-inch wheelbase chassis. The Series 10 eight shown weighed 4,715 pounds.

to the line. LaSalle prices were also reduced again, the Series 50 Coupe selling for only $1,255. Cadillac built 8,653 LaSalles, and LaSalle sales once more exceeded those of the more expensive Cadillac.

Sales of the prestigious V-16 continued to dwindle to insignificant levels. In 1935 they fell to their lowest point yet, with only 50 sold in the entire model year. The sixteen-cylinder Cadillac was now in its sixth year, but the ultra-luxury market for which it had been intended had all but vanished. Every one of these gigantic cars which came off the assembly line in Detroit must have cost General Motors a bundle. But the impressive V-16 was essential to maintaining Cadillac's image of prestige and engineering leadership.

In 1935, Cadillac got into the commercial car field in a big way. The division introduced a special, 160-inch wheelbase commercial chassis designed and built specifically for mounting hearse and ambulance bodies. This chassis was powered by Cadillac's V-8 engine. The funeral car and ambulance business would remain an important sideline at Cadillac for the next 45 years.

The biggest news in the automobile industry this year, however, was rival Packard's dramatic invasion of the medium price field with a new, smaller Packard called the 120. Powered by a 257 cubic inch, 110-horsepower straight eight, the new lower-priced Packard 120 was a stunning success with more than 25,000 sold in its first year on the market. There is no question that the introduction of the Packard 120 suppressed LaSalle sales and made incursions into the lower end of the Cadillac V-8 market. Some automotive historians contend that the arrival of the Packard 120 was the beginning of the end for Packard, and eventually killed off the LaSalle.

The pleasingly-proportioned 1935 Cadillac Five-Passenger Town Sedan was built in no less than six versions. These included the Series 10 by Fisher on 128-inch wheelbase for $2,495; Series 20 by Fisher on a 136-inch wheelbase for $2,695; Fleetwood Series 30 with flat windshield on the 146-inch chassis at $3,345; and the Series 30 with V-type windshield by Fleetwood for $3,795. There were also two V-12's, the Series 40 by Fleetwood with straight windshield at $4,045 and a V-windshield version with a $4,495 price tag. Fleetwood also offered a Style 5833-S Town Sedan on the Series 60 sixteen-cylinder chassis. This is the Series 10 eight.

1935

The largest cars in Cadillac's 1935 eight and twelve-cylinder model lineup were a series of seven-passenger sedans. At the top of the line were five and seven-passenger Town Cabriolets and a seven-passenger Limousine Brougham. The seven-passenger sedan line included a Series 20 Fisher Seven-Passenger Sedan on 136-inch wheelbase at $2,795, and a Fisher Seven-Passenger Imperial which sold for $2,945. The Series 30 line included a Fleetwood Seven-Passenger Sedan on a 146-inch chassis with flat windshield for $3,445 and a companion Seven-Passenger Limousine at $3,645. A Series 30 Fleetwood Seven-Passenger Sedan with "V" windshield listed for $3,895 and the Limousine for $4,095. The V-12 Series 40 Seven-Passenger Sedan of 140-inch wheelbase sold for $4,145, and the V-12 Seven-Passenger Limousine for $4,345.

Eighteen body styles, all by Fleetwood, were available on the V-12 Series 40 Cadillac chassis for 1935. Except for radiator and ventilator panel emblems, styling of the 1935 eights and twelves was identical. All V-12 models were built on the same 146-inch wheelbase chassis. This is the 1935 Cadillac Series 40 Convertible Coupe for two or four passengers, which was offered with V-type windshield only. Price started at $4,745 and in stock form this sporty model weighed 5,485 pounds.

The handsomely-styled 1935 Cadillac Series 40 twelve-cylinder Town Sedan was available with either a flat windshield or the increasingly-popular V-type. With straight windshield, the V-12 Series 40 Town Sedan for Five Passengers listed for $4,045. The Fleetwood Special Town Sedan, Style 5733-S with V-windshield, was considerably more expensive at $4,495. This body style was also available in the V-8 line. Note the big bullet taillights and center-hinged doors.

Fleetwood Style 5780 was an Imperial Convertible Sedan for Five Passengers on the twelve-cylinder Series 40 chassis for 1935. This V-windshield model with glass divider listed for $4,995. Like Cadillac's 1935 eight and sixteen-cylinder models, the V-12's were equipped with new, flat-style bumpers front and rear. This four-door model weighed an even 5,800 pounds.

1935

Cadillac formally entered the commercial car field this year with a special 160-inch wheelbase commercial chassis designed specifically for use by funeral car and ambulance builders. The long-wheelbase chassis utilized the "355-D" V-8 engine. In the funeral directing profession, the carved-panel hearse had made a dramatic comeback. The A.J. Miller Co. of Bellefontaine, Ohio built this richly-carved, six-panel draped hearse body with beavertail rear for the 1935 Cadillac commercial chassis.

The Cadillac Motor Car Division of General Motors published a special commercial chassis catalog for 1935, aimed at funeral directors and ambulance operators. This publication extolled the virtues of the new Cadillac long-wheelbase commercial chassis which was designed for the mounting of hearse and ambulance bodies. This is a limousine-style ambulance built by the Meteor Motor Car Co. of Piqua, Ohio on the 1935 Cadillac V-8 commercial chassis. From this point on, the funeral coach and ambulance business became an important sideline for Cadillac.

The prestigious Cadillac V-16, now in its sixth year, got a new model designation this year. Although still often referred to as the Series 452-D the 1935 Sixteen was officially listed as the Cadillac Series 60. The most noticeable exterior change was the replacement of the delicate biplane bumpers used in 1934 with more practical one-piece flat bumpers in 1935. This is the lithe-looking Series 60 Convertible Coupe for Two or Four Passengers, Fleetwood Style 5835, which sold for $7,700. A 2/4 passenger Stationary Coupe was also available, for $7,550. Cadillac V-16 prices were lowered by $200 this year, which in this price range really didn't mean all that much.

In addition to Cadillac-chassised hearses and ambulances, funeral directors usually owned several large sedans which were used as pallbearers' or family cars. In larger cities, limousine livery services maintained fleets of big sedans which they would rent out to undertakers in their area as needed. Starting in 1935, Cadillac actively cultivated this market with a series of special sedans aimed specifically at the funeral director and auto livery operator. These were called Cadillac Commercial Sedans, or Livery Limousines. They featured special interior upholstery and trim designed to stand up to hard public use. This is the 1935 Cadillac Special Livery Sedan, Fleetwood Body Style 6075-SL. Also available were two special Cadillac Livery Limousines, Styles 6075-L and 6075-LL.

All 1935 Cadillac V-16 models continued to use the huge, 154-inch wheelbase chassis introduced the previous year. Three open body styles were available, including two and five-passenger Convertible Coupes and a big Convertible Sedan. This is the 1935 Cadillac Series 60 Five-Passenger Convertible Coupe, Fleetwood Style 5885, only two of which were built this year. The elegant Victoria top nicely suits this very long chassis. Price was $8,150.

1935

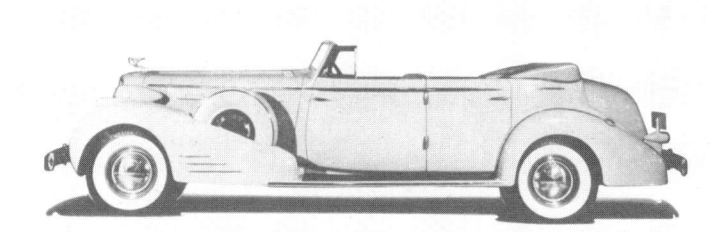

This was the largest production convertible built in the United States. The four-door Convertible Sedan body style seemed to suit the 154-inch Series 60 chassis especially well. In all dimensions and proportions, it was truly overpowering. Twenty-one of these impressive convertible sedans, Fleetwood Style 5880, were built between 1934 and 1937, only one of which did not have an Imperial division (Style 5880-S). Four of these big five-passenger cars were sold in 1935.

The 1935 Cadillac V-16 line included several monstrous five and seven-passenger sedans, all by Fleetwood. The Series 60 "Special Sedan for Seven Passengers", Style 5875-S, tipped the scales at a whopping 6,150 pounds and had a factory price of $7,550. The companion "Special Seven-Passenger Limousine", Style 5875 with Imperial division glass, cost $200 more. A mere 50 sixteen-cylinder Cadillacs were sold this year.

Although still referred to in some quarters as the Series 452-D (same as in 1934), the 1935 Cadillac V-16 was officially shown as the Series 60. The Series 60 sixteens and lower-priced Series 40 twelves continued to use composition roofs and did not have the all-steel "turret top". This is a catalog illustration of a 1935 Series 60 Town Cabriolet for Five Passengers, a formal, open-front town car. Fleetwood Style 5812, it was priced at $8,950 and weighed 6,310 pounds. Topping the Series 60 line for 1935 was a Seven-Passenger Town Cabriolet (Style 5825) and a Seven-Passenger Limousine Brougham, Style 5891. The Seven-Passenger Town Cabriolet was the most expensive model in the 1935 line, at $9,050.

For 1935, the LaSalle was redesignated the Series 50. All 1935 LaSalles sported a new V-type windshield. The wheelbase of 120 inches was one inch longer than the Series 350 LaSalle of 1934. Flat, single-piece bumpers replaced the unusual biplane bumpers used only on the 1934 models. Again, four body styles were offered, all by Fisher. The lowest-priced car in the line was the sleek Two-Passenger Coupe, which sold for $1,225. In standard showroom trim this model weighed 3,475 pounds.

All sixteen-cylinder Cadillacs of this era were huge cars. The five and seven-passenger sedans were almost sinister in appearance. They were certainly intimidating when parked next to lesser makes. This Series 60 Five-Passenger Sedan towers over the little English GM sedan next to it. Style 5830-S, the "Special Five-Passenger Sedan by Fleetwood" was priced at $7,400 and weighed 6,240 pounds. Note the massive, new flat-style bumper.

1935

At $1,325, the 1935 LaSalle Convertible Coupe for Two Passengers was priced only $100 higher than the lowest-priced car in the Series 50 line, the Two-Passenger Coupe. The well-equipped Convertible Coupe shown has white sidewall tires and metal-covered sidemount spares. This model was advertised as either a two or four-passenger car, depending on whether or not it was equipped with a rumble seat. Standard weight was 3,510 pounds, or 60 pounds heavier than the Coupe.

LaSalle's 1935 model lineup included an important, new body style. It was a Two-Door Touring Sedan for Five Passengers, an attractive two-door coach. Like most of GM's 1935 models, it featured the all-steel "turret top". The Two-Door Touring Sedan had a base price of $1,255 and weighed 3,620 pounds. Note the rear-hinged doors, sometimes referred to as "suicide doors". This model had a bustle-back trunk. The four-door sedan had a flat back.

LaSalle's sensational 1934 styling was carried almost unaltered into 1935. The dainty biplane bumpers were replaced with less attractive but far more practical single-piece flat bumpers, and all four 1935 models had stylish, new V-type windshields instead of the flat pane used in 1934. The Series 50 designation bestowed on LaSalles this year would be continued through 1940, when the fondly-recalled LaSalle came to the end of the line. This Convertible Coupe shows the front fender chevrons and LaSalle's classic thin, vertical grille design.

The 1935 LaSalle Series 50 Four-Door Touring Sedan had rear-hinged doors and GM's revolutionary new all-steel "turret top". Except for new, flat bumpers this five-passenger car was identical in appearance to the 1934 model. Price was $1,295. The stock Series 50 four-door sedan weighed in at 3,650 pounds. The Club Sedan of 1934 was discontinued, leaving this the only four-door model in the 1935 LaSalle line.

Viewed from the rear, the 1935 LaSalle Four-Door Sedan had a smooth, flat-back profile. Like the more expensive Cadillacs, the two and four-door Series 50 sedans—advertised as Touring Sedans—had two vertical bars in their rear windows. Note the dual trunk handles. The spare tire is carried in a compartment in the bottom of the trunk. The companion Two-Door Touring Sedan had a bustle-back trunk faired into the rear body sheet-metal. This streamlined body almost qualifies as a fastback.

1936

Nineteen Thirty-Six was another year of important product development for Cadillac. Once again, all Cadillac models (except LaSalle) were given new series designations.

The comprehensive 1936 Cadillac Motor Car Co. product line began with the Series 50 LaSalle. Next came an all-new, lower-priced Cadillac called the Series 60. The new Series 70 and 75 Cadillacs were V-8's. Series 80 and 85 Cadillacs were V-12's. The sixteen-cylinder Cadillac was renamed the Series 90—a designation that would remain until the V-16 was discontinued in 1940.

Perhaps the most important new product Cadillac took to market this year was the Series 60. This was an all-new, lower-priced Cadillac that was designed to compete, along with the LaSalle, with rival Packard's spectacularly successful 120 which had stormed the medium-priced field the previous year. The 120 brought the lustre and prestige of the Packard name within reach of a much larger segment of the car-buying public. Some automotive historians still insist that the introduction of a smaller, less-expensive Packard was the beginning of the end for the Packard Motor Car Co. But, like Cadillac, Packard had seen its principal market—that for large, expensive luxury cars—drastically reduced by the Depression. As a component of giant General Motors, Cadillac was in a much better position to weather the economic storm. But rival Packard was an independent, and the Packard 120 was a necessary survival step that undoubtedly preserved the company's future, at least for a while.

The Packard 120 was outselling the LaSalle by nearly five to one and was also taking sales from the lower end of Cadillac's V-8 line. The Series 60 Cadillac was brought in to strengthen Cadillac's representation in the medium-price market. Three models were offered, all with Fisher bodies on a new, short, 121-inch wheelbase. Series 60 Cadillacs shared bodies with the

Cadillac's new Series 60 price leader was designed to take on Packard's very successful 120. Built on a 121-inch wheelbase, the Series 60 was offered with three Fisher body choices. The Series 36-60 Touring Sedan had a factory price of $1,695 and weighed 4,010 pounds. The new Series 60 Cadillacs were powered by a 125-horsepower, 322 cubic inch version of Cadillac's all new monobloc V-8 engine.

For 1936, Cadillac again shuffled its model designations. Eight-cylinder models were now known as the Series 60, 70 or 75, depending on engine and wheelbase. The new low-line Series 60 was designed as Cadillac competition for Packard's extremely successful 120. Three Fisher bodies were offered on a new, short 121-inch Series 60 wheelbase. Series 60 Cadillacs were powered by a 125-horsepower, 322 cubic inch version of Cadillac's all-new monobloc V-8. The Series 60 Two-Passenger Coupe was the lowest-priced car in the 1936 Cadillac line, at $1,645. This body style was also available in a Fleetwood version in the Series 70 on 131-inch wheelbase, and as a V-12 Series 80, also on 131-inch wheelbase with Fleetwood coachwork. The Series 70 Coupe had a $2,595 price tag and the V-12 Coupe started at $3,295.

Cadillac's fresh 1936 styling include a new, narrow vertical grille design not unlike that used on the style-setting LaSalle. The new Cadillac grille featured a series of thin, horizontal bars. This is the Series 60 Five-Passenger Touring Sedan equipped with optional sidemounted spares. The sidemounts were half-buried in deep wells in Cadillac's new front fenders.

1936

smaller 1936 Buicks. Prices started at $1,645. LaSalle prices had been further reduced for 1936 and began at only $1,175.

For 1936, Cadillac had an all, new "third generation" V-8 engine. This 90-degree, L-head engine was smaller and more powerful than those used previously and was the forerunner of the modern, lightweight V-8 engines that would power Cadillacs for many years. The new V-8 was of monobloc construction—cylinder block and crankcase were a single iron casting—and featured downdraft carburetion. It was produced in two versions: Series 60 Cadillacs used a 125-horsepower, 322 cubic inch displacement version of 3-3/8-inch bore and 4-1/2-inch stroke. Series 70 and 75 cars had a more powerful 135-horsepower version that displaced 346 cubic inches. Stroke was identical, but the larger engine had a 3-1/2-inch bore.

Series 70 Cadillacs were built on a 131-inch wheelbase. Four Fleetwood body choices were offered. The Series 75 utilized a longer 138-inch chassis and offered a selection of 10 Fleetwood bodies. Series 80 Cadillacs were V-12's on the 131-inch wheelbase, with four Fleetwood body choices. Ten Fleetwood bodies were available for the Series 85 Twelve on 138-inch chassis.

Ten Fleetwood bodies were available on the generally-unchanged 1936 Cadillac Series 90 Sixteen, which continued on the gigantic 154-inch wheelbase chassis introduced for 1934.

The 1936 Cadillacs, advertised as the "Royal Family of Motordom", finally got hydraulic brakes, two years after LaSalle. Most models now had front-hinged doors. Cadillac's fresh, new styling included a narrow, vertical grille design not unlike that pioneered by the style-setting 1934 LaSalle. Twelve and sixteen-cylinder Cadillacs got the Unisteel "Turret Top" introduced on most GM cars for 1935.

LaSalle went into its third model year with little change. The porthole hood ventilators were gone, replaced by a series of horizontal louvers in a convex panel that ran the length of the hood. Again, only four models were offered, all on a 120-inch wheelbase. In addition to intense competition from Packard, the LaSalle got some additional competition from Ford's new Lincoln Zephyr. Introduced for 1936, the beautifully-styled Lincoln Zephyr was inspired by John Tjaarda's 1933 Chicago World's Fair show car. With prices starting at $1,275, the Zephyr was powered by a 110-horsepower V-12 engine of 267 cubic inch displacement. Wheelbase was 122 inches. Like the Packard 120, the Lincoln Zephyr was an instant hit with more than 15,000 sold in its first year on the market. Despite this strong competition, LaSalle sales jumped to just over 13,000 cars in 1936.

This would be the third and final year for the Oldsmobile straight-eight engine which had been used in the LaSalle since 1934.

Sales of sixteen-cylinder Cadillacs continued at low ebb. Only 52 Series 90 cars were sold in 1936, including three bare chassis supplied to outside coachbuilders. The flagship Sixteen was carried over from 1935 virtually without change. The 452 cubic inch V-16 was now in its seventh year. In April, 1936 Cadillac staged a special showing of V-16 models in the Cadillac-LaSalle showroom on the ground floor of the General Motors Building in Detroit. Four 1936 Series 90 Cadillacs were displayed in appropriately lavish surroundings. They included an Aero-Dynamic Coupe, a big seven passenger sedan, a limousine and a convertible sedan. These four cars had a combined value of more than $35,000—a fortune in 1936.

The industry was enjoying its best year since 1929. Cadillac sales doubled to 25,884.

To stimulate sales, the industry's new model announcement dates had been advanced two months into the fall, instead of early in January.

The 1936 Cadillac Series 70 offered a choice of four Fleetwood bodies on a 131-inch wheelbase. This is the Series 70 Stationary Coupe for two or four passengers, which had a base price of $2,595. This Fleetwood body style was also available as a Series 80 V-12, also on 131-inch chassis, with a factory delivered price of $3,295. Series 70 Cadillacs were powered by a new 135-horsepower, 346 cubic inch V-8 engine. The lower-priced Series 60 models used a 125-horsepower, 322 cubic inch engine. A rumble seat was standard equipment on this car.

One of the sportiest body styles in the much-changed 1936 Cadillac line was the two or four-passenger Convertible Coupe. This nifty body style was available as a Series 60 with a Fisher body on the 121-inch wheelbase at $1,725; as a Series 70 on the 131-inch chassis with a Fleetwood body, at $2,695 or as a V-12 Series 80 by Fleetwood, also on the 131-inch wheelbase with a $3,395 price tag. The Series 36-70 Convertible Coupe shown, with rumble seat enticingly open, is configured for four passengers and is equipped with optional white sidewall tires and sidemounts.

1936

Series 75 Cadillacs for 1936 were offered with a selection of 10 Fleetwood bodies, all on a 138-inch wheelbase. This is the Series 75 Convertible Sedan, a five-passenger car which sold for $3,395. The Series 75 utilized the same 135 horsepower, 346 cubic inch displacement V-8 engine used in the smaller Series 70 Cadillacs. This model did not have an Imperial glass division between the front and rear seats.

Cadillac's comprehensive 1936 model lineup included two series of V-12 cars. Series 80 Twelves included four Fleetwood body styles on a 131-inch wheelbase. Series 85 Cadillac V-12's offered a choice of no less than 10 Fleetwood bodies on a 138-inch chassis. This is the luxurious Series 85 Fleetwood Convertible Sedan with Imperial division, which sold for $4,095. Four convertible sedan models were available this year, including a V-8 Series 70 on 131-inch wheelbase with no glass division, with a factory price of $2,745; a V-8 Series 75 on 138-inch wheelbase for $3,395; the V-12 Series 80 on the 131-inch chassis at $3,445, and the Series 85 Twelve shown here.

In addition to a wide selection of standard body styles, Cadillac offered a series of special-order custom and semi-custom bodies by Fleetwood. This unusually attractive 1936 Cadillac Five-Passenger Touring Car does not appear on Cadillac's standard price list for this year. It appears to be a Series 80 V-12 on the 131-inch wheelbase. Note the large bullet-shaped taillights used from 1934 through the 1936 model year.

Five-passenger Town Sedans were offered only in the 1936 Series 75 and 85 Cadillac lines. Both rode on long 138-inch wheelbases and carried Fleetwood four-door bodies. The V-8 Series 75 Town Sedan had a factory list price of $3,145. The V-12 Series 85 Town Sedan cost $700 more, or $3,845. Cadillac's sixteen-cylinder Series 90 line for 1936 also included a Town Sedan, Fleetwood Style 5833-S. The Town Sedan's doors opened from the center, and this body style featured "blind", or windowless, upper rear quarters.

This is another Fleetwood special. It is a sleek five-passenger Coupe, probably a V-12 Series 80 on the 131-inch wheelbase. This car was photographed in Cadillac's factory garage. It lacks the metal cover for the sidemounted spare tire. Cadillac's eight and twelve-cylinder models for 1936 were externally identical except for the V-8 or V-12 emblems at the front of the hood ventilator panels.

1936

Cadillac's 1936 model offerings included two of these smart five-passenger Fleetwood Formal Sedans. The V-8 Series 75 version, Fleetwood Style 7519-F, had a base price of $3,395 and weighed 4,805 pounds. The Series 85 Twelve, Fleetwood Style 8519-F, weighed 5,115 pounds and had a factory-delivered price of $4,095. Both were built on the same 138-inch wheelbase chassis. The Series 75 Formal Sedan shown has a smooth back, padded, leather-covered roof, and external, folding luggage rack.

Five-passenger Touring Sedans were offered in Cadillac's Series 70 and 75 for 1936. The Series 70 Touring Sedan by Fleetwood utilized the 131-inch wheelbase chassis and had a factory price of $2,445. The Series 75 variation, on 138-inch chassis, was priced $200 higher, at $2,645. Both were V-8's. The Series 70 Touring Sedan shown weighed 4,670 pounds with the standard single spare residing in the trunk.

This well-equipped 1936 Cadillac Touring Sedan for Five Passengers sports dual sidemounted spare tires with metal covers and white sidewall tires, both popular Cadillac options. Cadillac's new Series 70 and 75 eight-cylinder models were powered by an all-new, "third-generation" monobloc V-8 engine. Cylinder block and crankcase were cast in one piece. This extremely efficient engine was built in two versions—a 125-horsepower engine of 322 cubic inch displacement for the Series 60 and a 135-horsepower, 346 cubic inch version for the Series 70 and 75.

Except for insignia, Cadillac's twelve-cylinder models for 1936 were virtually indistinguishable from the V-8's. This is a V-12 Series 85 Five-Passenger Sedan by Fleetwood on the long 138-inch wheelbase. The example shown has a smooth back and no trunk. A bustle-back trunk version was also available. All 1936 Series 85 Cadillac Twelves used the 138-inch chassis. This car has no sidemount spare cover.

This is what Cadillac's V-8 and V-12 standard sedans looked like when viewed from the rear. The backlight, or rear window, has two vertical dividers. In a General Motors program aimed at fewer parts and greater interchangeability between its various automotive divisions, some 1936 Buicks and Cadillacs shared the same bodies. This five-passenger Cadillac Touring Sedan was photographed on the roof of the GM Building in Detroit. Note that this car has been fitted with special chromed headlight shells. This was probably a styling experiment, as chromed shells were not listed in Cadillac's accessory book this year.

1936

Cadillac sedans with bustle-back trunk were listed as Touring Sedans. Fleetwood offered four Five-Passenger Touring Sedans on Cadillac chassis for 1936. These included a Series 70 on the 131-inch wheelbase which sold for $2,445; Series 75 on the 138-inch chassis at $2,645; Series 80 on the 131-inch wheelbase at $3,145 and a Series 85 Touring Sedan on the 138-inch wheelbase for $3,345. Series 70 and 75 cars were V-8's. Series 80 and 85 were Twelves. This car looks very dressy with its chrome hubcaps and wheel covers and wide whitewall tires.

Cadillac's senior eight and twelve-cylinder sedans could be ordered with or without a conventional bustle-back trunk. Some owners still preferred the old-style folding luggage rack on which a portable trunk was carried when touring. This is the 1936 Cadillac Fleetwood Seven-Passenger Imperial, a limousine offered in Series 75 and 85 on the 138-inch chassis. The 75 was a V-8 and the 85 a V-12. The Series 75 Fleetwood Imperial had a factory price of $2,995 and weighed 5,045 pounds. The V-12 Series 85 variant cost $3,695 and tipped the scales at 5,230 pounds.

Some of Cadillac's premium models bore catalog names that were nearly as long—and impressive—as the cars themselves. This, for instance, is the 1936 Cadillac Series 85 Twelve Seven-Passenger Touring Imperial by Fleetwood, a 138-inch wheelbase limousine with the conventional bustle-back trunk. The Series 85 Touring Imperial had a factory price of $4,445. An eight-cylinder Series 75 version was also available for a little less, or $3,695. Note the V-12 emblem on the upper right-hand side of the grille.

Big seven-passenger Fleetwood Touring Sedans were offered in the 1936 Series 75 and 85. The eight-cylinder Series 75 Fleetwood Touring Sedan for Seven Passengers had a factory list price of $2,795. The V-12 Series 85 version cost $3,495. Both of these models were on a 138-inch wheelbase and had the standard bustle-back trunk. Sidemounts were and added-cost accessory.

The market for ostentatious automotive luxury had been almost obliterated by the Depression, but there were still a few wealthy customers out there who insisted on exclusivity and conspicuous dignity. Consequently, Fleetwood continued to turn out a small number of formal, open-front town cars even during the worst years of the Depression. Things were looking up by 1936, and the market for individual custom and semi-custom bodies perked up a bit. This is a 1936 Cadillac Series 85 Twelve-Cylinder Fleetwood Town Cabriolet for Seven-Passengers, Style 8543, a 138-inch wheelbase car that sold for $5,145. An eight-cylinder version was also offered as a Series 75 for $4,445. It was Fleetwood Style 7543. This impressive V-12 had a padded rear roof, no trunk, but folding luggage rack.

1936

This Fleetwood semi-custom was not shown on Cadillac's standard 1936 price list. It is an attractive open-front Limousine Brougham, a once-popular formal body style that by now had all but disappeared. The weatherproof top has been buttoned into place over the chauffeur's compartment, and this car has a conventional bustle-back trunk. The chassis is a twelve-cylinder Series 85 on 138-inch wheelbase.

The 1936 Cadillac was a thoroughly-redesigned car. Styling included a narrow, vertical grille not unlike that on the LaSalle, new bodies and fenders. Headlights were mounted on stalks growing out of the sides of the radiator shell instead of on stanchions. Mechanical improvements included hydraulic brakes (two years after LaSalle got them) and an all-new, flathead monobloc V-8 engine. This shot shows the new grille with its fine horizontal bars. The same grille was used on both V-8 and twelve-cylinder models.

Cadillac's public relations program frequently involved loaning out appropriate cars for VIP's and special functions all over the country. This metallic silver 1936 Cadillac V-8 sedan was the official Silver Anniversary Car marking the 25th birthday of the Dixie Highway. The official sedan was photographed in the Michigan snow outside the Detroit plant before it headed south for the party.

The sixteen-cylinder Cadillac entered its third consecutive model year with no significant appearance or mechanical changes. But the prestigious V-16 was given a new series designation for 1936—Series 90. All Cadillac Sixteens were impressive cars, but the Series 36-90 Convertible Sedan was especially stunning. Only six of these majestic four-door convertibles, Fleetwood Style 5880, were built this year. List price was an equally impressive $7,850. A movable glass partition continued as standard equipment on this model. Without the Imperial division, it was shown as Style 5880-S. This 1936 Cadillac Series 90 Convertible Sedan was photographed in a showroom setting.

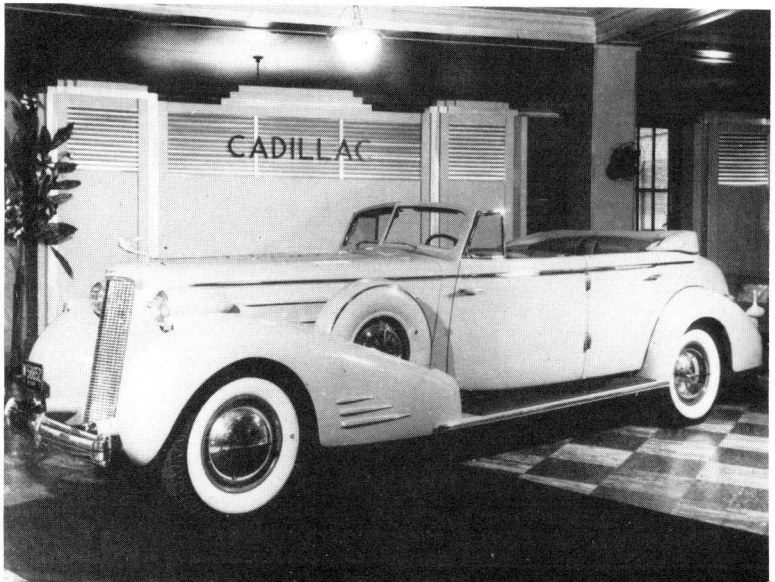

Cadillac continued to export a small number of chassis to Europe each year. Vanden Plas of Belgium designed and built this rakish Convertible Victoria body for the 1936 Cadillac V-8 chassis. Note the very heavy chrome molding which follows the contour of the rather long rear body. Barrel-type hinges are chrome-plated.

1936

In April, 1936, Cadillac staged a special salon exhibition of its sixteen-cylinder models in the Cadillac-LaSalle Showroom in the General Motors Building in Detroit. Four Series 36-90 Cadillacs—representing a good chunk of that year's V-16 production—were featured in this special display. One of these was the cream-colored Fleetwood Convertible Sedan being demonstrated here. The others included an Aero-Dynamic Coupe, a Seven-Passenger Sedan and a big Series 90 Limousine. Only 52 sixteen-cylinder Cadillacs were built this year, including 49 completed cars and three chassis supplied to outside body builders.

Only 10 Fleetwood body styles, all with V-type windshields, were offered on the Series 90 sixteen-cylinder chassis for 1936. All rode on the same tremendous 154-inch wheelbase. This is the Series 36-90 Seven-Passenger Limousine, Fleetwood Style 5875, which carried a list price of $7,550. Half of the Cadillac V-16's built this year were of this style. A companion Fleetwood Seven-Passenger Sedan, Style 5875-S, was priced $200 less. Two other seven-passenger cars in the V-16 lineup this year included the Imperial Cabriolet, Style 5875-FL ($7,850) and a Seven-Passenger Town Cabriolet, Style 5825, the most expensive model in the entire 1936 Cadillac line at $8,850. No flat-windshield models were offered this year.

While the lingering Depression had all but killed off the custom body business and was taking its toll of independent luxury car manufacturers, Fleetwood continued to offer Cadillac's customers a series of special order custom and semi-custom bodies for the V-16 chassis. Very few, however, were actually built. One of these was this special Seven-Passenger Town Cabriolet. This severely-formal, open-front town car, Fleetwood style 5725, had a blacked-out grille, painted instead of chrome hupcabs and wheel covers, and very little bright exterior trim. Only one was built at a cost of $8,500. The roof luggage rack was similar to those common on large cars in Europe at the time but rare in America. Note the smooth trunkless back and center-hinged doors.

Cadillac carefully controlled who could mount special or custom bodies on its premium sixteen-cylinder chassis, limiting this privilege almost exclusively to GM's own Fleetwood Body Division. This year, only three V-16 chassis were shipped to outside coachbuilders. Brunn and Company of Buffalo, N.Y., a firm better known for its custom creations for the Buffalo-built Pierce-Arrow, and the competitive Lincoln, created this striking open-front town car on the 1936 Cadillac Series 36-90 chassis. Cadillac continued to refer to this body style as a Transformable Brougham or Town Cabriolet, although this very formal, razor-edged style was also sometimes called a Panel Brougham. The result is a pleasing contrast in curved and straight body lines. Despite the extreme chassis length, it works. Elimination of the usual sidemounted spares emphasizes the impressive frontal dimensions of this car.

Here is a full-custom body under construction in the Fleetwood Body plant in Detroit. This would appear to be the only Style 5725-C Town Landau built this year. It was an imperial landaulet design with a fully-collapsible rear quarter. Fleetwood custom and semi-custom bodies received a great deal of meticulous hand-finishing with teams of experienced bodybuilders assigned to one-off or special bodies for days at a time. This one is still in the body-in-white stage prior to preparation for painting. An automotive trade journal of the day said this car cost its discerning owner an incredible $17,000.

1936

The new Cadillac commercial chassis was enthusiastically received by special body builders. Most of these extra-long wheelbase V-8 chassis were fitted with hearse or ambulance bodies, but some interesting other applications were to be found, too. A prominent funeral car manufacturer, The Eureka Co. of Rock Falls, Ill., built this special 1936 Eureka-Cadillac Police Patrol wagon for the City of Binghamton, N.Y. Note the screened side windows and rear handrails. Needless to say, this sinister conveyance was painted black—as in Maria!

The Eureka Co. used some colorful names for its funeral cars and ambulances. Premium hearses were Eureka Chieftains and ambulances were Blackhawks. This is a 1936 Eureka Blackhawk limousine-style ambulance on the Cadillac commercial chassis. The body was finished in a cream color with black fenders and top. Note the double belt molding and red flashing lights atop the front fenders.

Exactly 98 years after it was founded, the James Cunningham, Son & Co. of Rochester, N.Y., long a respected builder of fine carriages, then individual custom motor cars, ceased vehicle production in 1936. When the market for its very expensive custom-built passenger cars vanished in the Depression, Cunningham had to rely more and more on its hearse and ambulance business for survival. The company also built some town car bodies for the Ford V-8 chassis. But not even this helped. While the very high quality of its products was never questioned, Cunningham trailed its competitors when it came to styling. This 1936 Cunningham-Cadillac Style 374-A carved-panel hearse looks like a horse-drawn body that has been transplanted onto the new Cadillac Series 75 commercial chassis, but it is not: Cunningham actually built them this way. Cunningham is still in business today in upstate New York, in the electronics field.

Cadillac continued to offer a series of special sedans for funeral directors and limousine livery operators. These commercial sedans were trimmed and upholstered specifically for rental use. This is a 1936 Cadillac Commercial Seven-Passenger Sedan by Fleetwood. The interior was upholstered in broadcloth and whipcord. The car shown carried an exterior rear-mounted spare, and appears to be the only Cadillac style to do so this year.

In addition to the Seven-Passenger Commercial Sedan, Cadillac offered the funeral service trade a special Commercial Seven-Passenger Touring Limousine. This Fleetwood-bodied model had a conventional trunk and Imperial division for privacy for the rear-seat passengers. Larger funeral firms and limousine livery operators would rent cars like this out to funeral directors on an as-needed basis.

1936

The 1936 LaSalle, officially Cadillac's Series 36-50, showed little exterior change from the 1935 models. The porthole vents were gone, replaced by a cleaner-looking linear ventilator treatment. The attractive front fender chevrons remained for the third straight year. The lowest-priced car in the 1936 model line was the LaSalle Series 50 Stationary Coupe, still a most attractive car which sold for only $1,175. This model was sold as a two or four-passenger car, depending on whether trunk deck or rumbleseat version was ordered. Since no step-plates are visible, it is assumed that this is the two-passenger version.

The most expensive model in the 1936 LaSalle Series 50 line was this Convertible Coupe, a two or four-passenger car that sold for $1,255 in standard form. All 1936 LaSalles rode on the same 120-inch wheelbase used in 1935. The example shown sports metal-covered sidemounted spare tires buried deeply in the new front fenders. This would be the third and final year for LaSalle's Olds-built straight-eight engine.

LaSalle sales jumped to more than 13,000 in the 1936 model year, but continued to trail far behind those of the spectacularly successful Packard 120 which had invaded the medium-priced car field the previous year. The new, smaller Packard, which brought Packard prestige within the reach of people who had heretofore only dreamed of ever owning one, was outselling the LaSalle by nearly five to one. This is the Fisher-bodied Series 36-50 LaSalle Two-Door Touring Sedan which sold for only $1,185. The five-passenger coach sports optional white sidewall tires.

This 1936 LaSalle Touring Sedan has been turned into a rolling billboard. The automobile was a popular novelty advertising medium through the 1930's. LaSalle features extolled include knee-action shock absorbers, GM's Uni-steel turret top, a "peak-load" generator, luxurious interiors, low operating cost per mile and an advertised as-equipped price of only $1,175. LaSalle at this time was fighting a losing sales battle with Packard's extremely successful 120 which had stormed the medium-priced car market the previous year.

The 1936 LaSalle line continued with only four body styles, all by Fisher on a 120-inch wheelbase. All of GM's smaller sedans, including the 1936 Cadillac and LaSalle, featured very round side window openings. This is the 1936 LaSalle Series 36-50 Four-Door Touring Sedan, a five-passenger car which sold for $1,225. This would be the final year for the Olds-built straight-eight engine used in the LaSalle since the 1934 model year.

1937

This was to be a record year for Cadillac. Paced by record-breaking sales of its restyled LaSalle, Cadillac sold 46,152 cars during the 1937 model year, nearly double its 1936 sales which in turn were twice as high as they had been in 1935. It looked like the Great Depression was finally over. Prosperity had returned to what was left of the industry.

For 1937, Cadillac offered its customers seven different series of cars which ranged in price from just over $1,000 ($995 at one point) to more than $9,000. These included the handsomely restyled Series 50 LaSalle; the V-8 Series 60 Cadillac; an all-new, single model Series 65 Cadillac V-8; Series 70 and 75 Cadillac Eights; the V-12 Series 85, and the sixteen-cylinder Series 90.

The 346 cubic-inch-displacement monobloc V-8 engine introduced for 1936 was made standard in all 1937 Cadillac V-8 models. Horsepower remained at 135. The companion 125-horsepower, 322 cubic inch derivative of this engine which had been used in the 1936 Series 60 Cadillac was relegated to the LaSalle, finally replacing the Oldsmobile-built straight-eight which had powered all LaSalles built between 1934 and 1936. This would also be the last year for the Cadillac V-12 and the 90-degree, 452 cubic inch Cadillac V-16 engine.

Only 211 sixteen-cylinder Cadillacs were produced during the 1934-1937 model run, averaging a mere 50 sales a year. Although annual changes were minimal, each of these huge cars must have cost General Motors a bundle to build. However, they more than preserved Cadillac's public image of engineering leadership and prestige, and for these reasons alone have to be considered successful. There is little question that all of the other cars in the Cadillac line carried the mighty Sixteen. No independent manufacturer could have afforded them.

The 1937 LaSalle's fresh styling included a refined vertical grille design, massive new front fenders and a much stronger horizontal hood louver treatment. The big, bullet-shaped headlights were repositioned to give the car a lower look. Prices were LaSalle's lowest ever and Cadillac promoted the 1937 LaSalle as "completely Cadillac-built."

The Series 37-50 LaSalle's wheelbase was increased four inches to 124 inches, and the chassis frame was lowered and stiffened. A new body style—a very attractive Fisher four-door convertible sedan—was added to the 1937 line making five LaSalle body styles in all. LaSalle's strong performance image was further strengthened when Ralph de Palma paced the 1937 Indianapolis 500-miles Race in a LaSalle Convertible Coupe. De Palma also put the LaSalle through a 500-miles-plus durability run on the famed Brickyard. LaSalle sales rocketed to an all-time record 32,005. Despite its best sales year ever, the LaSalle was still handsomely outsold by Packard's seemingly invincible

The 1937 Cadillac was an extensively changed car inside and out. All 1937 Cadillac V-8's were powered by the 346 cubic inch, 135-horsepower engine that had been used in the 1936 Series 70 and 75. Series 60 models rode on a four-inch-longer 124-inch wheelbase. This is the 1937 Cadillac Series 60 Coupe, one of the four Fisher body styles offered in Cadillac's lowest-priced series. Model 6027, this attractive coupe with angled "B" pillar had a factory delivered price of $1,655.

For 1937, all of Cadillac's eight and twelve-cylinder models were given fresh, clean styling. The new look included a die-cast "eggcrate" grille and shapely bullet-formed front fenders. This is the sporty Series 37-60 Convertible Coupe, one of four models in the 60 Series. This two-passenger convertible had a factory price of $1,790 and weighed 3,745 pounds. Front and rear bumpers were also new and incorporated the Cadillac crest at their centers.

This elevated view of the 1937 Cadillac Series 60 Four-Door Touring Sedan shows Cadillac's fresh, new exterior styling. Note especially the large, redefined bullet-formed front fenders and elongated bullet headlights mounted high on the hood. Like the grille, the finely-detailed hood vent panels were also die-cast. This five-passenger sedan, on 124-inch wheelbase, had a factory delivered price of $1,760 and weighed 3,845 pounds.

1937

One-Twenty and the all-new six-cylinder Packard Model 115.

The 1937 Cadillacs got fresh, new styling including a new die-cast "eggcrate" grille and big, bullet-formed front fenders. Bumpers were new, too, with the Cadillac coat-of-arms at their centers. Series 60 and 65 Cadillacs had full-length, die-cast ventilator panels just below the chrome hood molding. Series 70, 75 and 85 models had much smaller hood vents positioned at the lower rear of the hood with four small horizontal trim pieces. On cars equipped with optional side-mounted spares, the hood vents were almost completely hidden making for a very clean frontal appearance.

There were four Fisher-bodied Series 60 Cadillacs, all n a 124-inch wheelbase that was four inches longer than the previous year. The only model offered in the new Series 65 was a Fisher four-door Touring Sedan on a 131-inch wheelbase.

Cadillac's Series 70 model lineup included four Fleetwood body styles on the 131-inch wheelbase chassis. A total of 11 Fleetwood body styles were available in the popular Series 75, all on the same 138-inch wheelbase. Six Fleetwood body choices were offered for the Series 85 Twelve, also on the 138-inch chassis. Except for grille and decklid insignia, Series 75 and 85 Cadillacs were virtually indistinguishable. This would be the seventh and final year for the 370 cubic inch, 90-degree Cadillac V-12 engine which had been introduced for the 1931 model year. Sales of V-12 Cadillacs had tapered off to nearly nothing: those who could afford this kind of automotive luxury invariably opted for the more prestigious and much more distinctive Cadillac V-16.

This year would also be the eighth and last year for the 452 cubic inch "first generation" Cadillac sixteen-cylinder engine. Eleven Fleetwood body styles were available for mounting on the huge, 154-inch wheelbase Series 90 chassis for 1937. The big Sixteen's styling had not been changed (except for bumpers) in four years. The Series 37-90 Cadillac Sixteens were identical in appearance to those introduced for 1934. Again, only 50 were built. Industry observers were confident that this would be the V-16's last year. The ultra-luxury market was gone. There was simply no need for such an ostentatious and expensive car.

Packard, which now dominated the medium-priced field with its extremely popular One-Twenty had intensified its grip on this segment of the market with the introduction of a six-cylinder junior Packard, the 115, for 1937. This car, too, was an immediate success.

Cadillac broadened its model range this year with the addition of an all-new Series 65. This series consisted of only one body style, a Fisher Four-Door Touring Sedan. The new Series 65 Cadillac rode on a 131-inch wheelbase, the same as the Series 70. Designated the Model 6519, it was priced at $2,190. This Series 37-65 five-passenger sedan was photographed on the roof of the GM Building in Detroit—a secure location for shooting prototypes. Like all 1937 Cadillac Eights, it was powered by the 135-horsepower, 346 cubic inch engine introduced the previous year.

Like the lower-priced Series 60, the 1937 Cadillac Series 70 line included four body styles. All were by Fleetwood. Series 70 Cadillacs were built on a 131-inch wheelbase. This is the handsome Series 37-70 Fleetwood Sport Coupe, a two-passenger model with a factory delivered price of $2,905. These folks apparently are about to embark on a spin around the roof of the General Motors Building in Detroit. The cars and models were later superimposed on more rustic backgrounds for advertising purposes. The Series 70 Coupe was Model 7057.

One of the prettiest cars in the entire 1937 Cadillac line was this impressive Convertible Sedan. This is the Series 37-60 Convertible Sedan by Fisher, a 124-inch wheelbase car that sold for $2,120. This well-equipped example has optional whitewall tires and metal-covered sidemount spares. A similar model joined the 1937 LaSalle line this year, also by Fisher. Fleetwood offered four-door convertible sedans in the 1937 Series 70 and 75 eight-cylinder Cadillac lines, and as a Series 85 Twelve. The Series 70 convertible sedan on 131-inch wheelbase had a factory delivered price of $3,060; the Series 75 on 138-inch chassis, $3,730 and the V-12 Series 85, also on 138-inch wheelbase, $4,450.

Series 70 Cadillacs had a different hood ventilator treatment than the lower-priced Series 60 and 65 models. Series 70, 75 and 85 Cadillac hoods had four short chrome streamers covering small vents at the lower rear of the cleanly-styled hood. Series 60 and 65 cars had larger die-cast vent panels mounted higher on the hood. This is the Series 37-70 Convertible Coupe by Fleetwood, Model 7067, which had a factory-delivered price of $3,005. All Series 70 Cadillacs utilized the 131-inch wheelbase V-8 chassis.

1937

This gentleman is about to enter the 1937 Cadillac Series 70 Four-Door Touring Sedan, a five-passenger car which sold for $2,695. Designated Model 7019, the car shown has optional white sidewall tires. This profile view shows the clean hoodline and stylish new bullet-formed front fenders. Note the concentric rings on the chrome wheel covers.

Shown with its massive hood raised is this beautifully restored 1937 Cadillac Series 70 Convertible Sedan with a body by Fleetwood. The very desirable car is owned by Odd Braathen of Minneapolis and is here attending the 1981 Concourse at Fort Snelling, Minn. This body was available in either fast-back style or in the more popular bustle-back trunk form seen here. Designated Model 7029, the Series 37-70 Convertible Sedan had a factory price of $3,060. The optional dual sidemounts and whitewall tires enhance its appearance.

Cadillac's Series 75 for 1937 offered the buyer a choice of no less than 11 body styles, all by Fleetwood and all on a long 138-inch wheelbase chassis. This is a 1937 Cadillac Series 75 Five-Passenger Sedan, Style 7519, which sold for $2,915. The car shown has a smooth back and no built-in trunk. Models incorporating the bustle-style trunk were listed as Touring Sedans: without, they were just Sedans. Note how deeply the sidemounted spare is buried in the new style front fender.

This elevated three-quarter rear view shows the standard bustle-back trunk on the 1937 Cadillac Series 75 Five-Passenger Touring Sedan. Note the two vertical divider bars and the Cadillac "V" emblem on the trunk lid. The Series 85 V-12 Sedan was identical except for insignia. The Series 37-75 Sedan had a factory delivered price of $2,915. The car in the background is a 1929 LaSalle, apparently intent on crunching the left rear fender of its younger brother.

Cadillac's restyled 1937 front end included a new die-cast grille, the first of a long series of finely-detailed "egg-crate" grilles that would identify Cadillacs off and on for many years. Note the three small horizontal chrome trim pieces on the catwalks between the lower grille and the massive, bullet-formed front fenders. Wide, new bumpers front and rear had a circular Cadillac emblem at their centers. This is the Series 37-75 Formal Sedan for Five Passengers. It was Fleetwood Style 7509-F, with closed upper rear quarters and a padded leather-covered top. Price for this formal four-door sedan was $3,785. This elegant body style was also available in the V-12 Series 85, also on 138-inch wheelbase, as Fleetwood Style 8509-F. The "F" indicates that this car does not have a glass division between the front and rear seats.

1937

This elevated view shows Cadillac's handsome, new 1937 front-end ensemble. The grille is a die-cast honeycomb, somewhat wider in appearance than the 1936 grille. The small, horizontal chrome trim pieces in the catwalk area between the fenders and grille emphasize the car's width. Dual sidemounted spare tires are half-buried in the massive new bullet-formed front fenders. Hood ventilators on Series 70, 75 and 85 Cadillacs were quite small and on cars equipped with optional sidemounts were almost totally hidden. Cadillac's styling was now very clean and integrated with rounded and curved surfaces everywhere. A full-length bright chrome molding extended from the front of the hood to the rear of the body at the beltline. A "V" emblem indicating the number of cylinders under the long, tapered hood was affixed to the upper right-hand side of the eggcrate grille.

This would be the seventh and final year for the Cadillac V-12. The 370 cubic inch twelve-cylinder engine, a derivative of the 1930 sixteen, had been introduced in 1931. Demand for the V-12 had been steadily falling, along with that for the companion V-16. But the Twelve was axed at the end of the 1937 model year while the V-16 was reincarnated for a production run of another three years. For 1937, Cadillac offered six Fleetwood body styles on the Series 85 V-12 chassis, all on 138-inch wheelbase. This is the formal open-front Fleetwood Town Car, Style 8543. A seven-passenger car, it sold for $5,575 and was available with or without quarter windows. It was listed as Fleetwood Style 8543. Note the padded rear roof and the compartment above the front of the fixed roof for storage of the weatherproof top for the chauffeur's compartment.

This is the 1937 Cadillac V-12 Series 85 Fleetwood Town Car with quarter windows with its weatherproof top buttoned into place over the chauffeur's compartment. These big town cars were seven-passenger jobs. Few were built. Note the small parking lights atop the front fenders. These were not used on any other 1937 Cadillacs. Also of interest are the dipped front door window sills and the extremely slender windshield pillars. This Series 37-85 Town Car has a luggage rack rather than the built-in bustle-back trunk.

Among the most practical and popular body styles in the Cadillac Series 75 line for 1937 were the big seven-passenger sedans. Four basic versions were offered, all by Fleetwood and all on the same 138-inch wheelbase chassis. The Series 37-75 Special Touring Sedan for Seven Passengers, Style 7523-S, was priced at $2,710; next came the Special Touring Imperial Sedan, Style 7533-S, at $2,910. The standard Seven-Passenger Touring Sedan, Style 7523 had a factory price of $3,070 and the Touring Imperial Sedan, Style 7533, carried a $3,270 price tag. In the Series 85 V-12 line, the Seven-Passenger Touring Sedan cost $3,790 and the companion Imperial Seven Passenger listed at $3,990. All of these models were virtually identical in appearance.

This is the limousine version of the 1937 Cadillac Fleetwood V-12 Town Car, one of six body styles available in the Series 85 twelve-cylinder line. This car has a bare metal top and full quarter windows instead of formal, closed quarters and a leather-covered, padded rear roof area. In earlier years Cadillac referred to this body style as a Limousine Brougham, but now showed its open-front models simply as Town Cars. Very few were being built now, and even then only on special order. The ritzy Grosse Pointe, Mich., setting is entirely appropriate, even though the recent Depression had placed many such mansions up for sale.

1937

In a period of less than seven years the American custom body business had been all but wiped out. With chilling impartiality the Depression killed off old, established coachbuilders and small custom body shops alike. By the end of the decade only a few remnants remained. One of these was the Rollston Co. of New York, which was reorganized and renamed Rollson in 1939. Rollston built this one-off, full-custom Town Cabriolet on a 1937 Cadillac Series 75 eight-cylinder, 138-inch wheelbase chassis. Note the flat, one-piece windshield, a style also favored by rival Derham of Pennsylvania. The hubcaps and wheel covers on this aristocratic Series 37-75 town car were painted the same color as the body.

A prominent funeral coach and ambulance builder, the A.J. Miller Co. of Bellefontaine, Ohio, built this special eight-door sightseeing bus on a much-lengthened 1937 Cadillac Series 75 chassis. The posh Broadmoor Hotel near Colorado Springs operated a large fleet of sightseeing limousines of this style to transport its guests up and down Pike's Peak. Stretched cars of this type were also used as airport limousines. This nicely restored Miller-Cadillac is one of hundreds of cars in the fabulous Harrah's Automobile Collection in Reno, Nevada. Its canvas covered top could be removed in good weather so that tourists could enjoy a convertible-like view of the Rocky Mountains.

The Meteor Motor Car Co. of Piqua, Ohio, designed and built this limousine style hearse body for the 1937 Cadillac commercial chassis. Beginning this year, only the Henney Motor Co. of Freeport, Ill. was permitted to mount funeral car and ambulance bodies on the Packard commercial chassis. All of the other makers were forced to use other chassis. Most settled on Buick, Cadillac and LaSalle commercial chassis. Ironically, by the time Packard ceased production of commercial chassis 17 years later, Cadillac dominated the small but prestigious professional car industry.

This limousine style ambulance is parked in front of the Meteor Motor Car Co. office in the small city of Piqua, Ohio. Meteor and rival A.J. Miller merged to form Miller-Meteor in 1956, but the company went out of business at the end of the 1979 model year. Note the large parking lights on the front fenders of this two-toned ambulance.

The sixteen-cylinder Cadillac's styling remained almost unchanged between 1934 and 1937. This would be the fourth and final year for the huge 154-inch wheelbase chassis. This impressive Seven-Passenger Limousine, Fleetwood Style 5875, was by far the most popular Series 90 body style this year with 24 examples built, accounting for almost half of the 50 Cadillac Sixteens produced in 1937. Factory delivered price was $7,900. Only two Style 5875-S Seven-Passenger Sedans were built this year. Price was $7,645. The big, chrome spinner hubcaps were still very much in vogue.

Powered by essentially the same 452 cubic inch engine introduced early in 1930, the sixteen-cylinder Cadillac was now in its eighth year. This would be the last year for the magnificent "first generation" V-16 powerplant: an all-new V-16 engine was in the works. For 1937, Fleetwood offered a selection of 11 body styles for the giant 154-inch wheelbase Series 90 Cadillac chassis. This is the 1937 Cadillac Series 90 Convertible Sedan, Fleetwood Style 5880. Only five of these monstrous five-passenger convertibles were built this year, bringing to 20 the number built since the V-16 was restyled in 1934. Factory delivered price was $8,205, and weight was 6,100 pounds.

1937

The Detroit Fire Dept. in 1937 took delivery of one of the largest and most impressive ambulances ever to see service in the United States. A gift of Detroit Fire Commissioner Paxton Mendelssohn, a wealthy fire buff who amassed his fortune working for the Fisher Bros., it was built on a special stretched Series 90 Cadillac V-16 chassis. The one-off custom body was built by the Meteor Motor Car Co. of Piqua, Ohio. In 1951, Commissioner Mendelssohn had this huge body transferred to a new Cadillac chassis. The ambulance continued to respond to multiple-alarm fires in Detroit into the early 1970's. It is seen here in front of Detroit Fire Dept. Headquarters shortly after it went into service. Note the frosted rear window glass. Warning equipment includes a set of whirling Buckeye Roto-Rays above the oversized windshield and a standard 12-inch fire apparatus bell on the right running board.

This would be the biggest year in LaSalle's history. But despite sales of a record 32,005 cars, LaSalle was still unable to overcome the commanding lead built up by Packard with its hot-selling junior series. The 1937 LaSalle was a much-changed car. Besides fresh styling, the Series 37-50 LaSalle rode on a four-inch longer, 124-inch wheelbase and was powered by a V-8 engine for the first time in four years. This is the snappy 1937 LaSalle Convertible Coupe for two or four passengers, which sold for $1,350. Model 5067, the Convertible Coupe weighed in at 3,715 pounds.

The lowest-priced model in the 1937 LaSalle lineup was the svelte Series 37-50 Coupe for two or four passengers. Model 5027, it sold for only $1,155. All five 1937 LaSalle models—a four-door convertible sedan was added to the line this year—were built on a new 124-inch wheelbase. The "B" pillar on this body style was angled forward slightly. Headlights were also positioned lower.

One of three closed body styles in the line was the 1937 LaSalle Two-Door Touring Sedan for Five Passengers, as illustrated in the year's sales literature. This two-door coach sold for a mere $1,275 and weighed 3,780 pounds. Three chrome chevrons—a LaSalle trademark since 1934—continued on the front of each front fender. Note the "LaS" emblem in the center of the front bumper and the small LaSalle nameplate at the rear of the hood above the ventilator louvers. This car was listed as Model 5011.

Four years after the introduction of the boldly-styled 1934 model, the LaSalle was still something of a style-setter. Cadillac's stylists were reluctant to tamper with the car's basic styling, especially the narrow vertical grille design. Consequently, the Series 37-50 LaSalle bore an unmistakable resemblance to its immediate predecessors. This is the 1937 Coupe. Note the single rear window divider and the long, clean rear deck.

1937

For 1937, the LaSalle retained its distinctive, narrow grille design. The hood now sported eight large vertical vents on each side, ornamented by six chrome streamers which gave the front end a very strong horizontal look. This is the popular Series 37-50 Four-Door Touring Sedan, a five-passenger car that had a factory delivered price of $1,320. It was listed as Model 5019 and weighed 3,810 pounds in stock form.

The LaSalle model line was expanded this year with the addition of an exciting new body style—a four-door Convertible Sedan. At $1,680 this was the most expensive car in the Series 37-50 line. It was Model 5049 and tipped the scales at 3,850 pounds. Why this gentleman has the top down on his 1937 LaSalle Convertible Sedan in the dead of winter is anybody's guess, but you have to admit it looks gorgeous. And the gals don't seem to mind a bit as they admire the frozen fountain in the background.

Here is the new 1937 LaSalle Convertible Sedan with its top up. In either configuration, this was an uncommonly handsome automobile. The car shown is also equipped with optional dual sidemounted spare tires. The four-door Convertible Sedan would remain among LaSalle's model offerings until this fondly-remembered make was discontinued at the end of the 1940 model year. LaSalle's sales rocketed to a record 32,005 cars this year but still trailed those of the Packard One-Twenty. And to make matters worse, Packard added a six-cylinder series to its junior line this year.

For 1937, the Oldsmobile-built straight-eight engine that had been used in the LaSalle since 1934 was replaced by a V-8. This eight was the same 322 cubic inch, 125-horsepower engine that had been used in the 1936 Series 60 Cadillac. For the first time in four years, the LaSalle was again "all-Cadillac." Legendary racing driver Ralph de Palma drove a 1937 LaSalle, that year's Indianapolis "500" Official Car, in a 500-mile-plus endurance run at the Indianapolis Motor Speedway. De Palma averaged 75.59 miles on this impressive performance run.

Cadillac broadened its considerable appeal to funeral car and ambulance builders with the addition of a LaSalle commercial chassis. This 160-inch wheelbase, V-8 chassis proved very popular with hearse and ambulance builders between 1937 and 1940. The A.J. Miller Co. of Bellefontaine, Ohio, built this limousine ambulance body on the new 1937 LaSalles Series 37-50 commercial chassis. Two Cadillac commercial chassis were also available to the trade.

1938

Nineteen Thirty-Eight was a milestone year for important new products at Cadillac. The company introduced the revolutionary Cadillac Sixty Special and took the industry by surprise with its announcement of a totally-new Cadillac V-16.

The stunning, new Sixty Special strongly influenced the look of GM and other American cars for years. Its bold styling advances marked a sharp break with the past. It was the first complete car styled by a talented young designer named William L. Mitchell, who had recently been made head of Cadillac's styling studio. Mitchell, in a distinguished design career that spanned more than 40 years, went on to style many outstanding cars for General Motors, including the Corvette Sting Ray and the 1963 Buick Riviera. But the Sixty Special was always one of Bill Mitchell's personal favorites.

The styling prototype that evolved into the milestone 1938 Cadillac Sixty Special was originally conceived as a LaSalle model, but as development progressed a decision was made to put it into producion as a new kind of smaller Cadillac. It made its bow as a supplemental model to the division's lowest-priced Cadillac line, the Series 60. The classic Sixty Special was unique in that it featured distinctive upper and lower body forms. The trunk was a gracefully curved extension of the rear body. The Sixty Special had no bright beltline trim and it was noticeably lower in appearance than most other cars of its day. But what made the Sixty Special's styling really daring was something it did not have—running boards. It was the first U.S. car to do away with running boards completely. All other cars eventually followed.

The Sixty Special's upper body or "greenhouse" had large side windows that were angled sharply at the corners and edged with thin, chrome frames. The windows were nearly flush with the car's sides. A massive horizontal-theme grille, squared-off pontoon fenders and a thin, flat roofline gave the Sixty Special a low, powerful stance.

It was available only in one body style, a four-door, five-passenger Touring Sedan that had been styled by Mitchell to look like a convertible. Wheelbase was 127 inches, three inches longer than other 1938 Cadillac Series 60 models. The new Sixty Special had its own, unique chassis frame with a high kickup at each end and a massive structural "X" member at its center which gave it the stiffest platform of any car. The result was a car three inches lower than other Cadillacs, but with no sacrifice in interior headroom. The same basic body with annual front-end styling changes was carried through the 1941 model year. The new Sixty Special was a resounding sales success, outselling all other 1938 Cadillac models. A total of 3,703 were sold in its first year at a factory delivered price of $2,090. The Sixty Special name was reserved for Cadillac's most distinctive four-door sedan model into the early 1970's.

All 1938 Cadillacs except the new Sixteen were powered by the same 346 cubic inch, 130-horsepower V-8 engine. The standard Cadillac line included the

From an historical point of view, the most important car in Cadillac's 1938 model line was the stunning, new Sixty Special. A truly epic design which broke sharply with then current passenger car styling concepts, the Sixty Special was styled by the gifted young William L. Mitchell, at the time the new head of Cadillac's styling studio. The boldly-styled 1938 Sixty Special influenced the look of American cars for years. It was a spectacular sales success, too, outselling all other 1938 Cadillac models. The Sixty Special's most significant styling features included the complete elimination of running boards, distictive upper and lower body forms, a thin, flat roofline and large, flush-mounted side windows edged with thin chrome frames. The Sixty Special was also a full three inches lower than the other 1938 Cadillacs. A massive, horizontal-theme grille, big bullet headlights nestled low between squared "suitcase" front fenders and the tapered hood and the absence of any bright trim at the beltline made the new Sixty Special a real standout. It was unlike anything else on the road, and it was a Cadillac. This photo says it all.

Viewed from any angle, the new Cadillac Sixty Special was a beauty. The wide rear window featured two thin vertical dividers, continuing a Cadillac styling motif employed since 1934. The 60-S had its own, distinctive "greenhouse" above the beltline with large side windows set in bright, chrome frames. The windows were sharply angled at the corners. But the 1938 Cadillac Sixty Special was especially notable for two things that weren't there—running boards. This marked a bold break with the past. With frontal styling changes, this same handsome body style was carried through the 1941 model year. It still looks good today. The Sixty Special was the first of many distinctive cars styled by Bill Mitchell for General Motors over a period of many years. Others included the 1963 Corvette Sting Ray, the 1963 Buick Riviera and the front-wheel-drive Cadillac Eldorado.

1938

Series 60 with four Fisher body styles on a 124-inch wheelbase; the new Sixty Special, also by Fisher; three Series 65 models by Fisher on a 132-inch wheelbase, and the Series 75 with a choice of 14 Fleetwood bodies all on the same 141-¼-inch wheelbase. The all new Series 90 Sixteens utilized the same Fleetwood bodies and 141-inch chassis as the Series 75 but had different front-end styling. The last-generation "Sixteen Fleetwoods" are described in a separate chapter.

Cadillac's eight-cylinder cars had two distinctively different grille designs. The Series 60 and new Sixty Special had a pleasing new front-end that featured a series of thin, horizontal grille bars that extended around the sides of the squared-off nose. Series 65 and 75 had tall, vertical die-cast "eggcrate" grilles. Inside, the gearshift lever on all models was relocated from the floor to the steering column. The Cadillac name appeared in large, bold script on front and rear bumpers on all models.

The 1938 LaSalle was given modest styling changes. Series 38-50 models had a new vertical die-cast eggcrate grille and heavy horizontal hood ventilator panels that extended the entire length of the hood from grille to cowl. Again, five Fisher body styles were offered, all on the same 124-inch wheelbase. LaSalles continued to use the 125-horsepower, 122 cubic inch V-8 engine. LaSalle's 1938 sales were disappointing, dropping to 15,575 or less than half the all-time record 32,005 sold the previous year. Packard's grip on this end of the market seemed unbreakable.

The 1938 model year was a poor one for the entire U.S. industry. The country had been hit by another recession. Motor vehicle production fell 40 per cent and auto sales were in their worst slump since the Depression. Cadillac's dealers delivered 24,843 cars this year—half as many as they had sold in 1937.

By the end of the year Pierce-Arrow had been liquidated, joining Franklin, Marmon, Peerless, Stutz, Cunningham and Duesenberg on the growing list of now-defunct luxury car manufacturers.

The stunning new 1938 Cadillac Sixty Special was offered in only one body style, a five-passenger Touring Sedan on 127-inch wheelbase. The Sixty Special's wheelbase was three inches longer than that of other 1938 Series 60 Cadillacs. Factory list price was $2,090. Model 6019-S, the Sixty Special sold like hotcakes: 3,703 were produced, outselling every other car in the Cadillac line. The Sixty Special was powered by the same 346 cubic inch V-8 engine used in all other 1938 Cadillacs except the Series 90 Sixteen. Standard weight was 4,170 pounds. A ribbed chrome rocker molding occupied the space where the running boards usually went.

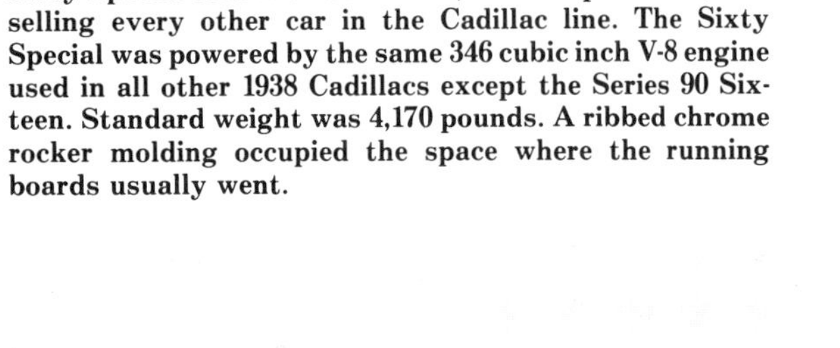

Most Cadillac Sixty Specials built between 1938 and 1940 were sold without sidemounted spare tires. But some customers insisted on having them, and sidemounts were optionally available. This is an early production 1938 Cadillac Sixty Special Touring Sedan with optional six-wheel equipment. Viewed in profile, the smooth sidemounted spare tire cover complements the Sixty Special's distinctive lines rather well. From the front they were a trifle heavy-looking. The boldly-styled Sixty Special influenced the look of GM's cars for years.

The Cadillac Motor Car Division built two of these smart 1938 Cadillac Sixty Special convertible sedans, which were driven by senior company executives. Actually, Bill Mitchell's revolutionary Sixty Special Touring Sedan was styled to look like a four-door convertible. The original Sixty Special prototype was conceived as a LaSalle model but as the project progressed it evolved into a new, smaller Cadillac. The Sixty Special had its own unique chassis frame with a high "kickup" at both ends, resulting in an extremely stiff platform and a car that was three inches lower than other Cadillacs. One 1938 Sixty Special Two-Door Coupe was also built.

In an effort to keep its doors open, the Brunn Co. of Buffalo, N.Y. offered a series of semi-custom bodies for mounting on standard Cadillac chassis. This old, established coachbuilder also designed and built a few full-custom bodies for the 1938-40 Cadillac V-16 chassis. Brunn offered this dashing Convertible Victoria body style for the all-new 1938 Cadillac Sixty Special chassis. It is not known if any of these were actually built.

1938

The lowest-priced Cadillac offered for 1938 was the Series 60 Two-Passenger Coupe, Model 6127. Body was by Fisher. Factory-delivered price was $1,695 and the Sixty Coupe weighed 3,855 pounds. The "B" pillar was angled forward slightly and the strong front-end ensemble was similar to that of the style-setting new Cadillac Sixty Special. Four Fisher body styles—Coupe, Convertible Coupe, Convertible Sedan and four-door Touring Sedan were offered on the 124-inch wheelbase Series 60 chassis for 1938. The Sixty Special was three inches longer.

Here is the 1938 Cadillac Series 60 Convertible Coupe all dressed up to go. Optional equipment includes new single-piece sidemount covers and white sidewall tires. The result is a very attractive car. The Series 60 Convertible Coupe was shown as Model 6167. Factory price was $1,815. The restyled 1938 Series 60 had a strong new horizontal grille design instead of the tall, vertical grilles favored on earlier Cadillac models.

Despite the drab-looking factory rooftop setting, we couldn't resist throwing this one in: the 1938 Cadillac Series 60 Convertible Coupe simply looks so attractive with its top down, and especially when it is equipped with side-mounted spares. Wheelbase was 124 inches and this two-passenger convertible weighed 3,845 pounds. Only 145 of these rather stubby-looking convertibles were sold this year.

This was Cadillac's lowest-priced 1938 sedan, the bread-and-butter Series 60 Five-Passenger Touring Sedan, Style 6119. It is seen here in base form without whitewalls, sidemounts or other accessories. The Series 38-50 Touring Sedan by Fisher had a factory-delivered price of $1,780 and 1,295 were sold. This plain-vanilla body was shared with other, lesser General Motors car lines as the corporation continued to move toward greater divisional standardization.

Rivalling the new Sixty Special in the glamour department was the 1938 Cadillac Series 60 Convertible Sedan. This most attractive Fisher four-door convertible, Style 6149, was built on the standard Series 60 wheelbase of 124 inches. It sold for $2,215 and only 60 were built. Note the sleek fastback styling. The more expensive 1938 Series 65 Convertible Sedan had the same smooth back. Sure it's winter, but you have to admit that this car looks gorgeous with its top and windows down.

Dual sidemounts really did something for Cadillacs of this era. A pair of those smooth, new single-piece spare tire covers greatly enhanced the front-end appearance of even the plainest four-door sedan. This is the 1938 Cadillac Series 60 Touring Sedan, Model 6119, with optional six-wheel equipment. Whitewall tires were also optional. In standard form this model weighed 3,940 pounds. This was by far the most popular of the four 1938 Series 60 body styles. Other models included a Coupe, Convertible Coupe and a four-door Convertible Sedan.

1938

For 1938, Cadillac offered funeral car and ambulance builders a special Series 60 commercial chassis. Wheelbase was 159 inches. A Series 75 commercial chassis and LaSalle commercial chassis were also available for mounting hearse and ambulance bodies. This was perhaps the most distinctive funeral car built on the Series 38-60 commercial chassis. It is a classic open-front Town Car Limousine Funeral Coach designed and built by the A.J. Miller Co. of Bellefontaine, Ohio. The ornamental grille in the rear side window opening is of interest: these were usually found only in ambulances or combination coaches —dual-purpose vehicles which could be used as either funeral cars or ambulances. It is doubtful that this formal car was ever used as an emergency ambulance.

The Superior Body Co. of Lima, Ohio, which once mounted most of its funeral car and ambulance bodies on Studebaker chassis, was now building principally on various GM commercial chassis, including Cadillac, LaSalle, Buick and Pontiac. This is a 1938 Superior-Cadillac limousine-style hearse on the Series 60 commercial chassis. Note the new "goddess" hood ornament which doubled as the latch for the 1938 Cadillac's new alligator hood. The "Cadillac" name now appeared in bold script in the center of front and rear bumpers in all series.

Cadillac's 1938 Series 65 included only three models, all with Fisher bodies and all on a 132-inch wheelbase. Series 65 Cadillacs featured the same front-end styling as the more expensive 1938 Series 75 models, which were on a longer 141-inch wheelbase and had Fleetwood coachwork. This is the Series 38-65 Touring Sedan for Five Passengers, a conventional four-door sedan which had a factory delivered price of $2,290. It was shown as Style 6519. The 346 cubic inch V-8 engine used in Series 65 models was rated at 130 horsepower.

The 1938 Cadillac Series 65 line included two four-door Touring Sedans. In addition to the standard five-passenger Touring Sedan, Style 6519, there was the formal Style 6519-F. The 6519-F had a partition with a glass division between the front and rear seats and thus qualified as a close-coupled limousine. The Series Sixty-Five Model 6519-F Touring Sedan sold for $2,360 or only $70 more than the standard sedan. Only 110 of these models were built compared with 1,178 Style 6519 sedans.

This is a three-quarter rear view of the 1938 Cadillac Series 65 Touring Sedan with Imperial division, Style 6519-F. The new four-door Fisher sedan body had an angled "C" pillar and split rear window treatment with a thick center divider. The sidemounted spare tires and white sidewall tires were extra-cost options. Wheelbase for this series was 132 inches.

Series 65 Cadillacs for 1938 were positioned between the low-line Series 60 (124-inch wheelbase) and the more expensive Series 75 models (141-inch) in Cadillac's comprehensive V-8 line. Series 65 cars had the same front-end styling as Series 75 models but utilized a 132-inch wheelbase chassis. This is the elegant 1938 Cadillac Series 65 Convertible Sedan with sleek fastback styling. The Fisher convertible sedan, Style 6549, sold for $2,605 and was the most expensive car in the Series 38-65 line. Weight was 4,580 pounds. Only 110 were built.

1938

Cadillac's most comprehensive model selection for 1938 was to be found in the medium-range Series Seventy-Five. A total of 14 body styles, all by Fleetwood, were offered in this series, all on the same 141-inch wheelbase. This is the 1938 Cadillac Series 75 Fleetwood Coupe for Two Passengers, Style 7557, which carried a factory delivered price of $3,280. The handsome Coupe weighed 4,675 pounds. Fifty-two were built. The Series 38-75 line also included a five-passenger coupe, Style 7557-B.

The 1938 Cadillac Series Seventy-Five Coupe for Five Passengers was identical in appearance to the slightly less expensive two-passenger stationary coupe. This car was Fleetwood Style 7557-B, and it sold for $3,380. The Series 75 Fleetwood Coupe shown is equipped with some popular options—white sidewall tires and smart, sidemounted spares. Note the strong vertical grille profile which was not unlike that on the top-line 1938 Cadillac Sixteen. This model weighed 4,775 pounds.

The beautiful 1938 Cadillac Series 75 Fleetwood Convertible Coupe had a long, pleasingly rounded rear deck. Style 7567, this model was marketed as a two-passenger car although three could be accommodated in reasonable comfort on the bench-type seat. New, larger pontoon fenders were squared at the rear and looked good with or without sidemounts. The weatherproof boot buttoned into place over the folded convertible top presents a tidy appearance. The Series 75 Convertible Coupe weighed 4,665 pounds.

Certainly among the most appealing of all 1938 Cadillac body styles was the dashing Series 75 Convertible Coupe for Two Passengers. Fleetwood Style 7567, the Convertible Coupe was priced identically with the five-passenger Series 75 Coupe at $3,380. Series 75 Cadillacs for 1938 had a massive, new vertical "V"-type die-cast eggcrate grille design. Three sets of short, horizontal ventilator louvers graced each side of the hood on all 1938 Cadillac V-8 models. Series 60 and 60 Special models had a different, horizontal grille treatment.

Most Fleetwood body styles offered in Cadillac's 1938 Series 75 line were also available on the all-new Series 90 Sixteens, which shared the same 141-¼-inch wheelbase chassis. The premium Sixteen Fleetwoods, however, had different styling forward of the cowl. This is the very impressive 1938 Series 75 Fleetwood Convertible Sedan, a five-passenger car which sold for $3,945. Note the expansive Victoria-type rear roof appearance. The Series 75 Fleetwood Convertible Sedan was shown as Style 7529.

1938

Fleetwood produced only 58 of these big four-door convertible sedan bodies for the 1938 Cadillac V-8 Series 75 chassis. Here is how this imposing car looked with its huge top let down, the center posts removed and the window glass lowered into the doors. The roomy Series 75 Convertible Sedan weighed 3,795 pounds. Convertible sedans were available in every one of Cadillac's 1938 car lines, from the LaSalle on up to the awesome V-16.

The elegant four-door Town Sedan was a popular Cadillac body style through the style-conscious decade of the 1930's. This is the 1938 Cadillac Series 75 Town Sedan by Fleetwood, Style 7539, a semi-formal, five-passenger car that carried a factory-delivered price of $3,635. Only 58 of these handsome Fleetwood Town Sedans were built on the V-8 Series 75 Cadillac chassis this year. The Town Sedan weighed an even 4,900 pounds. Note the smooth metal roof and the pleasingly rounded rear body contours.

This imposing car marked the beginning of one of the most famous and longest-running series of Cadillacs ever built—the prestigious Series Seventy-Five sedans and limousines which were produced under the Series 75 model designation through the 1976 model year, a span of an incredible 41 years! From 1977 on they were simply called the Fleetwood Limousine and Formal Limousine. This is the 1938 Cadillac Series 75 Fleetwood Touring Sedan for Seven Passengers, Style 7523, which sold for $3,210. A total of 380 were built this year. A Series 75 Fleetwood Seven-Passenger Imperial Touring Sedan, Style 7533, was also available, with Imperial glass division. The Seventy-Five Imperial carried a $3,360 price tag and 479 were built in the 1938 model year. Rounding out the line was a Seven-Passenger Formal Sedan with closed quarters and a padded top. In any configuration, these stately big sedans commanded respect, and usually got it.

Near the very top of Cadillac's 1938 V-8 car line was this impressive Series 75 Formal Sedan. This is the Series 38-75 Fleetwood Seven-Passenger Formal Sedan, Style 7533-F. The "F" suffix indicated that this car had an Imperial glass division between the front and rear passenger compartments. This big formal four-door sedan with leather-covered, padded top and closed rear quarter panels sold for $3,995. Only 40 were built. The 1938 Series 75 line also included a Fleetwood Five-Passenger Formal Sedan, Style 7559, which sold for the same price as the 7533-F. The Seven-Passenger Formal Sedan weighed 5,105 pounds.

The most dignified—and expensive—body style in Cadillac's 1938 Series 75 model offerings was the stately Fleetwood Town Car, Style 7553. Only 17 of these correctly formal, seven-passenger open-front Town Cars were built this year. Formal styling touches included closed rear quarters and a padded, leather-covered roof with small rear window. Fleetwood Town Cars had special windshields with much thinner pillars than other Series 75 models. At $5,115 this was by far the most expensive car in the 1938 Cadillac V-8 line. Wheelbase was 141-¼ inches. The removable top was stored in the trunk when not being used. Fleetwood formal four-door models had distinctive rear door uppers with integral ventilator panes and rounded rear frames.

1938

For fleet customers, Cadillac offered two large Business Sedans in its 1938 Series 75 line. These included the Seven-Passenger Business Sedan by Fleetwood, Style 7523-L, and a companion Seven-Passenger Imperial Business Sedan with glass division, Style 7533-L. These models, which were favored by limousine livery services and funeral directors, were sometimes configured as eight-passenger cars. The 7523-L sold for $3,105 and the 7533-L was priced $155 higher. Only 25 of each were sold in the 1938 model year, or just 50 in all.

Brunn and Company of Buffalo, N.Y. continued to design custom and semi-custom bodies for various Cadillac chassis. This is a Brunn Touring Cabriolet with collapsible rear roof for the 1938 Cadillac Series 75 chassis. This design was commissioned for the Capitol Cadillac Co. of Atlanta, Ga. Brunn did a number of these gorgeous semi-convertible bodies for mounting on other chassis, including Lincoln. Note the extra glass above the windshield header.

Cadillac exported only eight "CKD" (completely knocked-down) Series 75 chassis this year. One of these became this very interesting open-front Town Car by Franay of Paris. The two-tone exterior color is especially striking and this car's French styling is much more dashing than its staid-looking American counterpart by Fleetwood. Styling highlights include a raked, one-piece flat windshield and sweeping rear body lines. European coachbuilders sometimes referred to this formal body style as a Sedanca.

Cadillac continued to offer a series of special eight-passenger sedans intended for use by funeral directors and limousine livery operators. These cars featured special, hard-wearing interior upholstery and trim for public service. The Fleetwood-bodied Cadillac eight-passenger sedan on 141-inch wheelbase was quite popular with the trade. To further enhance their appeal, Cadillac marketed special-purpose variations of these big sedans. This is the 1938 Cadillac Fleetwood Sedan Ambulance. The right-hand "B" pillar could be removed and the right front passenger seat taken out to accommodate a standard wheeled ambulance cot. A locking device clamped the cot firmly in place during transit. Cadillac marketed Fleetwood sedan ambulances through its Commercial Car Department well into the 1940's. Some outside firms also did similar conversions of Cadillac Series 75 sedans.

This is the special 1938 Cadillac Fleetwood Invalid Chair Car, a special-purpose modification of Cadillac's big eight-passenger commercial sedan. Equipment in this package included a specially-upholstered wheel chair which locked into place in the roomy rear passenger compartment and a three-piece folding ramp used to get the wheel chair in and out of the car. The Invalid Car could be easily converted into a conventional sedan. Cadillac was now actively wooing the commercial car operator, especially funeral directors and limousine livery service operators.

This was a disappointing year for LaSalle sales. The U.S. economy was again in recession. Packard was still outselling all of its competitors with its phenomenally-successful junior cars, and the Lincoln Zephyr wasn't doing badly, either. LaSalle sales in 1938 dropped to less than half those of record 1937. Again, five Series 50 LaSalle body styles were offered, all on the same 124-inch wheelbase. This was the lowest-priced car in the line, the Series 38-50 LaSalle Coupe for two or four passengers. Model 5027, it weighed 3,745 pounds and was priced at a very reasonable $1,259. The 1938 LaSalle had a new vertical eggcrate grille, and the chevrons had finally been deleted from the front fenders.

1938

With its top up or down, the 1938 LaSalle Convertible Coupe was a most attractive car. The Series 38-50 Convertible Coupe, Model 5067, sold for only $1,420. Weight was 3,735 pounds in stock form shown here. Note the raised side window glass, the squared-off "suitcase" front fenders and the long, low-mounted headlights atop fluted vertical stanchions. Hood ventilator panels had a strong horizontal theme. This was a fairly popular model with 819 examples sold.

It's cherry blossom time in Washington, D.C. and this upwardly mobile young couple are touring the capital in their sporty new 1938 LaSalle Convertible Coupe. This car is loaded with extra-cost options, including dual sidemounted spare tires and white sidewall tires. The light body color enhances its sporty appearance. Despite modest restyling, sales of 15,501 LaSalles in the 1938 model year were less than half those of 1937—LaSalle's best year ever.

The 1938 LaSalle received modest exterior styling changes. These included a massive, die-cast vertical grille and long, bullet-shaped headlights set low in the front-end sheetmetal between the hood and squared pontoon front fenders. The circular "LaS" emblem on the upper right hand side of the grille was replaced with a golden "V-8". This is the Series 38-50 Two-Door Touring Sedan, Model 5011, which sold for $1,345. Sales of this model totalled 700.

Here's an over-the-shoulder view of the 1938 LaSalle cockpit. The gearshift lever on all 1938 Cadillacs from the LaSalle on up to the mighty V-16 was moved from the floor to a handy new location on the steering column just below the wheel. This Series 38-50 Convertible Coupe is equipped with optional dual sidemounted spare tires. The smoothly redesigned 1938 sidemounts hinged forward and presented a very tidy appearance. Note the windsplits atop the long, low-mounted bullet headlights.

By far the most popular model in the 1938 LaSalle line was the practical Four-Door Touring Sedan, Model 5019. Sales totalled 9,765 accounting for a high proportion of the 15,501 cars sold this year. The five-passenger Series 38-50 Touring Sedan sold for $1,385 and weighed 3,830 pounds. Note the massive horizontal hood ventilator panel which featured four chrome streamers that extended the full length of the hood from grille to cowl.

1938

Only 265 LaSalle four-door convertible sedans were sold in 1938. Again, this was an exceptionally attractive car with its long flowing fastback roofline and full upper rear quarters. The Series 38-50 Convertible Sedan shown here, Style 5049, carried a $1,825 price tag making it the most expensive car in LaSalle's 1938 model lineup. This is a 1938 LaSalle catalog illustration. The convertible sedan is equipped with optional sidemounts.

Like Cadillac, LaSalle offered the funeral directing profession specialized sedans for livery service. This is a sedan ambulance version of the Series 38-50 LaSalle Two-Door Touring Sedan. The right front seat could be removed to accommodate a full-sized ambulance cot for emergency service or for just transferring patients to and from hospitals—a courtesy service frequently provided by local funeral directors as a public relations gesture.

The LaSalle commercial chassis had become very popular with American funeral car and ambulance builders. Sayers and Scovill of Cincinnati, Ohio, one of the oldest and largest funeral car builders in the U.S., switched from Buick to LaSalle chassis exclusively for its products this year. One of S & S's most popular 1938 models was the Damascus Carved Panel Funeral Coach on a thinly-disguised Series 38-50 LaSalle 159-inch wheelbase commercial chassis. Note the extra chrome pieces on and above the grille. This extra trim was intended to mask the car's identity and eliminate "year marks", thus extending the car's styling life.

This year saw an important, new type of vehicle introduced to the funeral directing profession. It was the coupe-style Flower Car. These special long-wheelbase coupes were designed to lead funeral processions. Behind the coupe-type cab was an open well for displaying floral tributes. Up to now, large open touring cars or phaetons with their tops let down were used for this purpose. The coupe-style Flower Car caught on quickly and within a year or two all principal funeral car builders included them in their product lines. The Eureka Co. of Rock Falls, Ill. was one of the first to offer a full-sized funeral flower car. The Henney Motor Co. of Freeport, Ill., which built on Packard chassis exclusively, also came out with a flower car at about this time. This is a 1938 Eureka-LaSalle Flower Car. Customers could specify a folding convertible top: most went for fixed-roof jobs styled to look like convertibles. Full-sized flower cars were produced in small numbers until 1976. Flower car conversions of standard Cadillac Coupe de Villes are still available today.

The Eureka Company this year boasted the largest carved panels in the American funeral coach industry. This is a 1938 Eureka-LaSalle four-panel carved hearse undergoing road testing at the General Motors Proving Grounds near Milford, Mich. Note the richly-detailed drapery panels complete with imitation tassels and tieback cords. The carved-panel hearse had reached the peak of its popularity at this time. The owner's nameplate—O'Conner Service Co.—is mounted above an ornamental escutcheon plate on the front door.

1938-1940 V-16

The Cadillac Motor Car Company's management seemed to delight in occasionally taking the rest of the industry by surprise. They'd done it with the introduction of the industry's first mass-produced V-8 passenger car engine in 1914, the epic V-16 in 1930 and the companion V-12 a year later. Now they did it again with the totally-unexpected announcement of an all-new Cadillac V-16 luxury car line for 1938.

Most auto industry observers had written off the sixteen-cylinder engine as an automotive anachronism left over from the recent Depression. The market for such a complex and sophisticated automobile powerplant had all but vanished. The fact that Cadillac had managed to sell only 50 or so V-16's annually in recent years appeared to bear this out. Why, even Pierce-Arrow and Duesenberg were gone now, and rival Packard had turned to lower-priced cars and mass-production to stay alive. Yet, as the curtain went up on the big New York Automobile Show in October, 1937 Cadillac once again caught its rivals off guard with its public introduction of a totally-new sixteen-cylinder car program for 1938.

And once more, the hype that surrounded this announcement was more than just hyperbole. Cadillac revealed details of a completely new L-head V-16 engine, a unitized monobloc powerplant that was considerably smaller and lighter than the venerable 452 cubic inch engine which it replaced, which had powered Cadillac Sixteens from 1930 through the 1937 model year. With its cylinder banks splayed at an extremely shallow 135 degrees, compared to the narrow 45 degrees of the 452, the potent new V-16 was almost horizontally-opposed. Yet this new, smaller engine, which drew heavily on the enbloc technology incorporated into Cadillac's highly-regarded new V-8 engine for 1936, produced virtually the same 185 horsepower as the old V-16.

It was a "square" design of 3-¼-inch bore and stroke with separate distributors, downdraft carburetors and manifolding for each of its cylinder banks. The new engine was six inches shorter, 13 inches lower, 250 pounds lighter and had significantly fewer parts than the earlier V-16.

Cadillac's new short-stroke engine displaced 431 cubic inches -- 21 fewer than its predecessor. Compression ratio was 6.67:1. The 1938 engine was so flat that it could be tucked up under the firewall, permitting a smaller car but with no sacrifice in interior space.

As efficient as it was, the new Cadillac sixteen-cylinder engine lacked the aesthetic appeal of the high-styled, beautifully-detailed 452. Perhaps this was a concession demanded by the cost-conscious Nick Dreystadt who was determined to reduce the number of Cadillac's parts and to bring the division's costs under strict control. More than likely it was simply because the new V-16 sat so low in the chassis that one had to lean over a broad fender and peer down over a high hood side panel to even see it.

Fleetwood offered a full choice of 12 semi-custom bodies, all on the same 141-¼-inch wheelbase chassis. These ranged from two-passenger coupes and convertibles on up to town sedans, big convertible sedans and seven-passenger formal limousines and open-front town cars. Introductory prices ranged from $5,140 to $7,175 and remained unchanged through the 1940 model year.

Cadillac continued to designate its flagship Sixteens the Series 90. For 1938, the V-16 line was known as Cadillac's Series 38-90; for 1939, 39-90 and in its final year, 40-90. The final generation V-16 was produced over three model years with only minor cosmetic exterior styling changes, mainly in taillight and bumper details. Inside, Series 90 instrument panels were redesigned every year and were compatible with those used in the V-8 Series 75 Cadillacs. For 1938, the gearshift lever was moved from the floor to the steering column. All of the Fleetwood V-16 body styles were also available on the less-expensive eight-cylinder Series 75 Cadillac chassis which utilized the same 141-inch wheelbase as the V-16. Styling, however, differed forward of the cowl.

For 1939, all Sixteens got full-length chrome running board edge moldings, which were carried over into 1940. Also in 1939, the license plate was transferred from a stanchion atop the left rear taillight to a new position above the trunk lid handle in the center of the rear deck. For 1940, the long parking lamps atop the front fenders were chrome-plated. The 1940 models also boasted sealed-beam headlamps and directional signals.

Bright young GM stylist William J. Mitchell, who had designed Cadillac's stunning 1938 Sixty Special, was responsible for the 1938-1940 Series 90's styling. Mitchell gave the imposing Sixteen a massive, clifflike

Cadillac once again took the American automobile industry by surprise with its announcement in October, 1937 of a totally-new line of sixteen-cylinder cars for 1938. The Series 38-90 Cadillacs were new from road to roof and from bumper to bumper—including an all-new, flathead V-16 engine. Twelve body styles, all by Fleetwood, were offered on a new, shorter 141-¼-inch wheelbase. This is the 1938 Cadillac Series 90 Style 9067 Fleetwood Sixteen Convertible Coupe, only 10 of which were built in 1938. Factory list price for this sporty model was $5,440. Fleetwood's new Series 90 bodies were designed by GM stylist William Mitchell, who also styled the milestone 1938 Cadillac Sixty Special. He bestowed the prestigious Sixteen with a massive, vertical grille design and strong horizontal body lines.

1938-1940 V-16

frontal appearance with a nearly vertical die-cast eggcrate grille thrust forward of large, squared front fenders. Headlights were attractively faired into the sheetmetal between the front fenders and the long, tapered hood.

Optional sidemount wheel covers were smooth, single-piece units and proved to be very popular options with V-16 customers. The three, long horizontal hood louvers originally used on the 1933 Cadillac Sixteen were retained and were duplicated in miniature on the lower edges of all four fenders. These trim pieces were painted on 1938 Series 90 Cadillacs with a bright, chrome bead on the lower edge. They were fully chromed for 1939 and 1940. A heavy new "goddess" hood ornament doubled as the latch handle for the new rear-hinged alligator-type hood.

By 1940, the Sixteen's styling had fallen noticeably behind that of Mitchell's contemporary, very sleek V-8's. Besides, there was no longer any need for such an ostentatious car line. A mere 61 had been sold in 1940—shades of miniscule V-16 registrations during the depths of the recent Depression. The super-rich had dropped out of the market—lavish personal transportation was now clearly passe.

So the mighty V-16 was quietly and unceremoniously discontinued—along with the LaSalle—at the end of the 1940 model year. History had repeated itself. The original Cadillac V-16 was introduced just after the 1929 stock market crash and as the U.S. economy slid into a disastrous depression. Similarly, the all-new 1938 Cadillac V-16 line made its bow precisely as the American economy slumped into another recession. Strong auto industry sales of 1936 and 1937 were sharply reversed in model year 1938. It was another case of unfortunate timing.

In all, Cadillac built 4,403 sixteen-cylinder cars during this classic's spectacular 11-year model run. Two thirds of those had been built in the car's very first year! Factory records show production of 3,677 between 1930 and 1933; a mere 212 between 1934 and 1937 and 514 between 1938 and 1940. The final V-16 series saw 315 cars delivered in the 1938 model year, 138 in 1939, and only 61 for 1940.

While it is doubtful that General Motors made money from the V-16 program, this magnificent series undoubtedly did much for GM. It clearly established Cadillac as the new American luxury car leader and bestowed upon Cadillac an image, an aura of mystique that continues to surround the marque to this day. The Cadillac V-16 was a monumental piece of automotive engineering, the likes of which have not been seen since. Fortunately for all, numerous fine examples of each five V-16 series survive today in the hands of appreciative collectors.

The completely redesigned 1938 Cadillac V-16 model lineup included only two two-passenger models, a convertible coupe and this huge Sport Coupe, Fleetwood Style 9057. Like the convertible, this coupe was a massively-proportioned car, even though the new Series 90 Sixteen chassis was a full 13 inches shorter than the previous series. Weight was 4,915 pounds. The Sixteen Fleetwood Sport Coupe carried a factory list price of $5,340. Eleven were built for 1938, six the following year and only two in 1940 for a total of 19 in all. This V-16 Sport Coupe has no sidemounted spares. Most 1938-40 Sixteens carried them. There was certainly plenty of room in that tremendously long rear deck to store a spare.

Among the 12 Fleetwood body styles offered on the all-new 1938 Series 90 Sixteen chassis were two coupes—for two or five passengers. This is the Five-Passenger Sport Coupe, Style 9057-B. With a factory list price of $5,540 it was priced only $200 higher than the two-passenger model. Weight was 5,015 pounds. Eight of these sporty coupes were delivered in 1938. Five were built in the 1939 model year and only one was sold in 1940 for a total of 14. The new "goddess" hood ornament doubled as the latch handle for the Sixteen's rear-hinged "alligator"-type hood. The five-passenger Fleetwood Sport Coupe had a three-piece rear window. the two-passenger variation had a one-piece backlight.

The sixteen-cylinder Cadillac's massive chassis dimensions stretched two-passenger-model body proportions to incredible lengths. The Style 9067 Fleetwood Convertible Coupe is a good example. Note the extremely long rear deck which, incidentally, no longer housed a rumble seat. Ten cars of this style were built by Fleetwood in the 1938 model year. Seven were built in 1939 and only two in 1940, the V-16's final year, for a total of 19 in all. Cadillac also offered a V-16 Stationary Coupe for Two Passengers. Although sold as a two-passenger car, the Convertible Coupe came with a pair of folding, forward-facing auxiliary seats.

1938-1940 V-16

The Series 90 Cadillac Sixteen of 1938 was continued through the 1940 model year with only minor exterior changes. Chrome running board trim strips were added for 1939 and taillight and bumper details were altered from year to year. General Motors continued to exercise strict control over the release of V-16 chassis to custom body builders other than Fleetwood. This is the only full-custom Series 90 Cadillac built in 1940. It is a special Five-Passenger Coupe designed and built by Derham of Rosemont, Pa. The leather-covered roof is unusual for this body style and the rear deck contour somehow doesn't quite work. The three horizontal trim pieces used on the lower edges of all four fenders on production 1938-40 Sixteens were deleted on this custom. Note the ventilator panes in the rear side windows.

The very impressive 1938 Cadillac Series 90 Convertible Sedan by Fleetwood carried a factory list price of an even $6,000. It was designated Style 9029. Fleetwood produced only 19 of these beautiful four-door convertibles during the Sixteen's final three model years. This one was delivered without whitewall tires. This was the sixth consecutive model year in which the Sixteen's long hood featured three horizontal ventilation louvers—a styling motif originally introduced for 1933.

Once again, one of the most attractive body styles on the all-new Cadillac V-16 chassis was the versatile four-door convertible sedan. This model, Style 9029, was officially listed as the Series 90 Fleetwood Convertible Touring Sedan for Five Passengers. Only 19 were built between 1938 and 1940, most of them (13) in the first year. Four were delivered in 1939 and only two in the Sixteen's last year. Eleven were produced with an Imperial glass division between the front and rear seats. This Series 38-90 Cadillac V-16 Convertible Sedan was photographed on the roof of the GM Building in Detroit prior to the all-new Sixteen's public announcement in October, 1937.

With its huge, tentlike top lowered, the Cadillac V-16 Convertible Sedan by Fleetwood had the imposing appearance of a parade phaeton—a use to which these huge convertibles were not infrequently put. Again, the four-door convertible sedan body style was especially suited to the Series 90 Cadillac's long 141-¼-inch wheelbase. With the top up, rear roof proportions of this car were almost those of a convertible Victoria. The upper door center posts had to be removed and side door glass lowered to convert this car into a true convertible. The Style 9029 Convertible Sedan weighed 5,350 pounds.

Fleetwood designed and built two of these truly gigantic V-16 convertible sedans for the Roosevelt White House in 1938. Both were constructed on special 165-inch wheelbase Cadillac Series 90 chassis. They were designated Fleetwood Body Style 9006. Seven-passenger cars, they were intended for use as security cars by the White House Secret Service staff. President Franklin D. Roosevelt usually used the "Sunshine Special", a custom-built, armored 1938 Lincoln Model "K" which was replaced by another Lincoln—Dwight D. Eisenhower's "bubbletop", in 1951. The Cadillac V-16 security cars were later repowered with V-8 engines. Both of these cars are believed to still exist today. Note the big Federal coaster siren mounted between the fender and hood, and the body extension between the doors.

1938-1940 V-16

Here is one of the two special White House security cars with its huge top let down. These special convertible sedans could often be spotted in newsreel footage of President Roosevelt's—then Truman's—motorcades. These were cars of truly staggering proportions. Wheelbase was 165 inches—11 inches longer than the gargantuan 154-inch wheelbase used on 1934-37 Cadillac Sixteens. Note the grab handles on the windshield posts. Secret Service agents assigned to the White House security detail often rode the running boards in presidential motorcades.

The Fleetwood Sixteen Town Sedan was an especially handsome five-passenger car. Style 9039, the Town Sedan combined the formality of closed rear quarters with the practicality of four doors and a conventional bustle-back trunk. The Series 38-90 Fleetwood V-16 Town Sedan had a factory list price of $5,695 and weighed 5,140 pounds. All 1938-40 four-door V-16 Cadillacs had doors which opened from the center. The new 1938 Cadillac Sixteens sported one-piece, hinged sidemount wheel covers.

Among the most popular Series 90 Cadillac body styles for 1938 was the practical five-passenger sedan. Fleetwood offered this body style in two versions between 1938 and 1940—Style 9019, a Touring Sedan for Five Passengers and Style 9019-F, a Five-Passenger Imperial with glass division between the front and rear seats. The standard sedan had a factory list price of $5,140 and weighed 5,105 pounds. The Imperial, shown here, cost $5,215 or only $75 more. This Style 9019-F Imperial has no sidemounts, unusual for a formal car. Fleetwood built 48 of these big four-door sedans in 1938, including five Imperials. A total of 67 were delivered through 1940, including seven Imperials.

This elegant, formal four-door sedan body style was offered in five and seven-passenger variations on the all-new 1938 Cadillac V-16 chassis. The Series 38-90 Fleetwood Formal Sedan for Five Passengers, Style 9059, was priced identically with the seven-passenger model, Style 9033-F, at $6,055. Upper rear roof quarters were closed and the top was covered with a padded, leather material. This very stylish sedan was not offered with a divider partition between the front and rear seats. However, a Seven-Passenger Imperial, Style 9033, was offered as a conventional six-window model in the 1938-40 Series 90 line. The Five-Passenger Formal Sedan weighed 5,345 pounds.

Fleetwood built a total of 23 of these sleekly-styled five-passenger Series 90 Town Sedans between 1938 and 1940. Twenty were delivered in 1938, two in 1939 and only one in 1940. The rear window featured two vertical dividers, a styling feature first used in 1934. Note the V-16 emblem on the deck lid. This was Fleetwood Style 9039. From this angle, the car looks slightly Packardesque.

1938-1940 V-16

The Fleetwood Formal Sedan featured closed rear quarter panels and a thickly-padded, leather-covered roof. The roof had two full-length seams, visible on each side of the padded, frameless rear window opening. A total of 14 of these five-passenger Fleetwood Formal Sedans were built between 1938 and 1940, while another 30 seven-passenger Formal Sedans were produced in the same three-year production run. This example has the very popular optional sidemounted spare tires and wide whitewall tires. The principal difference between the five and seven-passenger Formal Sedans was a set of folding rear auxiliary seats in the latter.

Among the most popular of all 1938-1940 Series 90 Cadillac V-16's were the big, impressive seven-passenger sedans and limousines. The principal difference between the Series 90 seven-passenger sedans and the more formal limousine was an "Imperial division" with adjustable glass partition between the front and rear compartments. This is the 1938 Series 90 Fleetwood Seven-Passenger Touring Sedan, Style 9023, which carried a list price of $5,270. Weight was 5,185 pounds. A total of 92 of these huge sedans were built between 1938 and 1940, including 65 of these 1938 models. Eighteen were built in 1939 and only nine in the Sixteen's last year.

This is the 1938 Cadillac Series 90 Seven-Passenger Sedan by Fleetwood without the sidemounted spare tire option. The new Fleetwood bodies for the V-16 chassis were also available on the 1938 V-8 Series 75 Cadillacs which shared the same 141-inch wheelbase. Although the V-16 Cadillac's wheelbase was more than a foot shorter, Cadillac boasted that its new Fleetwood Series 90 bodies were actually larger than those mounted on the 154-inch 1937 sixteen-cylinder chassis. This view shows the new Sixteen's massive front-end ensemble, horizontal hood louvers and squared "suitcase" fenders. Remarkably, Series 90 prices remained unchanged from introduction in the fall of 1937 through the end of the 1940 model year.

By far the most popular of all "last generation" Cadillac Sixteens was the staid Seven-Passenger Touring Imperial Sedan, Fleetwood Style 9033. A total of 175 of these awesome V-16 sedans were built between 1938 and 1940, including no less than 95 of these 1938 models. The regal Fleetwood Touring Imperial had an adjustable glass division between the front and rear compartments and was intended to be chauffeur-driven. Most, of course, were. Factory list price was $5,420 and the Style 9033 sedan weighed a hefty 5,348 pounds. Strangely enough, Fleetwood's Style 9033-F was an altogether different-looking car, a seven-passenger Formal Sedan with closed rear quarters and a padded leather-covered roof and not offered with the Imperial division.

Exterior differences between 1938, 1939 and 1940 Cadillac V-16's were extremely minor, limited in the main to bumper and taillight detail changes. Inside, however, there was a redesigned instrument panel every year. Series 90 Cadillacs for 1939, got full-length chrome running board trim strips which were carried over onto the 1940 models. This is a 1939 Cadillac Seven-Passenger Sedan by Fleetwood. Note the larger, chromed taillight housing and the slender, vertical "V-16" insignia on the sidemount cover. The three horizontal chrome trim pieces on the lower edges of the fenders were now chromed.

1938-1940 V-16

After an 11-year production run, the mighty sixteen-cylinder Cadillac came to the end of the line. The prestigious V-16 was discontinued at the end of the 1940 model year. In its little more than a decade as the flagship of the General Motors passenger car fleet, the legendary V-16 had firmly established Cadillac as America's luxury car leader and had been largely responsible for creating the mystique which continues to surround the Cadillac name to this day. The 1940 Cadillac Sixteens were almost unchanged from the 1938 and 1939 models, but the 1940 Series 90 had chromed parking lamps atop their front fenders, sealed-beam headlamps and redesigned taillights. Only 61 Series 40-90 Cadillac V-16's were built before this truly classic car was quietly dropped. The Sixteen had done its job well, even if it never was a moneymaker. This is the 1940 Cadillac Series 90 Seven-Passenger Touring Imperial Sedan by Fleetwood, 20 of which were built accounting for nearly half of the final year's production.

Fleetwood built only 18 of these open-front town car bodies for the last-generation 1938-1940 Cadillac V-16 chassis. Style 9053, the Fleetwood V-16 Town Car at $7,175 was by far the most expensive model in the entire Cadillac line. Eleven of these 1938 models were built, including one which was delivered to the Vatican for Papal use. Five Series 90 Fleetwood Town Cars were built in the 1939 model year and only two for 1940. These severely formal Town Cars were seven-passenger jobs.

This is a 1938 Fleetwood catalog illustration of the Series 90 Cadillac Sixteen Town Car. In this view, the front roof section has been removed but the door glass left in the raised position. With its top in place, the Series 90 Town Car, Fleetwood Style 9053, was almost indistinguishable from the Style 9033-F Formal Sedan. Front door construction, however, differed. Cadillac V-16 customers could specify the conventional bustle-style trunk illustrated on this car or a fastback roofline. Most opted for the practical trunk.

Only half a dozen 1938-1940 Cadillac V-16 chassis were released to outside custom body builders for the installation of special custom bodies. General Motors had always preferred that its own Fleetwood Body Division have complete control over its V-16 body designs. Two bare Series 90 chassis were shipped in 1939. One of these was used as the basis for this highly individual full custom formal Town Car designed and built by Derham of Rosemont, Pa. for Mrs. H. B. DuPont of Wilmington, Del. Note the flat, single-piece windshield and the fender overhang behind the relatively short, angular body. This distinctive limousine was still being used by the DuPont family well into the 1960's.

An old-line carriage builder and one of the few custom coachbuilders to survive the Depression was the Derham Body Co. in the Philadelphia suburb of Rosemont, Pa. Derham became the only outside custom body house to do work for Cadillac in later years, building formal Fleetwood Seventy-Five Limousine conversions into the mid-1960's. Derham designed and built this special, individual open-front Town Car on a 1939 Cadillac Series 90 Sixteen chassis for a Mrs. David Haas, at the time a resident of the ritzy Waldorf-Astoria Hotel in New York. This very formal car has a one-piece, flat windshield instead of the standard Fleetwood "V" glass. The three chrome trim pieces appear to have been transplanted from the rear fenders onto the sidemount spare tire cover—however, the spare trim is much smaller and definitely custom made. A wide belt molding follows the contour of the unusual customized rear fenders and the rear roofline is of interest. Note the special molding behind the rear side window.

V-16

Few custom-body houses survived the depression. The few that did found little work. Demand for one-off, custom bodies was a phenomenon whose main life-span lasted from the mid-1920's to the mid-1930's, and by 1938 this facet of the automotive world had all but disappeared. But there still remained a few wealthy individuals and families who continued to demand something special when it came to choosing a new luxury car. For this small segment of the auto market, the Derham Body Co. of Rosemont, Pa. bravely offered a number of special body designs—some of them quite radical—for mounting on the 1938-40 Series 90 Cadillac V-16 chassis. Very few were actually built, and of those that were, most were extremely conservative formal town cars and limousines. One that was anything but conservative was this proposed fastback Sport Coupe with almost circular doors and extra glass above the windshield header. Looking far more like a proposal for a toy car than a body to set on the pretentious V-16 chassis, this multi-colored coupe would have had front end design resembling an oversized Ford of the day. Apparently no Cadillac sheet metal parts were to be used in this body, with the only hint to its heritage being the "V-16" design on the hubcaps or wheel covers.

Cadillac took the industry by surprise again with its introduction of a totally new, 135-degree sixteen-cylinder engine for 1938. This new V-16, which drew heavily on the design principles of Cadillac's all new V-8 engine introduced for 1936, replaced the venerable 90-degree, 452 cubic inch V-16 which had been utilized with only minor improvements between 1930 and 1937. The all-new 1938 V-16 was shorter and 250 pounds lighter than the engine it replaced but turned out the same 185 horsepower. Like the 1936 V-8 from which it was derived, the 431 cubic inch displacement 1938-40 V-16 was of monobloc, or "enbloc" design. It was a "square" design with 3-¼-inch bore and stroke. Unlike the first-generation V-16, the new engine was purely functional. No attempt had been made to "style" it, as was the beautiful 1930-1937 Sixteen. Perhaps this was because the new V-16 was mounted low in the chassis, almost out of sight. The new L-head V-16's two separate cylinder blocks each had their own distributors, downdraft carburetion and manifolding, and in effect, operated almost like two independent straight-eight engines sharing a common crankshaft.

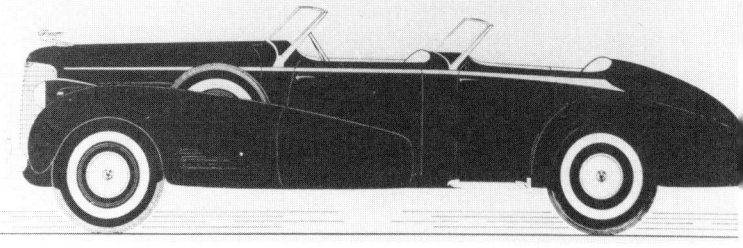

Another Derham proposal for the 1938-40 Cadillac V-16 was this interesting and attractive dual-cowl sport phaeton, which seems to trace its blood lines back to the much-admired dual-cowl phaeton of the first series V-16. Unlike the sport coupe, this car would have utilized a basic Series 90 front end, with Derham coachwork taking over aft of the cowl. Large custom pontoon front fenders all but swallowed the side-mounted spares, and unless an access door was provided somewhere, these heavy spares would have been a real joy to remove when needed. No running boards were used, but small step plates were to be provided for the tonneau passengers. It is not known what type of weather protection was planned for this car, or for that matter, if the design proposal was even carried far enough to take weather protection into account.

A third Derham proposal for the V-16 chassis that was never put into reality was this 5-passenger Touring Car. The vehicle, which at first glance looks like a long 7-passenger Parade Phaeton, had a relatively short-coupled passenger compartment, and a very long trunk deck. In a very advanced styling move for the era, the rear fenders were to be a part of the body sheet metal, and were up-swept at the trailing edges, in what probably would have been Cadillac's first tail fins. The long front fenders were also very advanced, in that they would have faded into the front door sheet metal—a styling move that didn't become popular on production vehicles until after World War II. Derham called this style its 5-passenger Victoria. On these designs, it is interesting to note that Derham leaned toward wide, chromed belt moldings. It's a pity that none of these Derham proposals were actually built. From a styling point of view, these very advanced designs would have been a knockout. After World War II, the concept of custom cars returned, but this time with the major hotspot being in the Los Angeles area, and the major customers being an off-shoot of the hot rod contingent.

The 1938 sixteen-cylinder Cadillac was truly new from road to roof, including its chassis. The new Series 90 Sixteen chassis, which was produced with only very minor improvements and changes from 1938 through the 1940 model year, was a massive, bridgelike affair with very deep side rails and a completely new, almost horizontally-opposed 135-degree V-16 engine mounted extremely low in the frame. Two huge air cleaners are mounted on the top of the engine. The same 141-¼-inch wheelbase was used to carry all 12 Fleetwood bodies. Note the huge radiator and the eight body mounts.

1939

For 1939, Cadillac went to market with five separate passenger car lines on four different wheelbases. An all-new Series 61 replaced the lowest-priced Cadillac Series 60, and all 1939 Cadillacs except the unchanged Series 90 Sixteens had dramatically-restyled front ends.

Gone was the vertical, clifflike look of Cadillac's 1938 eight-cylinder cars. The freshly-styled 1939s sported a bold, new sharply-pointed grille design. The attractive new grille was swept back slightly and consisted of a series of very fine horizontal chrome bars. These were complemented by equally thin, vertical grille bars in the wide "catwalk" area between the prowlike grille and the front fenders. The catwalk grilles were non-functional, although at one point consideration had been given to a catwalk-cooled car. Headlights were long, conical types and were mounted higher on the hood panel sides. The die-cast ventilator panels at the rear of the hood also had a finely-detailed horizontal motif. The total effect was a much sleeker-looking car.

Cadillac's lower-priced models utilized GM's new "B-O-P" bodies (Buick-Oldsmobile-Pontiac) which were also shared by the 1939 LaSalle. These bodies had significantly-increased glass area—about 27 per cent—which gave Cadillac's smaller cars a lighter, airier appearance. Running boards were optional on some models, a styling trend initiated by the pioneering 1938 Cadillac Sixty Special, and rumble seats had gone the way of the buggy whip.

In 1938, Cadillac had introduced a "Sunshine Roof" option for some of its closed bodies. This consisted of a retractable metal panel above the front seat, like those used at the time on some British and European cars. The GM "Sunshine Turret Top Roof" was offered on both Cadillac and LaSalle but was never a big seller. Only 1,500 sunroofs were sold on Cadillacs and LaSalles between 1938 and the end of the 1941 model year, when this option was discontinued.

The new Series 61 Cadillac line included four Fisher body styles on a 126-inch wheelbase which was two inches longer than the 1938 Series 60 which it replaced. Series 61 models were powered by the same 140-horsepower, 346 cubic inch V-8 engine, however.

The milestone Sixty Special also received Cadillac's new, V-type grille and front end. Rearward of the cowl this style-setting four-door Touring Sedan was generally unchanged. During the 1939 model year, a division-glass option was added. This partition did not have a header bar in the ceiling. The retractable glass divider travelled in narrow channels in the door posts. The Sunshine Roof option was also available on the Sixty Special which continued in a single body style on a 127-inch wheelbase.

A total of 14 Fleetwood body styles were available in the moderately-expensive Cadillac Series Seventy-Five. Wheelbase remained at 141 inches. The wide selection of distinctive Fleetwood bodies was identical with some of those available on the big Sixteen Fleetwoods which shared the same chassis.

The 1939 Series 75 Cadillacs also got the new, pointed grille. Body styles ranged from sporty two-passenger coupes and convertibles on up to impressive seven-passenger sedans and limousines, big formal sedans and stately open-front town cars. Like the Series 61 and Sixty Special, the Series 75 Cadillacs were powered by the 346 cubic inch V-8 engine which had been introduced three years earlier, but which had been uprated to 140 horsepower.

While generally advertised as a two-passenger car, the 1939 Cadillac Series 61 Convertible Coupe came equipped with two opera-type folding auxiliary seats designed to accommodate two extra passengers on short trips. Model 6167, the handsome Series 39-61 Convertible Coupe sold for only $1,770 and weighed 3,765 pounds. All Cadillacs except the style-setting Sixty Special sported full-length chrome belt moldings. Long, cone-shaped headlights sported horizontal trim pieces on their sides. Note the stone guards on the front of the rear fenders. Only 350 of these sporty ragtops were sold this year.

Cadillac once again shuffled its model designations, replaced the lowest-priced Series 60 with a new Series 61. Four Fisher body styles were offered on the 126-inch wheelbase Series 61 chassis which was two inches longer than the Series 60 which it replaced. The lowest-priced model in the new Series 61 Cadillac line was the Model 6127 Coupe for two or four passenger which had a factory-delivered price of $1,620. It was the second-best seller, with 1,023 delivered during the 1939 model year.

In a move toward greater standardization across General Motors' various car lines, four GM passenger car divisions this year shared the same basic "B-O-P" bodies. The initials stood for Buick-Oldsmobile-Pontiac, but the same smaller GM bodies were also used for the 1939 LaSalle and Cadillac's lowest-priced Series 61 models. The new bodies had much-greater glass area and an angled "C" pillar. By far the most popular model in the new 1939 Cadillac Series 61 line was the Four-Door Touring Sedan, Model 6119, of which 3,955 were built. An Imperial Sedan with glass division, Model 6119-F, was also offered. Base price for the five-passenger, four-door sedan was $1,680. Weight was 3,770 pounds.

This view of the 1939 Cadillac Series 61 Four-Door Touring Sedan shows the Cadillac's dramatically-restyled front end. The pointed grille featured a series of very fine horizontal bars which were duplicated in the catwalk area between the hood and fenders. Headlights were very long and were mounted higher on the hood. General Motors' new "Sunshine Roof" was offered on this style as Model 6119-A, but only 43 were sold. The Sun Roof, a retractable metal panel over the front seat area, had been introduced during the 1938 model year. Only 1,509 were produced between 1938 and the 1941 model year, about half of them in model year 1939.

Despite its sporty good looks, the convertible sedan body style was attracting few customers. Only 140 of these 1939 Cadillac Series 61 Convertible Sedans, Model 6129, were sold. The most expensive car in the Series 61 family, the Convertible Sedan at $2,170 was the only one priced above $2,000. Four-door convertible sedans were available in every one of Cadillac's four 1939 car lines (except the Sixty Special) from the Series 50 LaSalle on up to the prestigious Series 90 Sixteen Fleetwood. This style was offered with or without a bustle-back trunk. The Series 39-61 Cadillac Convertible Sedan weighed 3,810 pounds.

Cadillac's most popular hearse and ambulance chassis this year was the LaSalle, with 874 delivered to funeral car and ambulance builders. Next came the Series 61 commercial chassis shown here, which had the same 156-inch wheelbase as the LaSalle. A 161-inch Series 75 commercial chassis was also offered. A total of 237 Series 61 commercial chassis were shipped from the Detroit factory. This is a 1939 Superior-Cadillac Limousine Style Funeral Coach with optional sidemounted spare. The right-hand sidemount sometimes carried a folding, wheeled casket truck rather than a spare tire. The Superior Coach Corp. was located in Lima, Ohio.

1939

The 1939 Series 50 LaSalles also got a new look up front. The redesigned 1939 grille was a tall, very slender affair with a series of extremely fine horizontal bars. It was reminiscent of the narrow, vertical grille first used on the daring 1934 LaSalle. The new grille was flanked by wide catwalk grilles with vertical chrome bars. All five 1939 LaSalles rode on a new 120-inch wheelbase that was four inches shorter than the previous year. The LaSalle was powered by a 122 cubic inch, 130-horsepower Cadillac V-8 engine. LaSalle sales improved considerably, jumping to just over 22,000 in the 1939 model year, but still trailed far behind Packard. Despite beautiful styling, superb engineering and excellent value for the money, the LaSalle just couldn't seem to catch up with its rivals. Perhaps if the car had carried the Cadillac name it might have been different.

The American automobile industry was on the rebound from the 1938 recession. Cadillac's sales shot up to 35,582. It was a year of records: the industry built its 75 millionth car; Ford produced its 27 millionth vehicle, and Chevrolet built its 15 millionth car. Note that at this point, Ford had produced more than one-third of all cars ever built in the U.S.

But war clouds were gathering over Europe again. Nazi Germany was flexing its military muscle. In September, 1939 Germany invaded Poland and Great Britain and Canada declared war. The United States would not enter the conflict for another two years, but American industry was already girding for defense production. Cadillac in 1939 received its first defense assignment. The Detroit plants began to produce engine parts for the Allison V-1710 aircraft engine which would soon power several famous U.S. fighter aircraft.

Like all other 1939 Cadillacs except the Series 90 Sixteen, Cadillac's highly successful Sixty Special sported a new, pointed nose. The boatlike prow featured a series of fine, horizontal grille bars which were duplicated vertically in the "catwalk" area between the grille and front fenders. The hood ventilator casting also consisted of a series of fine, horizontal lines. Rearward of the cowl, the Sixty Special Four-Door Touring Sedan's styling was unchanged. Factory-delivered price was also unchanged at $2,090. The wheelbase continued at 127 inches. Designated the Model 6019, this five-passenger, Fisher-bodied car weighed 4,110 pounds. Note the long, tapered headlights which were mounted higher on the hood panel sides.

1939

The style-setting Cadillac Sixty Special continued to sell extremely well. More than 5,500 were sold in the 1939 model year compared with 3,703 in this milestone car's first year on the market. Here's the restyled 1939 Sixty Special in profile. The finely-detailed, pointed grille was swept back slightly. The die-cast vent panel at the rear of the hood side panel replaced the three sets of short horizontal hood louvers used on the 1938 model. A "Sunshine Roof" option was available this year, but only 280 were sold—225 Style 6019-SA's and 55 Style 6019-SAF's, the latter on the Imperial Sedan version with glass division.

While more than a few Cadillac Sixty Special customers ordered optional sidemounted spare tires, the car looked better without them. The sidemounts looked better in profile than head-on. Viewed from the front they made the otherwise clean-lined Sixty Special look bulky. Here is a well-dressed 1939 Cadillac Sixty Special, Model 6019, so equipped. Cadillac was beginning to blossom forth with some exciting two-tone color combinations, which especially suited the Sixty Special's distinctive upper and lower body shapes.

The 1939 Cadillac Series Seventy-Five model line included a very comprehensive selection of 14 body styles, all by Fleetwood and all built on the same 141-inch wheelbase V-8 chassis. Most of these Fleetwood bodies were also available on the V-16 Series 90 chassis of identical wheelbase. This is the 1939 Cadillac Series 75 Fleetwood Coupe, Style 7557, which sold for $3,280. It was listed as a two-passenger car but two more passengers could be accommodated in a pair of standard folding opera seats. The Series 39-75 five-passenger coupe, which was identical in outward appearance, was Style 7557-B. Only 36 Series 75 Fleetwood Two-Passenger Coupes were built during the 1939 model run.

The 1939 Fleetwood Five-Passenger Coupe was externally identical to the two-passenger version but sold for exactly $100 more. Factory-delivered price was $3,380. This model was listed as Fleetwood Style 7557-B. Weight was 4,695 pounds—exactly 100 pounds heavier than the Style 7557 Two-Passenger Coupe. Only 23 of these cars were sold this year. Interior appointments included a silk curtain which could be pulled down over the rear window. Note the massive "goddess" hood ornament which functioned as the hood latch.

The 1939 Cadillac Series 75 Fleetwood Convertible Coupe was a very sporty-looking car. This two-door convertible was sold as a two-passenger car but standard equipment included a pair of folding opera seats in the rear. Upholstery was genuine leather or Bedford cord fabric depending on customer preference. Only 27 of these Fleetwood Convertible Coupes, Style 7567, were sold in the 1939 model year. Factory-delivered price was $3,380 and standard weight was 4,675 pounds.

This is the Style 7567 Fleetwood Convertible Coupe with its top down. The folding top recessed completely into a well behind the front seat. A button-down dust boot protected the retracted top and gave the car a clean, tidy appearance. Sidemounted spare tires were not nearly as popular now, and most of these cars were sold without them. Some customers insisted on having them, however.

Four-door convertible sedans were available in every one of Cadillac's 1939 model lines except the Sixty Special. All were impressive, beautifully-proportioned cars. This is the 1939 Cadillac Series 75 Fleetwood Convertible Sedan, Style 7529. The five-passenger Series 39-75 Convertible Sedan had a factory-delivered price of $3,945 and weighed 5,030 pounds. This one has dual sidemounted spares and double-faced white sidewall tires. Just 36 of these classic convertible sedans were built this year.

The Series Seventy-Five Fleetwood Convertible Sedan was offered with or without a bustle-style trunk. Most customers preferred the trunk, and relatively few "fastback" convertible sedans were sold. Here is how the Style 7529 Fleetwood Convertible Sedan looked with the top up. Note the Cadillac name in small block letters on the finely-detailed, horizontal hood ventilator panel. The Cadillac name also appeared in bold script on both front and rear bumpers.

Fleetwood produced 51 of these five-passenger Town Sedan bodies for the 1939 Cadillac Series 75 chassis. The semi-formal Fleetwood Town Sedan had closed upper rear quarters and a metal top. This model, Fleetwood Style 7539, has the standard trunk. A much less popular fastback version was also offered. The Series 39-75 Fleetwood Town Sedan had a factory-delivered price of $3,635 and weighed 4,820 pounds. All four doors are equipped with no-draft ventilator panes.

Cadillac's senior V-8 line consisted of a series of big five and seven-passenger four-door sedans at the upper end of the Series Seventy-Five model range. This is the Fleetwood Five-Passenger Touring Sedan, Style 7519, equipped with the optional sidemounts and sporting accessory fog road lights of the era. Owned by Steve Scalzo of Minneapolis, the car is one of 543 such models built in 1939. Its base price then was $2,995. An Imperial Sedan for Five Passengers, designated Style 7519, was also offered, at a factory price of $3,155. Only 53 of these were built. The chauffeur-driven Imperial was identical to the Sedan on the exterior, but was equipped with a glass division between the front and rear compartments.

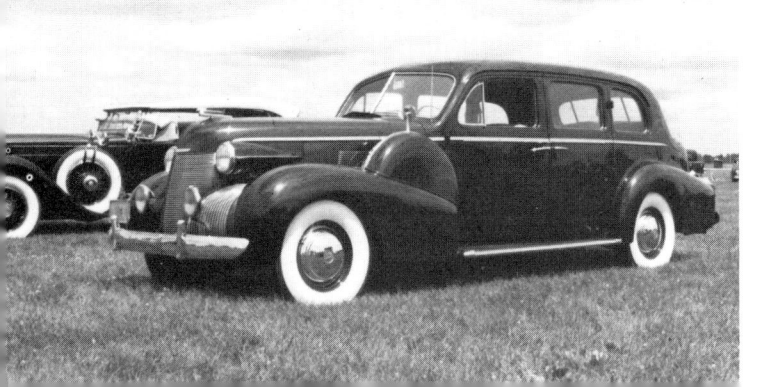

The big 1939 Cadillac Series 75 Seven-Passenger sedan was available in several versions. Style 7523 was the standard Series 39-75 Fleetwood Seven-Passenger Touring Sedan, 412 of which were sold with a factory-delivered price of $3,210. Style 7523-L was a Seven-Passenger Business Sedan aimed at the funeral service trade. The most popular model in the 1939 Series 75 line was the Fleetwood Seven-Passenger Imperial, Style 7533, which sold for $3,360. A total of 638 were delivered this year. The 7533 was generally identical to the 7523 but had a glass partition between the front and rear passenger compartments. Adding confusion to this nomenclature was the Style 7533-F, a much-different looking Fleetwood Formal Sedan with closed rear quarters and a padded leather roof. This is the Style 7523 Fleetwood Seven-Passenger Touring Sedan with optional sidemounted spare tires.

The elegant Fleetwood Formal Sedan was available in five and seven-passenger versions. A total of 97 were built, including 53 five-passenger Formal Sedans, Style 7559, and 44 seven-passenger models, Style 7533-F. The principal difference was a pair of folding opera seats in the seven-passenger job. These formal sedans had no header bar for the division glass. With the partition lowered, it was a conventional sedan. With the glass raised in its side channels it was a formal limousine. Note the leather-covered roof and absence of quarter windows. Both models were priced identically at $3,995.

1939

The most expensive car in Cadillac's V-8 offerings for 1939 was the regal Series Seventy-Five Fleetwood Town Car. Style 7553, this Fleetwood semi-custom was a very formal open-front town limousine of the now rare, classic style. Very few town cars were being built now. Only 13 Series 39-75 Town Cars were sold. This same Fleetwood formal body was also available on the V-16 chassis which shared the same 141-inch wheelbase with the Series 75. The Cadillac-Fleetwood Town Car was a seven-passenger model, with a factory-delivered price of $5,115. Weight was 5,095 pounds. All Series 75 Cadillacs had the same heavy chrome side molding.

The Cadillac Motor Car Division continued to market special livery-type limousines for funeral directors and commercial limousine operators. These big sedans were specially trimmed for hard commercial use. Sedan ambulance and invalid car versions were also available through Cadillac's Commercial Car Department in Detroit. In special sales literature prepared for the funeral trade, Cadillac sometimes advertised these business sedans as eight-passenger cars. This is the 1939 Series 75 Fleetwood Seven-Passenger Business Sedan, Style 7523-L, 33 of which were sold this year. Two Business Imperials, Style 7533-L with glass division, were also built in 1939.

This is a Derham-bodied full custom built on the 1939 Cadillac Series Seventy-Five eight-cylinder chassis. The style is a correctly-formal open-front town car. Note the smooth top, center-hinged doors and complete lack of bright body side moldings. The flat, single-piece windshield was something of a Derham town car trademark. This one is equipped with a sun visor. Derham continued to do small numbers of Cadillac Fleetwood Seventy-Five formal limousine conversions well into the 1960's, long after all of the other mainline coachbuilders had vanished. The Derham Body Co. was located in the Philadelphia suburb of Rosemont, Pa. Note that the division glass cannot be lowered, but instead consists of two sliding panes, which could be opened partially either for ventilation or to converse with the chauffeur. Also, unlike the Cadillac bodies which this year had all doors latching on the center pillar, this car has all doors hinged on that pillar.

The Cadillac, LaSalle and Buick funeral cars and ambulances built by The Flxible Co. of Loudonville, Ohio, are interesting because of their extra-high, built-up hoods. Flxible had to raise the height of the hood to meet the beltlines of its high-sided professional car bodies. These vertical extensions required special fillers for the hood side panels and extra grille pieces. This is a 1939 Flxible-Cadillac Limousine Funeral Car. Compare the height of the hood on this hearse with those on other 1939 Cadillac-chassised funeral cars and ambulances.

1939

The A.J. Miller Co. built this clean-lined limousine style ambulance on a 1939 Cadillac commercial chassis. Note the angled "C" pillar, frosted rear window glass, high window sill line and the siren with front flasher light mounted on the roof. This ambulance is on display on the showroom floor of a large Cadillac-LaSalle-Oldsmobile dealership. The sign in the window reads "Brogan". Hearses and ambulances were rarely displayed in passenger car dealer showrooms. Such vehicles were built to customer order and usually were delivered directly from the factory to the customer's business establishment.

Like its more expensive cousin, the Cadillac, the 1939 LaSalle got a nose job this year. LaSalle's new front-end included a tall, very narrow grille not unlike the one on the milestone 1934 LaSalle and a series of fine vertical grille bars in the "catwalk" area between the grille and front fenders. Big, conical headlights with horizontal trim on the sides were mounted higher up on the hood side panels. Running boards were optional this year. The 1939 LaSalle Series 50 Coupe for two or four passengers, Model 5027, was the lowest-priced car in the 1939 line, with a factory-delivered price of only $1,240. It was LaSalle's second-best seller with 2,525 delivered. The handsome Series 39-50 Coupe weighed 3,635 pounds.

For 1939, LaSalle continued to offer five Fisher body styles, all on a new, four-inch shorter 120-inch wheelbase. This is the 1939 LaSalle Series 50 Convertible Coupe, a two or four-passenger car listed as Model 5067. Factory-delivered price was $1,395. The LaSalle had a cleaner, more streamlined appearance this year. Greatly increased glass area made these cars look lighter and airier—especially the closed models. Note the finely-detailed die-cast ventilator panel at the rear of the hood. Side-mounted spare tires were available but these cars looked better without them.

The practical 1939 LaSalle Two-Door Touring Sedan was a slow seller this year. Only 977 Series 39-50 Two-Doors, Model 5011, were sold, compared with sales of more than 15,000 four-door sedans and 2,500 coupes. Included were 23 Two-Door Touring Sedans with the new "Sunshine Roof" option, Model 5011-A. The five-passenger Two-Door Touring Sedan weighed 3,710 pounds and sold for $1,280. The rocker molding treatment (running boards were now optional) was similar to that used on the style-setting Cadillac Sixty Special.

1939

The extensively-restyled 1939 LaSalle boasted a very substantial 27 per cent increase in glass area. This is very evident in the new General Motors "B" bodies shared with various other GM car divisions. The Series 39-50 LaSalle Four-Door Touring Sedan, Model 5019 was by far the most popular body style in the 1939 LaSalle line with 15,688 sold. This total included 380 Style 5019-A's with the new "Sunshine Roof" option. The five-passenger Four-Door Touring Sedan weighed 3,740 pounds and had a factory delivered price of $1,320. All 1939 LaSalles were powered by the 322 cubic inch, 125-horsepower V-8 engine introduced two years earlier.

Demand for the very attractively-styled LaSalle Convertible Sedan was beginning to taper off. Only 185 of these four-door convertibles, Model 5029, were sold this year. It was the most expensive body style in the 1939 LaSalle line with a price tag of an even $1,800. The tall, narrow grille featured a series of very thin horizontal chrome bars. The "catwalk" grilles were non-functional. The hood vent panel casting was also finely detailed with a series of thin, horizontal bars.

The stylish Flower Car had caught on with American funeral directors in a big way. All major funeral car and ambulance builders now offered them. Many were built on the LaSalle 156-inch wheelbase commercial chassis. This is a 1939 Meteor-LaSalle Flower Car, which utilized the entire upper body structure of the 1939 LaSalle Coupe. Note the drapes in the rear side windows. The rear deck could be adjusted to carry various floral arrangements in funeral processions. Most flower cars had a compartment below the flower deck that was large enough to carry a casket, permitting this versatile vehicle to be used as a hearse, flower, or utility car.

The old Cincinnati firm of Sayers and Scovill set a new style in formal funeral car styling with the introduction of its new "Victoria" landau on LaSalle chassis in 1938. This style, with closed upper rear quarters, padded leather-covered top and stylish carriage bows, soon became the most popular body style in the funeral coach industry and remains so to this day. Although some other funeral car builders—Eureka for one—had done a few custom-built landaus earlier, S & S popularized the style. This is the 1939 Sayers and Scovill Victoria on LaSalle chassis. For 1939, it was also available on the Cadillac commercial chassis. The company (Hess & Eisenhardt now) was still building the Victoria model on Cadillac chassis for the 1981 model year, an incredible 43 years later!

Sayers and Scovill of Cincinnati continued to disguise its LaSalle and Cadillac funeral cars and ambulances with a series of extra chrome bars on the grille and ornamental castings around the grille opening. The idea was to erase its model year—a ploy that didn't work very well. This is the 1939 S & S LaSalle "Masterpiece" Carved-Panel Hearse. All LaSalle identity has been replaced with S & S emblems, even on the hubcaps. Note the distinctive side window shape and the elegant coach lamp between the front door and the drape panel. The highly-ornamental carved-panel hearse had passed its peak in popularity and was definitely on the wane.

1940

For Cadillac and the American automobile industry in general, 1940 marked the end of one era and the beginning of another. It was the final year for the mighty Cadillac V-16, and for the venerable LaSalle. The 1940 model year was also the last in which side-mounted spare tires were available on any of the Cadillac Motor Car Division's cars.

The turbulent decade that had just ended had seen the zenith of the golden classic era, a peak that had been blunted by a crushing depression. Automotive ostentation was decidedly out of fashion now, and the once-flourishing custom body business had all but vanished. New kinds of cars were being designed and built for changing times.

For 1940, Cadillac went to market with seven car lines — five series of Cadillacs and two LaSalles. Cadillac dropped one series and added two new ones for 1940. LaSalle got a new premium model line. The lowline Series 61 Cadillac of 1939 was discontinued and replaced by an all-new Series 62. A new Series 72 Cadillac — built only during the 1940 model year — was introduced as a lower-priced companion for the big Series 75 sedans and limousines.

All V-8 Cadillacs for 1940, including Series 62, 72, 75 and the Sixty Special, sported massive new die-cast grilles. Cadillac's stylish pointed nose was retained but the horizontal grille bars and vertical catwalk cooling grilles were heavier now. Two horizontally-textured ventilator castings replaced the large single units used on each side of the 1939 Cadillac hood, and front and rear bumpers now bore the large, modern Cadillac script.

Like most 1940 U.S. cars, Cadillacs and LaSalles were now equipped with sealed-beam headlights. Parking lights incorporating new standard turn indicators were mounted on top of the 1940 Cadillac's front fenders. Running boards were going the way of the buggy whip: they were still optional on the 1940 Cadillacs and LaSalles, but customers could take or leave them at no extra cost.

The most stylish of all the 1940 Cadillacs were the dazzling new Series 62 models. This series included four sleek Fisher "torpedo" body styles on a new 129-inch wheelbase. These cars were wider, lower and much more integrated in appearance than the other cars in the line. They had no beltline trim and clearly reflected the strong styling influence of the earlier Sixty Special. Windshields were swept back at a 45-degree angle. The rear windows were curved, and door hinges were completely concealed. Initially, only a four-door sedan and two-door coupe were offered in this new series, but a two-door convertible coupe and four-door convertible sedan were added later in the model year.

The distinctive Cadillac Sixty Special was available in no less than four versions for 1940, all based on the same shapely four-door notchback body style. For the first time, Sixty Special bodies were now built by Fleetwood, thus commencing a long succession of Fleetwood Sixty Specials which would continue as Cadillac's premium four-door sedan well into the 1970's. The 1940 Sixty Special had new, flush-mounted taillights and revised bumpers. Sunroofs and side mounted spare tires were still optionally available on special order.

The 1940 Sixty Special line included, besides the standard five-passenger Touring Sedan, an Imperial Sedan with glass division and two formal open-front Town Cars.

The all-new Series 72 Cadillac boasted recirculating ball steering gear and set the style for Cadillac's big formal sedans and limousines for the rest of the decade. The "72" featured a fine chrome reveal molding around its side window openings and a broad belt molding into which the door handles were set. There were eight models, all on a 138-inch wheelbase, which was three inches shorter than the 1940 Series Seventy-Five. The Series 72 line included Fleetwood five and seven

The distinctive Cadillac Sixty Special entered its third year with a modest, but effective, grille change. Like all other eight-cylinder 1940 Cadillacs, the Sixty Special featured fewer but heavier grille bars and a return to the same style of multiple hood louvers that had been used on the original 1938 model. The Sixty Special's wheelbase continued at 127-inches and the factory-delivered price of $2,090 remained the same for the third consecutive year. This is the standard 1940 Cadillac Sixty Special Touring Sedan by Fleetwood, Model 6019-S. Of the 4,600 Sixty Specials built during the 1940 model year, 4,242 were of this model. Weight was 4,070 pounds.

The Cadillac Sixty Special was still an attractive and very stylish car. For the first time, Sixty Special bodies were now built by Fleetwood. Thus the 1940 model was the first of a very long line of Fleetwood Sixty Specials. No less than four Sixty Special models were available this year, all based on the same handsome four-door notchback sedan body. In addition to the standard Model 6019-S Touring Sedan, the Sixty Special family included a close-coupled limousine version with glass division. A total of 113 were built, including 110 Model 6019-FS Imperial Sedans and three Imperial Sedans with sun roofs, designated Fleetwood Style 6019-AF. Two 1940 Sixty Special town cars rounded out the line. The Imperial Sedan carried a factory list price of $2,230. Styling changes also included new taillights, bumpers with modern Cadillac script, and a narrow "V" emblem in the center of the rear decklid.

1940

passenger sedans, limousines and formal sedans. Except for the new grille with its stronger Cadillac identity, Series 75 models were largely carryovers. This would be the last year, however, that this series would include a complete range of models from coupes and convertibles on up to limousines and town cars.

The Cadillac V-16 supercar went out with a whimper. A mere 61 Series 90 "Sixteen Fleetwoods" were sold in the 1940 model year, bringing down the curtain on an illustrious 11-year model run that did much to elevate Cadillac to its position of leadership among luxury car builders. Much less complex cars with half as many cylinders offered virtually the same performance as the sixteen now, and the market for such conspicuous automotive luxury had disappeared. The last-generation sixteen-cylinder Cadillac was almost unchanged except for cosmetic detail changes between 1938 and 1940. The only noticeable changes on the rare 1940 Series 90 models were sealed-beam headlights and front fender-mounted parking and turn signal lights.

After a 14-year model run, the LaSalle went out with a real splash. For 1940 there were no less than nine LaSalle models in two separate price classes. The standard LaSalle Series 50 comprised five body styles on a new 123-inch wheelbase three inches longer than in 1939. A spectacular new premium Series 52 was added. Called the LaSalle Special, this series initially included only a four-door sedan and two-door coupe. A convertible coupe and four-door convertible sedan joined the line in the latter half of the 1940 model year.

The new LaSalle Specials were sensational. They utilized the same sleek GM "C" body used on the Series 62 Cadillacs. The high-styled Series 52 LaSalles handsomely outsold the standard Series 50 cars despite their higher price tags and are still considered by many to be the most beautiful LaSalles ever built. The 1940 LaSalle's 322 cubic inch V-8 engine was uprated to 130 horsepower by a one-eighth inch increase in the barrel diameter of its Carter downdraft carburetor.

But the successful new Series 50 LaSalles weren't enough. Although there was a 1941 LaSalle waiting in the wings, the decision was made to discontinue the LaSalle at the end of the 1940 model run. The first LaSalle had made its bow with much fanfare in 1927. Now, some 205,000 cars later, the last one rolled off Cadillac's Detroit assembly line in August, 1940 almost unnoticed. The LaSalle must be regarded as one of Cadillac's few real failures. It was never more than just a Cadillac cousin.

Packard had proved the wisdom of affixing a prestige nameplate to a lower-priced car with its extremely successful Package 120 in 1935. Lincoln did the same thing with the Zephyr. For 1941, Cadillac replaced the LaSalle with a new series of lower-priced cars carrying the magic Cadillac name: sales hit a new record. The division soon overtook Packard to assume leadership of the U.S. fine car field, which it has dominated ever since. While rumors persist that the fondly-remembered LaSalle nameplate will someday be revived for a new, smaller Cadillac, it hasn't happened — yet.

Oldsmobile introduced GM's new Hydra-Matic automatic transmission as an extra-cost option on its 1940 models. Ford placed the Lincoln Continental into limited production, offering a stunning convertible cabriolet and later a two-door coupe. Packard quietly dropped its V-12 and reshuffled its 1940 model line with new six-cylinder 110 junior series cars and the Super Eight Series 160 and 180.

This year's 41st National Automobile Show in New York would be the last. Europe was at war, and America was being inexorably drawn into the conflict.

General Motors' sporty "Sunshine Turret-Top Roof" option was introduced during the 1938 model year and was offered through 1941. It was never very popular, however, with only 1,500 sold over the four model years it was available. Half of those were sold in 1939. The Sunshine roof was patterned after those of some British cars and consisted of a sliding metal panel in the roof over the front seat. With the demise of the convertible in the late 1960's and early 1970's, the sun roof enjoyed a somewhat popular revival on many American cars large and small — including some Cadillacs. This is the Sunshine Roof, patents on which were held by GM's Ternstedt hardware division, as installed on the 1940 Cadillac Sixty Special. It was manually operated, and was also available on the LaSalle. A total of 230 Style 6019-SA Sixty Special sedans with the sun roof option were sold in the 1940 model year.

Two truly elegant close-coupled town car models were added to the Cadillac Sixty line for 1940. Only 15 of these very distinctive open-front town cars with Fleetwood semi-custom coachwork were built. They included nine cars with standard painted metal roofs, Style 6053-MB (for metal back) and six Style 6053-LB town cars with formal, leather-covered roofs. The metal back variation carried a factory price of $3,465 and the leatherback model was the most expensive car in the 1940 Sixty Special line, at $3,820. This is the handsome Style 6053-LB with the chauffeur's compartment roof removed. Standard equipment on these formal cars included a glass division. The Fleetwood Sixty Special Town Car weighed 4,365 pounds.

1940

The 1940 eight-cylinder Cadillacs — Series 62, 72, 75 and Sixty Special — entered a new decade with a revised grille design. The horizontal grille bars were heavier than the fine bars used on the 1939 models and the vertical chrome trim integrated into the catwalk area between the pointed front end and front fenders was also more prominent. Big, bullet-shaped headlight shells contained new sealed-beam headlamps and parking lights and new standard turn indicators were on top of each front fender. Large, modern Cadillac script appeared on front and rear bumpers for the first time. The big driving lamps on this car were extra-cost options.

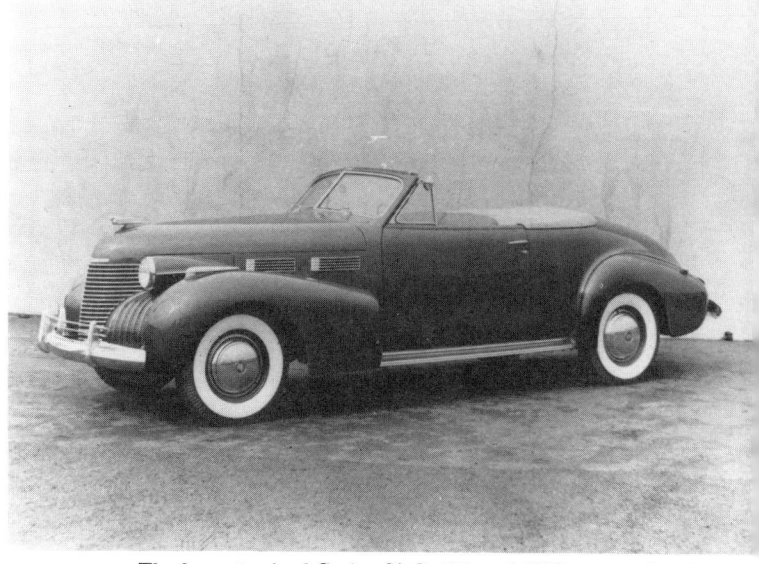

The lowest-priced Series 61 Cadillac of 1939 was replaced by a new Series 62 for 1940. The new Fisher-bodied Series 62 consisted of four models, all on a new 129-inch wheelbase that was three inches longer than the Series 61 which it replaced. The new Series 62 Cadillacs had softly rounded, cleanly-styled "torpedo" bodies that were much more integrated in design than Cadillac's other 1940 models. The sleek torpedo "C" bodies were also shared by the new premium 1940 LaSalle Series 52 Specials. This is the gorgeous 1940 Cadillac Series Sixty-Two Convertible Coupe, Model 6267. Factory price was $1,795. Note the optional bumper grille guard. The dashing Series 62 convertible couple weighed in at 4,045 pounds.

This is the very handsome 1940 Cadillac Series Sixty-Two Convertible Coupe with its top up. With its top up or down, the Series 40-62 convertible coupe turned heads wherever it went. It still does at shows today. Only 200 of these fleet-looking two-door convertibles were built this year. The Series 62 convertible could be ordered as either a two or four-passenger car. Note the tiny, slitlike rear window, which must have provided some fun when parallel parking in a tight spot.

A mere 75 of these impressive Cadillac Series Sixty-Two Convertible Sedans were built during the 1940 model year. Style 6229, the four-door convertible sedan had a factory list price of $2,195 and tipped the scales at 4,230 pounds. The interior included standard leather upholstery. The Series 40-62 Convertible Sedan was marketed as a five-passenger car. Few survive today, but those that do are highly regarded collector items in the upper price bracket.

Cadillac's new Series 62 models featured the sleek new Fisher GM "C" body. Contours were graceful and softly rounded. These cars had a cleaner, more integrated look than the other cars in the 1940 Cadillac line. The lowest-priced model in the Series 40-62 line was this Series Sixty-Two Coupe for Two or Four Passengers, Model 6227, which sold for $1,685. Weight was 3,940 pounds. A total of 1,322 were sold. Note the sharply-angled "B" pillar and distinctive upper and lower body forms. Two-tone exterior color choices also enhanced the strong good looks of these cars.

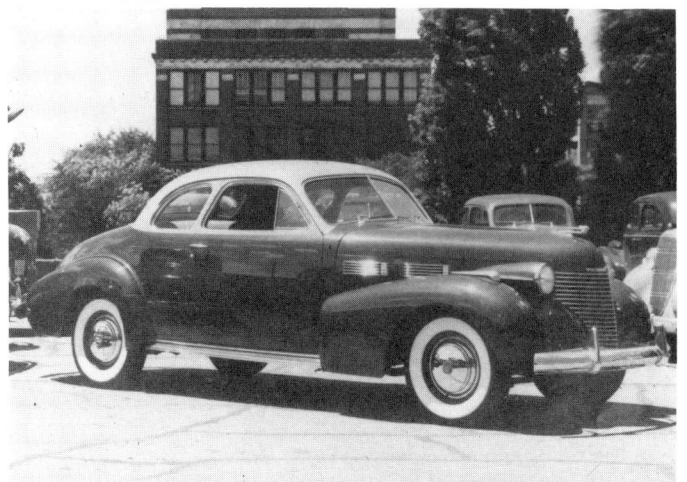

1940

The new Series 62 Cadillacs clearly reflected the styling influence of the 1938 Sixty Special, with distinctive upper and lower body forms and very little chrome trim. The new Series 62 Cadillacs replaced the former Series 61 at the bottom of the standard Cadillac line. By far the most popular body style in this new series was the Series Sixty-Two Four-Door Touring Sedan for Five Passengers, Model 6219, of which 4,242 were delivered during the 1940 model year. Factory-delivered price was $1,745. The handsome Series 62 Sedan weighed 4,030 pounds.

The 1940 model year would be the last in which Cadillac would offer side-mounted spare tires. This is the 1940 Cadillac Series Sixty-Two Four-Door Touring Sedan for Five Passengers equipped with optional six-wheel equipment. Other accessories on this very well-equipped Series 40-62 Sedan include a grille guard, fog lamps, white sidewall tires and two-tone paint treatment. The total effect is one of a very dashing family sedan indeed. Sidemounted spare tires dated back to the teens, when they were a practical necessity, and not an appearance item. Sidemounts reached their zenith in the style-conscious 1930's when most true classics — including Cadillacs and LaSalles — carried them. Various conversion builders have attempted, unsuccessfully, to resurrect the noble sidemount in recent years. Cadillac phased them out in style, as this photo attests.

Among the most unique of all production Cadillacs was the Series 72, which was built only during the 1940 model year. The Series 72 Cadillac was designed as a lower-priced companion car to the big Series 75 sedans and limousines. Eight models, all Fleetwood four-door sedan variations, were offered on a 138-inch wheelbase. Also offered was a Series 72 commercial chassis. A total of 1,526 Series 72 Cadillacs were produced. This one, the Series Seventy-Two Touring Sedan for Five Passengers, Style 7219, was the most popular with 455 delivered. It was also the lowest-priced car in the series with a factory-delivered price of $2,670. A Series 72 Imperial Sedan for Five Passengers with glass division, Style 7219-F, was also available at $2,740. An even 100 of the latter were sold.

At a glance, the new Series 72 Cadillac was nearly identical to the more expensive Series Seventy-Five. But the Seventy-Two's wheelbase was three inches shorter and the 72 had different beltline trim. The 72's windows also had thin chrome reveals and there were no vertical divider bars in the rear window. The author photographed this nicely-maintained 1940 Series 72 Cadillac at a service plaza near Arlington, Wash., in 1976. The Series 72's most easily recognizable styling feature are the large, rectangular taillight units mounted high on the sides of the bustle style trunk.

The 1940 Series 72 Cadillac's 138-inch wheelbase was three inches shorter than that of the more expensive Series 75. The moderately-priced Series Seventy-Two was built in both five and seven-passenger sedan body styles. This is the Series Seventy-Two Touring Sedan for Seven Passengers, Fleetwood Style 7223, which sold for $2,785. A Seven-Passenger Series 72 Touring Imperial with glass division, Style 7233, had a factory-delivered price of $2,915. The same style of chrome side molding, with a thicker dropped section on the doors, was used on both Series 72 and 75 Cadillacs. Door handles on the Series Seventy-Two were built into the chrome molding. On 1940 Series Seventy-Five Cadillacs they were mounted below.

1940

The new, one-year-only Series 72 Cadillac line included two big business, or livery, sedans. These were offered with seven or nine-passenger seating and were intended for use by funeral directors and commercial limousine operators. Fleetwood Style 7223-L was the standard seven-passenger business sedan, only 25 of which were sold at $2,690 each. An Imperial version with glass division, Style 7233-L, was also offered. Only 36 were built. Price was $2,825. All Series 72 Cadillacs were powered by the same 346 cubic inch V-8 engine used in Series 62, 75, and Sixty Special models.

This would be the last year in which Cadillac's premium V-8, the Series 75, would be offered in a complete range of body styles, from coupes and convertibles on up to big limousines and formal town cars. For 1940 Cadillac's Series Seventy-Five included 12 models, all with Fleetwood bodies and all built on the same 141½-inch wheelbase. The Series 40-75 line included two coupes, this two or-four-passenger model and a slightly more expensive five-passenger coupe. The Cadillac-Fleetwood Series Seventy-Five Coupe for two-or-four passengers, Style 7557, had a $3,280 price tag and weighed 4,785 pounds in stock form.

At the top of the new Series 72 Cadillac line were a pair of formal four-door sedans with closed upper rear quarters and leather-covered roofs. The 1940 Series Seventy-Two Formal Sedan for Five Passengers, Fleetwood Style 7259, was the most expensive car in the Series 72 line with a factory-delivered price of $3,695. Weight was 4,670 pounds. A Seven-Passenger Formal Sedan, Style 7233-F, was priced identically. Only 38 Series 72 Formal Sedans were built, including 20 Imperial sedans with glass division between the front and rear seats and 18 with an "X" division (without header). Note the uniquely-shaped rear doors equipped with ventilator panes.

This would appear to be a Fleetwood Custom, as no open-front town car body style was shown on the standard 1940 Cadillac Series 72 price list. This was likely a one-off special-order factory conversion of a Series 72 Formal Sedan executed for some high-ranking GM executive, as was sometimes done. In later years such conversions were handled outside, mainly by Derham of Rosemont, Pa. The bright chrome reveals around the side window openings were unique to 1940 Series 72 Cadillacs. They later showed up on Fleetwood Seventy-Fives.

The 1940 Cadillac-Fleetwood Series Seventy-Five Coupe for Five Passengers was priced an even $100 higher than the two-or-four-passenger model. Listed as Model 7557-B, it weighed 4,810 pounds and had a base factory price of $3,380. Production was miniscule, with only 27 Series 75 coupes built in the entire 1940 model year, divided into 15 Style 7557 two/four passenger coupes and 12 Style 7557-B's. Both were externally identical.

With its attractive, pointed nose and long, flowing rear deck, the 1940 Cadillac Series Seventy-Five Convertible Coupe was the sportiest-looking body style in the Series 75 line this year. Body was by Fleetwood. Only 30 Style 7567 convertible coupes were built for 1940. Base price was $3,380 — the same as the five-passenger coupe. This two-or-four-passenger car weighed 4,915 pounds.

1940

Cadillac advertised its premium Series 75 eight-cylinder cars as "Cadillac-Fleetwoods." All 12 Series 75 body styles were also available on the 1940 Series Ninety "Sixteen Fleetwoods," which shared the same 141-inch wheelbase chassis. Although small, sales of the imposing Series Seventy-Five Convertible Sedan were considerably higher than those of the less expensive two-or-four-passenger two-door convertible. A total of 45 Style 7529 four-door convertible sedans were sold during the 1940 model year compared with only 30 convertible coupes. The 1940 Series 75 convertible sedan had a factory delivered price of $3,945 and weighed 5,110 pounds.

The 1940 Cadillac Series 75 model lineup included this attractive five-passenger Town Sedan. Only 14 of these Fleetwood Style 7539 four-door sedans with closed rear quarters and plain metal roofs were built this year. The factory-delivered price of the Cadillac-Fleetwood Series Seventy-Five Town Sedan was $3,635, and this model weighed 4,935 pounds. A five-passenger car, the Series 75 Town Sedan was not available with a glass division.

The comprehensive 1940 Cadillac Series 75 model lineup included two six-window, five-passenger sedans. Style 7519 was the standard Touring Sedan for Five Passengers, a 4,900-pound car which sold for $2,995. A total of 155 were sold. Style 7519-F was a Five-Passenger Touring Imperial with glass division between the front and rear passenger compartments. Only 25 Style 7519-F's were built this year. The Imperial was 40 pounds heavier than the standard sedan.

By far the most popular body style in the entire Series Seventy-Five line for 1940 was the impressive Seven-Passenger Touring Imperial, Fleetwood Style 7533. Cadillac delivered 338 of these imposing limousines, which had an adjustable glass division between the front and rear passenger compartments as standard equipment. The example shown, from this year's Series 75 deluxe catalog, is equipped with optional sidemounted spare tires — the last year they would be offered. The Seven-Passenger Imperial weighed 4,970 pounds and had a factory delivered price of $3,360.

This was the style of car with which the distinguished Cadillac Series Seventy-Five name would soon become synonymous. The body style shown is the 1940 Cadillac-Fleetwood Series Seventy-Five Touring Sedan for Seven Passengers, Style 7523. Note the more massive grille design and revised hood louver treatment. The Series 75 Seven-Passenger Touring Sedan weighed 4,930 pounds and had a factory list price of $3,210. A total of 166 were built during the 1940 model year. An Imperial version with glass division was also available.

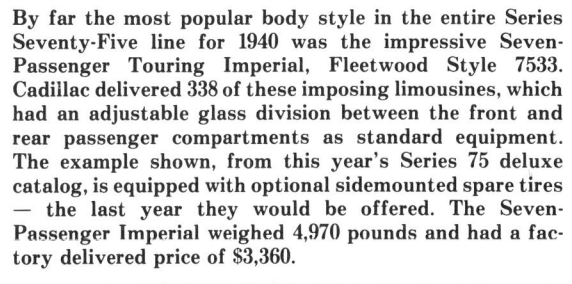

1940

At the upper end of the very comprehensive 1940 Cadillac Series 75 model line were two elegant formal sedans. This handsome Fleetwood body style was available in both five and seven-passenger versions. Style 7559 was the Cadillac-Fleetwood Series Seventy-Five Formal Sedan for Five Passengers, a 4,900-pound car with a $3,995 price tag. Style 7533-F was a seven-passenger version with Imperial glass division which sold for exactly the same price. A total of 48 five-passenger Series 75 Formal Sedans and 42 Seven-Passenger Formal Imperials were sold in model year 1940. Both models had closed upper rear quarters and leather-covered roofs.

The most expensive model in the 1940 Cadillac Series 75 line was the correctly formal Cadillac-Fleetwood Series Seventy-Five Town Car. Style 7553, it was a classic open-front seven-passenger car with a leather-covered roof and closed upper rear quarter panels. The Series Seventy-Five Town Car weighed 5,095 pounds. Price was a hefty $5,115. The car shown is equipped with optional side-mounted spare tires. A weatherproof cover could be buttoned into place over the chauffer's compartment. A few 1940 Series 75 Cadillacs were built with smooth backs instead of with the standard bustle-style trunk. Only 14 of these ritzy Series 75 Fleetwood Town Cars were sold this year.

For the undertaking and limousine livery trade. Cadillac continued to offer a special eight-passenger Series Seventy-Five Business Sedan. These big cars were equipped with special, heavy-duty upholstery and had two folding auxiliary seats. They were popular as pallbearer's and family sedans. Style-conscious funeral directors often operated carefully-matched fleets of late-model Cadillac-chassised hearses, flower cars, and limousines. This Series 75 Business Sedan is equipped with optional side-mounted spare tires.

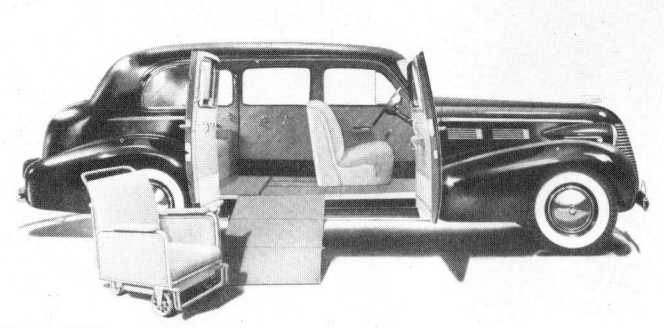

Cadillac's 1940 commercial car line included this Series Seventy-Five Invalid Chair Car, which was desinged for both private use or for use by livery or ambulance companies who had enough invalid transportation business to warrant such a vehicle. Included in the equipment was a special upholstered wheeled chair, designed along the lines of an overstuffed chair, and folding ramps to facilitate moving this chair (and its occupant) in or out of the car. Notice that there is no centerpost on the right side, the doors latching into the roof and floor when closed. This, obviously, allowed a large opening for easier movement of the wheeled chair. Standard front and rear seats were also provided, or a single front seat could be installed so that the wheeled chair could be positioned in the forward area.

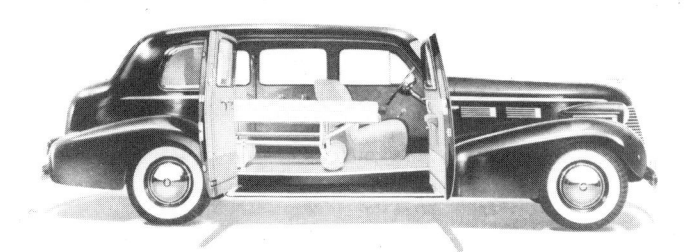

An Ambulance/Combination Car was part of the Cadillac Series Seventy-Five commercial car line. Sometimes refered to as "Sedambulances" these cars enjoyed some popularity in areas that couldn't afford a full-time, full-price ambulance. They were also used by livery firms which specialized in transporting bed-ridden invalids, where no great amount of medical equipment was necessary. As with the Invalid Chair Car, the doors latched together at the mid-point, with no centerpost blocking movement of the stretcher. This car had only one single front seat, but a full rear seat. As a combination car, these vehicles could be used either as an ambulance, or be incorporated into the funeral service with very little change in fittings.

1940

Only three 1940 Cadillac Series Seventy-Five 141-inch wheelbase chassis were released to outside coachbuilders. Cadillac's own Fleetwood Body Division continued to take care of most of its dwindling custom and semi-custom body requirements. Brunn and Co. of Buffalo, N.Y., designed and built this striking open-front town car with an all-aluminum body on a Series 40-75 Cadillac V-8 chassis for an Ohio steel executive. It was finished in black with a contrasting pale yellow side stripe. Owned at the time by Charles W. Coleman of Grosse Pointe Park, Mich., it was photographed by the author at the Classic Car Club of America Midwest Grand Classic at the Indianapolis Motor Speedway in July, 1963.

The old Derham Body Co. in the Philadelphia suburb of Rosemont, Pa., continued to build a small number of custom bodies for the Cadillac V-8 chassis. This style of collapsible landaulet with a distinctive flat, one-piece windshield was a Derham specialty. The rear roof could be let down and the cover over the chauffeur's compartment removed, but the center section was fixed. Derham also built at least one of these stately landaulets on a 1941 Cadillac Series 75 chassis. Owned by Roger Morello of Mason, Ohio, this impressive Derham landaulet was originally built for actress Barbara Hutton. It was photographed by Steve Hagy in 1980.

The Cadillac Motor Car Division of General Motors marketed three separate long-wheelbase commercial chassis in 1940. These included a 165-inch wheelbase Series 72 hearse and ambulance chassis; a 161-inch Series 75 chassis, and an extremely popular 159-inch LaSalle Series 50 commercial chassis. Only 275 Series 72 and 52 Series 75 commercial chassis were sold this year compared with a whopping 1,030 LaSalles. This is a 1940 Meteor-Cadillac Limousine Funeral Coach on the Series 72 commercial chassis. Note the ornamental coach lamp built into the rear of the front door. The Meteor Motor Car Co. was located in Piqua, Ohio.

The formal landau funeral car body style introduced by Sayers and Scovill in 1938 was sweeping the industry. By 1940 most American funeral car builders had added landau models to their product lines. The landau was rapidly replacing the ornamental carved-panel hearse which had dominated the scene through the 1930s. This is the 1940 Sayers and Scovill "Aristocrat" Landau Funeral Car on Cadillac chassis. This Cincinnati firm continued to mask the front end of its products in an attempt to disguise the chassis identity, a ploy which never really worked. Even the hubcaps bore "S&S" insignia instead of the Cadillac crest.

1940

The ornamental carved-panel hearse which had dominated the American funeral car industry through the 1930s was rapidly disappearing, replaced by the stylish new formal landau. The carved-drape hearse in the late 1930s was succeeded by a less ornate gothic-panel style. This is a 1940 Superior-Cadillac Gothic-Panel Carved Hearse. Note the downswept rear beltline, which worked quite nicely with the three-panel gothic style. The effect was that of a lighter-looking carved funeral car. Superior Coach Corp. was located in Lima, Ohio.

All principal American funeral car builders now offered coupe-style flower cars, many built on Cadillac and LaSalle commercial chassis. This is a 1940 Meteor-Cadillac Flower Car built on the 165-inch wheelbase Series 72 hearse and ambulance chassis. Meteor utilized the entire upper body structure of a standard Series 62 coupe for this model. The flower car shown had an open well for transporting funeral flowers. Other models featured an adjustable rear deck for flower display.

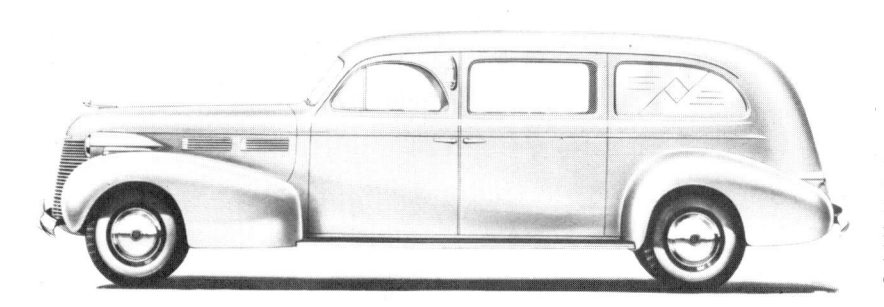

Except for hardware and trim, there was little exterior difference between ambulances and hearses of this period. Combination hearses and ambulances were common in smaller towns and cities. This is a 1940 Meteor-Cadillac straight ambulance. Note the light exterior color and frosted and etched rear window glass. The siren was mounted under the hood. Large emergency flasher lights had not yet come into serious use in most parts of the country.

Cadillac's monobloc 346 cubic-inch V-8 engine, introduced in the company's 1936 models, continued to power most of the division's cars including the 1940 Series 62, 72, 75 and Sixty Special. Now in its fifth model year, the 135-horsepower 346 had proved to be an extremely dependable design and a lively performer. For 1940 bore remained at 3½ inches and stroke 4½ inches. The 1940 LaSalle utilized a 130 H.P., 322 cubic inch derivative of this engine. The 1940 Series 90 Sixteen Fleetwoods were powered by the all-new 431 cubic inch V-16 introduced in 1938.

LaSalle for 1940 entered its 14th, and final, model year. But the LaSalle went out in style. Many Cadillac-LaSalle buffs consider the handsomely-restyled 1940 model the best-looking LaSalle ever built. The 1940 LaSalle's revised front-end styling included an extremely narrow grille and new front fenders with built-in sealed beam headlights. There were no less than nine 1940 LaSalle body styles in two separate series. This is the standard Series 50 Convertible Coupe for two or four passengers, Model 5067. Factory-delivered price was only $1,395 and the Series 40-50 convertible coupe weighed 3,805 pounds. A total of 599 were sold.

1940

For 1940, LaSalle offered two distinctively different series of cars. There were five standard Series 50 models and four all-new, premium Series 52 LaSalle "Special" body styles. The sleekly-styled Series 52 LaSalle Specials used the attractive new GM "torpedo" body which was lower, wider and more integrated in appearance than those used on the less expensive standard LaSalles. This new body was also shared with the 1940 Series 62 Cadillac. The LaSalle Series 52 Special Convertible Coupe and Convertible Sedan were added to the line midway through the 1940 model year. This is the Series 52 Special Convertible Coupe, Style 5267, which sold for $1,535 or only $140 more than the standard Series 50 convertible coupe. A total of 425 were built. Running boards were delete options on both Series 50 and 52.

The four-door convertible sedan was the least popular of the five body styles available in the standard 1940 LaSalle Series 50. Only 125 were sold. Style 5029, the Series 40-50 Convertible Sedan had a factory-delivered price of $1,800 and weighed an even 4,000 pounds in stock trim. All 1940 LaSalles were built on a three inch longer 123-inch wheelbase chassis and were powered by a Cadillac-built 322 cubic inch, 130-horsepower V-8 engine. With its top up or down, the LaSalle Convertible Sedan was a very good-looking car.

Midway through the 1940 model year, a convertible coupe and convertible sedan were added to the new Series 52 LaSalle Special line. A torpedo-bodied Series 52 coupe and sedan had been available from the start of the model year. The Style 5229 Series 52 Special Convertible Sedan was the most expensive car in the comprehensive 1940 LaSalle model line, with a factory-delivered price of $1,895. It was also one of the best-looking LaSalles ever built. Only 75 were produced, making it the rarest of all 1940 production LaSalles. Running boards were a delete option on this model and added 35 pounds.

This is the standard 1940 LaSalle Series 50 Convertible Coupe with four-passenger seating. Series 50 models had a chrome body side molding. The new premium Series 52 LaSalles were lower and did not. This car, with four stylishly-attired models, was posed against a drab railroad yard background atop one of the Cadillac plant buildings near downtown Detroit. Through advertising agency artistry, it was miraculously transposed in the 1940 LaSalle sales catalog into much more pleasant country surroundings. Black sidewall tires were standard and running boards continued as delete options which could be installed in the field.

Here's the handsome 1940 LaSalle Series 52 Special Convertible Sedan with its top down. A total of 13,750 LaSalle Specials were built in model year 1940, outselling the standard Series 50 LaSalle by more than 3,000. The Series 52 Special Convertible Sedan and companion Convertible Coupe had a brief, but successful, production run. This smartly-dressed young model is all set to take a spin around the roof of one of Cadillac's Clark Ave. buildings in a photo session during preparation of that year's sales catalogs.

1940

The lowest-priced car in the nine-model 1940 LaSalle line was the standard Series 50 Coupe for two or four passengers. Base price of this model was only $1,240. It was listed as Model 5027. Weight was 3,810 pounds and a total of 1,527 were sold. The Series 40-50 Coupe shown is equipped with running boards which added 45 pounds when specified. All 1940 LaSalles could be ordered with or without running boards. When not used they were replaced with a bright stainless steel sill molding.

For 1940 LaSalle offered its customers two price classes — the standard Series 50 and a new, premium Series 52 LaSalle Special. This is the beautifully-styled Series 52 LaSalle Special Coupe for two or four passengers. Designated Model 5227, it had a factory list price of $1,380 or only $140 more than the standard Series 50 coupe. Customers could take or leave running boards at no extra cost. An even 3,000 of these sleekly styled Series 40-52 Coupes were delivered during the 1940 model year, LaSalle's last.

Only two Series 52 LaSalle Special models were available at the beginning of the 1940 model year in October, 1939. They were closed body styles, a four-door touring sedan and the two-door coupe shown here. A convertible coupe and convertible sedan joined the Series 52 Special line later in the year. The popular, new Series 52 LaSalle Specials featured sleek "torpedo" styling. They used very little exterior trim and relied on smooth, integrated body forms and attractive exterior colors for their pleasing visual impact. They were knockouts and sold very well during their brief lives on the market.

A two-door sedan was available only in the standard LaSalle Series 50. It was not offered in the new premium Series 52. All Series 50 LaSalles boasted GM's Unisteel Turret-Top body construction with Hi-Test safety glass all around. The Series 40-50 Two-Door Touring Sedan for Five Passengers, Style 5011, accounted for only 366 sales. Factory delivered price was $1,280, and it weighed in at 3,760 pounds. Nine Series 50 two-door sedans with sun roofs, Model 5011-A, were also sold this year.

LaSalle's standard Series 50 model lineup for 1940 included five body styles, the most popular of which was the four-door touring sedan. Style 5019, the Series 50 Touring Sedan for Five Passengers weighed 3,790 pounds and had a factory list price of $1,320. A total of 6,698 Series 50 sedans were sold in the 1940 model year, including 140 with optional sunshine roofs. Another 132 were exported. The Series 40-50 sedan shown is equipped with running boards and optional white sidewall tires.

1940

The 1940 model year was LaSalle's last, and the last in which Cadillac would offer side-mounted spare tires on any of its cars. This is a very well-equipped 1940 LaSalle Series 50 Touring Sedan for Five Passengers, Model 5019. The sidemounts feature a stylized "LS" emblem in the center. This four-door, six-window sedan was by far the most popular model in standard Series 50. The double-faced white sidewall tires were also optional. The wheels are finished in a lighter color than the rest of the car.

This may not have been the prettiest LaSalle built in 1940, but it was certainly the longest! The builder of this eight-door "woodie" station wagon is not identified. It is likely that this super-stretch job was used for hotel or airline limousine service. The LaSalle commercial chassis was a big seller in 1940, far exceeding sales of both Cadillac funeral car and ambulance chassis. This monster must have been a bear to park.

The Flxible Co. of Loudonville, Ohio was one of several large funeral car builders that mounted hearse and ambulance bodies on the 1940 LaSalle commercial chassis. Others included Sayers and Scovill of Cincinnati, the A. J. Miller Co. of Bellefontaine, Ohio, Meteor Motor Car Co. of Piqua, Ohio, The Eureka Co. of Rock Falls, Ill., and the Superior Coach Corp. of Lima, Ohio. Flxible's professional car bodies had unusually high belt lines, so it was necessary for Flxible to raise the height of the LaSalle hood. This is a 1940 Flxible-LaSalle Style "AA" carved-panel hearse, a style which was fading from the scene now. Within two years the carved-panel funeral car would be gone, it's place taken by the more stylish new landau.

The all-new Series 52 LaSalle Specials for 1940 created the same kind of sensation the style-setting 1934 LaSalle had seven year earlier. With their smooth, uncluttered "torpedo" body shapes, the new 1940 LaSalle Series 52 Special sedan and coupe were immediate sales successes, handsomely outselling the less expensive Series 50 LaSalles. Initially, only the Series 52 sedan and coupe were available. A convertible coupe and convertible sedan were added to the line midway through the 1940 model year. By far the most popular of all nine 1940 LaSalle models was the beautifully-styled Series 52 Special Touring Sedan for Five Passengers, Style 5219, of which 10,118 were sold. Full rear fender skirts enhanced the Special's looks even further. The Series 52 Sedan had a factory-delivered price of $1,440 and weighed 3,900 pounds. The new torpedo bodies were significantly lower and wider than those used on the standard Series 50 LaSalle.

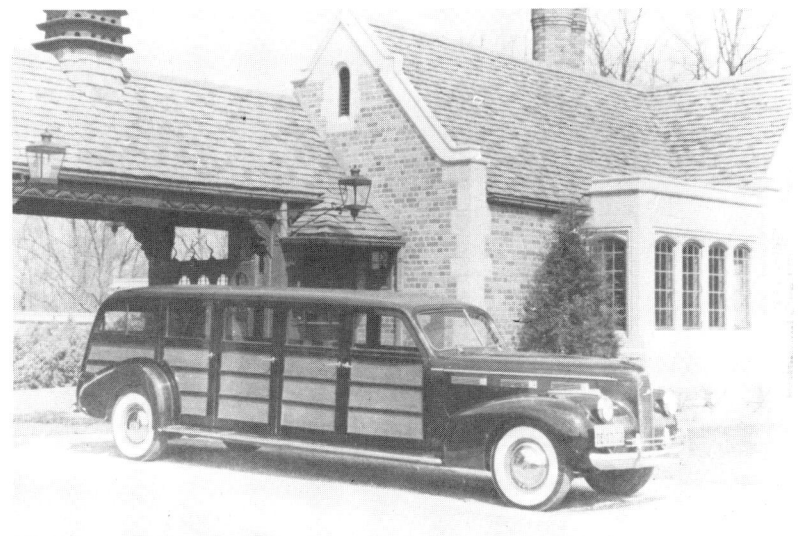

The 1940 LaSalle Series 50 commercial chassis was extremely popular with American funeral car and ambulance builders. A total of 1,030 were shipped to various hearse and ambulance builders, far more than the 327 Series 72 and 75 Cadillac commercial chassis sold that year. This coupe style flower car on the Series 40-50 LaSalle commercial chassis was built by the A. J. Miller Co. of Bellefontaine, Ohio. It utilized the entire upper body structure of the 1940 LaSalle Series 50 Coupe. These impressive cars were used to lead funeral processions.

1941

The 1941 model year holds a special place in Cadillac's long and proud history. Cadillac sales set a new record. It was a year of major styling change, and more than any other, firmly established Cadillac's strong and extremely successful styling identity. General Motors' new Hdra-Matic automatic transmission was offered in Cadillacs for the first time. The LaSalle and V-16 were gone, and 1941 would be the last full model year before U.S. motor vehicle production was again cut off by another world war.

All of Cadillac's extensively-restyled 1941 cars got a massive new rectangular grille and front-end ensemble. The new grille was a broad, vertical design with forward-thrusted center section and eight horizontal chrome bars. This handsome grille set the style for Cadillac's frontal appearance for many years to come and was widely imitated by other cars, including GM's own Chevrolet. For the first time a valance panel filled the space between the bottom of the grille and the front bumper. A grille guard was built into the front bumper between the vertical bumper guards.

Headlights were built into wide new front fenders. Rectangular parking and turn signal lamps were incorporated into the upper outer edges of the grille. Provision was made for fog lamps in openings below the headlights. When optional fog lamps were not specified, these openings were covered by circular chrome ornaments with a "V" emblem in the center.

The high, new coffin-nose hood had no side panels. Finely-textured horizontal ventilator castings were located on the rear sides of the hood. The Cadillac name appeared in small block letters above the grille header and beneath a bold, new Cadillac crest surmounted by a pair of tall, unpraised wings. A new "flying lady" hood ornament/latch handle was set well back from the front of the hood. Fenders were squared, and except for those on the Sixty Special all were ornamented with three horizontal chrome trim spears. Most of the 1941 models were equipped with fender shields, or skirts, the lower center of which bore a round Cadillac emblem. The spare tire was carried vertically in the trunk: sidemounted spares were no longer available. The gasoline filler opening was cleverly concealed under the swing-up left rear taillight.

Nineteen models were offered in six series, three of them new. All all-new lowline Series 61 Cadillac replaced the LaSalle, a factor which contributed to Cadillac's vastly-improved 1941 sales. There has been a Series 61 in 1939, along with a lower-priced Series 50 LaSalle, but none in 1940. The 1941 line included a new mid-range, single body style Series 63 six-window sedan. The Series 72 of 1940 was dropped, replaced with a new long-wheelbase, lower-priced companion car to the Series 75 called the Series 67. From 1941 on, the distinguished Series Seventy-Five name would be reserved for Cadillac's most expensive big formal sedans and limousines. The Fleetwood name was used only on the Sixty Special and Seventy-Five.

Series 61, 62, 63 and Sixty Special Cadillacs for 1941 were all built on the same 126-inch wheelbase. Series Seventy-Five models rode on a 136-inch wheelbase. The new Series 67, which shared its single six-window body with the 1941 Buick Limited, boasted Cadillac's longest wheelbase for this year, 139 inches.

Now in its fourth year, the Cadillac Sixty Special also got the handsome new grille and front-end. Long, sweeping front fenders extended back into the front doors. Many Cadillac historians consider the suave 1941 model the best-looking Sixty Special ever built. Wheelbase was reduced one inch and Sixty Special prices were increased for the first time since this style-setting car was introduced for the 1938 model year.

All 1941 Cadillacs were powered by the same 346 cubic inch, L-head V-8 engine, which had been uprated to 150 horsepower. For the first time since 1926, Cadillac had reverted to a one-name, one-engine marketing strategy.

The new Series 61 Cadillacs had sleek new Fisher two and four-door fastback bodies and were the most aerodynamically-styled Cadillacs since the limited production Aero-Dynamic coupes of 1934-37. Series 62 models retained the same torpedo, or "projectile-styled" bodies which had been introduced for 1940, and which worked well with the new grille and front end. Styling of the Series Seventy-Five models was much more angular and conservative than the rest of the cars in the line and received only minor styling changes through the remainder of the decade.

The Imperial glass division in Series Seventy-Five limousines was now electrically-operated, controlled by switches in the rear seat armrests. Series 61 and 62 Cadillacs were available in both standard and Deluxe versions. This would be the last year in which Cadillac would offer a four-door convertible sedan (Series 62 Deluxe). The division's convertible tops were now raised and lowered with vacuum assist. Series 61, 62

Many Cadillac enthusiasts consider the 1941 Cadillac Sixty Special the most beautiful ever built. The massive, new horizontal grille and long, flowing front fenders which extended into the doors complemented perfectly the Sixty Special's crisp upper and lower body forms. This would be the fourth and final year for Bill Mitchell's classic four-door sedan which had been originally introduced in the fall of 1937 for the 1938 model year. The 1942 Sixty Special would be a totally different car. The Sixty Special's wheelbase was reduced one inch to 126 inches for 1941 and the price was increased for the first time since this car was announced three years earlier. The factory-advertised price went up $105 to $2,195. A total of 4,100 Sixty Special were built during the model year. In all, 17,900 of these handsome sedans were produced between 1938 and 1941.

1941

and 67 models had concealed running boards. Running-boards were no-cost options on other models.

The most important new Cadillac convenience option this year was General Motors' Hydra-Matic automatic transmission, which had been introduced as an option on the 1940 Oldsmobile. The automatic transmission did away with gear shifting and the clutch pedal. It cost $110 extra and was ordered by 30 per cent of Cadillac's customers the first year it was offered.

The Cadillac Motor Car Division chalked up sales of a record 66,130 cars during the booming 1941 model year — 20,000 higher than the previous all-time sales record established in 1937.

Halfway through the prosperous 1941 model year, rival Packard stirred up some excitement with its beautifully-styled new medium-priced Packard Clipper. While the American economy was buoyant, war clouds were on the horizon. Plants everywhere were already gearing up for war production. No one knew how long the boom — and peace — would last.

Bill Mitchell's epic Cadillac Sixty Special seemed timeless. Four years after it was introduced, this car was still a style-setter. It was the most-imitated design of its day. Most of the 4,100 Sixty Specials delivered in 1941 — 3,693 — were of this style, the standard Fleetwood Sixty Special Touring Sedan for Five Passengers, Style 41-6019. Another 185 with sun roofs, Style 6019-A, and 220 Imperial Sedans with a formal glass division, Style 6019-F, were also sold. The Sixty Special Imperial Sedan was priced at $2,345 or $150 higher than the standard 41-6019.

The glorious custom-body era was over, even at Fleetwood. Only one 1941 Cadillac Sixty Special Town Car was built this year. It was a one-off show car designated Style 6053-LB (for leather back), which made its splashy debut in New York. This Fleetwood custom creation sported a daring white leather-covered roof. Derham, which built about half a dozen similar cars this year, executed a small number of open-front Cadillac town cars on special order for individual customers in later years, but this was the last one to come from Cadillac-Fleetwood. It is believed that this car survives in the hands of a West Coast collector.

The 1941 Cadillac Sixty Special shown here is a special semi-custom. It has a leather-covered roof and heavy Series 75 belt moldings. It has no hood ventilator louvers, Cadillac emblem on the front of its hood nor Fleetwood script nameplates on the lower edges of the front fenders as used on standard production Sixty Special. Fleetwood and GM Styling turned out a number of Sixty Special customs, about a dozen in all, between 1938 and 1941. Most were done for high-ranking GM executives. Some of these cars had stretched and lowered bodies. Others had special styling accents like dummy landau irons, special wheel covers and V-16-style fender spears.

When the curtain went up on Cadillac's completely restyled 1941 models, there was no LaSalle. The LaSalle's place at the bottom of the Cadillac Motor Car Division's price structure was filled by an all-new Series 61 Cadillac which was offered in sleek two and four-door fastback body styles on a 126-inch wheelbase. This is the extremely attractive 1941 Cadillac Series 61 Five-Passenger Coupe, which was available in Standard and Deluxe versions. This Standard Series 61 Coupe, Model 6127, was the lowest-priced Cadillac in many years with a base price of only $1,345. It was also the hottest-selling car in the entire 1941 Cadillac line with 11,812 delivered. The Deluxe version, Model 6127-D, was priced $90 higher at $1,435. A total of 3,015 Deluxe Coupes were sold this year.

1941

The sleek, new Series 61 Cadillacs were the most aerodynamically-styled Cadillacs built since the limited production Aero-Dynamic Coupes of 1934-37. Their sweeping fastback rooflines strongly influenced the appearance of all American cars through the 1940's. The Sixty-One's belt molding swooped down at the rear, creating an ideal color break for some striking two-tone paint combinations. The triangular-shaped rear side window also dipped down slightly at the rear. Side window openings were encircled with a continuous bright chrome molding.

The 1941 Cadillac Series 61 four-door sedan was also a racy-looking fastback. Like the companion Series 61 two-door coupe, standard and deluxe versions of the sedan were available. This is the standard 1941 Cadillac Touring Sedan for Five Passengers, Model 6109, the second-best selling car in the line this year with sales of 10,925. Factory advertised price was $1,445. The Series 61 Deluxe Sedan, Model 6109-D, cost $1,535 or only $90 more. A total of 3,495 Deluxe Sedans were sold in the 1941 model year. All 1941 Cadillacs were powered by the same L-head 346 cubic inch V-8 engine which by now was putting out 150 horsepower.

Styling of the new Series 61 two-door coupes and four-door sedans was quite similar, with the same sweeping fastback roofline, angled rear body pillar, continuous chrome molding encircling the side window openings and a slight dip at the rear of the triangular rear side window. The Sixty-One's belt molding swept down to the rear bumper. A total of 29,250 Series 61 Cadillacs were sold in 1941 which helped to propel the division's sales to a new record. In just one year the new Series 61 cars had demonstrated the wisdom of dropping the LaSalle and replacing it with a lower-priced car carrying the magic Cadillac name.

Brunn and Company, an old-line custom body builder located in Buffalo, N.Y., was well-known for its conservative custom and semi-custom bodies created for Pierce-Arrow and Lincoln chassis. But Brunn turned out some rather radical designs in addition to its bread-and-butter formal jobs. One of these was this flambouyant conversion of a 1941 Cadillac Series 61 Deluxe Coupe. The front fenders were extended halfway to the front doors (shades of what was to come in 1942!) and the rear side windows were closed in. The upper quarters were spanned by an oversized carriage bow. A bold side spear and stark two-tone paint job added even more dazzle, as if it were needed. A portion of the standard chrome window molding was retained.

Coachcraft Ltd., a custom body shop located in Los Angeles, designed and built this special station wagon on a 1941 Cadillac Series 61 chassis. Cadillac woodies are few and far between. This one was built for cowboy movie star Charles Starrett. Note the absence of side slats on this wood-panelled body and the full-length roof luggage rack.

1941

The Series 62 Cadillac line for 1941 offered six models in four body styles, all on the same 126-inch wheelbase shared by the Series 61, 63, and Sixty Special. This is the very attractive 1941 Cadillac Series 62 Deluxe Coupe for Five Passengers, Style 6227-D. At $1,510 the Deluxe Coupe was priced $90 higher than the standard coupe. A total of 1,900 Series 62 Deluxe Coupes were sold this year, only 85 fewer than the standard model. Wheel shields, or skirts, were standard on the Deluxe. Note the thin, angled "B" pillar and pleasing two-tone paint job.

Cadillac's 1941 Series 62 coupes and sedans were offered in both standard and Deluxe versions. This is the standard Series Sixty-Two Coupe for Five Passengers, Model 6227, which sold for $1,420. The Deluxe model cost $90 more. Weight was 3,950 pounds. Cadillac's dealers delivered 3,885 Series 62 Coupes during the 1941 model year, of which 1,985 were the price-leading standard model. This coupe is equipped with optional fog lamps mounted in the openings provided for this purpose below the headlights.

The 1941 Cadillac Series 62 Convertible Coupe was available only in a Deluxe version. Shown as Model 41-6267-D, it weighed, 4,055 pounds and carried a factory advertised price of $1,645. The Series 62 Deluxe Convertible Coupe for two or four passengers represented exceptional value for the money. A respectable 3,100 were sold in the 1941 model year. This would be the last year that Cadillac convertible coupes would have fully-closed rear roof quarters. Subsequent models would have quarter windows.

The milestone 1941 model year would be the last in which Cadillac would offer its customers a four-door convertible sedan. Only 400 were built. Consequently, Cadillac's last production convertible sedan is much in demand by collectors today. Even with its top up it was a very beautiful car. The 1941 Cadillac Series 62 Convertible Sedan had a factory advertised price of $1,965 making it the most expensive model in the series. Like the companion convertible coupe, the 1941 convertible sedan was available only in a Deluxe version. It was shown as Model 6229-D.

By far the rarest of all 1941 Cadillac Series 62 body styles was the Model 6229-D four-door convertible sedan. Only 400 of these gorgeous convertibles were built in the entire 1941 model year. Running boards were optional at no extra cost. Three convertible top color choices were offered and complemented the leather upholstery. Cadillac's last convertible sedan was marketed as a full six-passenger car. In stock form the convertible sedan weighed 4,230 pounds. Cadillac's convertible tops were now raised by vacuum, which finally took the muscle work out of raising and lowering the top.

1941

Cadillac's restyled Series 62 models for 1941 retained the sleek Fisher torpedo bodies introduced the previous year. Combined with Cadillac's massive new front-end ensemble they were extremely good-looking cars. The 1941 Series 62 four-door sedan was offered in both standard and deluxe versions. This is the 1941 Series Sixty-Two Deluxe Touring Sedan for Five Passengers, Model 6219-D, which sold for $1,535. Cadillac's dealers sold 7,754 of these beauties, many with the attractive optional two-tone paint treatment shown on this car. The standard Series 62 four-door sedan, Model 6219, was priced a mere $40 less, at $1,495 and outsold the Deluxe model by 258 units.

Very few Cadillac station wagons have been built over the years and most of these were one-off specials done by various conversion shops. J. T. Cantrell and Son Co., of Huntington, L.I., N.Y. was for many years a major builder of wooden station wagon bodies, supplying them to several major auto manufacturers. Cantrell built this conventional style "woodie" on a 1941 Cadillac chassis. The doors open from the rear and are mounted on barrel hinges. This station wagon body is very similar to those offered on the 1941 Chevrolet chassis, built by Ionia.

Cadillac introduced three new series for the big 1941 model year, a lowline Series 61 which replaced the LaSalle and two special six-window four-door sedans Series 63 and 67. Actually a revival of a series that was originally offered in 1939 but not 1940, the Series 61 Cadillac was produced through the 1951 model year. But the Series 63 and the larger Series 67 models were built only during model years 1941 and 1942. This is the 1941 Cadillac Series 63 Touring Sedan for Five Passengers, Style 6319. A total of 6,800 Series 63 Cadillacs were built during the two-year model run, including 5,050 of these 1941 models. Factory advertised price was $1,695. Fender shields, or skirts, were standard.

The handsome new Series 63 Cadillac utilized GM's new six-window "B" body which was mounted on the same 126-inch wheelbase chassis used for the 1941 Series 61, 62 and Sixty Special models. This was the only Series 63 body style offered. The Series 63 Cadillac's sleek notchback styling included concealed running boards. Weight was an even 4,100 pounds. Note the two vertical dividers in the rear window and the bright molding enclosing all three side window openings in place of a belt molding. The triple horizontal fender spears were used on all 1941 Cadillacs except the Sixty Special.

Another new addition to the Cadillac family for 1941 was the long-wheelbase Series 67, a big four-door sedan built only in the 1941 and 1942 model years. The new Series 67 Cadillac occupied approximately the same market niche as the now-defunct Series 72, which had been offered only during model year 1940. The new Series 67 rode on the longest passenger car wheelbase in Cadillac's stable this year — 139 inches. The 67 was thus three inches longer than the top-line Series Seventy-Five but cost $400 less. Only 900 were built this year. Prices ranged from $2,595 for the standard 1941 Cadillac Series 67 Touring Sedan for Five Passengers on up to $2,890 for the Sixty-Seven Touring Imperial for Seven Passengers.

1941

Four Series 67 models were offered for 1941, all using the same six-window, four-door body shell and long 139-inch wheelbase. The Series 67 line included a Touring Sedan for Five Passengers, Style 6719, which sold for $2,595; a Five-Passenger Touring Imperial, Style 6719-F ($2,745); a Touring Sedan for Seven Passengers, Style 6723 which carried a price of $2,735, and the top-line Seven-Passenger Touring Imperial, Style 6733, which sold for $2,890. Weights ranged from 4,555 pounds to 4,705 pounds. Production included 315 Five-Passenger Sedans, 95 Five-Passenger Imperials, 280 Seven-Passenger Sedans, and 210 Seven-Passenger Imperials.

The 1941 Fleetwood Seventy-Five was available in five and seven-passenger sedan versions, with or without an electrically-operated Imperial glass division. The division glass was raised or lowered by switches mounted in the rear seat armrests. Imperial models had leather-upholstered front compartments and were intended to be chauffeur-driven. Most were. The 1941 Series Seventy-Five Imperial Sedan for Five Passengers sold for $3,150. A total of 132 were built this year.

Cadillac sales literature of this period was beautifully illustrated and is much in demand by automotive collectors today. This is a catalog illustration of the 1941 Cadillac Fleetwood Seventy-Five Seven-Passenger Touring Sedan, Style 7523. This body style was also offered with an optional Imperial glass division for chauffeur driving. Cadillac sold 405 of these big seven-passenger sedans. Factory advertised price was $3,140. The windsplit flowing back along the front fender from the headlight bezel remained a distinctive Cadillac front-fender styling touch for an incredible 17 years, making its final appearance on the 1958 models.

With its long wheelbase, fender skirts, concealed running boards and smartly angled "C" pillar, the new Series 67 Cadillac was an impressive, fine-looking car. Although it was designed as a lower-priced companion car to the premium 1941 Series Seventy-Five, the Sixty-Seven's wheelbase was three inches longer. The Series 67 Cadillac was available in five and seven-passenger versions. The principal difference was in the seating arrangement. Formal glass divisions were offered in both five and seven-passenger models for chauffeur driving. The Series 67 Cadillac shared its body shell with the 1941 Buick Limited.

Commencing with the 1941 model year, the distinguished Series Seventy-Five name was reserved exclusively for Cadillac's most expensive long-wheelbase formal sedans and limousines. While the 1941 Series Seventy-Fives got Cadillac's massive new grille and front-end ensemble, their styling was much more angular and restrained than the other cars in the line. But the imposing Fleetwood Seventy-Fives exuded class and were very well received. They also set the style for Cadillac's most exclusive models for the rest of the decade. Except for grille and trim revisions, the 1949 Fleetwood Seventy-Fives were quite similar in appearance to the 1941s. Eight five and seven-passenger models were offered for 1941, all using the same basic Fleetwood four-door sedan body on a 136-inch wheelbase. This is the 1941 Cadillac Fleetwood Seventy-Five Touring Sedan for Five Passengers, Style 7519, of which 122 were built. Factory price was $2,995.

This factory photograph proves that there almost was a 1941 Series 72 Cadillac. But the "72" was built only for the 1940 model year. The car shown went into production as the 1941 Series Seventy-Five, but it borrowed the 1940 Series 72's broad belt molding with integral door handles and the bright chrome reveals around the six side windows. This body style would remain generally unchanged through the 1949 model year. The car shown is the 1941 Fleetwood Seventy-Five Imperial Sedan for Seven Passengers, Style 7533, of which 757 were built this year, making it the most popular model in the high-priced Series Seventy-Five line. Factory price was $3,295. Note the broad, new grille design and dual-faced white sidewall tires.

The most expensive cars in the 1941 Cadillac model lineup were two stately Fleetwood Series Seventy-Five Formal Sedans. These were built in five and seven-passenger versions. Both featured closed upper rear quarters and leather-covered roofs. This is the 1941 Cadillac Series Seventy-Five Formal Sedan for Five Passengers. Now owned by J. Bruce Burgan of Minneapolis, the car had an original base price of $3,920. Designated Style 7559, it had a production run of only 75. This model had the "X"-type glass division with no header: the glass, when raised, fit snugly up against the interior headlining, effectively deadening conversation between the front and rear compartments. Note the rounded door upper frame with ventilator pane, which was similar to that used on the 1940 Series 72.

1941

At $4,045 F.O.B. Detroit, the Series Seventy-Five Seven-Passenger Imperial, Style 7533-F was the most expensive model in the entire 1941 Cadillac line. Ninety-eight of these severely formal limousines were sold this year. The impressive Fleetwood Formal Sedans had closed upper rear quarters and padded, leather-covered roofs. These distinctive cars were built by Fleetwood, although Derham later turned out similar conversions of standard Series Seventy-Five sedans and limousines. Style 7533-F had a conventional division partition between the front and rear compartments. The division glass was now raised and lowered electrically rather than by the hand crank fitted previously.

During the 1941 model year, Cadillac-Fleetwood built 60 of these special Series Seventy-Five Business Sedans, which were designed for the funeral service trade. This is the Style 7523-L Nine-Passenger Business Sedan, 54 of which were sold at $2,895 a copy. This was thus the least expensive car in the high-priced Series Seventy-Five line. Also offered was the Style 7533-L Business Imperial with formal division, which sold for $3,050. Only six were built. Cadillac continued to publish special commercial car and chassis sales brochures. This 1941 Cadillac Business Sedan is posed against an appropriate cemetery background.

As the tempo of defense production increased, motor vehicle manufacturers large and small began to receive orders for specialized vehicles for the armed forces. A prominent funeral car and ambulance manufacturer, The Eureka Co. of Rock Falls, Ill. built this special 15-passenger bus on a 1941 Cadillac commercial chassis. It was one of many such "stretches" Eureka eventually did for the U.S. Army. These coaches were used to transport personnel to and from vital war plants. Note the clearance lights on the roof.

All of the 1941 Cadillacs featured a bold, new front end ensemble that was to set the style for future Cadillacs for many years to come. The massive new rectangular grille had eight horizontal bars and built-in parking lamps. Headlights were built into the new front fenders and provision was made for a pair of optional fog lamps. For the first time there was a valance between the bottom of the grille and the front bumper. A new style Cadillac crest with a pair of vertically-raised wings was affixed to the front of a high coffin-nose hood. The Cadillac name appeared in small block letters between the grille header and crest, and the hood was surmounted by a new "flying lady" ornament/latch handle. The total visual effect was one of massiveness and handsome good looks that set the 1941 Cadillac far apart from all other U.S. cars.

1941

The Superior Coach Corporation of Lima, Ohio completely redesigned its funeral coach and ambulance bodies this year. Superior's new look included a belt line that swept down at the rear, just like the new style-setting Series 61 Cadillacs. The huge rear side doors were 55 inches wide. This is a catalog illustration of the 1941 Superior-Cadillac "Briarcliff" Side-Loading Limousine Funeral Coach. Flower trays are mounted in the rear side windows behind the velvet drapes.

Cadillac offered hearse and ambulance manufacturers two commercial chassis for 1941. These included a 163-inch wheelbase Series 62 commercial chassis, 1,475 of which were sold, and a Series 75 chassis on an identical wheelbase. Only 150 Series 75 commercial chassis were shipped. This is a 1941 Meteor-Cadillac Landau Funeral Coach on the Series 62 commercial chassis. Note the unusual front door window shape with built-in coach lamp and the Series 75-style reveal molding in the rear side window. The landau was now the preferred funeral car body style. The Meteor Motor Car Co. was located in Piqua, Ohio.

The coupe-style Flower Car was now a popular versatile vehicle in many funeral home fleets. In addition to transporting the flowers at the head of the funeral cortege, the flower car doubles as a hard-working service vehicle between funerals, carrying caskets, chairs and other equipment and generally doing the work of a conventional service car. This is a 1941 Meteor-Cadillac Flower Car. It is shown with the weatherproof boot buttoned into place over the stainless steel flower deck. This impressive professional car used the entire upper body structure of a Series 62 Coupe.

This is what the Cadillac commercial chassis looked like when it left the Clark Avenue plant in Detroit on its way to various hearse and ambulance builders in the U.S. midwest, most of them in Ohio. The special long-wheelbase commercial chassis was shipped with the hood, rear fenders and cowl components in a large carton strapped to the bare frame. The independent professional car builders designed and built their own bodies for mounting on this special chassis. The hearse and ambulance business was an important sideline for Cadillac for many years and did much to enhance the marque's image of prestige. Truck buffs will appreciate the heavy-duty 1940 Chevrolet tractor being used to haul these chassis. Note that in this era, the car carrier trailers were only large enough to hold three commercial chassis, and in fact, even had to be modified at the forward end to allow the lead chassis to ride well ahead of the cab. A medium-sized truck in its era, this vehicle could probably carry or pull no more weight than one of today's heavy-duty fifth-wheel pick-up units.

Here's a 1941 Superior-Cadillac in uniform. Funeral car and ambulance builders were beginning to receive sizeable orders for special military ambulances from the U.S. armed forces. This special 1941 Cadillac field ambulance with blanked-out rear windows and roof clearance lights was built for the U.S. Army Medical Department. Some of these ambulances eventually saw service overseas, though most were used on large military bases in the U.S.

1941

John J. C. Little was a small, independent Canadian funeral car and ambulance conversion builder who operated a busy shop in the southwestern Ontario town of Ingersoll. Mr. Little did this interesting ambulance conversion of a standard 1941 Cadillac Series 62 four-door sedan for John Labatt Ltd., a major Canadian brewery. A new roof and upper rear quarters were grafted onto the rear. Note the big driving lamp and the front fender guides.

Yes, there was a 1941 LaSalle — almost! Although the LaSalle was discontinued at the end of the 1940 model year, there was a restyled 1941 model waiting in the wings. But it never made it into production. This is a prototype 1941 LaSalle premium series four-door notchback sedan which would have utilized the same GM torpedo "C" body used on the 1940 LaSalle Series 52 Special and the 1940 Cadillac Series 62. It would have been little more than a lowline 1941 Cadillac with a vertical instead of horizontal grille motif.

Perhaps the most elaborate of the numerous one-off funeral cars and ambulances built in the small but busy shop of John J. C. Little in Ingersoll, Ontario, was this special gothic-panel hearse built to order for funeral director Russel Needham, of Chatham, Ontario. This car was a straight side-loader. It had no rear door. It had blue stained glass in its hand-carved wooden side window openings and a Henney side-loading casket table. A conversion of a standard 1941 Cadillac Series 62 coupe, it was around (although not in funeral service, until the early 1970's. The large rear side doors are almost completely concealed by the heavy Gothic panels.

Here is another 1941 LaSalle proposal. This model, had it made it into production, would either have shared its sleek four-door fastback sedan body with the 1941 Series 61 Cadillac, or it would have taken the place of the "61" in Cadillac's 1941 line. As it happened, Cadillac dropped the LaSalle and replaced it with a new, lower-priced Cadillac called the Series 61. There had been a Series 61 Cad in 1939 but not in 1940. The 1941 LaSalle would have retained its traditional, narrow vertical grille design but would have had six prominent "mailslot" air intakes in the area between the grille and front fenders. Note the two trim pieces above the slots between the headlights and grille and the three-bladed spinner hubcaps. In a nifty flashback to the classic 1927-1932 LaSalle, a "LaS" badge was mounted in the center of the grille guard between the bumperguards on this full-sized styling model.

This is yet another 1941 LaSalle styling exercise. As it turned out, this good-looking four-door sedan based on the new General Motors "B" body showed up in production as the 1941 Series 63 Cadillac. Note the three horizontal trim pieces on the rear of the fenders: these were used on all 1941 Cadillacs except the Sixty Special. The LaSalle name appears in small block letters on the rear of the front fenders, above the horizontal trim. The full rear fender skirts on most 1941 Cadillacs were embellished with a circular Cadillac crest. This photo was snapped under the floodlights in a GM styling studio in Detroit's New Center area at precisely 2:15, but we don't know if that was a.m. or p.m. The lights burned late as a new-model launch approached.

1942

Cadillac's 40th anniversary year would prove to be an eventful — and abbreviated — one. Cadillac had enjoyed all-time record sales in the 1941 model year, and while material shortages and uncertainty about the war raging in Europe clouded the outlook, the American automobile industry anticipated continued robust sales of new passenger cars in the 1942 model year.

Although Cadillac had fielded a much-changed series of cars for 1941, the 1942 model year was another one of significant styling advances. The 1942 Cadillacs sported glittering new grilles and long, sweeping pontoon fenders. Aerodynamically-styled fastbacks were the rage and set the style for the rest of the industry for the remainder of the decade. Sales of Cadillacs equipped with optional Hydra-Matic automatic transmissions, (introduced the previous year) doubled to 60 per cent of production in the shortened 1942 model year.

All of the 1942 Cadillacs had bright, new grilles. The new grille was larger and even more massive than the one of the previous year. Six heavier chrome bars spanned almost the entire width of the car. New vertical front bumper guards were topped by horizontal chrome bullets which would soon become a Cadillac front-end styling hallmark. The new grille, with annual refinements, would distinguish Cadillacs through the 1947 model year. Redesigned taillights followed the contour of the rear fenders.

But Cadillac's big styling ace this year was in the fenders. Series 61, 62, 63, 67 and Sixty Special models had long, flowing front fenders that extended well back into the front doors.

These bold new fenders were elliptical in shape instead of squared off at the rear and imparted an impression of length and lowness. The rear fenders were similarly rounded and on four-door models also extended into the doors. Rear fender skirts were now standard on all models.

Cadillac's "Fightin' '42s" were offered in 22 body styles in six series. The most changed of these was the all-new Fleetwood Sixty Special. The 1942 Sixty Special bore no resemblance to the classically-styled William Mitchell design that had been so successful between 1938 and 1941. The new car was much more similar in styling to the other cars in the 1942 line, but was still distinctive. Wheelbase was stretched to 133

The 1942 Cadillac Fleetwood Sixty Special was a completely new car which bore no resemblance to the classic Bill Mitchell design that had been so successful from 1938 through 1941. The all-new Fleetwood Sixty Special, however, was still a very distinctive premium four-door sedan that stood out from all other models in the 1942 Cadillac line. It was seven inches longer and one inch lower than the 1941 Sixty Special. Wheelbase was now 133 inches, only three inches shorter than the Fleetwood Seventy-Fives. Styling differences that set the 1942 Cadillac Fleetwood Sixty Special apart from the other 1942 Cadillacs included special wide door center pillars, heavy chrome window reveals, fine chrome louvers on the upper rear quarters behind the rear doors, and a series of bright chrome louvers on the lower portion of both front and rear fenders. Note the extra chrome molding above the windshield on this pre-production model. Possibly it was intended for a leather-covered roof not used on production cars.

The massive, new grille introduced on Cadillac's 1941 models was refined and enlarged for 1942. The revised 1942 grille design spanned almost the entire width of the car's front end. It featured six heavier horizontal chrome grille bars and 15 thin vertical ones. Round parking lights were built into the outer ends of the upper grille. The openings for optional fog lamps below the headlights were now rectangular. When fog lamps were not ordered, these openings were occupied by horizontally-textured chrome inserts. The prominent Cadillac crest with upraised wings continued on the front of the high, coffin-nose hood. With annual modifications, this style of bright, massive grille was used on all Cadillacs through the 1947 model year.

There were two 1942 Cadillac Fleetwood Sixty Specials. The standard model, Style 6069, was a five-passenger sedan which carried a substantially-higher 1942 price of $2,435. Also offered was an Imperial Sedan with glass division, Style 6069-F, which sold for $2,589. Cadillac built only 1,875 Sixty Special before production was halted by the war, including 1,684 standard 6069s and 190 Imperial Sedans. Note the two vertical dividers in the rear window, a feature first used on the 1934 Cadillacs, and the Fleetwood script on the rear deck. As might be expected, the Sixty Special had a lavishly trimmed and appointed interior.

1942

inches. The Sixty Special was seven inches longer and one inch lower than before. The 1942 Sixty Special had extra-wide center door pillars. The side windows were edged with heavy chrome reveals. A series of short, vertical chrome louvers ornamented the lower half of both front and rear fenders, and fine chrome louvers were mounted on the rear quarters just behind the doors. The all-new Sixty Special boasted substantially improved interior roominess and was offered in two models — a standard Fleetwood Sixty Special and an Imperial sedan version with glass division. Interiors were appropriately lavish.

One of the most significant new models this year was the beautiful Series 62 Sedanet. This was actually a Cadillac two-door coupe with a full rear seat and long side windows. The Sedanet, however, had a long, sweeping fastback roofline and a downswept chrome belt molding that, combined with a nearly flat backlight, gave the Sedanet a racy look that strongly influenced the look of all U.S. cars through the 1940s.

The Series 62 convertible coupe was another new style-setter. This model also had a full rear seat. For the first time, the convertible top had wind-down quarter windows. Offered only in a Deluxe version, it was the only convertible in the 1942 Cadillac line — the convertible sedan had been dropped. The wheelbase of Series 62 models was increased three inches, to 129 inches.

Series 61 Cadillacs were available only in two body styles, both torpedo-styled fastbacks. The Deluxe versions offered in 1941 were discontinued. Series 63 and 67 models were largely carryover cars but both got Cadillac's bright, new grille design. This would be the second and final year for both.

> This is a rare 1942 Cadillac Fleetwood Sixty Special custom, an open-front town car conversion by Derham. The rear roof is leather-covered and the removeable weatherproof curtain is in place over the chauffeur's compartment. This car is finished in a striking red with matching red leather roof front and rear. Note the tiny Derham script nameplates at the rear of the hood, which replaced the triangular shield emblem used earlier. Cadillac-Fleetwood built its last town car the previous year. From 1942 on, Derham produced small numbers of Cadillac customs and semi-customs including formal limousine conversions and a few town cars and convertible sedans. This Sixty Special Town Car was on the business end of a tow bar when Steve Hagy caught up with it on a highway in Ohio in 1979.

Fleetwood Seventy-Five Cadillacs retained their conservative 1941 styling. While they did get the new grille, this series carried over the squared-off front fenders and triple chrome fender trim spears of 1941. A circular chrome Cadillac crest with red insert was mounted on the lower edge of the standard fender skirts. The Series Seventy-Five line included five and seven-passenger sedans and limousines, two Fleetwood formal sedans with closed upper rear quarters and leather-covered roofs, and two big nine-passenger Business Sedans for the funeral service trade. Cadillac offered no open-front town cars this year, but Derham did a small number of town car conversions of standard Series Seventy-Five limousines on special order.

All of the 1942 Cadillacs were powered by the same 150 horsepower, 346 cubic inch V-8 engine. All series rode on strong X-type chassis frames. Inside, there was a new T-grip handbrake handle, all-weather ventilation, and improved brakes. Series 62 and Sixty Special models had redesigned instrument panels.

Just as the 1942 model year was shifting into high gear, history intervened. On December 7, 1941 the Japanese launched the infamous attack on U.S. military installations at Pearl Harbor, Hawaii. Within hours the United States had declared war on both Japan and Nazi Germany. The changeover to all-out war production began immediately.

In the weeks before civilian motor vehicle production ceased for the duration, Cadillac and other American automakers were subject to increasingly severe restrictions affecting vital materials. This resulted in thousands of "blackout" models — cars with painted grilles, bumpers and trim instead of chrome. Cadillac built 2,150 "blackout" 1942 models before civilian car production ended on February 4, 1942. The division had built only 16,511 cars in the war-shortened 1942 model year. There would be no more for another three years.

Seven weeks after the last cars had come off the line, the Cadillac assembly lines in Detroit were turning out tanks. Cadillac was at war.

> Only one Cadillac Sixty Special chassis was released for the installation of full-custom bodywork during the abbreviated 1942 model year. The final result was this stunning dual cowl, four-door convertible sedan which was designed and built by General Motors Styling. Construction began in January, 1942 and the instrument panel bears a plate signed by GM Styling Chief Harley J. Earl. The rear cowl and windshield are removeable and the car has a full convertible sedan top with conventional side glass. Wheelbase is standard 1942 Sixty Special 133 inches. This most unusual Cadillac apparently spent most of its life out on the West Coast. It was purchased in Ohio in 1978 by Cadillac enthusiast John Linhardt of Jamaica, N.Y., who still owns, and treasures, it. Note the sweeping, full-length fenders and total absence of bright exterior trim. The body design bears an amazing resemblance to the 1940 Chrysler Newport parade phaeton.

1942

The 1942 Cadillac Series 61 line included only two models — a two-door coupe and a four-door sedan. Both retained the sweeping fastback profile that made these cars Cadillac's top sellers in 1941. The Deluxe models, however, were discontinued. The Series 61 Cadillacs utilized GM's new Fisher "B" body and both body styles rode on a 126-inch wheelbase. This is the 1942 Cadillac Series 61 Coupe for Five Passengers, Model 6107. Factory price was $1,560. Only 2,740 were built before civilian production ended in February, 1942.

The 1942 cars were enthusiastically received by a nervous American public. Europe was embroiled in an escalating war, and Americans everywhere were wondering how long the United States could remain out of the conflict. The 1941 model year had been a highly successful one and U.S. auto manufacturers were hopeful for even higher sales in 1942. Cadillac began production of its 1942 models on October 1, 1941 in the face of material shortages and increased defense production. The splashy 1942 model announcements would be the last for another three years. This is an auto show display of 1942 Cadillacs. The car in the foreground is a Series 61 four-door sedan. Clockwise from the left are a Series 67 sedan, a Fleetwood Seventy-Five Formal Sedan and Limousine, a Series 62 Sedan in the middle and a Series 63 Sedan. The car at the extreme right is equipped with optional fog lamps.

Cadillac's 1942 Series 61 offerings included only two body styles, a two-door coupe and this attractive four-door fastback sedan. Only 5,700 Series 61 Cadillacs were produced before 1942-model production was halted by the war. Of these, 3,194 were this Series 61 Four-Door Sedan for Six Passengers, Model 6109. No Deluxe sedan was offered this year. Factory list price for the Series 61 Sedan was $1,647. Both 1942 Series 61 cars retained their angled rear door pillars and bright chrome moldings that encircled the side window openings. The more expensive Series 62 models had vertical door and window posts.

The new 1942 Cadillac Series 62 Five-Passenger Coupe, or "Sedanet", was even more aerodynamic in form than the torpedo-styled Series 61 models of 1941. The beautiful Sedanet, which was actually a coupe with a full rear seat, featured a long, sweeping fastback roofline that was complemented by a chrome belt molding that swept down to the base of the rear decklid. The backlight was nearly flat and had two chrome dividers. The long, narrow rear side window also dipped down slightly at the rear. The sleekly-styled Sedanet set the style for the rest of the decade. It was produced through the 1947 model year, and there were fastbacks in the 1948 and 1949 Cadillac lines. Wheelbase was 129 inches, three inches longer than in 1941. Standard and Deluxe models were offered, at $1,667 and $1,754 respectively. Only 1,045 Series 62 Sedanets were built in the 1942 model year, including 515 of these standard Model 6207 Coupes and 530 Model 6207-D Deluxe Coupes.

Three Series 62 Cadillac body styles were offered for 1942 — the racy new Sedanet Coupe, a four-door sedan and a convertible. The most popular of these was the Series 62 Five-Passenger Sedan, Style 6269. A total of 1,780 were built, compared with 1,743 Deluxe Sedans, Style 6269-D. The standard four-door Series 62 sedan had a factory price of $1,754. The Deluxe cost only $84 more, or $1,836. All 1942 Cadillacs came with rear fender skirts as standard equipment. Two-tone paint was a popular option: standard two-tone color combinations were offered at no cost. Special two-tones cost an extra $12.50 to $25.00 depending on color.

1942

The most dramatic styling change on the 1942 Cadillacs were bold new bullet-shaped fenders on Series 61, 62, 63, 67 and Sixty Special models. The long front fenders extended well back into the front doors. On four-door models the rear fenders extended forward into the door panels. Heavy chrome moldings protected front and rear fenders at their centerlines. The three horizontal chrome spears used on all fenders on the 1941 models were now found only on Series Seventy-Five Cadillacs. The new pontoon fenders went especially well with the softly-rounded upper and lower body forms of the 1942 Cadillac Series 62 Four-Door Sedan. Note the two vertical rear window dividers and the Cadillac emblem on the rear decklid.

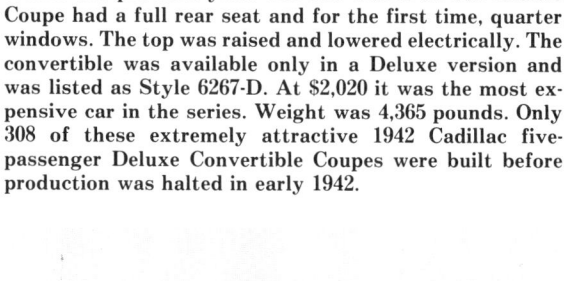

One of the most-changed cars in the 1942 Cadillac line was the Series 62 Convertible Coupe. This was the only soft-top model offered: Cadillac had built its last convertible sedans the previous year. The new Series 62 Convertible Coupe had a full rear seat and for the first time, quarter windows. The top was raised and lowered electrically. The convertible was available only in a Deluxe version and was listed as Style 6267-D. At $2,020 it was the most expensive car in the series. Weight was 4,365 pounds. Only 308 of these extremely attractive 1942 Cadillac five-passenger Deluxe Convertible Coupes were built before production was halted in early 1942.

This would be the second, and final, year for the Series 63 Cadillac. The Series 63 was again available in only one model, a four-door, five-passenger six-window notchback sedan. Wheelbase was 126 inches, the same as the Series 61. Factory advertised price was $1,882. This model was listed as Style 6319. A total of 1,750 Series 63 Sedans were built before production was prematurely terminated by the war. In all, Cadillac produced 6,800 Series Sixty-Three Sedans in the 1941 and 1942 model years. Note the fender-mounted accessory radio antenna.

Series Seventy-Five Cadillacs for 1942 retained their conservative 1941 body styling but got Cadillac's bright, new grille. All Series Seventy-Five models had carryover squared front fenders instead of the sleek new pontoon fenders that extended into the doors on the other 1942 Cadillacs. The three chrome fender spears, also a 1941 carryover, appeared only on 1942 Seventy-Fives and the commercial chassis, as did the ventilator panel at the rear of the hood sides. This vent casting had a pointed front for 1942. Eight Series Seventy-Five models were offered this year, all variations of the same Fleetwood four-door sedan body for five or seven passengers, and all on the same 136-inch wheelbase. This is the 1942 Cadillac Series Seventy-Five Touring Sedan for Five Passengers, Style 7519, of which 205 were built. Price was $3,306. Style 7519-F had an Imperial division and cost $3,459. Only 65 were sold.

The 1942 model year would also be the last for the Series 67 Cadillac, which was built only in 1941 and 1942. The Series 67 was a lower-priced companion to the expensive Series Seventy-Five sedans and shared its four-door body with the Series 90 Buick Limited. The Series 67 rode on the longest wheelbase in the entire Cadillac line — 138 inches, or two inches longer than the Seventy-Five. Four Series 67 models were offered; Style 6719, the standard Series 67 five-passenger sedan which sold for $2,896; Style 6719-F with formal division, $3,045; Style 6723 Series 67 Seven-Passenger Sedan, $3,045; and Style 6733, Series 67 Imperial Sedan for Seven Passengers, with division, $3,204. An even 700 were built. The seven-passenger sedan was the most popular, with 260 delivered. The five-passenger sedan was next, with 200 sold, followed by the Seven-Passenger Imperial (190) and 50 five-passenger Imperials.

Two seven-passenger sedans were included in the 1942 Fleetwood Seventy-Five line. Style 7523 was the standard Seven-Passenger Sedan, 225 of which were delivered at $3,459. The most popular car in the series was the Style 7533 Seven-Passenger Imperial, 430 of which were sold at $3,613. The 1942 Series Seventy-Five line also included two nine-passenger Business Sedans aimed at the funeral service trade. This is the Style 7523-L Nine-Passenger Business Sedan, 29 of which were built. Price was $3,152. Only six nine-passenger Business Imperials, Style 7533-L, were delivered this year at a price of $3,306 each.

1942

Cadillac's most expensive 1942 production models were a pair of formal sedans at the top of the prestigious Fleetwood Seventy-Five series. Offered in five and seven-passenger versions, these impressive formal sedans had closed upper rear quarters and padded, leather-covered roofs. Their rear windows were smaller and did not have vertical dividers. Only 60 Style 7559 Fleetwood Five-Passenger Formal Sedans with "X" division (no ceiling header) and 80 Style 7533-F Seven Passenger Formal Sedans with Imperial division were built in the shortened 1942 model year. The five passenger formal sedan carried a price of $4,330 and the seven-passenger version $4,484, the most expensive of all. This 1942 Fleetwood Seventy-Five Formal Sedan is of interest — it sports postwar stainless steel running boards, later rear bumpers, and has no (standard) rear fender skirts or the chrome trim spears used on front and rear fenders on all 1942 Seventy-Fives.

The big news in the U.S. funeral car industry this year was a Flxible innovation. The Flxible Co., of Loudonville, Ohio, came out with a totally new concept in funeral cars which it called, appropriately, the Flxible Innovation. Flxible's new hearse design combined the low, coupe-type styling of a flower car with the formal dignity of a carved-panel hearse. The Innovation had a smooth, rounded rear deck that was not designed to carry flowers. Flxible built two of these unusual funeral coaches on 1942 Cadillac commercial chassis, and at least one other without carved sides on a 1942 Buick. But the Second World War intervened and this revolutionary concept was not revived when Flxible resumed peacetime funeral car and ambulance production in 1946. This was the last year Flxible was to build on the Cadillac chassis. From 1946 on the company used Buick chassis exclusively.

The Proctor-Keefe Co., a large truck body builder in Detroit, built two of these special Cadillac rescue squad cars for the Detroit Fire Department. The custom Proctor-Keefe squad bodies were mounted on 1942 Cadillac Series 75 commercial chassis. This one was wrecked in a collision in 1948. It's mate was retired in 1954. Note the roof-mounted Roto-Rays, the Federal coaster siren on the left front fender and the bell on the other. Squad 4's rig stands beside a B-24 Liberator bomber and a P-39 Airacobra fighter plane at a prewar air show display at Detroit City Airport.

Cadillac and Fleetwood discontinued the production of open-front town cars in 1941. But custom coachbuilder Derham continued to provide such formal transportation for those who absolutely had to have one. Derham built small numbers of such custom conversions of standard Cadillacs for some years, although Town Cars had disappeared by the late 1940's. This is a Derham Town Car conversion of a 1942 Cadillac Fleetwood Seventy-Five Formal Sedan, one of six similar town cars done by Derham this year. This custom body house also did at least one town car conversion of a 1942 Fleetwood Sixty Special.

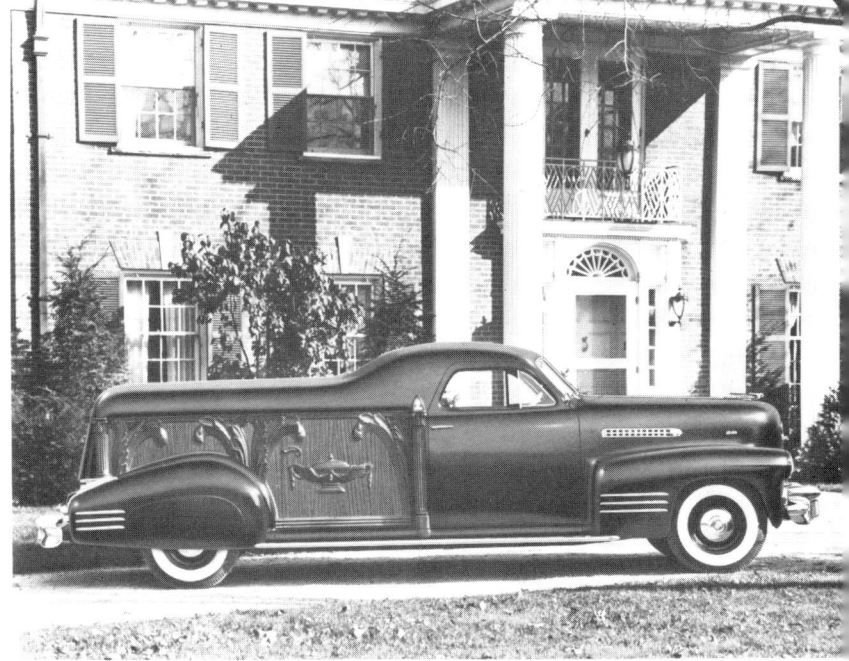

The Cadillac Motor Car Company built only 425 long-wheelbase commercial chassis before 1942-model production was cut off following the Japanese attack on Pearl Harbour. All were Series 75 hearse and ambulance chassis with 163-inch wheelbase. Funeral directors who were unable to purchase new motor equipment before production was halted were forced to take extremely good care of their existing equipment, as no one knew how long it would be before new vehicles would again be available. The A. J. Miller Co. of Bellefontaine, Ohio built this landau-style funeral car on the 1942 Cadillac commercial chassis. Note the fender skirts, small coach lamps and diminutive carriage bow. The 1942 Cadillac's new bullet-shaped bumper guards are shown to advantage in this photo.

1943 - 1945

The Cadillac Motor Car Division of General Motors made an important and distinguished contribution to the American war effort during the dark days of the Second World War. Again, history repeated itself: just as had happened during the Great War of 1917-1918, Cadillac's vast manufacturing facilities were quickly converted to defense production after the United States was drawn into the conflict late in 1941.

Cadillac had actually begun the production of war materiel — engine parts for the GM-built Allison V-1710 aircraft engine — in 1939, more than two years before Pearl Harbor. By 1941 Cadillac was already gearing up for war production. Because of a chronic shortage of Continental tank engines, the U.S. Ordnance Department was working closely with Cadillac Engineering to adapt the highly-regarded Cadillac V-8 engine as a military vehicle powerplant. Early tests were encouraging. A punishing 500-mile run in a converted tank proved beyond a doubt the Cadillac engine's unique suitability for this purpose.

When the United States finally entered the war in December, 1941, Cadillac immediately began the conversion of all of its plants to vital war work. Cadillac's principal wartime assignment was the production of light tanks. Only 55 days after the last Cadillac passenger car for the duration was driven off the Clark Avenue assembly line in Detroit, the first M-5 tank was rolled out in April, 1942. It was the first of thousands of tanks and specialized armored vehicles to pour out of the Cadillac plant in Detroit over the next few years and which saw action on virtually every battlefront in Europe and the Pacific.

A 13.6-ton tank, the nimble M-5 carried a 37-millimeter main gun in its turret. It was powered by two Cadillac V-8 engines coupled to a pair of GM Hydra-Matic automatic transmissions. Each engine-transmission combination powered one track. Steering was accomplished by manipulating two levers. Early in 1943 Cadillac began production of an improved version, the M-5-A1, and the M-8 howitzer motor gun carriage. The M-8 was equipped with a 75-millimeter gun in an open turret and utilized basically the same undercarriage as the M-5.

In 1944, Cadillac began production of a larger, faster and more sophisticated battle tank, the famous M-24 Chaffee, which replaced the M-5. The M-24 was also powered by dual Cadillac engines linked to Hydra-Matic transmissions. The 18-ton M-24, also classified as a light tank, carried a 75-millimeter gun. In 1945 the M-19 anti-aircraft motor gun carriage joined the M-24 on the Cadillac tank plant assembly line. The M-19 mounted two 40-millimeter anti-aircraft guns. Cadillac built-and-powered light tanks saw action on many battlefronts and served with the armies of the United States, Canada, England and France. The M-24 was also used in Korea and ultimately evolved into the M-41 Walker Bulldog, which saw service into the 1970s.

The last tank, an M-24, clanked off the Cadillac assembly line on August 24, 1945. In all, Cadillac produced 12,230 military vehicles during the Second World War.

In addition to tanks, motor gun carriages and anti-aircraft units, Cadillac produced millions of precision aircraft engine parts for another General Motors division, Indianapolis-based Allison. Cadillac's contribution consisted of crankshafts and camshafts, connecting rods, piston pins and complex reduction gear assemblies for the Allison V-1710 aircraft engine. This engine powered several famous American fighter aircraft including the Curtiss P-40 Hawk, the Bell P-39 Airacobra and later P-63 Kingcobra versions of the North American P-51 Mustang, and the twin-engined Lockheed P-38 Lightning.

The twin-boomed P-38 exerted a strong influence on postwar Cadillacs. Prior to the United States' entry into the war, GM styling chief Harley Earl sent some of his stylists out to Selfridge Field near Detroit to look at an early production Lightning. The long twin tail booms with their side-mounted air scoops and vertical tail surfaces provided the inspiration for the controversial tail fins which sprouted on the 1948 Cadillacs, and which were carried to outrageous extremes a dozen years later.

During the war, Cadillac's employees received numerous awards for proficiency in war production. One of these was the Army-Navy "E" for excellence. The coveted "E" flag flew proudly with the stars and stripes over the Clark Avenue complex until final victory was won in August, 1945.

It is likely that Cadillac assembled some cars for use by the U.S. armed forces during the war. The big Fleetwood Seventy-Five was a popular staff car preferred by several American generals, including General Douglas MacArthur. Once again, history had repeated itself. Senior American military officers had exhibited a clear preference for Cadillac staff cars in the First World War.

On October 17, 1945, the first postwar Cadillac passenger car came down the line. It was a warmed-over 1942 model with minor detail changes. Cadillac built only 1,142 cars in 1945. The division had in hand orders for more than 100,000 cars. The postwar sales boom had already begun.

Cadillac goes to war! In less than two months, Cadillac in early 1942 converted from civilian passenger car to all-out war production. Cadillac's principal wartime products were light tanks, notably the M-5 and M-42, and motor gun carriages. Although civilian cars would not be available for another three years, Cadillac continued to advertise with patriotic emphasis on the division's important contribution to the American war effort. This wartime ad shows a Cadillac-built and powered tank in action. The censorship board carefully removed or obscured all relevant details.

1943 - 1945

Just 55 days after the last car rolled off Cadillac's Clark Avenue assembly line in Detroit, the division in April, 1942 began production of light tanks. For the next two year's Cadillac's principal product was the M-5, which was succeeded in 1944 by the larger and more powerful M-24 tank. These tanks were powered by a pair of modified Cadillac V-8 engines coupled to GM Hydra-Matic automatic transmissions. The M-5 light tank was armed with a 37-millimeter main gun in its turret.

This was Cadillac's 1944 model — the potent new M-24 Chaffee light tank which succeeded the M-5. The new M-24 was also powered by two Cadillac V-8 engines coupled to Hydra-Matic automatic transmissions, one for each track. Steering was done by manipulating twin levers. One of the fastest battle tanks of its day, the M-24 saw action on all battle fronts and continued in service with several armies well beyond the Second World War. The M-24's turret was equipped with a 75-millimeter gun. Note the machine gun mounted behind the top hatch.

In 1943, Cadillac began production of the M-8 howitzer motor carriage, which was developed off the M-5 tank chassis. It, too, was powered by dual Cadillac engines. The pugnacious-looking M-8 mounted a 75-millimeter gun in its open turret, as well as a machine gun.

M8 75mm HOWITZER MOTOR CARRIAGE

Another Cadillac contribution to the U.S. Army's Second World War arsenal was the M-19 anti-aircraft gun motor carriage. Production of the M-19 began at Cadillac in 1945 only a few months before hostilities ended. Like the M-24 tank, the M-19 was powered by twin Cadillac V-8 engines. This is the M-19-A1, which carried two 40-millimeter anti-aircraft guns.

The last M-24 tank clanked off the Cadillac Motor Car Co. assembly line in Detroit on August 24, 1945. Less than two months later the first postwar Cadillacs were being driven off the line. Cadillac produced more than 12,000 military vehicles and millions of precision aircraft engine parts during World War II. For its significant contributions to the American war effort, Cadillac received numerous production awards including the coveted Army-Navy "E" Award for excellence in defense production.

As all-out war approached, the United States Government snapped up whatever it could get from civilian ambulance manufacturers for service at American military establishments all over the U.S. and abroad. The Superior Coach Corp., Lima, O., delivered a fleet of 1941 Superior-Cadillac limousine style ambulances to the Medical Department of the U.S. Army. Most of these commercial-chassis ambulances were assigned to various army camps and bases around the country. Military field ambulances were more utilitarian. Note the blanked-out side windows and small standard hubcaps. Whitewall tires and wheel trim rings were considered unnecessary civilian frills.

Between 1942 and 1945, Cadillac built thousands of tanks and motor gun carriages for the United States Government. All were powered by the rugged Cadillac V-8 engine. In 1941, U.S. Ordnance Dept. engineers, working with Cadillac, adapted the 346 cubic inch Cadillac V-8 as a military vehicle powerplant. The engine was beefed up and teamed with a GM-built Hydra-Matic automatic transmission. Cadillac-powered tanks and gun carriages acquired an excellent reputation for outstanding performance and reliability on all war fronts. Although the factory discouraged the practice, more than a few of these sturdy tank engines found their way under the hoods of Cadillac passenger cars after the war. Numerous components, however, were not interchangeable with civilian Cadillac engines.

Beginning in 1939 and continuing through the Second World War, Cadillac produced millions of precision engine parts for the Allison V-1710 aircraft engine. This engine powered several famous U.S. fighter planes including the Curtiss P-40 Hawk, the Bell P-39 Airacobra and P-63 Kingcobra; versions of the North American P-51 Mustang, and the twin-boom Lockheed P-38 Lightning. The twin-engined Lightning, shown downing a German bomber in this wartime Cadillac ad, was the inspiration for the style-setting tail fins which sprouted on postwar Cadillacs beginning in 1948, and which reached ridiculous heights a decade later.

With their supply of new motor equipment indefinitely cut off, funeral directors and ambulance operators were forced to take extremely good care of their prewar equipment. No one knew when new automotive equipment would again be available. Hearses, ambulances and limousines were pampered and meticulously maintained through the war years. Even service parts were hard to come by. Funeral directors who had been used to ordering new vehicles every year simply had to make do with what they had at the time of Pearl Harbour. It was a four-year-old 1941 Miller-Cadillac Funeral Coach identical to this one that transported the remains of the late President Franklin D. Roosevelt from his summer home in Warm Springs, Ga., where he died in April, 1945 to the local railway station for the sad journey back to Washington.

Less than two months after tank production ended, the first postwar Cadillac was driven off the assembly line on October 17, 1945. It was a Series 62 four-door sedan, the only 1946 Cadillac body style offered for several months after peacetime production resumed. Cadillac built only 1,142 cars in 1945, but had orders for 100,000 more from a car-hungry public. Here, Cadillac executives greet the first regular production 1946 Cadillac as it came off the line. The Cadillac plant had not built a civilian car in more than three years.

1946

For the first time in more than three years, new passenger cars were once again rolling off Detroit's assembly lines. After years of wartime restrictions and shortages, during which no new cars were available to the American public, U.S. automobile manufacturers were swamped with orders for new postwar cars.

The Cadillac Motor Car Division of General Motors had ended tank production in late August, 1945. Less than two months later, on October 17, the first postwar Cadillac came down the line. It was greeted with appropriate fanfare as it was driven off. For the next four years it was virtually a seller's market. Car-hungry Americans who had "made do" long enough, eagerly snapped up anything on wheels as long as it was new. With relatively few cars to sell, dealers added to their profits by aggressively merchandising optional equipment and accessories. Waiting lists were the norm if you wanted a new car of any make—including Cadillac.

The huge conversion from war back to peacetime production was a slow and gradual one. Immediate postwar production was frequently interrupted by shortages of vital raw materials, particularly sheet steel. Whitewall tires were almost impossible to get, and more than a few 1946 Cadillacs were shipped from the factory with temporary wooden bumpers, which were later replaced with chrome ones in the field.

Cadillac's 1946 models were little more than mildly warmed-over 1942 cars. The 1942 model year had barely reached stride when production was cut off by the war. Styling was identical to Cadillac's 1942 models with bright chrome grilles, big pontoon fenders and flowing fastback rooflines on many models.

The new postwar Cadillacs, however, did show some changes. The revised grille design featured six heavier horizontal bars which gave the car a somewhat lower appearance. Parking and turn signal lamps and the openings for optional fog lamps were still incorporated into the outboard edges of the eggcrate grille. All of the 1946 Cadillacs sported massive new three-piece front and rear bumpers which extended around the corners of the car for greater sheet-metal protection. And for the first time, the soon-to-be-world-famous Cadillac "V" emblem and crest appeared on the front of the hood and on the trunk. The Cadillac name was carried in small block letters above the horizontal chrome trim on the front fenders.

All of the 1946 Cadillacs were powered by the "Battle-Proved" or "Victory" 346 cubic inch V-8 engine which had benefitted from numerous improvements developed and incorporated during the war. The Cadillac V-8 had proved to be a rugged and extremely dependable tank powerplant throughout the war, and had been in production continuously since 1935, when it had been introduced for Cadillac's 1936 cars. Similarly, the Hydra-Matic automatic transmission which had been teamed with the Cadillac V-8 in thousands of tanks and military vehicles had been beefed up and improved through its unique wartime service.

For 1946, Cadillac offered 11 models in four series. The prewar Series 63 and 67 cars were gone, as was DeLuxe trim in Series 61 and 62 models. Only the Series 62 convertible came in deluxe form. For the first several months after peacetime production was resumed, just one Cadillac body style was available. It was the C-bodied Series 62 four-door sedan, which accounted for nearly half of all the Cadillacs built that year. Other models and series were not phased into production until well into 1946.

The lowest-priced Series 61 line consisted of only two models, a two-door coupe and four-door sedan. Both were prewar fastbacks on a 126-inch wheelbase. There were eventually three Series 62 models—the handsome two-door coupe or Sedanet; the four-door sedan and the sporty Convertible Coupe. There was only one Fleetwood Sixty Special in the 1946 model lineup, a four-door touring sedan on the same 133-inch wheelbase

Cadillac's 1946 models were little more than warmed-over 1942 cars with minor styling changes. The revised 1946 Cadillac grille design had six heavier horizontal grille bars which gave the car a slightly lower appearance. New three-piece front and rear bumpers extended around the corners of the car providing additional protection. Only one Fleetwood Sixty Special was offered for 1946, the standard Style 6069 five-passenger sedan. The Sixty Special retained its pre-war 133-inch wheelbase and exterior styling, but the small vertical chrome louvers were gone from the lower edges of front and rear fenders. The fine chrome louvers remained on the upper rear quarters just behind the rear door. Factory price F.O.B. Detroit was $3,054, and 5,700 were delivered to a car-hungry public during the 1946 model year.

The Derham Body Co. of Rosemont, Pa., one of the very few custom body builders still in business after the war, continued to produce small numbers of semi-customs on the Cadillac chassis, picking up where Fleetwood left off in 1941. This is a 1946 Cadillac Fleetwood Sixty Special Town Car by Derham. A glass division has been fitted between the front and rear compartments. Open-front town cars of this style were intended to be chauffeur-driven. Note the lack of white sidewall tires and the light-colored rear roof roof covering. A weatherproof curtain covered the chauffeur's compartment in less than ideal weather. Formal open-front town cars were now quite rare.

1946

The lowest-priced car in the 1946 Cadillac line was the sleek Series 61 Coupe for Five Passengers. A five-passenger fastback that had originally been introduced for 1941, the Series 61 Coupe was still a sporty-looking car. Only 800 of these fastbacks coupes, Model 6107, were produced during the 1946 model year making this the rarest of the Fisher-bodied series built that year. Base price was $2,022. The example shown has the standard small hubcaps without wheel trim rings.

Although introduced a full five years earlier, for the 1941 model year, the Cadillac Series 61 Coupe for Five Passengers was still a good-looking car. Exterior changes for 1946 included a "V" emblem on the rear deck and hood, the Cadillac name in small block letters above the horizontal chrome molding on the front fenders and a revised grille design. Bumpers extended around the corners of the car for greater protection. Only two Series 61 Cadillacs were offered for 1946, this two-door coupe and a four-door sedan, both on a 126-inch wheelbase.

Cadillac's lowest-priced car line, the Series 61, offered only two body styles for 1946—a four-door sedan and a two-door coupe. Both were sleek fastbacks built on the same 126-inch wheelbase chassis with Fisher bodies. This is the 1946 Cadillac Series 61 Five-Passenger Sedan, Model 6109, which sold for $2,146. Weight was 4,145 pounds. Of the 3,001 Series 61 Cadillacs delivered in the 1946 model year, 2,200 were four-door sedans like this one. The Series 61 Sedan looked much like it had before the war. The grille was revised, however, to give the car a lower look. The three side windows were still encircled by a bright chrome molding. The Cadillac name appears in small block letters on the front fenders, and "V" emblems were added to the front of the hood and the rear decklid.

introduced for 1942. The small vertical chrome louvers were gone from the Sixty Special's fenders, and no division glass was offered.

The Series Seventy-Five Fleetwoods retained their staid 1941 styling, with high, slab sides and squared-off fenders. The Seventy-Fives, however, did get the new 1946 grille and bumpers and were equipped with full-length stainless steel running boards. The Series Seventy-Five five and seven-passenger sedans and limousines would continue almost unchanged through the 1949 model year.

Cadillac offered no less than five Series Seventy-Five models, all on the same 136-inch wheelbase. The hood ventilator panels used in 1941 and 1942 were gone, as were the closed quarter panel, leather-topped Formal Sedans. Derham did a number of formal sedan and limousine conversions of postwar Series Seventy-Five Cadillacs.

Cadillac also continued to market two big Fleetwood Seventy-Five nine-passenger Business Sedans designed for funeral and limousine livery service. There was also a special 163-inch wheelbase Series 75 commercial chassis for mounting hearse and ambulance bodies.

Popular Cadillac options this year included Hydra-Matic automatic transmission ($176.49 extra); five wheel trim rings ($8.84), and bright license plates frames which sold for $3.39 a pair. This was also the first year Cadillacs were equipped with negative-ground batteries.

There were some important personnel changes on Clark Avenue this year. John F. Gordon succeeded Nicholas Dreystadt as Cadillac Motor Car Division General Manager. Dreystadt, who had followed the colorful L.P. Fisher in 1934, went to sister GM division, Chevrolet. Edward N. Cole became Cadillac's new Chief Engineer, succeeding Mr. Gordon, who went on to eventually become President of General Motors.

Cadillac built a respectable 29,194 cars during the 1946 model year which ended in late December. The American automobile industry officially observed its Golden Jubilee with a 12-day celebration in Detroit. The future again looked bright.

> A car-starved American public, weary of war and wartime shortages, eagerly snapped up anything on wheels when the 1946 models hit the showrooms late in 1945 and into 1946. Production was still hampered by material shortages as U.S. industry converted back to peacetime production. For those without connections and unable to obtain a new car, there were always the enticing sales brochures to drool over. This is a page from the 1946 Cadillac catalog. The car is a Series 61 Four-Door Sedan. About all you could do at this time was get your name on the waiting list... and hope.

Cadillac's mid-range Series 62 offerings for 1946 included three body styles, all on the same 129-inch wheelbase. All were listed as five-passenger cars. Except for minor grille and cosmetic changes, they were nearly identical in appearance to the pre-war 1942 models. But a car-hungry public snapped them up, anyway. This is the very attractive 1946 Cadillac Series 62 Coupe, or Sedanet, also sometimes called a Club Coupe. Style 6207, it sold for $2,249. A total of 2,323 were delivered in the 1946 model year. Note the long, sweeping fastback roofline which is still complemented by a downward curved chrome belt molding. Two-tone exterior color combinations were still popular in the immediate postwar years. This Series 62 Sedanet has full wheel covers but no whitewalls.

1946

For the first few months after Cadillac resumed civilian passenger car production in October, 1945, the only body style available was the Series 62 Sedan. This simplified production and enabled Cadillac to gear up for volume production. Other models and body styles were not introduced until well into the 1946 model year. This is the 1946 Cadillac Series 62 Sedan for Five Passengers, Model 6269. It was by far the most popular model offered this year, with 14,900 produced. Most auto manufacturers, including Cadillac, were still affected by steel and material shortages which limited production and equipment choices.

The undisputed glamour queen of the 1946 Cadillac line was the dashing Series 62 Convertible Coupe, which was not introduced until late in the 1946 model year. Model 6267-D, the convertible was available only in a well-equipped DeLuxe version, hence the "D" suffix in the model number. A five-passenger car, the Series 62 convertible was the most expensive car in the series with a factory list price of $2,521. Only 1,342 Series 62 convertibles were built in the 1946 model year. This elevated dockside pose shows the 1946 Cadillac's refined front-end ensemble to advantage, including the soon-to-be-famous Cadillac "V" emblem and crest on the hood. This "V" crest was also used on the trunk. Note the smoother hood ornament and bullet-topped bumper guards.

Here's how the 1946 Cadillac Series 62 Convertible looked with its top up. The 1946 convertible was nearly identical to the 1942 model. Only 1,342 of these sporty convertibles were built in the 1946 model year, and production did not begin until halfway through the model run. All 1946 Cadillacs were powered by the "Battle Proved" 346 cubic inch V-8 engine that had been introduced in the 1936 models more than a decade previously.

Although hampered by material shortages, Cadillac managed to build a respectable 29,194 cars during the 1946 model year. Of these, nearly half were Series 62 four-door sedans like this one. In fact, for several months after civilian production was resumed late in 1945, this was the only model Cadillac offered. For 1946 all Cadillacs got a revised grille design, new three-piece wraparound bumpers and "V" emblems on the front of the hood and the rear decklid. Note the rubber stone shield on the lower leading edge of the rear fender, a carryover from 1942. These were replaced with stainless steel stone shields in 1947.

Cadillac's most exclusive and expensive production cars, the regal Fleetwood Seventy-Fives, would show little exterior change for the next four years. The 1946 models, in fact, were little more than warmed-over versions of cars that had originally been introduced for 1941. The most noticeable exterior change for 1946 was broad, stainless steel running boards which extended onto the lower edges of the front and rear fenders. The Cadillac name appears in small block letters just behind the front wheel opening. The 1946 Seventy-Fives also got the same revised grille design and new bumpers used on all other 1946 Cadillacs. This is the 1946 Fleetwood Seventy-Five Sedan for Five Passengers, Style 7519, which listed for $4,238. Only 150 were built. The five-passenger sedan version was not available with a glass division. The seven-passenger model was.

As in previous years, Cadillac's Fleetwood Seventy-Five sedans for 1946 were built in both five and seven-passenger variations. The standard seven-passenger sedan for 1946 was Style 7523, of which 225 were built this year. Factory price was $4,415. Style 7533 was a Seven-Passenger Imperial with formal glass division, which sold for $4,609, making it the most expensive of all the 1946 Cadillacs. A total of 221 Imperials were sold in the 1946 model year. Two big Series Seventy-Five Business Sedans rounded out the line. All five 1946 Fleetwood Seventy-Fives were built on the same 136-inch wheelbase chassis. The ventilator panels on the rear hood sides on the 1941 and 1942 models were not used on postwar cars.

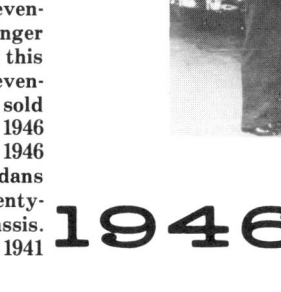

1946

The funeral directing profession still represented a large and important specialized market for Cadillac. The division's postwar 1946 product line included a special hearse and ambulance commercial chassis and two big nine-passenger business sedans designed for the unique requirements of funeral directors and limousine livery operators. The two business sedans included the Fleetwood Seventy-Five Nine-Passenger Style 7523-L, only 22 of which were built in 1946, and Style 7533-L, the Nine-Passenger Imperial Business Sedan with glass division. Only 17 Style 7533-L Business Imperials were sold. The standard nine-passenger business sedan was priced at $4,093 and the Imperial sold for $4,286. Here, a fleet purchaser looks over a 1946 Fleetwood Seventy-Five Business Sedan finished in grey-and-black livery in the Cadillac executive garage in Detroit.

The Hess and Eisenhardt Co. of Cincinnati, Ohio, builders of S & S funeral cars and ambulances, decided to enter the burgeoning airline limousine business this year. Rival Henney had been fairly successful in the airline limousine business before the war with a series of eight-door ground transportation limos built on stretched Packard chassis. Hess and Eisenhardt wanted in on this promising postwar market. This is a prototype six-door airline surface transportation coach built by Hess and Eisenhardt on the 1946 Cadillac Series Seventy-Five commercial chassis. It is a trifle hearselike in appearance but more integrated than other airline limos marketed at the time. Full-scale production did not begin until the following year.

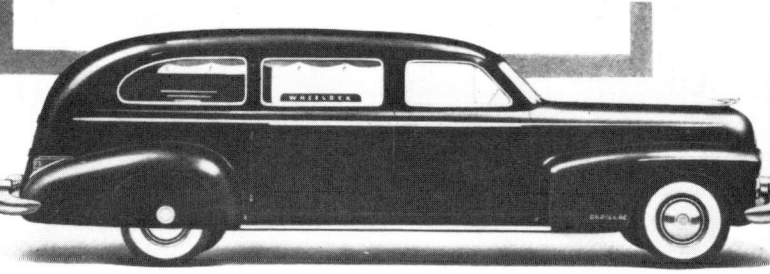

America's funeral directors, who had carefully nursed their pre-war funeral car fleets through the Second World War years, now badly needed replacement equipment. Their well-used prewar vehicles were literally wearing out, and those fortunate enough to get on waiting lists for the first postwar funeral cars and ambulances still faced long and uncertain waits. Some funeral car manufacturers were delivering their first postwar hearses late in 1945. Nearly all, except industry giant Henney, were back in full production in 1946. This is the 1946 Eureka-Cadillac "Chieftan" Three-Way side-loading limousine hearse on Cadillac Series Seventy-Five commercial chassis, produced by the Eureka Co. of Rock Falls, Ill. Like most American passenger cars, most immediate postwar funeral coaches and ambulances were basically carryover 1942 models that had been given minor cosmetic changes.

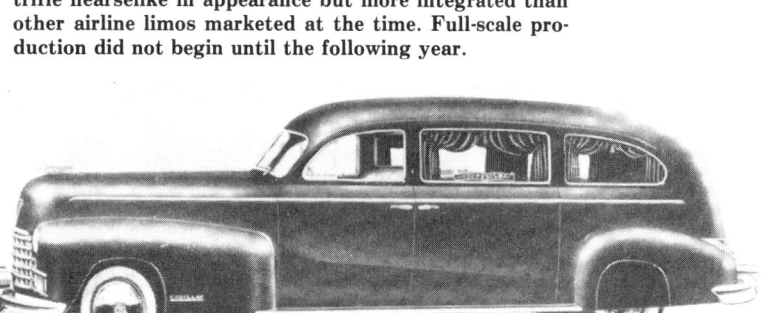

Maurice Schwartz, formerly of Bohman and Schwartz fame, continued to do specials and customs in the Pasadena, Calif. shops after he and Bohman ended their partnership in 1945. Schwartz did this very long custom six-door woodie wagon on a stretched 1946 Cadillac chassis. It was used by one of the major movie studios to haul actors and actresses between the studio and shooting locations. Note the conventional Series Seventy-Five style trunk.

Most of America's funeral car and ambulance builders found a large pent-up demand for their products when peacetime production resumed in 1946. Cadillac was able to deliver only 1,292 of its 1946 Series 75 commercial chassis, which were shared by The Superior Coach Corp. of Lima, Ohio; The Eureka Co. of Rock Falls, Ill.; Hess & Eisenhardt (S & S) of Cincinnati, Ohio; the Meteor Motor Car Co. of Piqua, Ohio, and the A.J. Miller Co. of Bellefontaine, Ohio. The Henney Motor Co. of Freeport, Ill., continued its alliance with Packard. This is the 1946 Superior-Cadillac Limousine Funeral Coach which was offered in both end-loading and side-servicing versions. The 1946 Cadillac commercial chassis had a 163-inch wheelbase.

1947

Cadillac entered the booming 1947 model year with essentially the same product lineup it had taken to market for 1942, some five years earlier. Even though its products were becoming dated as the first new postwar designs took shape on GM's drawing boards, this was no deterrent to a car-starved American public. Every new Cadillac that rolled out of the plant in Detroit was eagerly snapped up by a grateful customer. It was still a seller's market. If you wanted a new luxury car, you either knew someone or your name went on a long waiting list. The black market flourished in new passenger cars, parts and accessories.

The 1947 Cadillacs were little-changed refinements of the 1946 models, which in turn were warmed-over 1942 cars. Changes were few and mainly cosmetic. There were 11 models in four series on four different wheelbases. All 1947 Cadillacs were powered by the same 346 cubic inch V-8 engine which had powered Cadillacs since 1936. The current war-improved version was now rated at 150 horsepower and had hydraulic valve lifters with hardened ball seats. About the only significant functional change in the 1947 Cadillacs were standard Hydra-Lectric power window lifts in the Series 62 convertible and all Fleetwood Seventy-Five models.

The 1947 Cadillac's revised grille design featured five heavier, stamped grille bars instead of the six die-cast metal ones used previously. The header bar extended onto the front fenders, and the new grille, which continued the classic Cadillac eggcrate theme introduced for 1942, gave the car a wider, slightly lower appearance. Parking and turn signal lamps or optional fog lamps were housed in a rectangular panel in the lower outer edges of the grille. The high coffin-nose hood was ornamented with a new "V" crest, which featured the Cadillac emblem in a field of cloisonne. A new winged emblem appeared on the rear deck lid of all models except the 1947 Fleetwood Sixty Special, which retained its Fleetwood script. The small block letters previously used on the front fenders were replaced with stylish new Cadillac script nameplates, and the black rubber stone shields used on 1946 models were replaced with bright stainless steel stone shields.

But the most significant exterior styling change on the 1947 Cadillacs were the hubcaps. New premium wheel covers available on all series ($25.08 for a set of four) were bright, chrome-plated "sombreros" which quickly became another Cadillac style setter, widely imitated later on other cars and very popular with car customizers and hot-rodders. The new wheel covers featured a red Cadillac crest insert in the centers. Cadillac used these stainless steel "sombrero" wheel discs well into the 1950s.

The lowest-priced Series 61 included a coupe and sedan on a 126-inch wheelbase. Three Series 62 body styles included a coupe, four-door sedan and convertible on a 129-inch wheelbase. There was just one Fleetwood Sixty Special, still on its exclusive 133-inch wheelbase. At the top of the line were five Fleetwood Seventy-Fives in five, seven and nine-passenger versions all on the same 136-inch wheelbase chassis. Also offered was a 163-inch Series 75 commercial chassis. All body sheet-metal was pre-war carryover. By far the most popular model in the 1947 line was the Series 62 four-door sedan, of which more than 25,000 were sold.

Cadillac prices increased sharply this year, ranging from $2,200 for the Series 61 five-passenger coupe on up to $4,887 for the Fleetwood Seventy-Five Seven-Passenger Imperial Limousine. Typical optional equipment prices were $186.34 for the increasingly popular Hydra-Matic automatic transmission (92 per cent of Cadillac sales this year) and $73.65 for a Cadillac radio.

Cadillac delivered 61,926 cars during the 1947 model year—double 1946 sales and the division's best year since 1941. Cadillac's 1947 model year sales, in fact had reached more than 90 per cent of their record pre-war levels. The factory still had unfilled orders for more than 96,000 Cadillacs and was gearing up for the day when it could deliver 100,000 cars a year.

For the second time, Cadillac's retail new car deliveries had exceeded those of faltering Packard. Cadillac had quietly surged past its old rival and was about to assume permanent supremacy in the U.S. luxury car field. Packard was left with no "senior" cars and had to rely on a limited line of mid-range Clippers to take it through the immediate postwar years. Lincoln's V-12 Continental was not doing very well, either. With a strong, comprehensive product line and a booming new car market, Cadillac assumed its fine car market leadership virtually by default.

The Packard Motor Car Company built its 1-millionth car this year, and the U.S. automobile industry produced its 100-millionth. The newly-formed Tucker Corporation had created a sensation with its three-eyed Tucker Torpedo, which Preston Tucker promised would soon be in volume production.

Production of Cadillac's 1947 models ended in mid-January, 1948. A frenetic three-week new-model changeover began immediately to clear the way for Cadillac's first all-new postwar cars. Another triumph was at hand.

The 1947 Cadillacs showed little change from the 1946 models, which in turn were really 1942 carryovers. The lowest-priced 1947 Cadillacs were the Series 61 coupe and sedan which utilized a 126-inch wheelbase chassis. Both Series 61 models retained the sweeping fastback rooflines which had made them real style-setters before the war. All Cadillac prices were increased significantly this year. This is the 1947 Cadillac Series 61 Coupe for Five Passengers, Style 6107, which sold for $2,200. Weight was 4,145 pounds. A total of 3,395 Series 61 Coupes were sold this year in what was still a seller's market.

1947

Series 61 Cadillacs were easily recognizable by their forward-angled window pillars and long, sweeping fastback rooflines. Cadillac's lowest-priced series, the Sixty-One was available in only two body styles, a coupe and this four-door, five-passenger sedan. The 1947 Series 61 Sedan, Style 6109, showed little change from the 1942 model. There was still an enormous pent-up demand for new cars, however—especially Cadillacs—and 5,160 Series 61 Sedans were delivered in the 1947 model year. Price was $2,324 and the sedan weighed 4,225 pounds. New stainless steel stone shields replaced the rubber ones used previously.

The 1947 Cadillacs received minor exterior styling changes but were essentially the same cars that had been introduced in the fall of 1941 as 1942 models. After five years they were beginning to look a trifle dated, but a car-starved public snapped them up as fast as they came off the Clark Ave. assembly line. For 1947 there was a modified grille design with five heavier horizontal bars instead of the six used previously. A new "V" emblem with the Cadillac crest in a field of cloisonne graced the front of the hood, and a winged emblem appeared on the rear decklid of all but Sixty Special models (which retained the Fleetwood script). New Cadillac script nameplates replaced the small block letters formerly used on the front fenders and stainless steel stone shields replaced the black rubber ones used on 1946 models. But the most significant styling innovation this year was the bright chrome "sombrero" wheel covers available on all models. This is the 1947 Cadillac Series 62 Five-Passenger Club Coupe, or Sedanet.

The Series 62 Club Coupe, or Sedanet, was originally introduced as a 1942 model. Five years later it was still a most attractive car that was being widely imitated. Model 6207, the Series 62 Coupe for 1947 was priced at $2,446 or $197 higher than the 1946 version. In all, 7,245 of these sleek fastbacks were built during the 1947 model run, which began in early January of that year. Styling changes were minor—a modified grille design, new Cadillac script nameplates on the front fenders, stainless steel stone shields on the rear fenders and bright, chrome-plated wheel discs.

Once again, the practical Series 62 Four-Door Sedan was by far the most popular single body style in the entire 1947 Cadillac line. A total of 25,834 Series 62 Sedans, Model 6269, were sold in the booming 1947 model year. A five-passenger sedan, the Series 62 four-door had a factory price of $2,523 which was $199 higher than that of the comparable 1946 model. The revised 1947 Cadillac grille is shown to advantage here. It featured five horizontal bars instead of the six used previously. The header bar extended into the front fenders. This grille followed the classic Cadillac grille theme introduced for 1942. Parking or fog lamps were incorporated in a panel in the outer, lower edges of the bright eggcrate grille.

Viewed in profile, the 1947 Cadillac Series 62 Sedan was almost indistinguishable from the 1942 and 1946 models. But the big chrome wheel discs, Cadillac script nameplates on the front fenders and new stainless steel stone shields identify this as a 1947 model. The Model 6269 Series 62 Sedan was by far the most popular of all 11 Cadillac body styles offered this year. All three Series 62 Cadillacs—Coupe, Sedan and Convertible—were built on the same 129-inch wheelbase.

1947

The Cadillac Series 62 Convertible Coupe remained one of the most beautiful of all postwar U.S. cars. Little changed from 1942, it was still an extremely attractive car. For 1947, the Series 62 Convertible's appearance was enhanced by a revised grille design and bright chrome-plated stainless steel wheel covers. At $2,902 the sporty Convertible Coupe was the most expensive model in the Series 62 line. A total of 6,755 of these gorgeous convertibles were produced this year. They are still avidly sought by collectors today.

Cadillacs have always been synonymous with the rich and famous. Celebrities of all persuasions have flaunted their wealth and power by driving or riding in "The Standard of the World". By 1947, Cadillac had quietly overtaken Packard to assume permanent leadership in the luxury car field in America. Show business figures were especially eager to be photographed with America's four-wheeled symbol of success. Hence, cowboy movie star Roy Rogers mounted the hood of a 1947 Cadillac for a promotional wave of his good guy's white hat. Roy's V-8-engined steed is equipped with optional fog lamps, a $30.12 extra.

For 1947, Cadillac's distinctive Fleetwood Sixty Special was again available in just one model, a well-equipped and sumptuously-appointed four-door, five-passenger sedan, Style 6069. Except for minor exterior details, the 1947 Fleetwood Sixty Special looked almost the same as the 1942 model. Wheelbase continued at 133 inches, but the Sixty Special's price was hiked $141 to $3,195. Exterior appearance changes for 1947 included a revised grille design with five heavier horizontal bars, a new "V" hood emblem, bright metal stone shields on the rear fenders instead of the rubber ones used on 1946 models, and new Cadillac script nameplates on the front fenders which replaced the small block letters used previously. Another major styling change on most 1947 Cadillacs was bright stainless steel "sombrero" wheel covers. Despite its somewhat dated appearance, the 1947 Cadillac Sixty Special sold extremely well. A record 8,500 were delivered in the 1947 model year. Note the series of fine louvers on the upper rear roof quarters, a 1942 carryover.

Even with its top up and without whitewall tires, the 1947 Cadillac Series 62 Convertible Coupe was a sporty-looking car. Leather trim and Hydro-Lectric window lifts were standard. The five-passenger Series 62 Convertible, Model 6267, was still the only ragtop available in the Cadillac family. The bright, new "sombrero" wheel covers looked especially good on this car and were used on Cadillacs well into the 1950s. They got their nickname because of their resemblance to the familiar Mexican headgear. The chrome-plated, stainless steel wheel discs had a red Cadillac crest insert in their centers.

The prestigious Fleetwood Seventy-Five went into the 1947 model year with only minor changes. These included Cadillac's revised five-bar grille, new script nameplates on the lower edges of the front fenders above the stainless steel running board extensions, and a new winged crest on the rear decklid. Hydro-Lectric window lifts were now standard on Series Seventy-Five Cadillacs as well as on the Series 62 Convertible. Five Series Seventy-Five models were offered, all on the same 136-inch wheelbase. The most prolific of these was the Style 7533 Seven-Passenger Imperial Sedan, of which 1,005 were built. Next came the Style 7523 Seven-Passenger Sedan, with 890 delivered. Also sold in 1947 were 300 Style 7519 Five-Passenger Sedans, 135 Style 7523-L Nine-Passenger Business Sedans, and 80 style 7533-L Nine-Passenger Business Imperials. Prices ranged from $4,471 for the Series Seventy-Five Five-passenger Sedan on up to $4,887 for the Seven-Passenger Imperial—the most expensive car in the entire 1947 Cadillac stable. This is the Style 7523 Seven-Passenger Sedan.

1947

Ford's Lincoln may have been America's "Car of State", but it seems that there have always been at least a few Cadillacs parked in the White House garage. Here, United States President Harry S. Truman is about to enter a Cadillac Fleetwood Seventy-Five Limousine at 1600 Pennsylvania Ave. in Washington, D.C. The U.S. Secret Service has long relied on a fleet of specially-equipped Cadillac "security" cars to provide protection for the Chief Executive. President Truman's limo has no whitewalls but is equipped with fog lamps and the stylish new chrome wheel discs which, at $25.08 for a set of four, were a bona fide bargain for the American taxpayer in this instance.

Among the more interesting specials constructed on the 1947 Cadillac chassis was this custom six-door station wagon built by Maurice Schwartz, of Pasadena, Calif. It was commissioned by cowboy star Gene Autry, and was built on the 163-inch wheelbase Cadillac commercial chassis used for hearse and ambulance work. Note the standard small hubcaps and the roof luggage rack. Schwartz, who terminated his partnership with Bohman after the war, did a number of Cadillac customs for the elite of Hollywood's movie colony.

With the demise of the custom body business in the years preceding the Second World War, Cadillac relied on a few outside coachbuilders for its limited custom and semi-custom body requirements. Fleetwood had built its last custom bodies in 1942. The Derham Body Co. of Rosemont, Pa. continued to do custom body work for Cadillac's customers, primarily formal limousine conversions of standard Fleetwood Seventy-Five sedans and limousines. This is a good example of the kind of work Derham was doing after the war. This is an elegant Fleetwood Seventy-Five Formal Sedan with closed rear quarters, ornamental landau bows and a striking white fabric-covered roof. The author photographed this 1947 Cadillac Style 7533 Limousine at a Classic Car Club of America Midwest Grand Classic at Greenfield Village in Dearborn, Mich. about 1963.

The Hess & Eisenhardt Co., builders of Sayers & Scovill funeral cars and ambulances, continued to dabble in the commercial limousine business in the years immediately following the Second World War. This is the Hess & Eisenhardt Style 47500 Airline Coach built on the 1947 Cadillac Series 75 commercial chassis, the same long-wheelbase chassis the company used for its hearse and ambulance bodies. This nine-passenger car had an oversized trunk to carry the luggage of eight passengers. It was photographed in a cemetery, which is not surprising considering the Cincinnati manufacturer's long ancestry in the funeral car field.

This is one of the more unique vehicles to come out of the Hess and Eisenhardt Co. plant in Cincinnati, a firm long known for its unusual special-order automotive creations. It is a Sayers & Scovill Police Patrol Wagon built on the 1947 Cadillac Series 75 commercial chassis. The body shell is the same one H & E used for its S & S hearses and ambulances. Wheelbase was 163 inches. Of interest here are the screened rear windows, rear handrails and tunnel roof lights with red flashers. The extra chrome trim on the nose of the hood was an S & S trademark for many years. Where this one-off Caddy paddy wagon was delivered is not known.

1947

Builders of the well-known Sayers & Scovill funeral cars and ambulances, the Hess & Eisenhardt Co. of Cincinnati produced small numbers of specialized limousines and passenger coaches between 1946 and 1948. A few of these were special sightseeing coaches built for large resort hotels. These special sightseeing buses had canvas roll-back roofs so their passengers could enjoy the view of the mountains and sky. The coach could also be used as ground transportation between the hotel, airport or railroad depot between sightseeing runs, hence the roof luggage rack. This is an eight-door sightseeing and ground transportation coach built on a stretched Cadillac Series 75 commercial chassis. Note the Fleetwood Sixty Special louvers on the rear roof panel, the clearance lamps above the windshield and the canvas sun roof. Hess and Eisenhardt produced small batches of similar limousines during the 1950s.

With the return to peacetime production, several American funeral car builders were again producing coupe-style flower cars. Hess and Eisenhardt, builders of S & S hearses and ambulances, and The Eureka Co. were the first funeral car builders to offer postwar flower cars. This is a 1947 Sayers and Scovill "Florentine" deck-type flower car built on the 1947 Cadillac Series 75 commercial chassis. Wheelbase was 163 inches. The huge chrome script nameplate on the front door identifies it as one built for the Washington, D.C. funeral firm of W. W. Chambers.

There was still a huge backlog of orders for new funeral cars and ambulances to replace the well-worn units which had been carefully maintained and nursed through the Second World War years. New ambulances were especially hard to come by, and more than a few older-model hearses were converted to do ambulance work until replacements could be secured. This is a 1947 Meteor-Cadillac Limousine Ambulance which saw service with the Asbury Park, N.J. Fire Department. Note the tunnel roof lights and the fire extinguisher carried in a well in the right front fender. The Meteor Motor Car Co. was located in Piqua, Ohio.

Cadillac supplied 2,423 of its 163-inch wheelbase commercial chassis to U.S. funeral car and ambulance manufacturers during the 1947 model run. Principal commercial chassis customers were the Meteor Motor Car Co. of Piqua, Ohio; A. J. Miller Co. of Bellefontaine, Ohio; Superior Coach Corp. of Lima, Ohio; Hess & Eisenhardt Co. of Cincinnati, Ohio, and The Eureka Co. of Rock Falls, Ill. Closely tied to Packard, industry giant Henney was unable to get back into postwar production until the 1948 model year. This is the 1947 Meteor-Cadillac Combination Funeral Car and Ambulance. Note the red lenses in the fog lamp openings and the removable ambulance grille in the rear window. This combination coach is obviously more hearse than ambulance.

1948

Cadillac's first all-new postwar cars ushered in one of the most successful styling innovations in the history of the American automobile—tail fins. Controversial from the beginning, Cadillac's towering tailfins focused attention on the rear of the car as never before and touched off an industry styling craze that reached ludicrous heights a decade later before receding into the fenders by 1965. But more than anything else, the tail fin solidified Cadillac's henceforth-unmistakable product identity.

The origins of these curious appendages (Cadillac taillights began to sprout blades, or fins, in the late 1930s) can be traced back to the day before the Second World War when General Motors' styling chief Harley Earl sent a group of his stylists out to Selfridge Field, near Detroit, to study the then top-secret Lockheed P-38 twin-engined fighter plane. The Lightning's vertical stabilizers grew out of the ends of two long, slender tail booms with side-mounted airscoops. Intrigued by the P-38's sleek, aerodynamic design, the stylists went back to their drawing boards and translated some of the Lightning's radical features into styling concepts for possible cars of the future. Next came a series of three-eighths scale models known as the "Interceptor" series. But the war intervened and development work along these lines was not resumed until after the war, when William J. Mitchell returned to head up the Cadillac styling studio. Some of the daring aerodynamic concepts of the pre-war Interceptor project were incorporated into the division's first all-new postwar cars, scheduled for the 1948 model year.

In the beginning, not even Cadillac called them fins. Cadillac's 1948 advertising simply referred to... "rudder-type styling". The division's traditionally conservative dealers, however, were aghast when they were first exposed to them, but heaved sighs of relief when the unique tail fins were accorded overwhelming approval by a style-conscious public. Recalling Cadillac's 1915 "Penalty of Leadership" advertising, the fins were also the butt of more than a few jokes.

The totally-redesigned 1948 Cadillacs went into production in February, 1948 but did not go on sale in dealer showrooms until the following month. Actually only six of Cadillac's eleven 1948 models got the fresh new styling. Series 61, 62 and Fleetwood Sixty Special models reflected the handsome, new look but the staid Series Seventy-Fives retained their now-archaic styling which dated back to 1941. Another two years would pass before the Seventy-Five was redesigned in keeping with the rest of the line. Cadillac's 1948 product line included four series, the same as in 1947 but on only three wheelbases. And while the restyled cars boasted the freshest styling in the industry, all 1948 Cadillacs continued to use the same old 346 cubic inch V-8 engine that had been around since 1936.

The 1948 Cadillacs used only five basic body shells, half as many as in 1941. Series 61 and 62 coupes and sedans shared the same Fisher "C" bodies. Series 62 models were three inches shorter than the previous year and utilized the Series 61's 126-inch wheelbase. There were only two Series 61 models, a fastback coupe

The most startling styling innovation on the all-new 1948 Cadillac was the one which established Cadillac's product identity for years to come—tail fins! The totally-new 1948 Cadillac's rear fenders terminated in saucy, streamlined sheet metal humps which housed the taillights. Cadillac's epic styling gimmick brought new identity to the rear of the car and made the vehicle look lower. Tail fins became Cadillac's predominant styling theme for the next 14 years, reaching ludicrous heights a decade later, then gradually receding into flat rear fender lines. The 1948 Cadillac's "fishtail" precipitated numerous jokes and a styling revolution: Eventually, every car built in the U.S. sprouted tail fins, to the point where they became automotive styling cliches of dubious taste. The left taillight popped up to reveal the fuel filler opening, as many an astonished service station attendant soon discovered.

Production of Cadillac's first all-new postwar cars began in the Clark Avenue plant in Detroit in February, 1948. They went on sale the following month. Because of the extensive new-model changeover and delayed new-model launch, the 1948 model run lasted only nine months. But Cadillac managed to sell 52,706 of its 1948 models. Actually, only six of the division's eleven 1948 models got fresh, new styling. All five Fleetwood Seventy-Five models were carried over unchanged for another two years. In this in-plant photo, a Series 61 Coupe comes off the assembly line, followed by a Fleetwood Seventy-Five sedan.

1948

and a notchback sedan, the latter the first since 1941. The three Series 62 body styles included the fastback Coupe, or Sedanet; an extremely popular Four-Door Sedan and a sporty Convertible Coupe. The redesigned Fleetwood Sixty Special retained its own, exclusive 133-inch wheelbase and distinctive styling and was again available in just one four-door sedan body style. Rounding out the line were five carryover Fleetwood Seventy-Fives, all on the same 136-inch wheelbase with five, seven and nine-passenger seating arrangements in the same high, slab-sided sedan body style. The Seventy-Fives were virtually unchanged from 1947.

Cadillac's attractive new styling featured strong horizontal body lines which extended the front fenderline along the full length of the car, and terminated in those fishtail fins which sprouted from the tops of the rear fenders and housed vertical beacon-type taillights. The perky tail fins made the car look lower from any angle, even when viewed from the front.

The restyled Cadillacs also had greatly-increased glass area with thin pillars, curved two-piece windshields and curved backlights. The new grille was lower, wider and simpler in appearance with three heavy horizontal bars and seven vertical ones which successfully perpetuated the bright eggcrate theme introduced in 1942. Some detail touches were distinctly familiar, however, including Cadillac script nameplates on the front fenders, "V" emblems on the hood nose and rear deck and large, chrome "sombrero" wheel discs. The hood was lower and flatter and the rear decklid was counterbalanced. Most models rolled on larger low-pressure 8:20 x 15-inches tires. Bumpers were more massive and curved and continued to be protected by two bullet-tipped vertical guards.

The new Fleetwood Sixty Special was now nearly as long as the Seventy-Five and was given its own distinctive styling touches. These included a simulated vertical airscoop on the leading edge of the rear fenders just like on the tail booms of the P-38); a wide bright sill molding, fine chrome moldings around the side window openings, and a series of chrome louvers on the rear roof panel just behind the rear doors. Series 62 models had stone shields behind the front wheel openings and bright rocker sill moldings: the lower-priced Series 61 did not. Series 62 models had three horizontal chrome trim pieces below each taillight. Parking/turn signal/fog lamps on all of the restyled cars were built into the front fenders below the headlights. The windsplit first used on 1941 models now swept all the way back from the headlight rim to the front door.

The new 1948 Cadillacs also had a redesigned instrument panel that was used for just one year. All driving instruments and gauges were housed in a large rainbow-shaped pod that straddled the steering column directly in front of the driver. The new dash was curved, contributing to increased interior room. Some models—the Series 62 convertible and Fleetwood Seventy-Five sedans and limousines—offered Hydra-Lectric power windows and seats (standard in Sixty Special). A record 97 per cent of Cadillac's customers ordered the Hydra-Matic automatic transmission option this year.

Because of the late new-model launch, Cadillac's 1948 model run lasted less than nine months. But the 1948 models were enthusiastically received, and a total of 52,706 were delivered in the abbreviated model year.

In mid-1947 the Packard Motor Car Co. had introduced its first all-new postwar cars. The long-awaited Twenty-Second Series Packards sold very well, with 98,897 delivered by the end of the long 1948 model run. Ford dropped its slow-selling Lincoln Continental at the end of the 1948 model year, and after an unprecedented 13 years, Cadillac's venerable 346 cubic-inch flathead V-8 came to the end of the line. Cadillac's new high-compression overhead-valve engine was ready.

Cadillac's first all-new postwar cars were introduced to an enthusiastic public in March, 1948. The completely restyled 1948 Series 61, 62 and Fleetwood Sixty Special featured a lower, wider grille, controversial "fishtail" fins and much-improved all-around visibility with curved windshields and backlight glass. The two Series 61 models remained the lowest-priced cars in the 1948 line, even though all Cadillac prices went up substantially again this year. Both Series 61 body styles used the same 126-inch wheelbase chassis employed the previous year. This is the 1948 Cadillac Series 61 Two-Door Coupe, Style 6107, which except for exterior and interior trim was identical to the Series 62 Coupe. The Series 61 Coupe weighed 4,070 pounds and had a factory price of $2,728. A total of 3,521 of these fastback coupes were built this year.

Cadillac's Series 61 and 62 models for 1948 shared the same basic body shells but featured different interior and exterior trim. The easiest way to tell them apart is by their front fenders. The lowline Series 61 had no stone shields behind the front wheel opening and no chrome rocker sill molding. Series 62 models also had higher-level interior trim. This is the 1948 Cadillac Series 61 Four-Door Sedan, Style 6169, of which 5,081 were delivered in the shortened 1948 model year. Factory price was $2,833. The new 1948 Cadillacs were marketed as six-passenger cars. Earlier models were designed to carry only five passengers. Substantially-increased interior room made this important specifications change possible. For the first time since 1941, the Sixty-One Sedan was once again a notchback.

1948

Although it was a completely restyled car, the 1948 Cadillac Series 62 Coupe retained the classic aerodynamic fastback profile of the 1942 Sedanet. Cadillac's bold, new 1948 styling included long, flowing front fenderlines which faded into pleasingly-contoured rear fenders topped by those controversial "fishtail" fins. The Series 62 Coupe shared its body shell with the lower-priced Series 61 Coupe but had a higher level of interior and exterior trim. Of the three Series 62 body styles offered for 1948, the six-passenger Coupe was the least popular. A total of 4,764 Series 62 Sedanets were sold in the abbreviated 1948 model run. Price was increased to $2,912 and weight was 4,125 pounds. Perhaps because of its P-38 inspiration, GM seemed to favor aviation backgrounds when photographing the 1948 Cads.

The 1948 Series 62 Cadillac was three inches shorter and two inches narrower than the comparable 1947 model but boasted more interior room. Consequently, Series 61 and 62 models were listed as six-passenger cars, instead of five-seaters as previously. This is a three-quarter rear view of the 1948 Series 62 Coupe or Sedanet, taken at Hershey, Pa., in 1972. Note the three horizontal chrome trim pieces below each taillight used only on the 62 series, the "notched" side window shape and retention of the two vertical dividers in the curved rear backlight. The front fender windsplit now extended from the side of the headlight almost to the front door.

The practical Series 62 Four-Door Sedan was again by far the most popular of all body styles offered by Cadillac in the 1948 model year. Of the 52,706 Cadillacs produced during the nine-month model run, 23,997 were Series 62 Sedans. Style 6269, the six passenger Series 62 Sedan, sold for $2,996. Weight was 4,180 pounds. This view shows the 1948 Cadillac's handsome new grille design which continued the bright eggcrate grille theme first introduced for 1942. Note the parking/turn signal lamps built into the leading edges of the front fenders and the massive new bumpers with bullet-tipped vertical guards.

The totally-restyled 1948 Cadillac Fleetwood Sixty Special retained its individuality, and even managed to keep its exclusive 133-inch wheelbase. The new Sixty Special featured aerodynamically-inspired simulated air-scoop vertical side moldings and a wide, bright rocker sill molding. Also retained was the thin chrome edging around the side window openings. Style 6069-X, the Fleetwood Sixty Special had Hydro-Lectric power windows and a power-operated front seat as standard equipment. Weight was 4,370 pounds. The Sixty Special was marketed as a five-passenger car. Factory delivered price was hiked to $3,820. Barely visible are the series of fine louvers on the upper rear roof, a Sixty Special styling touch first seen in 1942.

The Series 62 Convertible Coupe remained the glamour queen of the 1948 Cadillac line. Sales of the sporty soft top in fact, exceeded those of the fastback Series 62 Sedanet. A total of 5,450 Series 62 Convertibles were delivered during the nine-month 1948 model run, compared with 4,764 Sedanets. At $3,442 the six-passenger Convertible Coupe was the most expensive body style in the series. The convertible was listed as Style 6267-X, the "X" suffix indicating that this model was equipped with standard Hydro-Lectric power windows and seat. In stock form the Series 62 Convertible weighed 4,450 pounds.

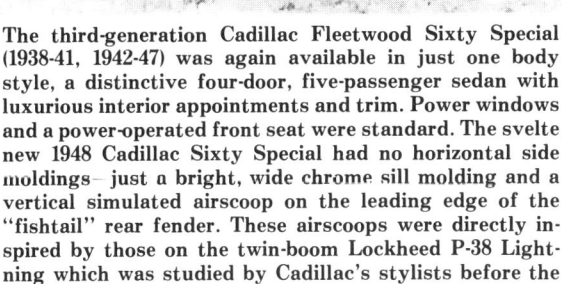

1948

The third-generation Cadillac Fleetwood Sixty Special (1938-41, 1942-47) was again available in just one body style, a distinctive four-door, five-passenger sedan with luxurious interior appointments and trim. Power windows and a power-operated front seat were standard. The svelte new 1948 Cadillac Sixty Special had no horizontal side moldings—just a bright, wide chrome sill molding and a vertical simulated airscoop on the leading edge of the "fishtail" rear fender. These airscoops were directly inspired by those on the twin-boom Lockheed P-38 Lightning which was studied by Cadillac's stylists before the war.

The most expensive model in Cadillac's 1948 line was the stately Fleetwood Seventy-Five Seven-Passenger Imperial Sedan, Style 7533-X, of which 382 were built. Factory price was $5,199. The "X" suffix indicated that this model was equipped with standard power windows, including the adjustable division glass in the partition between the front and rear compartments. Front compartments of these impressive limousines were upholstered in leather. The two 1948 Fleetwood Seventy-Five Business Sedans included Style 7523-L, a nine-passenger Business Sedan, and the Style 7533-L Nine-Passenger Business Imperial, which had a manually-operated division glass. The "L" stood for "livery", as these models were intended for funeral or commercial limousine livery service. Ninety 1948 Business Sedans were sold at $4,679, along with 64 Imperial Business Sedans which sold for $4,868.

Some of the most flamboyant styling ever seen on a Cadillac chassis came, not from an American coach-builder, but a French one. Jacques Saoutchik, of Paris, was noted for his truly exotic custom body designs. Saoutchik employed sweeping, curved body lines highlighted by broad chrome accents. The stunning drophead coupe shown here features finely-detailed canework on its sides and great expanses of chrome on the leading edges of both front and rear fenders. A wide one-piece curved windshield, two-tone paint and distinctive bright grille design contribute to the overwhelming appearance of this Saoutchik creation, which was built on a 1948 Cadillac chassis. The stock curved Cadillac instrument panel was one of the few original components left in the car. Only five stripped Cadillac chassis left the Detroit plant this year, including one Series 61, two Sixty-Twos and two Seventy-Fives.

Not all of the 1948 Cadillacs got the division's bold, new styling. All five Fleetwood Seventy-Five models carried over their severe 1947 styling, which dated back to 1941. The most prestigious (and expensive) of all of Cadillac's cars, in fact, retained this boxy, slab-sided styling for another two years. But Cadillac's affluent customers didn't mind. The impressive Seventy-Five exuded prestige and boasted an incredibly roomy and comfortable interior. It was available with seating for five, seven or nine passengers. This is the Style 7519-X Fleetwood Seventy-Five Sedan for Five Passengers, 225 of which were sold. Price was $4,779. Also delivered were 499 Seven-Passenger Sedans, Style 7523-X, at $4,999. Rounding out the line were an Imperial Limousine and two big Business Sedans.

The Hess & Eisenhardt Co., in the Cincinnati suburb of Blue Ash, O., continued to build a line of airline, hotel ground transportation and sightseeing coaches on the Cadillac commercial chassis. For 1948, this special hearse and ambulance chassis was redesignated the Series 76 and carried over Cadillac's 1947 styling. This is a six-door airline coach on the standard 163-inch wheelbase commercial chassis. An even longer eight-door "stretch" was also available. Note the Sixty Special louvers behind the rear door.

1948

Hess & Eisenhardt's eight-door airline ground transportation coach was built on a stretched Cadillac Series 76 commercial chassis. The 1948 Cadillac commercial chassis used carryover 1947 front-end sheet metal and rear fenders. Funeral car and ambulance manufacturers did not get Cadillac's all-new styling until the following year. Hess & Eisenhardt's airline limousines featured smooth, clean styling with up-to-date fastback rooflines. The company built these special-purpose coaches from 1946 through 1948.

By the time peacetime funeral car production resumed in 1946, the ornate carved-panel hearse had been replaced by the formally-styled Landau. This is the 1948 Meteor-Cadillac Landau Funeral Coach, Model 480-L, which had a factory driveaway price of $5,589 in standard rear-loading form. The side-servicing version, Model 480-LH with manually-operated casket table was priced at $6,516. The most expensive Meteor-Cadillac Landau, the 480-LHE, had an electrically-operated table and cost $7,057 at the Piqua, O. plant. Note the unique front door window shape incorporating a built-in coach lamp and the long, slender ornamental landau bow on the blind quarter.

All U.S. funeral car manufacturers continued to build long-wheelbase service cars. The service car was the workhorse of the funeral home motor fleet, doing all the heavy hauling and saving wear and tear on the funeral director's front-line hearse. This is the 1948 Meteor-Cadillac Service Coach, Model 482, on the Cadillac Series 76 commercial chassis. Factory driveaway price was $5,174—lowest in the Meteor Motor Car Co.'s 1948 product line. Note the large chrome-plated wreath ornament and coach lamps.

Coupe-type flower cars were still very much in demand by American funeral directors. This is a 1948 Eureka-Cadillac Flower Car built on the 163-inch wheelbase Series 76 commercial chassis. The example shown has an open rear well instead of an adjustable deck. The 1948 Cadillac commercial chassis carried over 1947 styling. The front fenders were squared off just ahead of the front door. This well-equipped flower car sports the chrome "sombrero" wheel covers and white sidewall tires.

Eureka also continued to offer its Cadillac ambulances in both landau version and the limousine style shown here. The limousine version cost $5,863, with the pod lights and the combination warning lights and siren on the roof being optional at extra cost. Other extras on this model are the small cast AMBULANCE sign at the base of the windshield, and the red lenses over the Cadillac parking lights. Blackwall tires were standard fare.

1949

Nineteen Forty-Nine was an extremely eventful year for the Cadillac Motor Car Division of General Motors. Sales set a new all-time record. Cadillac built its one-millionth car. A beautiful new body style, the pillarless hardtop coupe, was added to the line late in the 1949 model year, and Cadillac set a new standard for the industry with its first all-new engine in 12 years.

Cadillac's new overhead-valve, short-stroke, high-compression V-8 engine was an engineering triumph that ranked with its first V-type engine of 1915 and the epic 16 and 12-cylinder Cadillac engines of the 1930s. It was a brilliantly designed, extremely efficient passenger car powerplant that set the pattern for the entire industry. The 331 cubic inch engine delivered smooth, quiet high performance with fuel economy that was outstanding for a luxury car.

Development work on this milestone engine had begun in the late 1930s. The three men most directly responsible for its design were Harry F. Barr, the staff engineer in charge of Cadillac's engine design at the time, who went on to become Engineering Vice-President of the entire Corporation; Edward N. Cole, Cadillac's Chief Engineer who eventually became President of General Motors, and Cadillac Motor Division General Manager John F. "Jack" Gordon who also later ascended to the GM presidency.

Among the new engine's most significant innovations were its unique "slipper" pistons, credited to piston specialist Byron Ellis. The lower sides of the pistons were cut away so the piston nested between the crankshaft counterweights at the bottom of the stroke. This permitted the use of shorter connecting rods and a significant reduction in engine size and weight. This "compact" V-8 weighed 188 pounds less than the old 345 cubic inch flathead engine it replaced and was four inches shorter and nearly four inches lower. Yet it put out 160 horsepower, 10 more than the old engine, despite its smaller displacement. The new 331 engine also boasted a five-main-bearing crankshaft. The OHV engine had a 3-3/16-inch bore and 3-5/8-inch stroke resulting in an "oversquare" configuration.

The new engine had a 7:5-1 compression ratio but was designed to operate on much-higher future compression ratios as the use of new high-octane gasolines increased. Compression was eventually boosted to 10:5-1 by 1959. The 1949 engine proved to be a remarkably adaptable design. It continued in production with various modifications and improvements through the 1967 model year, by which time its block had been bored out to 429 cubic inches. Cadillac's sister division, Oldsmobile, came out with a short-stroke, high-compression 303 CID "Rocket" V-8 at about the same time. The U.S. auto industry eventually standardized on this type of engine, which dominated the scene into the early 1980s.

The other major product news at Cadillac this year was the introduction in July of the "pillarless" two-door hardtop body style. Cadillac's first hardtop (there had actually been one available in 1928) was called the Coupe de Ville. A Series 62 model, the stunning new Coupe de Ville combined the sporty appearance of a convertible with the practical advantages of a closed car. The idea is said to have originated with the wife of a GM executive who liked the open feel of a convertible but hated getting her hair mussed up. With the versatile hardtop she got the best of both worlds. Side and quarter windows retracted into the body giving the hardtop a spacious, open look. But GM's new hardtop body style was not exclusive to Cadillac: Buick introduced a Roadmaster Riviera Hardtop Coupe and Oldsmobile a Holiday Hardtop Coupe toward the end of the 1949 model year.

Cadillac's extremely successful 1948 styling was only slightly changed for 1949. The 1949 grille was wider and lower and consisted of two massive horizontal chrome bars and five short vertical ones. The center grille bar and parking light bezels extended around the corners of the front fenders. Cadillac's controversial "fishtail" fins had been well-received by an increasingly style-conscious public. In a major mid-year running change, a new, squared rear decklid like the one used on the Sixty Special was made standard on Series 61 and 62 models to increase trunk capacity.

Cadillac followed its styling triumph of 1948 with an equally impressive engineering breakthrough for 1949—an all-new overhead valve V-8 engine which set a new standard for the U.S. automobile industry. Cadillac's lowest-priced offerings for 1949 were again two Series 61 models, a Coupe and Sedan on a 126-inch wheelbase. This is a catalog illustration of the 1949 Cadillac Series 61 Coupe for Six Passengers, Style 6107, which except for exterior trim was identical in appearance to the more expensive Series 62 Coupe, or Sedanet. Price was $2,814 and the fastback Coupe weighed 4,070 pounds. Production totaled 6,409.

Of the 22,148 Series 61 Cadillacs built in the booming 1949 model year, 15,738 were Style 6169 Four-Door Sedans. The price-leader Series 61 Cadillacs did not have the front fender stone shields or bright rocker sill moldings found on the more expensive Series 62 models. This Sixty-One is equipped with still hard-to-get optional white sidewall tires. The six-passenger Series 61 Sedan had a factory price of $2,919 and weighed 4,145 pounds. The principal styling change this year was a wider, lower and more massive grille. Note the bright "sombrero" wheel covers.

1949

The 1949 Cadillac hood extended farther forward and lost its 1948 "lip" around the grille. Cadillac's breakthrough 1948-49 styling was used for only two years, but there was a different instrument panel each year. The "rainbow" cluster of the 1948 model was replaced with a more conventional hooded dash in the 1949 Cadillacs. By the end of the 1949 model year, 98 per cent of Cadillac's customers were opting for the Hydra-Matic automatic transmission.

Except for the late addition of the new Coupe de Ville, Cadillac's 1949 model lineup was unchanged from the previous year. There were four series on three wheelbases. Series 61 and 62 models shared the same 126-inch wheelbase. The lowest-priced Series 61 included only a coupe and convertible: Series 62 offered a coupe (Sedanet), sedan, convertible and hardtop. Again, there was only one Fleetwood Sixty Special, a premium four-door sedan on its own exclusive 133-inch wheelbase. The Fleetwood Seventy-Five carried over its now-archaic 1948 styling which actually dated back to 1941! The big Seventy-Five sedan and limousine had utilized the same stern automotive architecture clear through the 1940s and were not restyled until the 1950 model year. But the 1949 models did get Cadillac's new OHV engine. As in 1948, there were five models with seating for five, seven or nine passengers. Two models had a standard glass division.

The Cadillac commercial chassis was also redesigned this year and its exterior styling was now compatible with the rest of the division's cars, except for the carry-over Seventy-Five. Wheelbase of the special hearse and ambulance chassis, now designated Series 86, remained at 163 inches.

On November 25, 1949 Cadillac produced its 1,000,000th car. The milestone car, and the last Cadillac built during the 1940s, was a Coupe de Ville hardtop. It had taken Cadillac 47 years to turn out one million cars. It would take only nine years to build the second million.

In January, 1949 General Motors staged a lavish "Transportation Unlimited" show in the Grand Ballroom of New York's ritzy Waldorf-Astoria Hotel. This show, which was repeated in Boston, set the stage for GM's later Motoramas. The GM-only "Transportation Unlimited" show was a throwback to the glittering Automobile Salons and New York Auto Shows of the 1920s and 1930s. In its seven-car display in the Waldorf, Cadillac exhibited four special show cars created especially for this event. Three were modified Fleetwood Sixty Specials. One was the first true Coupe de Ville hardtop, a pillarless two-door on a 133-inch wheelbase Sixty Special chassis. The fourth car was a Series 62 convertible with a "westernized" interior. These were the first in a long succession of special Cadillac dream cars created to awe auto show crowds and gauge reaction to experimental features and styling ideas.

Sales of 92,554 cars in the booming 1949 model year set a new all-time record for Cadillac, toppling the previous record which had stood since 1941. "Motor Trend" magazine voted the 1949 Cadillac its first "Car of the Year." The U.S. automobile industry racked up sales of more than 6.2 million units, its highest in two decades. Introduction of the industry's 1950 models, including another completely-restyled line of Cadillacs was moved up to the late fall of 1949.

Cadillac continued to favor airport and aviation settings when photographing its postwar products, particularly the sleek Series 62 Coupe, or Sedanet. Here, two serious businessmen admire a Series 62 Coupe parked on the tarmac next to their DC-3. Surprisingly, the fastback Coupe was the least popular of the three standard Series 62 body styles (the new Coupe de Ville Hardtop was not introduced until late in the 1949 model year). Sales totaled 7,515 compared with deliveries of 8,000 Series 62 Convertible Coupes and a whopping 37,617 Four-Door Sedans. Style 6207, the Series 62 Coupe weighed 4,125 pounds and was priced at $2,992—$73 higher than the comparable 1948 model.

The overwhelming choice of Cadillac buyers in the record-setting 1949 model year was the handsome Series 62 Touring Sedan for Six Passengers. Of the 55,643 Series 62 Cadillacs delivered in the 1949 model year, 37,617 were four-door sedans like this one. Factory price was $3,076, or $80 higher than the 1948 model. Weight was 4,180 pounds. This car was photographed during a shooting session outside the Clark Avenue plant in Detroit. The sheets of white paper are carefully placed to throw light on the chrome. Cadillac's advertising agency then magically transplanted the car into a more idyllic setting for the 1949 sales catalog.

1949

A spectacular new body style joined the Cadillac model lineup late in the 1949 model year. It was a "pillarless" two-door hardtop coupe which coined a now-legendary Cadillac model name—Coupe de Ville. Only 2,150 Coupe de Villes were built this year, making it the rarest and most desirable of all 1949 Cadillacs. Style 6237-X, the stunning new Coupe de Ville carried a factory price of $3,497—$26 less than the 1949 Series 62 Convertible. At 3,857 pounds, the new hardtop was also the lightest car in the four-car series. The "X" indicated that the Coupe de Ville was equipped with standard power seat and windows. Cadillac enthusiasts consider the 1949 Coupe de Ville one of the most beautiful production Cadillacs ever built.

Cadillac's Series 62 model lineup for 1949 initially included three body styles, all on the same 126-inch wheelbase. A two-door hardtop coupe was added late in the model year. This is the sporty Series 62 Convertible Coupe for Six Passengers, Style 6267-X, which reached new heights of popularity this year, outselling even the aerodynamically-styled Series 62 Sedanet. The "X" suffix indicated that this model is equipped with standard power windows and front seat. Price was $3,523 making it the most expensive model in the series. Even the new Coupe de Ville Hardtop sold for less. An even 8,000 Series 62 Convertibles were built in the 1949 model run.

The Series 62 Convertible Coupe was an especially attractive car, even with its top up. Hydraulically-operated windows, including rear quarter windows, and a power seat were standard in this model, the most expensive in the Series 62 line. This car is equipped with the standard small hubcaps and optional wheel trim rings ($9.50 for a set of five), rather than the more popular chrome "sombrero" wheel discs. The three horizontal chrome trim pieces used below the taillights on the 1948 Series 62 models were deleted in 1949. The bright stone shields behind the front wheel openings and chrome rocker sill moldings set Series 62 Cadillacs apart from the less expensive Series 61 models.

Model year 1949 was a milestone one for Cadillac. The division set a new all-time sales record, a beautiful new hardtop body style was added to the line, and the company produced its 1,000,000th car. The 1,000,000th Cadillac built since the first single-cylinder Model "A" spluttered to life in 1902 was driven off the Clark Avenue assembly line in Detroit with appropriate ceremony on November 25, 1949. The millionth Cadillac was also the last one built in the epochal decade of the 1940s. Cadillac Motor Car Division General Manager John F. Gordon, left, and General Sales Manager Donald E. Ahrens pose with the milestone car, a 1949 Series 62 Coupe de Ville, in the executive garage. The third man is unidentified. It took Cadillac 47 years to produce its first million cars, and less than nine years to build its second million!

Cadillac's most distinctive sedan was still the elegant Fleetwood Sixty Special, which went into the 1949 model year with an all-new overhead valve V-8 engine and modest styling changes. Series 61, 62 and Sixty Special models got a wider, lower grille with two massive horizontal chrome bars. Parking lamp bezels now wrapped around the fenders to the front wheel openings. The impressive Fleetwood Sixty Special retained its exclusive 133-inch wheelbase. Only one body style, a formal four-door sedan, Style 6069-X, was offered. The 1949 model was advertised as a six-passenger car for the first time. Price was $3,859—$39 higher than in 1948. Sales nearly doubled to 11,399 this year. In addition, one full-custom Coupe de Ville was built on a Series Sixty Special chassis this year. It was a special show car designated Style 6037-X.

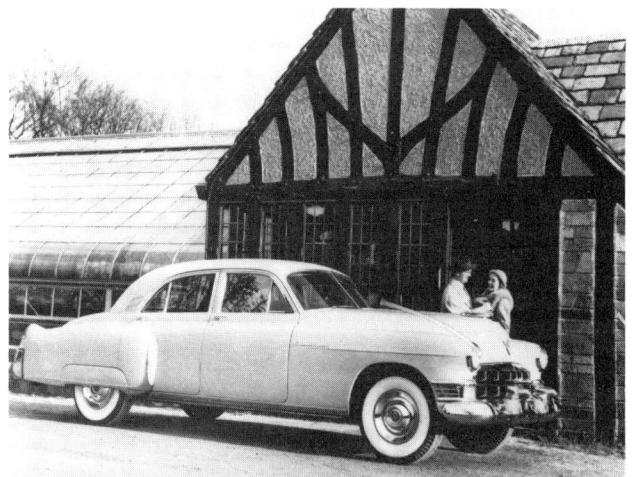

1949

The prestigious Fleetwood Seventy-Five sedans and limousines carried over their now-antiquated styling for yet another year. This would be the last. The imposing Fleetwood Seventy-Fives had changed little through the entire decade of the 1940s. The basic body shell dated clear back to 1941. But for 1949, the Seventy-Fives did get Cadillac's all-new OHV engine. The Series Seventy-Five model lineup included three big four-door sedans: Style 7519-X, Five-Passenger Sedan ($4,865) of which 220 were built; Style 7523-X, Seven-Passenger Sedan ($5,011) of which 595 were produced, and the Style 7523-L Nine-Passenger Business Sedan ("L" for livery), only 35 of which were built. Price of the latter was $4,691. The five-bar 1948 grille was carried over without change.

The staid architecture of Cadillac's most exclusive—and expensive—cars changed little through the entire decade of the 1940s. For 1949, the imposing Fleetwood Seventy-Five sedans and limousines carried over their towering, slab-sided looks for one more, final, year. Yet their faithful, well-heeled customers still loved them. Two formal limousines were included in the five-model Series Seventy-Five line for 1949. They included the Style 7533-X Seven-Passenger Imperial ($5,211) the most popular model in the series with 595 delivered, and the Style 7533-L Nine-Passenger Imperial Business Sedan ($4,890), a mere 25 of which were sold for funeral and livery service. This commercial limo is equipped with the small standard hubcaps and black sidewall tires.

Although the Fleetwood Seventy-Five sedan and limousine were not restyled until the 1950 model year, there once was a program to introduce a redesigned Seventy-Five for 1949. This program was dropped, however, and the 1948 model was carried over for one last year without styling changes. This GM Styling sketch of a proposed 1949 Fleetwood Seventy-Five sedan/limo turned up in the Clark Ave. archives. At least one full-custom Seventy-Five was built during the 1949 model year and incorporated some of the basic elements of this design. Of interest are the stone shields on the lower edges of the front fenders, which are similar to the running board extensions used on 1946-49 Seventy-Fives. The roofline is of the notchback sedan style. The rear side window contour did not use the graceful "Florentine curve" seen on many later Cadillac closed cars, including Seventy-Fives built from 1950 through the 1958 model year.

Only one of these custom-built Fleetwood Seventy-Five long-wheelbase sedans was produced in 1949, possibly as a styling proposal for a redesigned Seventy-Five that never made it into production. The 1948 model was carried over for one last year and the big Seventy-Five sedan and limousine were completely redesigned for 1950. This car is more or less 1949 stock forward of the front doors, but it has a number of Fleetwood Sixty Special styling touches including the wide chrome sill molding and vertical "air scoop" rear fender trim. The roof was padded and covered with a leatherlike material with two full-length seams. This one-off special was built for a Mr. Henry J. Taylor. Little else is known about this unique Cadillac.

Here is a three-quarter rear view of the one-off 1949 Cadillac Fleetwood Seventy-Five built by General Motors Styling and Fleetwood for a "Mr. Henry J. Taylor". At a glance it looks a little like a 1950 model, but the trim used makes it look more like a stretched Sixty Special rather than a Seventy-Five. Note the padded, leather-covered roof and the small "Frenched" rear window. Of special styling significance is the eloquent "Florentine curve" which sweeps around the outer edge of the rear quarter window and blends into the chrome belt molding. Note also the small, round backup lights and the separate antenna for the rear compartment radio. There were no Fleetwood nameplates on the exterior of this interesting custom.

1949

In January of 1949, General Motors staged a special "Transportation Unlimited" show in the grand ballroom of the Waldorf-Astoria Hotel in New York. This lavish show was a throwback to the exclusive Automobile Salons and New York Auto Shows of the 1920s and set the stage for GM's later Motoramas. Cadillac displayed seven cars in its "Transportation Unlimited" exhibit, four of which were special show cars. One of these was the first Cadillac Coupe de Ville: the production Coupe de Ville pillarless hardtop was not introduced until very late in the 1949 model year. The Coupe de Ville show car was built by Fleetwood on the 133-inch wheelbase Sixty Special chassis. It had a one-piece curved windshield, simulated air scoops on the rear fenders and bright chrome scuff plates below the doors. The one-piece windshield showed up on production Cadillacs the following year. This one-off show car, designated Style 6037-X, was later used by GM President Charles E. Wilson, who presented it to his secretary in 1957. It's still around out on the West Coast.

Of the four special show cars created for Cadillac's exhibit at GM's 1949 "Transportation Unlimited" show in New York, three were specially trimmed and equipped versions of the Fleetwood Sixty Special. This is the "Caribbean," which was finished in a dazzling "Caribbean Daybreak Metallic." The only custom exterior touches were leather-covered sills, or saddles, on the doors beneath the side windows. The interior was lavishly upholstered in lush French broadcloth and matching leathers.

The Cadillac "Embassy" was another specially-equipped and trimmed Fleetwood Sixty Special created for GM's 1949 "Transportation Unlimited" spectacular in New York. It featured vertical rear fender "air scoops" like those on the companion Fleetwood Coupe de Ville show car, and heavy horizontal rear fender moldings. The "Embassy" also had a leather-covered roof and a formal small rear window treatment. The front compartment was upholstered in leather and the rear passenger compartment in gray broadcloth. Other equipment included a glass division with a clock mounted in a glass header, and an umbrella, tool kit and two-way radio carried in the front doors. The dignified "Embassy" was actually a practical—for New York City—close-coupled limousine not unlike the Sixty Specials offered between 1939 and 1942.

We've all heard of "stretched" Cadillac customs, but here's a "condensed" one. This unusual three-passenger coupe was built by Pasadena, Calif. coachbuilder Maurice Schwartz on a much-shortened 1949 Cadillac Fleetwood Seventy-Five chassis. Note the extremely long rear deck, the golf club compartment door and the prewar-style coupe roof with forward-angled "B" pillar. The Seventy-Five's distinctive stainless steel running boards have been retained, as have the two vertical dividers in the rear window. This rather handsome coupe was custom-built for W.A. Woodward, of Oregon.

One of four special Cadillac show cars created for the General Motors "Transportation Unlimited" extravaganza in New York, the "El Rancho" was a specially-trimmed 1949 Cadillac Series 62 Convertible Coupe. Exterior appearance was stock, but the interior was done up in a western motif with saddle leather upholstery, leather-covered steering wheel and instrument panel and gun holsters in each front door pocket. The floor was covered in deep cowhide carpeting. Interior hardware was silver-plated and richly engraved. The four specials exhibited at this show were the first of a long line of Cadillac experimentals designed to dazzle the public and gauge reaction to new features and styling concepts. Many ideas first tried on show cars eventually showed up on production cars.

1949

The Derham Body Co. of Rosemont, Pa. built this huge convertible sedan for a Middle East customer who wanted something different, and got it. This car was constructed on a 1949 Cadillac Series 86 commercial chassis. Wheelbase was a gargantuan (for a passenger car) 163 inches. It was even more impressive with the top down. This four-door convertible sedan was equipped with extra-wide tires all around for desert travel.

Pasadena, Calif. coachbuilder Maurice Schwartz designed and built this unique 1949 Cadillac station wagon for Miguel Aleman, the President of Mexico. It was built on a stretched Series 62 coupe chassis. The fastback roofline has been preserved and the car is equipped with a roof luggage rack. The rear doors have ventilator panes just behind the "B" post. Note how high it is sprung. This had to be about the classiest "woodie" around at the time it was built.

Cadillac's 1949 commercial chassis got its first restyling since the war. Its exterior styling was now compatible with all of the division's passenger cars except the Fleetwood Seventy-Five, which carried over its 1948 styling. The all-new Series 86 commercial chassis had the same 163-inch wheelbase as before but was powered by Cadillac's all-new OHV high-compression V-8 engine. This is a 1949 Superior-Cadillac Model 1486 Ambulance, which sold for $6,085. It was photographed in the Superior Coach Corp. factory yard in Lima, O. prior to shipment. Note the spotlight mounted above the windshield and the coaster siren on the left front fender.

The Cadillac commercial chassis, designed especially for hearse and ambulance bodies, was completely redesigned for 1949. The new Series 86 commercial chassis now had the same styling as the rest of Cadillac's cars except the carryover Seventy-Fives. This is a 1949 Meteor-Cadillac End-Loading Limousine Funeral Coach, Style 86-490, which had a factory driveaway price at Piqua, O. of $5,942. Note the extremely high windshield glass and the distinctive Meteor Motor Car Co. coach lamp built into the front door. Cadillac shipped 1,861 Series 86 commercial chassis to funeral car and ambulance builders this year, considerably fewer than in 1948.

From 1949 through the 1956 model year, a distinctive styling feature of the landau funeral coaches built by the A. J. Miller Co., Bellefontaine, O., was an oval "porthole" window just behind the rear side door. This window was set at an interesting angle and was a delete option. Most customers wanted it, however. The standard Miller-Cadillac Landau Hearse on the Series 86 commercial chassis was priced at $5,700. When equipped with Hydra-Matic and an electrically-operated three-way casket table the price rose to $7,050. The 1949 commercial chassis featured Series 62 front fender stone shields and bright sill moldings.

1950

Cadillac sales reached a significant new plateau this year. The division sold more than 100,000 cars in a single year for the first time in its 48-year history. Cadillac had shifted into high, finally overtaking old rival Packard for good. The "Standard of the World" has dominated the U.S. luxury car market ever since. Within a few years, once-mighty Packard would be gone altogether.

The Cadillac Motor Car Division entered the second half of the century with a completely-restyled line of cars. Every model, from the lowest-priced Series 61 Coupe on up to the stately Fleetwood Seventy-Five Limousine had been totally redesigned. Cadillac would use this same basic styling for the next four years, through model year 1953. This year saw the first appearance of the graceful "Florentine curve" on Cadillac coupe rooflines and on the upper rear quarters of the redesigned Seventy-Five.

Gone was the sleek, taut styling of 1948 and 1949. The restyled 1950 Cadillacs were much heavier and bulbous in appearance, with swollen bodies and turtle-like rooflines. There wasn't a fastback to be found: this aerodynamic body style was a phenomenon of the 1940s, but it was destined for a revival on some lesser makes some 15 years later. Cadillac was careful, however, not to tamper with its incredibly successful basic styling elements. These included a massive, chrome-plated eggcrate grille and those sweeping fishtail fins.

The front fender line extended through the doors, dipping slightly before breaking into curved rear fenders which were capped on their leading edges by a vertical chrome "air intake" molding. All 1950 Cadillacs had a new, one-piece curved windshield. Rear windows wrapped around to meet the "C" pillar. Heavy Cadillac "V" emblems ornamented the front of the hood and rear decklids. Rooflines were broad and flat.

Cadillac's 1950 model lineup included nine models in four series on four wheelbases—essentially the same as in 1949. The new Series 61 Cadillacs were built on a four-inch shorter 122-inch wheelbase. These models used a new GM "B" body which it shared with sister GM divisions Buick and Oldsmobile.

There were four Series 62 body styles, all on the same 126-inch wheelbase used previously. Series 62 Cadillacs had bright rocker sill moldings and chrome extensions on the lower edges of their rear fenders between the fender skirts and rear bumper. This trim was not used on the less-expensive Series 61 cars. The redesigned Series 62 models used the new GM "C" body shell. Body styles included a Two-Door Coupe, two-door Coupe de Ville Hardtop, a Four-Door Sedan and a sporty Convertible Coupe. This would be the first time a pillarless hardtop was offered for a full model year: the style-setting Coupe de Ville had been introduced late in the 1949 model year.

The 1950 Fleetwood Sixty Special was also completely redesigned. It was built on a new 130-inch wheelbase that was three inches shorter than the 133-inch chassis exclusive to this model between 1942 and 1949. Again there was just one body style, a premium four-door sedan with exceptional rear seat legroom and a long list of standard comfort and convenience appointments. But the Sixty Special had all but lost its once-distinctive appearance. Except for a series of eight vertical chrome louvers on the lower edge of the rear doors, it looked almost identical to the Series 62 sedan. That didn't deter any customers, however. More than 11,000 were delivered in the 1950 model year, setting a new record for this model.

The most-changed of all the 1950 Cadillacs was the regal Fleetwood Seventy-Five, which got its first complete restyling in a decade. The all-new Seventy-Five utilized an exclusive "D" body and was built on a 10-inch longer 146-3/4 inch wheelbase. There were just two models now instead of the previous five—a Seven-Passenger Sedan and an Imperial Limousine with glass division. The impressive new Fleetwood Seventy-Fives had styling compatible with all other Cadillacs for the first time since 1941.

All 1950 Cadillacs had a redesigned front suspension which featured direct-action shock absorbers mounted inside the coil springs. Other mechanical improvements included a carburetor idle heating system that eliminated humid weather stalling and numerous refinements in the optional Hydra-Matic automatic transmission.

The entire Cadillac line was completely restyled for 1950. The same basic styling, with minor grille and exterior detail changes, would be used for the next four years, from 1950 through 1953. Again, the lowest-priced cars in the line were two Series 61 Cadillacs, a two-door Coupe and a four-door Sedan. Both utilized GM's new "B" body and were built on a four-inch shorter 122-inch wheelbase. The sleek fastback coupe was gone, replaced by this heavy-looking notchback. Style 6137, the 1950 Cadillac Series 61 Coupe had a factory-delivered price of $2,761 and weighed 3,829 pounds. Of the 26,772 Series 61 Cadillacs sold this year, 11,839 were coupes like this one. Note the wide whitewalls (still optional equipment) and the full chrome wheel discs.

Series 61 Cadillacs were still easily identified by their minimal chrome trim. The price-leader Series 61 models had no chrome rocker sill moldings or chrome extensions on the lower edge of the rear fenders between the fender skirt and the rear bumper. This is the 1950 Cadillac Series 61 Sedan, Style 6169, of which 14,619 were built. Shipping weight was 3,822 pounds and the 61 Sedan had a factory-delivered price of $2,866. The 1950 Series 61 Cadillacs shared this new General Motors "B" body shell with Oldsmobile and Buick. Note the vertical "C" pillar and wrap-around rear window. A total of 14,619 Series 61 Sedans were built in the 1950 model year.

1950

All 1950 Cadillacs were powered by the same 160-horsepower, 331 cubic inch OHV engine introduced the previous year. This extremely potent V-8 quickly attracted the attention of auto racing enthusiasts. Before long Cadillac engines were being wedged into every conceivable type of competition machinery, from street rods to Grand Prix cars. The 1950 Cadillac Series 61 Coupe with the high-compression, short-stroke 331 engine and three-speed manual transmission was reputed to be the fastest production car built in America. Despite their considerable bulk, these cars had surprisingly good handling. The 1949-1950 Cadillacs and Oldsmobile Rocket 88s were the first true postwar American "muscle" cars.

The most spectacular demonstration of Cadillac high performance took place at this year's 24 Hours of LeMans, in France. To the astonishment of spectators and other competitors, American sportsman Briggs Cunningham entered a pair of 1950 Cadillacs in this premier motor racing event. One was an aerodynamically-bodied racing car named "Le Monstre", an acknowledgement of its massive size compared with contemporary road racing machinery of its day. The other was a stripped-down but otherwise virtually stock Series 61 Coupe. The Frick-Tappet modified engines in both were fitted with four-barrel carburetion. The Coupe finished a very respectable 10th overall, even coming in ahead of the streamlined "Le Monstre", which finished 11th. A Cadillac-powered English Allard came in third. And a Cadillac won that year's grueling Pan-American Road Race.

General Motors staged another dazzling GM-only auto show. The 1950 edition, called the "Mid-Century Motorama", opened in New York's Waldorf-Astoria Hotel on January 15. This year's show was seen only in New York. Another three years would pass before GM took its Motoramas on the road. The focal point of Cadillac's exhibit at the Mid-Century Motorama was an oppulently-appointed Series 62 Convertible. Called the "Debutante", it was finished in a rich, buffed yellow, and its interior was upholstered with 187 Somali leopard pelts. All interior fittings and hardware were gold-plated. This rolling jewel case was valued at a then-fantastic $30,000! It is likely that this baroque show car provided the inspiration for "The Solid Gold Cadillac" of later stage and movie fame.

Cadillac's 100,000th car of the 1950 model year came off the assembly line in Detroit on November 16, 1950. Model year sales hit the six-figure mark for the first time, setting a new all-time record of 103,857. The division's sales had nearly doubled in just two years. There were some important personnel changes on Clark Ave. this year, too. Donald E. Ahrens succeeded John F. Gordon as Cadillac Motor Car Division General Manager. James M. Roche, a future GM Chairman, was appointed General Sales Manager, and Charles F. "Fred" Arnold became Chief Engineer.

Sales, and prospects, had never been better. But once more there were ominous clouds on the horizon. War broke out in far-off Korea in June, 1950. American armed forces were mobilized. And Cadillac found itself back in the tank manufacturing business again.

Cadillac's Series 62 model lineup for 1950 included four different body styles, all on the same 126-inch wheelbase. Styling was completely fresh. Series 62 Cadillacs used GM's new "C" body. This is the 1950 Cadillac Series 62 Coupe, Style 6237, of which 6,434 were built. All 1950 Cadillacs had a bright, vertical simulated "airscoop" molding on the leading edge of the rear fender. Series 62 Cadillacs had bright sill moldings and chrome trim on the lower edge of the rear fenders between the fender skirts and rear bumper. The Series 62 Coupe was generally identical in appearance to the more expensive Coupe de Ville Hardtop. The Series 62 Coupe sold for $3,150.

This would be the first year that a Cadillac hardtop was available through a complete model year. The stylish, new Coupe de Ville had been introduced late in the 1949 model year. This is the restyled 1950 Cadillac Coupe de Ville Two-Door Hardtop, Style 6237-D, which carried a factory-delivered price of $3,523, only slightly higher than the style-setting 1949 model. Only 4,507 Coupe de Villes (or Coupes de Ville for the purist) were built in the 1950 model year. Standard weight was 4,074 pounds. This beautiful, new body style lent itself particularly well to two-tone color combinations. The Coupe de Ville was traditionally a Series 62 body style.

The sportiest of all nine 1950 Cadillac models was still the sassy Series 62 Convertible Coupe. Like all of the other completely-restyled 1950 Cadillacs, it featured a curved, one-piece windshield and an even more massive grille which carefully perpetuated Cadillac's bright-eggcrate front-end styling. The rear of the car was still dominated by those sweeping "fishtail" fins. The Series 62 Convertible utilized the new GM "C" body and was listed as Style 6267. Production totaled 6,986—considerably higher than the Coupe de Ville Hardtop. Factory-delivered price was $3,654. This was undoubtedly the best-looking of Cadillac's restyled 1950 cars.

1950

By far the most popular single model in the 1950 Cadillac line was the utilitarian Series 62 Four-Door Sedan. Of the 59,818 Series 62 Cadillacs built this year—more than half of the division's production—41,890 were Series 62 four-door sedans. Style 6219, the Series 62 Sedan had a base price of $3,234. Weight was 4,102 pounds. Series 62 Cadillacs for 1950 utilized the same 126-inch wheelbase used in 1949. The small quarter-windows actually make this a six-window sedan. The "C" pillar treatment is totally different from the lower-priced Series 61 sedan, which had a vertical third pillar and wraparound rear window with two chrome dividers.

Just as police agencies routinely photographed citizens who have brushes with the law, General Motors Photographic takes "mug shots" of its new cars so that no detail, no matter how small, goes unnoticed. This profile shot of a pre-production 1950 Series 62 Cadillac Four-Door Sedan was taken at the General Motors Proving Ground at Milford, Mich., near Detroit. It graphically presents Cadillac's heavy, new profile and more extensive use of exterior chrome trim, particularly around the upper body or "greenhouse". Note the drab blackwall tires and standard small hubcaps.

Although they were still superbly-engineered cars that combined luxury with surprisingly high performance, the 1950 Cadillacs sacrificed something with their puffy, heavy-looking new styling. Gone were the lithe, taut lines of 1948 and 1949. All nine 1950 models used basically the same styling with variations in body shells, wheelbases and trim. The Fleetwood Sixty Special retained its own exclusive wheelbase—130 inches for 1950, three inches shorter than the 133 inches used between 1942 and 1949. The Sixty used a stretched "C" body with a rear deck nine inches longer than the C-bodied Series 62 Coupe. Style 6019, the Fleetwood Sixty Special had a factory list price of $3,797 and weighed 4,136 pounds. It was a big seller, with a record 13,755 delivered. Note the eight vertical chrome louvers on the lower edge of the rear door which set the Sixty Special apart from all other Cadillac models.

What more typified automotive accessories of the early fifties than sun visors and spotlights? Here is a well-equipped 1950 Cadillac Fleetwood Sixty Special Four-Door Sedan. This car is also equipped with optional fog lamps, which fit into openings in the outer ends of the new grille. When fog lamps were not used, these openings were filled with chrome concentric-ring trim pieces. Cadillac's 1950 grille was a more massive rendition of the same design used in 1949. The bullet-tipped vertical bumper guards had been a Cadillac styling feature since 1942. Note the heavy Cadillac "V" hood emblem, which was also used on the rear decklid of most models.

The most changed of all the 1950 Cadillacs were the big Fleetwood Seventy-Fives, which got their first complete restyling since 1941. There were now just two Fleetwood Seventy-Five models—a seven-passenger sedan and a formal limousine. The all-new Seventy-Fives were built on a 146-¾ inch wheelbase that was more than 10 inches longer than the 1941-49 models. A total of 1,459 Fleetwood Seventy-Fives were built during the 1950 model run, including 716 Style 7523-X Seven-Passenger Sedans and 743 of these Style 7533-X Seven-Passenger Imperials with glass division. The Seven-Passenger Sedan sold for $4,770 and at $4,959 the companion Imperial Limousine was the most expensive car in the entire 1950 Cadillac line. Only one 7523-L business sedan was sold this year.

1950

For the first time since 1941, the Fleetwood Seventy-Five's styling was now compatible with all of Cadillac's other cars. The completely redesigned and restyled 1950 Cadillac Fleetwood Seventy-Five was the only car that used GM's new "D" body, actually a much-stretched "C" body exclusive to this series. Fleetwood had long abandoned the custom-body business, but Derham of Rosemont, Pa. continued to do small batches of Cadillac semi-customs, mostly formal limousine conversions based on the Series Seventy-Five. This 1950 Derham-Cadillac has closed rear roof quarters and a white, fabric-covered roof. Note the special chrome roof and window edge moldings. These Derham formals had very small rear windows for privacy.

The 1950 model year marked an important new milestone for Cadillac. The division sold more than 100,000 cars for the very first time, this time overtaking rival Packard for good. Cadillac's sales had virtually doubled in just two years. The 100,000th Cadillac built in the record 1950-model production run came off the Clark Avenue final assembly line on November 16, 1950. It was a Style 6019 Fleetwood Sixty-Special Sedan. Here, Cadillac executives greet the milestone car as it is driven from the end of the line to the shipping department. Note the protective paper taped to the front bumper to protect it from in-transit scratches.

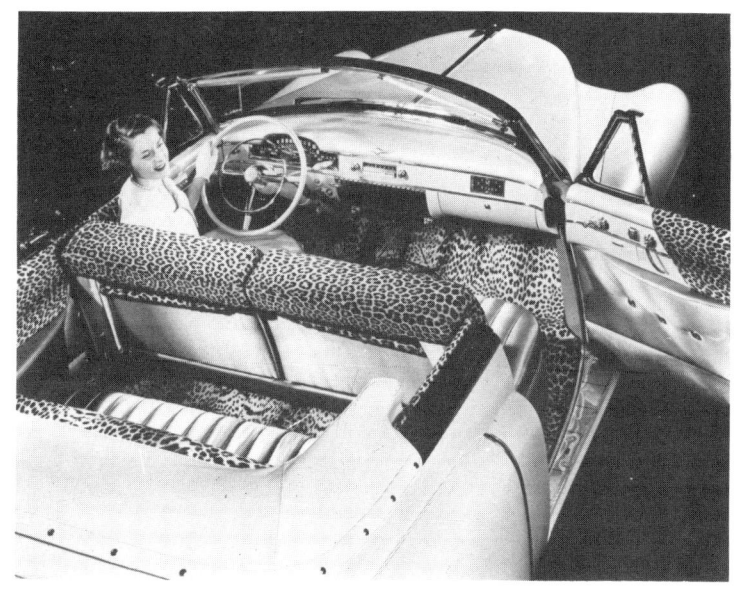

General Motors staged another dazzling GM-only auto show this year. The "Mid-Century Motorama" was again held at the Waldorf-Astoria Hotel in New York, in January of 1950. Cadillac created a special car called the "Debutante" for this show. A specially trimmed-and-painted Series 62 Convertible, the yellow-buffed Debutante may well have provided the inspiration for "The Solid Gold Cadillac" of stage and movie fame. The Debutante Convertible's interior was upholstered with 187 leopard pelts. All interior hardware, including the ignition key and chain, were gold-plated. Special show cars became a major feature of later GM Motoramas. This year's show was held in New York only. The Motoramas didn't really hit the road until 1953.

American sportsman Briggs Cunningham astounded everyone when he entered two Cadillacs in the 1950 LeMans Grand Prix in France. One was a nearly-stock 1950 Cadillac Coupe. The other was a Cadillac chassis that had been fitted with a special aerodynamic racing body. This special was named "Le Monstre" (The Monster), and it finished a very respectable 11th overall—right behind the Series 61 Coupe! Cadillac's new 331 cubic inch, overhead valve V-8 engine would soon earn even greater laurels in the legendary Cunningham C-1s and Cadillac-Allard J-2.

Cadillac's potent new OHV engine quickly attracted the attention of U.S. racing car builders. Even in stock form, the 1949 Cadillac was capable of 100-mile-an-hour top speeds and offered surprisingly good handling for a luxury car. Within a matter of months, Cadillac engines were being wedged into every imaginable type of competition car, from street rods to the fantastic British-built Allard which was cleaning up on the American road-racing scene. But the most remarkable competition performance of the year was turned in by an almost-stock 1950 Cadillac Series 61 Coupe, which U.S. sportsman Briggs Cunningham entered in the 24 Hours of LeMans. This Cadillac came in 10th overall, outrunning even a special aerodynamically-bodied Cadillac also fielded by Cunningham. A Cad-Allard came in third. The sight of that big, fat Caddy hurtling around the course among all those tiny purebred racing machines must have been truly memorable. By the end of the day a lot of smirks had disappeared from a lot of faces, and the symbol of American material success had acquired new lustre.

1950

Cadillac's Series 86 commercial chassis for 1950 also got the division's all-new styling. Just over 2,000 were delivered to five of the six major funeral car and ambulance manufacturers in the United States. Industry giant Henney continued to use the Packard commercial chassis exclusively. This is a 1950 Superior-Cadillac Limousine Funeral Coach with a manually-operated, side-loading casket table. The Superior Coach Corp. of Lima, Ohio was quickly rising to a position of leadership among American professional car builders. Note the Superior-Cadillac script on the front fender.

Cadillac's Series 86 commercial chassis had a new 157-inch wheelbase, six inches shorter than in 1949. The commercial chassis came with only the cowl and front end. Front doors and rear fenders were shipped loose. Specialized hearse and ambulance builders designed and built their own funeral coach, ambulance, service and flower car bodies for this special chassis. Each manufacturer incorporated his own styling touches. The Eureka Co. of Rock Falls, Ill. used a different rear door treatment, with the rear door cut into the rear fender. Most other U.S. funeral car builders lined up the "D" pillar with the vertical chrome rear fender trim. This is a 1950 Eureka-Cadillac Limousine Ambulance. Note the tunnel roof lights and Federal C-5 siren on the roof. Cadillac built 2,052 Series 86 commercial chassis this year.

Bedecked with every conceivable type of warning light system, and sporting a fire extinguisher built into the front fender, this Meteor ambulance must have been an impressive sight indeed. Even the coach lamps on the B-pillar and the parking lights have been equipped with red lenses. This style Meteor-Cadillac limousine ambulance carried a standard price tag of $6,634, exclusive of lights and other extra fittings. It was designated Meteor Model 501. This year Cadillac offered fog lights combined with the parking lights, and the red lenses on this model would indicate that these were probably refitted by Meteor to act as flashers as well as operate as parking lights.

The coupe-style flower car was now at its peak in popularity. Although this type of specialized funeral car had been around since the late 1930s, Superior Coach Corp. did not offer one until the 1949 model year. Superior called its Cadillac flower car the Coupe de Fleur, or "flower coupe," a name which is still reserved for Superior-Cadillac flower cars to this day. The 1950 Superior-Cadillac Coupe de Fleur had a factory driveaway price of $5,970. When equipped with an optional hydraulically-operated flower deck, the base price went up to $6,050. The rear window is the same used on the Series 61 Sedan.

Continuing as one of the last vestiges of the carved funeral car was this S & S service car. A quite elaborate vehicle for service car work, this model carried stamped metal plates in the rear window areas, which simulated the carved wood panels of an earlier era. The front door carries the owner's initial plaque or coat of arms, while the hood bears the traditional chromed S & S date marks that were supposed to hide the identity of the chassis.

1951

Totally redesigned the previous year, the 1951 Cadillacs received only minor styling changes. These included small "eggcrate" grilles in the outer ends of the massive main grille, below the headlights, and new bumpers with much-larger, forward-thrusted "bullet" bumper guards.

Cadillac's 1951 model lineup remained unchanged from 1950. Again, there were nine models in four series on four different wheelbases. Now in its third model year, the 331 cubic-inch Cadillac V-8 engine was also unchanged, although some improvements were made in the optional Hydra-Matic automatic transmission. A minor interior change included a switch from gauges to red warning lights for amperes and oil pressure.

This would be the final year for the lowest-priced Series 61 Cadillac. The Sixty-One was quietly dropped from the line May 1, 1951. The reason: low sales. Only 4,700 Series 61 Cadillacs had been ordered since the first of the 1951 model year. Cadillac had offered Series 61 models in 1938 and 1939, but none in 1940. A new Series 61 replaced the LaSalle at the bottom end of Cadillac's product lineup for the 1941 model year. The 1951 Series 61 line, on the short 122-inch wheelbase introduced for 1950, included only two body styles—a two-door Coupe and a four-door Sedan. At $2,804 the Series 61 Coupe was the lowest-priced 1951 Cadillac available.

Series Sixty-Two for 1951 included four different body styles on a 126-inch wheelbase. They were a two-door Coupe; the Coupe de Ville Two-Door Hardtop Coupe; a sleek Sixty-Two Convertible Coupe, and the practical Four-Door, Five-Passenger Sedan which accounted for more than half of Cadillac's entire 1951 model year production. Series 62 models continued to use chrome rocker sill moldings and bright rear fender trim not found on the less-expensive Series 61 Cadillacs.

The relatively expensive Fleetwood Sixty Special for 1951 was also virtually unchanged, except for the minor grille and bumper revisions, as were the long-wheelbase Fleetwood Seventy-Five sedan and limousine. This would be the last year Cadillac would offer a specially-trimmed Fleetwood Seventy-Five Business Sedan for funeral service and commercial limousine livery operators. These commercial four-door sedans had been available since the mid-1930s. Henceforth, funeral directors and limousine rental companies would have to be content with Cadillac's standard production seven-passenger sedans and limousines.

No history of Cadillac would be complete without special mention of the division's prominent presence in motor racing in the early 1950s. Production Cadillac passenger cars of this era—especially the basic Series 61 Coupe with the 331 OHV engine and stick shift—were among the first true American "muscle" cars. The compact, high-performance Cadillac V-8 quickly acquired an excellent reputation on the race track. Two of the most famous sports cars of their time were powered by Cadillac engines and transmissions.

The 1951 Cadillacs showed very little exterior change from the previous year. The only visible styling changes were larger front bumper "bullets" and small eggcrate grilles in the outer ends of the grille below the headlights. The lowest-priced car in the 1951 Cadillac line was the Series 61 Coupe for Five-Passengers, style 6137, which weighed 3,795 pounds and had a factory-delivered price of $2,809. Only 2,400 Series 61 Coupes were built before this series was quietly discontinued midway through the 1951 model year. As in 1950, the price-leader "61" had no rocker sill moldings or bright rear fender trim.

This would be the last year for the Series 61 Cadillac. The "61" had replaced the LaSalle as Cadillac's lowest-priced car in the 1941 model year, although there had been a Series 61 Cadillac (along with the Series 50 LaSalle) in 1938 and 1939. Only 4,700 Series 61 Cadillacs had been built in the 1951 model year before the series was quietly discontinued on May 1, 1951. Of these, 2,300 were Series 61 Four-Door Sedans. Style 6169, the Sixty-One Sedan sold for $2,917 and had a shipping weight of 3,827 pounds. Two-tone paint combinations were still very popular with Cadillac's customers.

The 1951 Cadillac Sedan 62 model lineup continued to offer four body styles, all on the same 126-inch wheelbase. The least expensive of these was the Series 62 Coupe for Five Passengers, Style 6237, of which 10,132 were built during the 1951 production run. Standard weight was just over 4,000 pounds—4,002 to be precise. The Sixty-Two Coupe sold for $3,369 in base form. This model was nearly identical in appearance to the Coupe de Ville two-door hardtop, which was priced $399 higher. All Series 62 Cadillacs had bright rocker sill moldings and chrome trim on the lower edge of the rear fenders.

1951

One was the legendary Cadillac-Allard. The other was the equally fast Cunningham.

Sydney J. Allard, a major English Ford distributor, had built his first Allard sports cars in the mid-1930s. Some of these inevitably found their way to U.S. shores where, when fitted with big American engines, delivered fantastic performance. The famous Allard J-2 was powered by the new overhead valve, short-stroke, high-compression Cadillac engine that had been introduced in 1949. For a time, this was an almost-unbeatable combination. Cadillac-Allards came in first, second and third at Watkins Glen. The Allard J2X had an improved front suspension system with a split axle, and later models were powered by Chrysler's new Hemi-head V-8. Others used Mercury or Lincoln V-8s.

Sportsman Briggs Cunningham, who had entered a pair of Cadillacs in the 24 Hours of LeMans the previous year, began building his own sports-racing cars in Palm Beach, Fla. The highly-successful Cunningham C-2 also used a Cadillac powertrain. Later C-3 Cunninghams had Chrysler Hemis.

When war broke out in Korea in June, 1950 Cadillac was again called upon to build tanks. This time, however, there was no interruption in civilian passenger car production. Cadillac was assigned the responsibility for the launch and operation of a huge new Tank Ordnance Plant in Cleveland, Ohio. Former Cadillac Chief Engineer Edward N. Cole was picked to head up this major defense production project. Cadillac built more than 3,700 M-41 Walker Bulldog light tanks in its 2½-million square foot Cleveland Tank Plant between 1951 and 1955. But the M-41, also sometimes referred to as the T-41, was not powered by Cadillac engines: these tanks used Continental or Lycoming six-cylinder engines. Typical of the Cadillac Motor Car Co.'s exemplary Second World War performance, the first Walker Bulldog rolled off the line three months ahead of schedule.

Cadillac sold 110,340 cars in the 1951 model year, setting another all-time record. The Packard Motor Car Company introduced a completely restyled 1951 model line, but it wasn't enough. Packard never overcame Cadillac's commanding sales lead. Chrysler came out with its new 180 HP engine with hemispherical combustion chambers—the famed "Hemi-head" V-8.

The 1952 models were on the way. Cadillac had now been in business for half a century, an event that would not pass unnoticed on Clark Avenue.

All 1951 Cadillacs carried over their fresh 1950 styling with minor changes to the grille and much-larger front bumper "bullets". There was new trim below the fin-mounted taillights. Backup lights were still housed in separate, small round units in the lower rear body below the trunk lid. The smartest-looking car in the 1951 Cadillac model lineup was again the Series 62 Convertible Coupe, Style 6267, which was also the most expensive body style in this four-car series. Factory-delivered price was $3,908. Long, flowing fender lines and the stretched "C"-body rear deck gave this car a low, road-hugging look.

By far the most popular Cadillac of all was the practical Series 62 four-door, five-passenger Sedan. A total of 54,596 Series Sixty-Two Sedans were sold this year, accounting for more than half of all 1951 Cadillac production. Style 6219, the Series 62 Sedan carried a factory-suggested price of $3,458 with standard equipment and weighed 3,983 pounds. The upper body structure included tiny quarter windows. Cadillac advertised this practical car as...."The ideal family sedan". Note the two-tone paint, wide whites and chrome "sombrero" wheel covers.

Cadillac sold more than 23,000 two-door coupes this year. These sales were split almost equally between the Series 62 Coupe and the more expensive Series 62 Coupe de Ville Two-Door Hardtop. Significantly, the sporty pillarless hardtop outsold the Coupe for the first time this year: a total of 10,241 Coupe de Villes were delivered, compared with 10,132 Series 62 Coupes. The name "Coupe de Ville" appeared in small gold script on the rear roof pillars for the first time. Interiors offered a selection of two-tone combinations in cloth and leather. The 1951 Cadillac Coupe de Ville, Style 6237-DX, weighed 4,077 pounds and had a factory price of $3,768.

Deliveries of the premium Cadillac Fleetwood Sixty Special set another record this year. A total of 18,631 Sixty Specials were sold in the 1951 model year, easily toppling the previous year's record 13,755. The well-equipped Fleetwood Sixty Special continued on its own, exclusive 130-inch wheelbase. Listed as Style 6019, it weighed 4,155 pounds and had a factory price of $4,060. Except for miniature grilles below the headlights and larger front bumper guards, styling was unchanged from 1950. The eight small vertical chrome louvers on the lower edge of the rear door (or front of the rear fenders, depending on how you look at it) were a Sixty Special styling touch first seen on the 1942 model.

1951

At the top of the 1951 Cadillac model lineup were two big Fleetwood Seventy-Fives—a seven-passenger sedan and a formal limousine, both built on a 147-inch wheelbase. This is the 1951 Cadillac Fleetwood Seventy-Five Seven-Passenger Sedan, Style 7523-X, which had a factory-delivered price of $5,098. A total of 1,090 were built. Also sold this year were 30 Style 7523-L Business Sedans. This would be the last year Cadillac would offer a specially-trimmed Business Sedan for funeral service and commercial limousine livery operators. The big Seventy-Five Sedan weighed 4,610 pounds.

Cadillac's largest and most expensive cars, the stately Fleetwood Seventy-Five Sedan and Limousine, were completely redesigned for 1950, so received minimal changes for 1951. The only visible exterior changes were larger front bumper bullets and small "eggcrate" grilles under the headlights. The most expensive model in the 1951 Cadillac line was the Fleetwood Seventy-Five Seven-Passenger Imperial Sedan, Style 7533-X, which had a factory price of $5,300. Weight was 4,641 pounds. The optional Hydra-Matic automatic transmission added another 90 pounds. A total of 1,085 of these formal limousines with glass division were sold in the 1951 model year, slightly less than half the Series Seventy-Five production.

Cadillac convertibles were always in demand when VIPs came to town. So it was that General Motors of Canada Ltd. provided a special Series 62 Convertible Coupe for her Royal Highness the Princess Elizabeth, now Queen Elizabeth II, and her husband, Prince Phillip, on their 1951 Canadian Royal Tour. Modifications included masts to fly the royal ensign and the Union Jack on the front of the car. This photo was taken at the Canadian National Exhibition Grandstand in Toronto. The author was in the honor guard as a Cub Scout and clearly remembers this sparkling, black car.

Cadillac's customers long lamented the fact that there was no station wagon in the company's otherwise comprehensive model lineup. But GM's product planners knew there was insufficient volume to justify the costly tooling required to build one. Several outside coachbuilders, however, did produce small numbers of Cadillac station wagon conversions based on the Series 62 four-door sedan. This 1951 Cadillac Station Wagon was built by Coachcraft of Los Angeles for Minneapolis contractor Merril M. Madsen. It was designed by Philip Wright. The power-operated rear window retracted into the tailgate. The styling is nicely compatible with the chassis.

No history of Cadillac would be complete without special reference to the fantastic Cadillac-Allard sports cars which dominated the American road-racing scene in the early 1950s. The Cadillac-engined Allard J-2 was the hairiest—and fastest—sports car of its day. The Allard Motor Co. Ltd. of London, England designed and built the car, which was powered by the 331 cubic inch Cadillac V-8 engine and Cadillac transmission. The J-2 was succeeded by the split-axle J2X in 1951. Later, Allards were powered by Chrysler Hemi and Mercury engines. The Cadillac-powered J-2 and Briggs Cunningham's equally-impressive Cadillac-powered C-2s left indelible impressions on U.S. auto racing history. Surviving examples are prized collectors' items today.

1951

The most unusual Cadillacs turned out during the 1951 model year were a fleet of special "harem cars" custom-built for King Ibn Saud, of Saudi Arabia. King Saud ordered 20 of these unique six-door limousines, which were designed and built by the Hess & Eisenhardt Co. of Cincinnati on the 1951 Cadillac Series 86 commercial chassis—the same 157-inch wheelbase chassis the company used for its hearses and ambulances. Each of these limousines was designed to transport six of the King's wives across the desert in air-conditioned comfort and privacy. Each cost a whopping (for the time) $12,500. The rear windows were equipped with one-way mirror glass so the rear seat passengers could see out without being seen. Hess & Eisenhardt completed this big order in 1952. It is not known if any of these unique six-door sedans still exist.

All of the major U.S. funeral car builders—Superior, Meteor, Miller, S & S, and Eureka—were now offering stylish deck-type flower cars on the Cadillac Series 86 commercial chassis, the same 157-inch wheelbase chassis they used for their hearses and ambulances. The Henney Motor Co. remained loyal to Packard and offered a coupe-style flower car on a special long-wheelbase Packard commercial chassis. This is an elevated view of a 1951 Superior-Cadillac Coupe de Fleur with hydraulically-adjustable stainless steel flower deck. Note the Series 61 rear window glass. The Superior-Cadillac flower car had a factory-delivered price of $6,470 at Lima, Ohio.

Cadillac built 2,960 Series 86 commercial chassis for U.S. funeral car and ambulance manufacturers during the 1951 model year. One of the company's major customers was the Meteor Motor Car Co. of Piqua, Ohio, which built this 1951 Superior-Cadillac End-Servicing Landau Funeral Coach, Style 86-510-L, which sold for $6,675. This model was also available with a manual or electrically-operated side-servicing casket table. Note the smooth metal roof and the coach lamp mounted on the extra-wide "B" pillar.

The long-wheelbase funeral service car had become something of a rarity by this time. Once a standard piece of funeral home rolling stock, the luxury chassised service or first-call coach was being displaced by less-expensive Pontiac or Chevrolet service cars, or by station wagons or converted panel delivery trucks. This is a 1951 Superior-Cadillac Service Car, which had a factory-delivered price at Lima, Ohio of $6,470. Note the modernistic wreath emblem and chrome streamers. Service cars were used for hauling caskets and equipment, saving wear and tear on the funeral director's first-line hearses. The small chrome pieces on the upper front door are holders for the funeral firm's nameplates.

War erupted in Korea in June, 1950 and Cadillac soon found itself back in the tank business. But this time civilian passenger car production was not directly affected, although there was some material rationing and shortages. Cadillac was responsible for operation of the huge Cadillac Ordnance Tank Plant in Cleveland, Ohio. Former Cadillac Chief Engineer Edward N. Cole was picked to launch this major operation. Cadillac built more than 3,700 M-41 Walker Bulldog light tanks in the Cleveland plant between 1951 and 1955. Many of these modern tanks saw service in the Korean conflict, which ended in a truce in 1953. Hundreds of others were used by foreign armies all over the world well into the 1970s. The M-41 (sometimes called T-41) mounted a 76-mm gun and was powered by Continental or Lycoming tank engines.

1952

The Cadillac Motor Car Division observed its 50th Anniversary in 1952. Half a century had passed since Henry Martyn Leland breathed life into the first single-cylinder Model "A" Cadillac in the fall of 1902. Before the 1952 model year was out, Cadillac would have produced its 1,300,000th car. The Master of Precision and the father of interchangeability would undoubtedly have been proud to see how his "Standard of the World" had steadily ascended to its position of leadership among American fine-car manufacturers.

The company's 1952 models were proudly introduced as the "Golden Anniversary Cadillacs". Styling changes were minimal this year, but the 1952 Cadillacs boasted numerous important engineering improvements and innovations.

Cadillac's 1952 model count fell to its lowest level in many years. Only seven body styles were offered, in three series and on three separate wheelbases. The Series 61, which had replaced the LaSalle in 1941 and had been Cadillac's lowest-priced series for a decade, was discontinued halfway through the 1951 model year leaving the Series 62 Coupe the least-expensive car in the line.

There were four Series 62 Cadillacs for 1952, all on the same familiar 126-inch wheelbase. Body styles included a Two-Door Coupe, the Coupe de Ville Two-Door Hardtop, a sporty Convertible Coupe and the extremely popular Series Sixty-Two Four-Door Sedan. Again, there was just one Fleetwood Sixty Special, a luxuriously-equipped premium four-door sedan built on its own exclusive 130-inch wheelbase. The sumptuously-appointed Sixty Special was intended to be owner-driven. At the top of the 1952 Cadillac product line were the stately long-wheelbase Fleetwood Seventy-Five Sedan and Limousine. Cadillac's modestly-higher 1952 prices ranged from $3,542 for the Series Sixty-Two Coupe on up to $5,572 for the Fleetwood Seventy-Five Imperial Sedan.

One had to look very carefully to spot the minor 1952 styling changes. Up front, small, horizontal winged emblems replaced the mini-eggcrate grilles used on the 1951 grille, below the headlights. A new, wider "V" emblem and crest appeared on the front of the hood and on the rear decklid of all models except the Sixty Special, which got its Fleetwood script back for the first time in several years. The Series 62 Sedan had a higher rear deck which provided increased trunk space. Backup and turn signal lights were now tidily incorporated into the base of the fin-mounted taillights. The 1952 Cadillac was equipped with a new four-muffler dual exhaust system. Tailpipes terminated in narrow horizontal slits in the outer ends of the redesigned rear bumper.

While the Golden Anniversary Cadillacs showed little exterior change from the previous year, there were major mechanical advances under the sheetmetal. Horsepower of the 331 cubic-inch engine was increased a substantial 30 BHP, to 190 at 4000 RPM. The V-8 engine was now equipped with a Rochester four-barrel Quadra-Jet downdraft carburetor and improved manifolding. This four-barrel carburetor

Cadillac observed its 50th Anniversary in 1952, so the division's 1952 models were proudly promoted as the Golden Anniversary Cadillacs. With the price-leader Series 61 discontinued halfway through the 1951 model year, the Series 62 Coupe became Cadillac's lowest-priced car. The 1952 Cadillac Series Sixty-Two Coupe, Style 6237, had a factory-delivered price of $3,542. A total of 10,065 were delivered in the model year. Standard weight was 4,174 pounds. Styling changes were absolutely minimal: new winged emblems replaced the miniature eggcrate grilles below the headlights for 1952. This Series 62 Coupe is equipped with an accessory sun visor color-keyed to the car's roof color.

As in 1951, the sporty Cadillac Coupe de Ville Two-Door Hardtop outsold the less expensive Series 62 Coupe. These cars were almost identical in appearance. The Coupe de Ville had small, gold script at the base of the roof pillar; the Coupe did not. Factory advertised delivered price of the Series 62 Coupe de Ville for 1952 was $3,962 or a fairly hefty $420 higher than the Coupe. At 4,205 pounds the hardtop weighed only 31 pounds more than the Coupe. This Coupe de Ville, complete with spotless wide whites, is displayed in an attractive indoor rock garden setting at a show. Coupe de Ville interiors featured leather combinations in light matching tones. Cadillac delivered 11,165 Coupe de Ville Hardtops this year, a new high for this body style.

1952

setup was adopted for all large U.S. eight-cylinder passenger car engines for years to come.

An improved Dual-Range Hydra-Matic automatic transmission offered manual control of special third and fourth gear ranges for congested urban or open-highway driving. This automatic transmission was made standard equipment in all Series 62 Cadillacs as well as the Sixty Special. It was optional only in the big Seventy-Fives. Saginaw power steering was offered for the first time as an extra-cost option on all 1952 Cadillacs. The hydraulic power steering gear took over when three or more pounds of "pull" were exerted on the steering wheel. Cadillac claimed that its new power steering option eliminated as much as 75 per cent of normal steering effort.

Because of these numerous engineering advancements, Motor Trend Magazine presented Cadillac with a special "Car of the Year" Engineering Achievement Award.

There was no General Motors Motorama in 1952 for the second straight year. But Cadillac did commission one special dream car for the auto show circuit. It was a specially-modified Series 62 Convertible. This car had a sweeping panoramic windshield with vertical corner posts, vertical chrome simulated air intakes on the leading edge of the rear fenders and dual radio antennas built into the top of the tail fins. This special Cadillac show car was the direct predecessor of the first Eldorado convertible, which went into limited production the following year.

Also shown at auto shows this year was the Cadillac "Townsman", a specially-trimmed Fleetwood Sixty Special Sedan finished in a lustrous Nubian Black with a golden, fabric-covered roof.

War raged in Korea. Cadillac's big Cleveland Tank Plant was in full operation. While civilian passenger car production was not seriously affected, there were some shortages. Cadillac's 1952 sales literature listed optional white sidewall tires—"when available".

Cadillac delivered 90,259 cars in the 1952 model year. It was the first year in several in which the division's sales did not set new records.

The practical Series 62 Four-Door Sedan again far outsold all other 1952 Cadillac body styles combined. A total of 42,625 were built in the 1952-model production run. Base price was $3,962, and the Sixty-Two Sedan weighed 4,151 pounds. Series 62 Sedan interiors were upholstered in two-tone combinations of dark broadcloth and cord, or patterned cloth of a lighter shade. The upper door panels were trimmed in dark broadcloth in six-inch squares. These two-tone color options were available in combinations of gray, brown, green or blue. A pushbutton radio was a $112 option. Power steering was available on the 1952 Cadillacs for the first time.

The sporty Series Sixty-Two Convertible Coupe remained the glamour queen of the Cadillac line for 1952. Style 6267, the 1952 Cadillac Series 62 Convertible weighed 4,419 pounds and had a factory price of $4,110. An even 6,400 of these attractive five-passenger convertibles were built during the 1952 production run. It was the most expensive body style in the four-car Series 62 line. The Convertible Coupe's interior was upholstered in leather in either solid or two-tone color choices. Hydra-Matic automatic transmission became standard on all Series 62 models as well as the Sixty Special.

You had to look very closely, but there were more changes on the rear of the 1952 Cadillacs than there were up front. A new, wider Cadillac "V" emblem and crest appeared on the hood and rear deck of all 1952 Cadillacs except the Sixty Special, which resumed the use of the decklid Fleetwood script. Directional signals and backup lights were now incorporated into the base of the fin-mounted taillights giving the car a tidier rear-end appearance. Exhaust ports on the 1952 cars were horizontal slits in the outer ends of the rear bumpers. The Series 62 Sedan also got a higher rear deck line which increased trunk capacity. Note the two vertical chrome bars in the wraparound rear window, a Cadillac styling element that dated all the way back to the 1934 models. These would soon disappear.

1952

Except for minor cosmetic styling details, the Cadillac Fleetwood Sixty Special looked almost unchanged between 1950 and 1953. There were many significant mechanical improvements in the 1952 model, however, including a more powerful 190-horsepower engine; a standard equipment Dual-Range Hydra-Matic automatic transmission with manual control over third and fourth gears; a four-barrel carburetor and, for the first time, power steering. Styling changes included new wing emblems below the headlights in the outer ends of the grille and the return of the distinctive Fleetwood script on the rear decklid. Style 6019, the 1952 Cadillac Fleetwood Sixty Special sold for $4,269. Weight was 4,258 pounds. Sales of 16,110 Sixty Specials were 2,000 below the 1951 record for this model. The Sixty Special was built on its own 130-inch wheelbase and used a modified GM "C" body with longer rear fenders and deck.

The largest, heaviest and most expensive cars in the 1952 Cadillac model line were the impressive Fleetwood Seventy-Fives. Again, there were only two models in this expensive series, both on a 147-inch wheelbase and sharing the same long four-door "D" body shell. Surprisingly enough, an automatic transmission was still optional equipment on these luxurious cars. This is the 1952 Cadillac Series Seventy-Five Sedan for Seven Passengers, Style 7523, which sold for $5,360. The big Seventy-Five sedan weighed 4,699 pounds. Exactly 1,400 Series Seventy-Five Sedans were built in the 1952 model year along with 1,800 Seventy-Five Limousines with partition and glass division.

At the very top of Cadillac's seven 1952 model offerings was the regal Fleetwood Seventy-Five Imperial Sedan for Seven Passengers. A formal Limousine with a glass division between the front and rear compartments, the Imperial Sedan was intended to be chauffeur-driven. The front compartment had a fixed-position seat and was upholstered in leather. The division glass could be hydraulically raised or lowered. The rear passenger compartment, with two auxiliary seats which folded into the partition when not in use, was upholstered in plush broadcloth. The stately Fleetwood Seventy-Five Limousine had a factory price of $5,572 and weighed a hefty 4,734 pounds. An even 1,800 were built this year—400 more than the lower-priced Seventy-Five Sedan.

This was a most important year in Cadillac's history. The division observed its 50th anniversary in 1952. The 1952 models were proudly promoted as the "Golden Anniversary Cadillacs". Although styling changes were minimal, the 1952 models incorporated many significant engineering changes including a more powerful engine, power steering, four-barrel carburetor and an improved Dual-Range automatic transmission. Here, Cadillac Motor Car Division General Sales Manager James M. Roche, left, and General Manager Donald E. Ahrens pose with a fire-engine red 1952 Cadillac Series 62 Convertible Coupe on the Canadian side of the Detroit River. The photo was taken in Windsor, Ontario to get the Detroit skyline in the background. The big "50" was ornamented with 60 Cadillac "V" emblems and crests.

1952

Coachcraft of Los Angeles developed a very attractive station wagon conversion of the standard Cadillac Series 62 Sedan, and actually built and sold a few. The Coachcraft Caddie wagon was offered with either plain metal sides or as a high-styled "woodie". This is the wood-trimmed version, which featured modest paneling on the upper body between the beltline and the fender break line. Note the ventilator pane built into the rear door window and the airline-type rear cargo compartment drapes to shield passengers and luggage from the sun.

Cadillac built only one special show car for this year's auto show circuit. There was no 1952 Motorama. This modified Series 62 Convertible Coupe was the forerunner of the limited-production Cadillac Eldorado convertible which was introduced the following year. Special styling features on this Cadillac "dream" car included a sweeping panoramic windshield (which eventually appeared on all GM cars), vertical chrome fake air intakes on the leading edge of the rear fenders, dual radio antennas built into the tops of the tail fins; a modified rear bumper with big jet-pipe exhausts, and special wheel covers. The interior was specially trimmed, too, and the top cover boot fit almost flush with the rear deck. The production 1953 Eldorado had cut-down doors, a flush-fitting metal top cover and chrome-plated wire wheels.

American funeral car manufacturers showed considerable resourcefulness in designing special bodies for their funeral cars and ambulances. This 1952 Meteor-Cadillac coupe-style flower car is a good example. The Meteor Motor Car Co. of Piqua, Ohio used the entire upper roof section of a Series 62 Coupe to build this impressive flower car. The stainless steel deck could be hydraulically raised or lowered to accommodate various floral arrangements. Banked with flowers, these cars were truly impressive sights as they slowly led funeral corteges between church and cemetery. More often than not, they preceded a Cadillac-chassised hearse and a string of Fleetwood Seventy-Five sedans or limousines.

With the United States involved in a war in Korea, U.S. funeral car and ambulance builders received sizable orders for special ambulances and medical units for the armed forces. The Superior Coach Corp., of Lima, Ohio built this special 1952 Superior-Cadillac ambulance for the U.S. Air Force Medical Service. Note the blanked-out center side window and the spotlight in the roof above the windshield. This unit was painted Air Force Blue with regulation yellow lettering. Ambulances of this type were generally used at air bases, with much more utilitarian 4 x 4s in service on actual battle fronts.

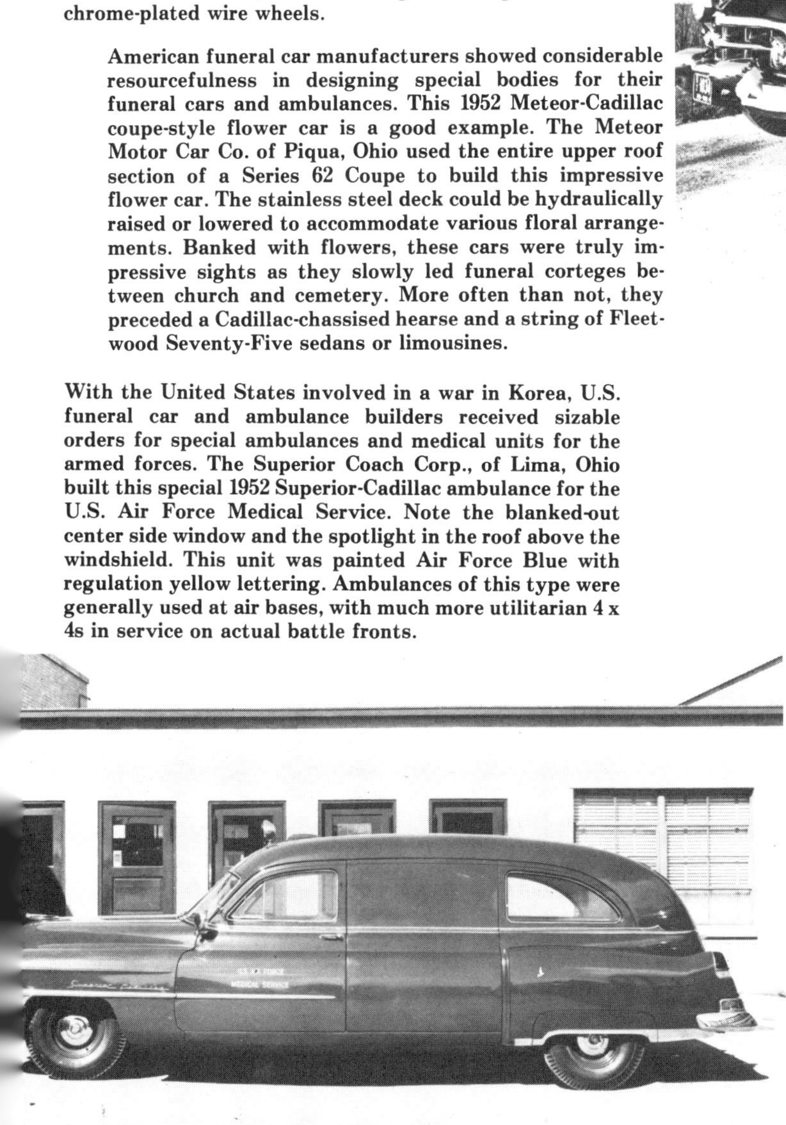

The Superior Coach Corporation of Lima, Ohio was now the second-largest builder of funeral cars and ambulances in the U.S., right behind the Henney Motor Co. Henney built exclusively on the Packard commercial chassis. Cadillac sold 1,694 Series 86 commercial chassis to funeral car and ambulance builders in the 1952 model year. This is a 1952 Superior-Cadillac End-Loading Limousine Funeral Coach. These coaches often fulfilled dual roles as combination funeral cars and ambulances. The window drapes could be quickly removed and casket rollers reversed to quickly convert this coach from a funeral car to an emergency ambulance.

1953

While styling remained basically unchanged for the fourth consecutive year, the 1953 Cadillacs boasted many significant engineering and equipment changes. There were luxurious new comfort and convenience options, and a spectacular new model joined the line.

Cadillac's 1953 model count was increased to eight with the surprise addition of a new, ultra-luxury convertible. This dazzling new model also saw the first use of a now-legendary Cadillac model name—Eldorado. Actually a limited-production version of the Series 62 convertible with several special styling features, the fully-equipped 1953 Cadillac Eldorado Convertible Coupe came with an equally impressive price tag: $7,750. The Eldorado cost nearly twice as much as the 1953 Series 62 Coupe and was priced more than $2,000 higher than even the Fleetwood Seventy-Five Limousine, traditionally Cadillac's most expensive production car.

The sporty new Eldorado's distinctive styling features included a low wraparound windshield like the one used on the 1952 show convertible; cut-down door sills that dipped at the ends of the doors, and a flush-fitting metal panel that completely concealed the convertible top when it was lowered, giving the car an exceptionally low and tidy look. The Eldorado stood only 58.8 inches high. Chromed wire wheels were standard equipment. Only 532 Eldorado convertibles were sold in the car's first year on the market. The new Cadillac Eldorado was featured at this year's GM Motorama and was one of the first "dream" cars that could be actually bought by the public, albeit for an appropriately stratospheric price.

The Eldorado was one of three limited-production convertibles introduced by General Motors during the eventful 1953 model year. Sister divisions Buick and Oldsmobile also got premium soft-tops with Eldorado-type styling. The most popular of this trio was Buick's Skylark, 1,690 of which were sold. Next came the Eldo, with 532 sales. Rarest of the three was the Oldsmobile Fiesta, only 490 of which were delivered. Chevrolet introduced its fiberglass-bodied, two-seat Corvette sports car this year, too.

All of the 1953 Cadillacs, including the Eldorado, were powered by the same 210-horsepower, 331 cubic inch V-8 engine. Horsepower was boosted a truly remarkable 32 per cent, or 50 horsepower, in one quantum leap from 160 to 210. Compression ratio went up to 8:25-to-one. Surprisingly, fuel economy was actually improved this year. All Cadillacs now had a 12-volt electrical system, a real necessity with the recent proliferation of power options. The most important new option available this year was factory air conditioning, a $619.55 option in all closed-body cars. Supplied by GM's Frigidaire Division, this system lowered interior temperature, and maintained it, at any level of comfort desired by the owner. The bulky evaporator unit and blower were mounted in the trunk.

Another new Cadillac option this year was an "Autronic Eye" system that automatically dimmed the headlights. The Autronic Eye unit itself was an outer space-type pod mounted on the left-hand side of the top of the instrument panel.

Chrome wire wheels were also a new Cadillac accessory this year. Price was $325 for a set of five, and they were available on all of the division's 1953 models.

Cadillac entered its second half-century with essentially the same styling it had used since 1950. The 1953 model year would be the fourth, and last, for this series of bodies. The most noticeable exterior change for 1953 was the replacement of the large bullet-shaped front and rear bumper guards with Cadillac's first massive grille "bombs" which doubled as bumper guards. Parking or fog lamps returned to their old locations in the outer ends of the grille, below the headlights. The lowest-priced car in the 1953 Cadillac line was the Series 62 Coupe, Style 6237, which had an advertised price of $3,571. Weight was 4,230 pounds. A total of 14,353 Series 62 Coupes were built this year. This photo was taken at the eastern entrance of the then relatively unique Pennsylvania Turnpike.

It was a neck-and-neck sales race again between the Series 62 Coupe and the considerably more expensive Series 62 Coupe de Ville Two-Door Hardtop. Both models were nearly identical in appearance. The Sixty-Two Coupe had fixed window pillars, and the Coupe de Ville had golden script calling out the model name at the base of the roof pillars. When the figures were tallied up at the end of the 1953 model run, the Style 6237-D Coupe de Ville had outsold the Coupe by 197 units—14,550 Coupes de Ville compared with 14,353 Series 62 Coupes. The 1953 model year was the Cadillac pillarless hardtop's fifth. The Coupe de Ville, at $3,994, was priced $423 higher than the Coupe and weighed 110 pounds more. The 1953 Coupe de Ville shown is equipped with optional fog lamps which fit into the parking light openings under the headlights.

1953

Typical 1953 Cadillac option prices included: vanity mirror, $1.85; pair of license plate frames, $4.28; white sidewall tires, $47.77; fog lamps, $36.91; Autronic Eye, $53.36; chrome wheel discs, $28.40; power window and seat regulators, $138.64; power steering, $176.98; automatic heating system, $119.00, and a signal-seeking radio with pre-selector, $120.60.

Cadillac's 1953 model lineup included five Series 62 body styles; a Two-Door Coupe; Coupe de Ville Two-Door Hardtop; Convertible Coupe; Four-Door Sedan, and the new highline Eldorado Convertible Coupe which came with a long list of standard-equipment features including whitewalls; wire wheels; power seat, windows and steering; signal-seeking radio; fog lights; automatic heating system; windshield washers, and license plate frames. There was still just one Fleetwood Sixty Special, and in the Fleetwood Seventy-Five series an Eight-Passenger Sedan and Imperial Limousine. Wheelbases were unchanged from the previous year.

The only noticeable styling change was the replacement of the bullet-shaped bumper guards with massive chrome "bombs". These huge horizontal pods were mounted in each side of the grille and doubled as effective bumper-and-grille guards. Smaller horizontal "bombs" were mounted above the rear bumper. The Sixty Special got broad, chrome trim on the lower edges of its rear fenders. All 1953 Cadillacs had golden-tone front and rear "V" emblems except the Sixty Special and Seventy-Fives which had Fleetwood script.

The 1953 GM Motoramas featured two new Cadillac experimental show cars. One was a small (for Cadillac) two-seater named the LeMans, which Cadillac described as a "sports prototype". This fiberglass-bodied, 115-inch wheelbase car was powered by a special 250-horsepower version of Cadillac's standard OHV engine, and was said to be a respectable performer on the road if not on the race track. The low-slung LeMans incorporated many styling features which showed up on the production 1954 Cadillacs, including its entire front end and grille. The LeMans had a prominent chrome air intake built into its rear fenders, a series of small vertical louvers on its rocker sills instead of a chrome molding, and turbine wheel covers.

The other 1953 Cadillac "idea" car was the Orleans, a forerunner to the Sedan de Ville pillarless four-door hardtop introduced three model years later, for 1956. The very stylish Cadillac Orleans featured center-opening doors. Buick and Oldsmobile introduced four-door hardtops during the 1955 model year. After its glittering New York opening, the General Motors Motorama went on the road this year, visiting six major cities and drawing more than one million people. The last one was held in 1961.

General Motors' passenger car production was seriously threatened by a totally-unexpected event this year. On August 12, 1953 the huge GM Hydra-Matic Division automatic transmission plant in the Detroit suburb of Livonia, Mich. was destroyed in a disastrous fire. Three employees died and damage exceeded $35 million—the largest industrial fire loss in U.S. history up to that time. The almost-new plant was leveled in a few hours, cutting off much of GM's supply of automatic transmissions.

Production experts were flown into Detroit from all over the country. Repairable machinery was salvaged from the still-smoking ruins. The corporation made emergency arrangements to lease space from Kaiser Motors in a former bomber plant near Willow Run, Mich. Meanwhile, plans were made to hurriedly adapt other GM transmissions for temporary use in various GM car lines, including Cadillacs. Consequently, for the last few months of the 1953 model year, Cadillacs were equipped with Dynaflow automatics supplied by Buick. Oldsmobile also used Buick Dynaflows in some of its late-production 1953 models. Chevrolet supplied some Powerglide automatics to Pontiac. But by the time the 1954 models went on sale, Hydra-Matic was back in business. This disaster had a traumatic effect on GM and on the entire industrial world.

The Packard Motor Car Co. also introduced a stunning, new limited-production convertible this year, the Packard Caribbean, as well as a series of long-wheelbase limousines with Henney bodies. In England, Frederick S. Bennett drove his 1903 single-cylinder Cadillac in a re-enactment of the 1903 Thousand-Mile Trial. Even though Bennett was more than 80 years old and the Cadillac—the first one imported—had more than 200,000 miles on it, the venerable one-lunger finished the run with a respectable average of 21.2 miles an hour.

Cadillac built 109,651 cars in the 1953 model year, nearly 20,000 more than in 1952 but slightly below the 1951 record of 110,340 cars.

This was a big year for Cadillac convertibles. Not only were sales of the Series 62 Convertible their highest ever, but Cadillac's soft-top offerings were enhanced by the addition of an extremely expensive new sports model—the 1953 Cadillac Eldorado Convertible. This is the standard Series 62 Convertible, Style 6267-X, which sold for $4,143. Weight was 4,500 pounds. A record 8,367 Series 62 Convertibles were sold in the 1953 model year, eclipsing the previous record 8,000 sold in 1949. The Series 62 Convertible shown is equipped wtih Cadillac's new, optional chrome wire wheels. Seven new leather interior choices in solid or two-tone combinations were offered in the Series 62 Convertible this year. Note how high the door sills look when compared with the new Eldorado.

1953

Still the most popular of all Cadillac body styles was the four-door, five-passenger Series 62 Sedan. Factory delivered price was $3,666 and the 1953 Sixty-Two Sedan weighed 4,225 pounds. Cadillac's dealers delivered 47,316 of these practical, roomy four-door family type sedans which were listed as Style 6219. The famous Cadillac "V" emblem was finished in a golden shade this year, front and rear, instead of the usual chrome. The most visible exterior styling change this year was the two huge chrome "bombs" built into the outer ends of the grille opening. These big, bright bullets provided more than adequate protection for the front end.

Sales of the distinctive Cadillac Fleetwood Sixty Special soared to a new record in 1953. Cadillac's dealers delivered an even 20,000 of these expensive premium sedans this year. The 1953 Fleetwood Sixty Special was given modest styling changes, including a revised grille design and broad chrome moldings on the lower edges of the rear fenders. The eight small vertical chrome louvers which set the Sixty Special apart from all other Cadillac models since 1950 were retained. Style 6019, the 1953 Sixty Special had a factory-delivered price of $4,304. Weight was 4,415 pounds. Air conditioning was optionally available in all Cadillacs this year for the first time. Note the small A/C air scoops at the base of the roof.

A new styling option which proved quite popular with Cadillac's customers this year was sporty chromed wire wheels. These bright chrome wheels, which added 30 pounds to the weight of the car, were available on all 1953 Cadillacs. Price was $325 for a set of five. They were standard equipment on the impressive new Eldorado Convertible. This is a 1953 Cadillac Fleetwood Sixty Special Four-Door Sedan equipped with these accessory wheels, posing at the 1972 antique car meet at Hershey, Pa. Note the extensive use of chrome trim on this model, which still had its own exclusive 130-inch wheelbase.

A legendary new model name joined Cadillac's 1953 model lineup—Eldorado. A Spanish word that translates to "the golden one", the Eldorado name would henceforth be reserved for some of Cadillac's most exclusive and expensive cars. The Eldorado name was first used on a special limited-production Series 62 convertible introduced during the 1953 model year. The dazzling, new Eldorado was a feature attraction at GM's 1953 Motoramas. But what made this car different from previous "dream" cars was that this one could actually be bought—for a whopping $7,750! Only 532 Cadillac Eldorado Convertible Coupes were built in this model's spectacular first year on the market. The swank, new Eldorado cost nearly twice as much as a Series 62 Coupe and $3,607 more than the standard Series 62 Convertible.

Cadillac's prestigious new Eldorado Convertible Coupe, Style 6267-S, had several distinctive styling features. These included a daring new wraparound windshield like the one used on the 1952 show car, cut-down, dipped door sills and a flush-fitting metal cover which completely concealed the convertible roof when it was lowered. Chrome wire wheels were also standard. For a very steep $7,750 one could now buy a genuine GM Motorama "show car". Only 532 of these stunning convertibles were built in the Eldorado's first year making it among the most desirable of all postwar Cadillacs. Weight was 4,800 pounds—300 more than the standard Series 62 convertible which looked rather plain in comparison. This is how the 1953 Eldo looked with its top up.

Cadillac's spectacular new Eldorado Convertible was one of four special limited-production convertibles introduced by General Motors during the 1953 model year. Buick came out with a Skylark Convertible, Oldsmobile with a Fiesta Convertible and Chevrolet introduced its fiberglass-bodied Corvette sports car. The product planners on West Grand Blvd. actually called the hefty Eldorado, Skylark, and Fiesta "sports cars," which was stretching the point: the two-place Corvette with its Blue Flame Six barely even made it. But the Cadillac Eldorado got worldwide exposure when U.S. President-Elect Dwight D. Eisenhower rode in one in his inaugural parade in Washington, D.C. in January, 1953. Here, "Ike" stands up and waves at the crowds lining the parade route.

1953

With the introduction of the exclusive new Eldorado Convertible, the Fleetwood Seventy-Fives were no longer the most expensive models in Cadillac's eight-car product line. The 1953 Cadillac Fleetwood Seventy-Five Sedan for Eight Passengers, Style 53-7523, had a factory-advertised price of $5,604—more than $2,000 lower than the 1953 Eldorado. The 1953 Fleetwood Seventy-Fives got the redesigned grille with prominent chrome bumper "bombs" but was otherwise unchanged in appearance for the fourth straight year. But there were numerous mechanical and equipment changes, including a boost in horsepower to a very potent 210 from 160 in 1952. Cadillac delivered 1,435 Series Seventy-Five Sedans this year.

Cadillac built only 765 Fleetwood Seventy-Five Imperial Sedans in 1953, less than half as many as in the previous year. The companion Series Seventy-Five Sedan outsold the limousine by a margin of two-to-one. The Eight-Passenger Imperial Sedan, or Limousine, had a 1953 factory price of $5,817 and weighed 4,900 pounds. Style 7533, its appearance was basically unchanged since 1950. The small chrome-edged rear window was a Fleetwood Seventy-Five styling element for many years, befitting this formal body type. A glass division was standard on the Series Seventy-Five Imperial. The horizontal rear bumper "bombs" were scaled-down replicas of those built into the massive 1953 Cadillac grille. This is an aft view of the author's 1953 Imperial Limousine, which was originally owned by the Eaton's Department Store family of Toronto.

Two new Cadillac show cars were unveiled at the 1953 General Motors Motoramas, which went on the road for the first time this year after a glittering New York opening. One of these "dream" cars was the 1953 Cadillac LeMans, which GM billed as a "sports prototype". Built on a short 115-inch wheelbase, the LeMans had a fiberglass body and was powered by a 250-horsepower Cadillac V-8 with 9-to-1 compression ratio. This show car incorporated many styling features, including its entire front end, which showed up on the 1954 Cadillacs. The rear end of this LeMans was restyled with cleaner tail fins the following year. Factory records indicate that at least three of these little two-seaters were actually built. One was sold to a Mr. Floyd D. Akers of Washington, D.C. in June, 1955. Another was built for a Mr. Goodman of Fisher Body, and yet another was turned over to GM's Engineering Department. Despite its heavy appearance, the LeMans was a lively performer.

Cadillac's other 1953 show car was the Cadillac Orleans, a forerunner of the next major innovation in U.S. passenger car body styles, the pillarless four-door hardtop. Buick and Oldsmobile both introduced four-door hardtops during the 1955 model year, but Cadillac's version—the Sedan de Ville—would not make its debut until Model Year 1956. The stylish Cadillac Orleans show car was a much-modified Series 62 Sedan with a redesigned roof and center-opening doors. It was equipped with the production 1953 Cadillac's new chrome wheel discs.

Prosperity had returned to the land, and with it a demand for something special—something really out of the ordinary in personal transportation. What more exclusive could you drive than a Cadillac station wagon, especially since the factory in Detroit didn't even offer one? Well-known industrial designer Brooks Stevens executed this nicely-done Cadillac station wagon, which had full nine-passenger seating. Note the rear wood paneling and the curved upper window line which is notched at the rear, with triangular corner window panes.

1953

The Cadillac Motor Car Division shipped just over 2,000 long-wheelbase commercial chassis this year. One of these was used to build this interesting 1953 Meteor-Cadillac Flower Car. Note the two-door coupe roof and rear window. The upper roof quarters have been blanked out and ornamented wtih small landau bows. This coupe-style flower car has a rear well instead of an adjustable deck. It was photographed in front of the Meteor Motor Car Co. offices in Piqua, Ohio.

Cadillac's special long-wheelbase commercial chassis was given a new series designation in 1952. It was now listed as Model 8680-S. The Detroit Plant turned out 2,005 of these 157-inch wheelbase hearse and ambulance chassis during the 1953 model year. This is a 1953 Superior-Cadillac Automatic Side-Servicing Landaulet which sold for $7,906 at the factory in Lima, Ohio. It is equipped with Cadillac's new chrome wheel discs. Note the tasseled draperies and the funeral firm's nameplates in the rear side window. A well-equipped hearse of this era also had a "Funeral Coach" sign mounted on the top of the dash.

King Ibn Saud of Saudi Arabia had become a good Cadillac customer. Two years earlier he had ordered a fleet of 20 custom-built Cadillacs to transport his many wives in royal comfort. But for really important state occasions, the Palace rolled out King Saud's special parade car, a monstrous four-door convertible sedan designed and built by Saoutchik of Paris. This 24-foot-long parade phaeton was built on a stretched 1953 Cadillac chassis. Features included fixed glass division pillars, bodyguard running boards which extended out from the sides of the car and the royal crest on the grille and hood. Flag masts are mounted on each front fender. Note the standard 1953 Cadillac wheel covers.

The latest innovation in rescue ambulances was the high-headroom body style. Although custom high-headroom ambulances had been built before, Hess and Eisenhardt, builder of S & S funeral cars and ambulances, introduced a production high-headroom ambulance in 1950. Within a year or two, most major ambulance builders had also added them to their lines. This is a 1953 Superior-Cadillac Rescue Ambulance. The upper roof was a huge, single piece of fiberglass. Note the roof vent and built-in tunnel type red warning lights. This professional high-body ambulance was built for the Milwaukee County Institutions.

1954

The totally-new 1954 Cadillacs received the Cadillac Motor Car Division's first complete restyling in four years. Cadillac's impressive new look was pure 1950s—longer, lower and wider. Wheelbase was increased three inches on all three familiar series—Sixty-Two, Fleetwood Sixty Special and the aristocratic Fleetwood Seventy-Five.

Cadillac's fresh, new 1954 styling featured GM's bold, "panoramic" wraparound windshield, just like the one used on the 1952 experimental show car and on the limited-production 1953 Eldorado. All of the 1954 Cadillacs had high, slab sides; broad, flat rooflines and long, wider rear decks. Coupe rooflines terminated in a delightfully graceful Florentine curve at their bases and had wide wraparound rear windows which complemented the new windshield. A full-width cowl ventilator intake was built into the base of the windshield.

The 1954 cars also had a new grille design that introduced Cadillac's famous "dollar grin". This new grille had distinctive upper and lower sections and was Cadillac's first integrated bumper/grille. It incorporated Cadillac's first real "Dagmars" — big upswept chrome bullets which took their colorful nickname from a well-endowed television actress of the day. These massive horizontal chrome bullets doubled as ornaments and functional grille guards. The grille itself had a fine, horizontal "eggcrate" texture that would distinguish Cadillac front ends for the next four years.

Once again there were a total of eight body styles. There were five in the very comprehensive Series 62 range, all on the same new 129-inch wheelbase.

The lowest-priced Cadillac offered for 1954 was the Series 62 Coupe, which was virtually indistinguishable from the more expensive Coupe de Ville, as both were now pillarless two-door hardtops. The only way to tell a 1954 Coupe de Ville from the Series 62 Coupe of the same year is by the tiny gold "Coupe de Ville" script on the sill below the rear quarter windows. Next came the standard Series 62 Convertible Coupe and the ever-popular Sixty-Two Sedan, the best-selling Cadillac of them all.

The 1954 Eldorado Convertible Coupe was much less distinctive than the original 1953 version. It utilized the same body shell as the standard 62 convertible, but came loaded with comfort and convenience equipment that cost extra on other models. Special 1954 Cadillac Eldorado styling touches included broad, ribbed metallic panels on the lower rear fenders and gold Cadillac crests on the door sills, lower rear fenders and wheel covers. Chrome wire wheels were also standard. The Eldorado's price was reduced $2,012 this year and 2,150 were built — nearly four times as many as the previous year.

There was still just one Fleetwood Sixty Special, a six-window, four-door sedan that came with a long list of standard luxury features. Wheelbase was back to its traditional 133 inches, last used in 1949. At the top of the 1954 Cadillac model lineup were the Fleetwood Seventy-Five Eight-Passenger Sedan and Imperial Sedan, which also got Cadillac's attractive new styling. Wheelbase was increased to 149-¾ inches on these impressive cars.

The horsepower race was on. The 1954 Cadillac engine put out 230 brake horsepower at 4400 RPM, 20 more than in 1953. Compression ratio remained at 8:25-to-1. Displacement was still 331 cubic inches, the same as when this engine was introduced six years earlier.

There were some other important mechanical changes and improvements this year, too. Power steering and windshield washers became standard equipment on all 1954 Cadillacs. Bendix Hydrovac power brakes were optionally available for the first time, as was a four-way adjustable power front seat. There were refinements in the Saginaw power steering,

The 1954 Cadillacs were totally new and marked the division's first complete restyling since 1950. Cadillac's new look was longer, lower and wider. It was difficult to distinguish the 1954 Series 62 Coupe from the more expensive Coupe de Ville, as both were now pillarless two-door hardtops. This is the 1954 Cadillac Series Sixty-Two Coupe, the lowest-priced car in the 1954 Cadillac line with a factory advertised price of $3,837. Standard shipping weight was 4,347 pounds. A total of 17,460 Series 62 Coupes, Style 6237, were sold this year outselling the companion Coupe de Ville by 290 units. Note the graceful Florentine curve in the rear roofline.

The completely restyled 1954 Cadillac featured the division's first combined bumper/grille. Principal elements included a fine "eggcrate" texture, integral parking/turn signal or fog lamps and a pair of upswept bumper guard "bullets". This new grille design featured an upper and lower section separated by a bar in the center. This is the attractive 1954 Cadillac Series Sixty-Two Coupe de Ville, Style 6237-D, which sold for $4,261 and weighed 4,409 pounds. Cadillac's dealers sold 17,170 of these two-door hardtops. Note the elegant Florentine curve in the rear roofline. The only way to tell the 1954 Coupe de Ville from the lower-priced Series 62 Coupe was the small gold script on the rear window sill and the more luxurious interior in the Coupe de Ville.

1954

including a lower steering ratio which meant less wheeling in tight turning and parking situations. Improvements were made in the GM Frigidaire air conditioning system which was optionally available in all closed models but was still unavailable in Cadillac convertibles.

Cadillac exhibited a record three new experimental show cars at the 1954 General Motors Motorama. After the usual gala New York opening, this industrial extravaganza was again taken on the road and repeated in Miami, Los Angeles, and San Francisco. Cadillac's advanced-concept "dream" cars included the swank Park Avenue, a four-door luxury personal sedan; the El Camino, a small two-passenger Coupe, and the La Espada, a companion two-seat Convertible Roadster.

The smartly-styled Park Avenue was the direct predecessor of the very expensive Cadillac Eldorado Broughams built in small numbers in 1957 and 1958. The El Camino and La Espada had fiberglass bodies and were built on short 115-inch wheelbases, compared with 133 inches for the Park Avenue. The product planners on Clark Avenue were beginning to hear rumors about an ultra-expensive new Lincoln specialty car said to be on the drawing boards in Dearborn. Suddenly, the promising Park Avenue "dream" car project assumed great importance, and urgency. The Ford Motor Company formally announced its Continental Mark II program in October for introduction the following year. The scramble for automotive exclusivity at any cost was on.

Cadillac opened a new three-story plating building at its main plant complex in Detroit this year. A total of 96,680 cars were built in the 1954 model year, not a record, but the division had quietly produced its 1.5 millionth car.

Intent on garnering a larger share of the booming American luxury car market, Chrysler Corporation registered its Imperial as a separate nameplate. The Imperial now stood alone; Chrysler up to now had marketed its most expensive cars as Chrysler Imperials.

While the "Big Three" automakers prospered, the sun was slowly setting on floundering Packard. A dynamic new president, James Nance, had been brought in the previous year to clean house. In 1954, Packard merged with Studebaker. It was all downhill from there.

The 1954 Cadillac model lineup again included two convertibles, the standard Series 62 Convertible Coupe and the considerably more expensive Eldorado. This is the standard 1954 Series 62 Convertible, Style 6267, which sold for $4,404. A total of 6,310 were built this year. The example shown is equipped with optional chrome wire wheels. All 1954 Cadillacs had GM's bold, new panoramic wraparound windshield with vertical "dog leg" side pillars that intruded slightly into the front passenger compartment. Cadillac's trademark tail fins were higher and more squared-off now. This same basic styling would be used for the next three years.

The premium Cadillac Eldorado Convertible was no longer the most expensive car in the line. For 1954, it was little more than a gussied-up Series 62 Convertible. But it still cost $1,300 more than the Sixty-Two soft-top—$5,738 compared with the 62's $4,404 base price. The 1954 Eldo, however, was priced more than $2,000 lower than the more distinctive 1953 Eldorado. Special styling touches on the Style 6267-S Cadillac Eldorado for 1954 included broad, ribbed chrome pannels on the lower rear fenders, fiberglass convertible top cover, gold Cadillac crests on the door sills and rear fenders and a lavishly-appointed and equipped leather interior. Chrome wire wheels were also standard. Buyers could choose between four special Eldorado colors: Aztec Red, Azure Blue, Alpine White, or Apollo Gold. A total of 2,150 were built—four times as many as in 1953.

Again in 1954, the six-passenger, four-door Series 62 Sedan topped all other Cadillac body styles in sales. A total of 33,845 were produced this year. Style 6219, the Series 62 Sedan had a factory-advertised price of $3,932 and weighed 4,330 pounds. Styling was all new with slab sides and broad, flat roof and rear deck lines. All 1954 Cadillacs had the same style of vertical chrome "air intake" molding extending from the window sill to the rocker molding, a continuation of the exterior trim theme used on the 1950-53 models. Note the air intake vent built into the base of the windshield. All eight 1954 Cadillacs were powered by the same uprated 230-horsepower, 331 cubic inch V-8 engine, and all five Series 62 models rode on a three-inch longer 129-inch wheelbase.

1954

Like all of the other 1954 Cadillacs, the distinctive Fleetwood Sixty Special was completely redesigned and restyled. But somehow it came off looking much like it had for the previous four years. The eight small chrome louvers on the lower rear doors were carefully retained, but otherwise this premium four-door sedan looked nearly identical to the Series 62 Sedan. For the first time since 1949, the Sixty Special was built on its traditional 133-inch wheelbase, three inches longer than the one used from 1950 to 1953. There was still only one Sixty Special body style, a very well-equipped four-door, six-window sedan. Factory price was $4,683 this year. Sales fell off significantly this year, from 20,000 in 1953 to 16,200. This model still carried the style designation of 6019.

Cadillac unveiled a record three experimental show cars at the 1954 GM Motorama. Two of these were small two-seaters, but the one that had the greatest impact on Cadillac's future product plans was the Park Avenue, a four-door luxury sedan that ultimately evolved into the ultra-luxury Cadillac Eldorado Broughams of 1957 and 1958. The 1954 Cadillac Park Avenue Sedan was built on the same 133-inch wheelbase as the production Fleetwood Sixty Special but had an over-all length of 230.1 inches, which was three inches longer. The Park Avenue was powered by the same 230-horsepower V-8 engine used in all 1954 Cadillacs. The smart Park Avenue introduced many styling elements eventually used on the next-generation 1957 Cadillacs, including the notched roofline, large front-wheel cutouts and "reversed" tail fin shape. After exhibition at numerous auto shows all over the country, the Park Avenue was turned over to Cadillac's Engineering Department for experimental purposes in April, 1955.

With the Eldorado downgraded and lowered $2,000 in price, the Fleetwood Seventy-Five Sedan and Limousine once again became the most expensive, and exclusive, models in the Cadillac product line. Like all other 1954 Cadillacs, the Series Seventy-Fives were completely restyled. Rooflines continued the classic 1950-53 profile, but everything else was new. This is the 1954 Cadillac Fleetwood Seventy-Five Sedan for Eight Passengers, Style 7523, which had a factory-advertised price of $5,874. Only 889 were built. Note the air conditioning intake scoops at the base of the roof, and the "Autronic Eye" automatic headlight dimmer visible through the windshield.

The largest, heaviest and most expensive car in Cadillac's totally-restyled 1954 model lineup was the formal Fleetwood Series Seventy-Five Imperial Sedan for Eight Passengers. Style 7533, the Fleetwood Imperial Limousine had a fixed partition with adjustable division glass and was designed for chauffeur operation. The imposing Cadillac Fleetwood Limousine weighed just over 5,000 pounds and had a starting price of $6,090. Only 611 were built this year. Total Series Seventy-Five production for 1954 was an even 1,500 cars.

Cadillac's other two 1954 idea cars were a pair of small two-seaters, one a coupe and the other a convertible. The coupe, shown here, was the Cadillac El Camino which was built on a 115-inch wheelbase, the same as the 1953 LeMans show car. Over-all length was 200 inches. The El Camino was powered by the same 230-horsepower V-8 used in all 1954 Cadillacs. The same style of rear fender tail fin showed up on the 1955 Cadillac Eldorado Convertible, the 1956 Eldorado Seville Hardtop and across the line on the 1958 models. The El Camino and La Espada were the first Cadillacs that had quad headlights.

1954

Cadillac's trio of 1954 show cars included two small two-seaters. One was the El Camino Coupe and the other was a convertible roadster named the La Espada. Both had fiberglass bodies and rode on 115-inch wheelbases with an over-all length of 200.6 inches. This is the 1954 Cadillac La Espada Sports Convertible. Note the metallic trim panels which flowed from the front wheel openings into the middle of the doors. These cars had sharply-raked 60-degree windshields and daggerlike tail fins that showed up later on production Cadillacs. Show cars like the La Espada, El Camino and Park Avenue were actually rolling laboratories which conditioned future customers to new styling concepts planned for the near-future. The El Camino name was later used on a 1959 Chevrolet pickup based on that year's passenger car, but the La Espada nameplate never made it onto a production GM vehicle.

The pillarless hardtop was the hottest thing in U.S. passenger car body styles during the Fabulous Fifties. General Motors had introduced its two-door hardtop coupe in 1949. The 1953 Cadillac Orleans show car was the first four-door hardtop sedan built in America. Buick and Olds introduced four-door hardtops in the 1955 model year and Cadillac came out with its Sedan de Ville Hardtop Sedan the following year. American funeral car and ambulance builders knew a good thing when they saw it, so it came as no surprise when the Superior Coach Corp. of Lima, Ohio announced the first pillarless hardtop hearse. Like a true hardtop, it had no "B" or even "C" pillar. This model actually went into production in 1955 and 1956 as Superior's "Beau Monde" and was available as a funeral coach, flower car or combination. This is the one-off 1954 prototype configured as a floral coach.

The long, low coupe-style Flower Car still enjoyed great popularity among American funeral directors. This is a deck-type flower car built on the 1954 Cadillac commercial chassis by the Meteor Motor Car Co of Piqua, Ohio. Meteor and other funeral car builders offered two styles of flower cars—a "western" type with low rear deck, and an "eastern" version favored in the northeastern part of the country. The "eastern" model had a higher rear deck to accommodate a casket in the stainless steel compartment below the flower deck. This is Meteor's "western" model, which sold for $7,308. The "eastern" style flower car cost $7,705. Cadillac supplied 1,635 Series 8680-S commercial chassis this year.

The wraparound rear window look was "in". Not even the funeral car and ambulance manufacturers could resist this latest innovation in automobile styling. The Hess and Eisenhardt Co., builders of S & S hearses and ambulances, introduced full wraparound rear window styling on its 1953 models. The other major professional car builders quickly followed. This is a 1954 Meteor-Cadillac Rear-Loading Limousine Funeral Coach with full-view styling. All U.S. funeral car builders redesigned their products this year for compatibility with Cadillac's all-new 1954 styling.

1955

Cadillac passenger car sales rocketed to a new all-time high this year as the U.S. automobile industry shattered all previous production and sales records.

The 1955 Cadillacs—except for the Eldorado—got modest styling changes, but horsepower and compression ratios were again increased and the restyled Eldorado Convertible got its own, special high-performance engine.

The 1955 Cadillac model lineup consisted of the same eight familiar body styles on three different wheelbases. Again, there were five Series 62 models on the same 129-inch wheelbase. These included the Series 62 Two-Door Hardtop Coupe, the lowest-priced car in the line; the more luxurious Coupe de Ville Two-Door Hardtop; the standard Sixty-Two Convertible Coupe; the popular Series 62 Four-Door Sedan and the exclusive Eldorado Convertible Coupe, which was still a member of the Series 62 family.

The 1955 Cadillac Eldorado got dramatic new rear-end styling this year. Gone were the traditional Cadillac fishtail fins, replaced by a pair of sweeping, pointed tail fins right off a rocketship. This very flashy Cadillac convertible also had full rear wheel cutouts without fender skirts. Round brake, turn signal, and backup lights were paired in clusters at the base of the long tail fins. The Eldorado name appeared above a gold "V" emblem on the rear deck.

Other special Eldorado styling touches included metallic appliques on the upper door and quarter window sills and new standard equipment "Sabre Spoke" cast aluminum and steel wheels, instead of the chromed wire wheels which had distinguished the Eldorado since it was introduced in 1953. The 1955 Eldo also got its own, exclusive hopped-up engine this year, too. It was a 270-horsepower version of Cadillac's 331 cubic inch V-8 equipped with dual four-barrel carburetors. This high-performance engine was optionally available in other 1955 Cadillacs. The Eldorado's price went up only incrementally this year, while most other Cadillac model prices were reduced. A record 3,950 Eldorado Convertibles were sold in the car's third year on the market.

The premium Cadillac Fleetwood Sixty Special Four-Door Sedan got minor styling changes for 1955. Both the Sixty Special and the Series 62 Four-Door Sedan now had the elegant Florentine curve rear rooflines previously used only on two-door coupes. The Sixty Special's distinctive vertical chrome louvers were repositioned from the bottom of the doors to a new location at the extreme ends of the rear fenders. There were 12 of these chrome accents now instead of eight, and a wide, chrome molding appeared on the lower edge of the rear fenders between the fender skirt and the rear bumper. The Sixty Special continued on its own 133-inch wheelbase in a single lavishly-equipped four-door sedan model.

The Fleetwood Seventy-Five Sedan and Limousine carried over their fresh 1954 styling but got Cadillac's revised 1955 grille design. The new grille had a heavier horizontal eggcrate texture and two longer, single-piece chrome bullets joined by a V-shaped center blade. Series 62 and Sixty Special models got six short vertical chrome louvers, three on each side of the rear license plate holder just below the deck lid. These new trim pieces made the rear of the car look wider. A remote-control deck lid release was added to Cadillac's options list this year.

The standard engine in all 1955 Cadillacs, except Eldorado, was a 250-horsepower, 331 cubic inch V-8. Horsepower was boosted 20 from 1954 and compression ratio went up to 9.0-to-1. Manifolding was also improved and despite the greater horsepower, fuel economy was actually improved again. Hydra-Matic automatic transmission became standard equipment on all Cadillacs. A quarter of a century would pass before a stick-shift Caddy was again available. Gray-tinted "E-Z Eye" glass replaced the green-tinted band at the top of Cadillac's wraparound windshield.

Cadillac showed three more all-new experimental show cars at the 1955 General Motors Motoramas. One

The 1955 Cadillacs showed little outward change from the all-new 1954 models. Styling modifications included a heavier eggcrate grille texture and larger, single-piece front bumper bullets. Series 62, Eldorado and Sixty Special models had a new chrome side molding treatment that swept up vertically at the beginning of the rear fenders. The lowest-price car in Cadillac's 1955 line was again the Series 62 Coupe, Style 6237, which sold for $3,568—$269 less than the comparable 1954 model. Most 1955 Cadillacs went into the showrooms with lower prices. Sales of the Sixty-Two Coupe increased sharply this year, to 27,879 from 17,460 in 1954.

The stylish Cadillac Coupe de Ville was gaining on the Series 62 Sedan in popularity among Cadillac's customers. Sales nearly doubled this year, from 17,170 in 1954 to 33,000. This model's $3,964 factory-advertised price was $297 lower than the previous year, which might have had something to do with it. Style 6237-D, the Coupe de Ville weighed 4,424 pounds, 66 more than the less expensive Series 62 Hardtop. The Series 62 Coupe and Coupe de Ville were almost indistinguishable, except for tiny gold "Coupe de Ville" quarter window sill script and upgraded interior trim in the Coupe de Ville.

1955

was the prototype of the ultra-luxury Eldorado Brougham which went into limited production in 1957 and 1958. The other two were less practical sports models which resurrected, at least temporarily, the warmly-remembered LaSalle name. All that these three idea cars had in common were their quad headlight units which became standard throughout the industry three years later.

Like the 1954 Park Avenue, the prototype Eldorado Brougham was a sleekly-styled pillarless four-door town sedan of generally smaller exterior proportions than current-production Cadillacs. The elegant Brougham stood only 54 inches high, seven inches lower than standard models and at 210 inches from bumper to bumper was a full 13 inches shorter. It had center-opening doors and a brushed, stainless steel roof—a favorite styling feature of GM styling chief Harley Earl that made it onto the production Broughams. The interior had specially-designed lounge seats and among other personal luxuries, a vanity case. The brochure handed out at the auto shows where the Eldorado Brougham was exhibited promised that the car would be placed into limited production the following year. It didn't make it, although Cadillac came up with a spectacular open-front Brougham Town Car to keep the public interested.

The two small sports cars included a two-seat convertible roadster and a four-door hardtop sedan, both called LaSalle II. These advanced idea cars were powered by experimental V-6 engines. In addition to these three "dream" cars, Cadillac exhibited a series of specially trimmed and painted standard production cars for the 1955 auto show circuit. These feature cars included the "St. Moritz", a customized Eldorado Convertible; a "Celebrity" Coupe de Ville Hardtop, and the "Westchester", a close-coupled limousine with padded, black vinyl roof and division glass based on the standard Sixty Special Sedan. These cars had exotic paint jobs and special interior trim. But it was the Brougham that really captured the public's fancy.

The Ford Motor Co. unveiled its long-awaited $10,000 Continental Mark II at the Paris Auto Show in October, 1955. Cadillac's response would not be ready for another two years—at precisely the same time the 3,000-car Mark II program ended.

Cadillac's advertising at this point in the division's history exuded status and snob appeal. The cars were portrayed against rich, velvety backgrounds and glittering jewels from Tiffany's, Van Cleef and Arpels, and Harry Winston's. A Cadillac motor car was irrefutable evidence that one was conspicuously successful, or had truly "arrived." One high-toned magazine ad depicted a stern corporate boardroom scene. The simple underline read: "A Meeting of . . . Cadillac Owners!" Cadillac's advertising, created by the Detroit agency of McManus, John and Adams, was enormously successful. A fire-engine red Cadillac convertible was the consolation prize on television's infamous "$64,000 Question" quiz show, prompting service station attendants everywhere to ask anyone who happened to drive up in one, "What question did you miss, stupid?"

Cadillac sold a record 140,777 cars in the memorable 1955 model year. Calendar year production exceeded 150,000 for the first time. The U.S. auto industry built an all-time record 9.2 million vehicles in 1955, including 7.9 million cars. The Packard Motor Car Company introduced its last line of all-new cars and finally offered a V-8 engine. In a nostalgic tribute to its proud heritage, Packard created a beautiful show car called the Request. Its principal styling feature was a narrow vertical grille reminiscent of the classic Packard radiator shape of two decades earlier. The next year would be Packard's last.

Cadillac's 1955 model lineup again included two convertible coupes, the standard Series 62 Convertible and the much flashier, more expensive Eldorado. Factory-advertised price of the 1955 Series 62 Convertible was reduced $327, from $4,404 in 1954 to $4,097. Sales went up significantly, from 6,310 in 1954 to 8,150 in 1955. The Series Sixty-Two Convertible weighed 4,627 pounds. All Series 62 Cadillacs and the Sixty Special had three vertical chrome louvers on either side of the license plate, below the rear decklid. The massive 1954-55 rear bumper incorporated large, round exhaust ports which jutted out at the base.

Unlike the rest of the 1955 Cadillacs, the Eldorado Convertible was given major styling changes. The 1955 Eldorado was the first Cadillac to abandon the famous fishtail fins, which had highlighted this marque's styling since 1948. The 1955 Eldo got boldly-styled new "rocketship" tail fins which eventually appeared on all Cadillac models, in 1958. The distinctive 1955 Eldorado rear end also had larger rear wheel openings without fender skirts, and brake and back-up lights housed in dual pods built into the base of the pointed tail fins. New "Sabre Spoke" wheels replaced the chrome wire wheels which were standard on Eldorado up to now. There was a metallic applique on the upper body sills and the Eldorado name appeared in script on the rear decklid. This was the year the Eldo got its own high-performance engine, a 270 horsepower version of Cadillac's 331 CID V-8 with dual four-barrel carburetion. Price went up only $76 this year, to $5,814. A record 3,950 Style 6267-S Eldorado Convertibles were sold in the 1955 model year.

1955

The graceful "Florentine curve" was extended to Cadillac four-door sedan rooflines this year. Both the Series 62 Sedan and the Fleetwood Sixty Special got this elegant rear roof styling. The roof side molding flowed smoothly down into the wraparound rear window molding which encircled the upper rear deck. The six-passenger Series 62 Four-Door Sedan, Style 6219, remained the most popular of all eight 1955 Cadillac body styles. Production totaled 44,904. The Series Sixty-Two Sedan's price was lowered $180 this year, to $3,657. Shipping weight was 4,370 pounds. Hydra-Matic automatic transmission became standard equipment on all 1955 Cadillacs.

The Fleetwood Sixty Special received modest styling changes for 1955. Like the less costly Series 62 Sedan, the premium Sixty Special now had a Florentine curve in its rear roofline. The vertical chrome louvers used on the bottom of the rear doors from 1950 through 1954 were repositioned at the ends of the rear fenders on the 1955 model. There were now 12 instead of eight. A broad chrome molding was used on the lower edge of the rear fenders between the fender skirt and rear bumper. At $4,342 the Sixty Special for 1955 was priced $341 lower than the previous year's model. A total of 18,300 Style 6019 Fleetwood Sixty Special Four-Door Sedans were built this year. This view shows the 1955 Cadillac's revised grille design with heavier horizontal eggcrate texture and two huge chrome bullets.

Except for a revised grille design, the prestigious Fleetwood Seventy-Five Sedan and Limousine for 1955 carried over their 1954 styling. The vertical rear fender molding and horizontal chrome trim that terminated at the ends of the front doors were retained: all other 1955 Cadillacs got a new side molding treatment that swept up at right angles at the rear fenders. The two Fleetwood Seventy-Five models offered for 1955 included the Style 7523 Eight-Passenger Sedan which sold for $5,694, and the Style 7533 Eight-Passenger Imperial Sedan, a formal limousine with glass division and a factory price of $5,895—highest in the 1955 Cadillac line. Production included 1,075 Series Seventy-Five Sedans and 841 Imperial Limousines.

The Hess and Eisenhardt Co. of Cincinnati, a well-known funeral car and ambulance builder which also did some custom car work, built this very special Fleetwood Seventy-Five Limousine for Mrs. Mamie Eisenhower, wife of the President of the United States and First Lady of the land. Special styling modifications included elimination of the "C" pillars for an open "hardtop" look in the rear, and a Series 62 Coupe rear window and Florentine curve roofline. The "Sabre Spoke" cast aluminum and steel wheels are the same as those used on the 1955 Eldorado. This custom-built limousine also had a retractable rear compartment roof panel to permit VIP passengers to stand up for parade appearances.

1955

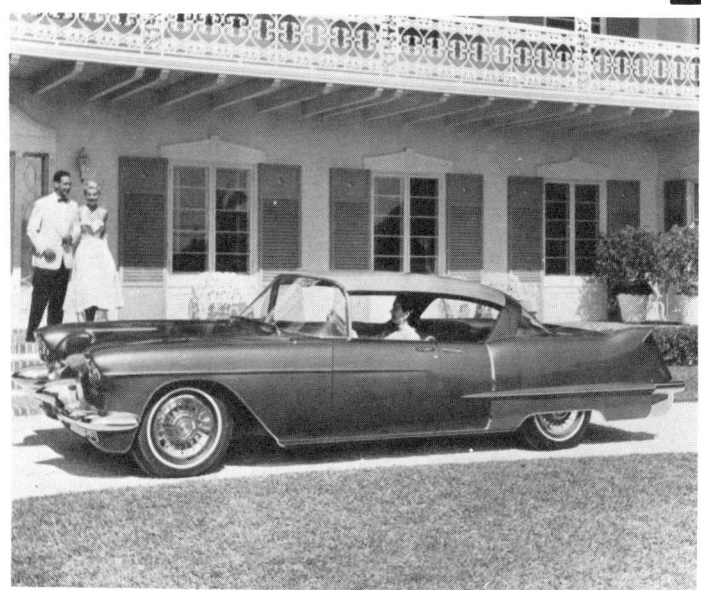

The first Cadillac Eldorado Brougham was a one-off show car which made its public debut at the 1955 GM Motorama. This car was the direct predecessor of the ultra-expensive, limited-production Eldorado Broughams built in 1957 and 1958. A pillarless four-door hardtop with center-opening doors, the experimental 1955 Cadillac Eldorado Brougham incorporated some of the basic styling elements of the 1954 Park Avenue show car, including large front wheel cutouts, a notched rear roofline and forward-swept tail fins. Known at General Motors Styling as the XP-38, the experimental Brougham had quad headlight units and a brushed stainless steel roof. Some of this car's styling features were later seen on the production 1957 Cadillacs.

Fifteen years had passed since Cadillac had dropped the LaSalle. Despite the fact that it was one of Cadillac's very few failures, the high-styled LaSalle was fondly remembered by many and there was strong sentiment to revive this once-proud name. The LaSalle nameplate was briefly resurrected in 1955. It appeared on two experimental Cadillac show cars both called LaSalle II, at the 1955 GM Motoramas. One was a small four-door hardtop sedan. The other was a two-seat convertible roadster. Both were powered by compact V-6 engines. This is the 1955 LaSalle II Four-Door Hardtop Sedan. Note the slotted, vertical grille design and the concave aluminum panels behind the front wheel opening, a styling feature later seen on production Chevrolet Corvette sports cars.

One of the stars of the 1955 GM Motorama, Cadillac's stunning Eldorado Brougham show car was a full seven inches lower and more than a foot shorter than current production Cadillacs. The sensational Eldorado Brougham bristled with advanced engineering features, including air suspension. This dazzling show car was the predecessor of the limited-production Eldorado Broughams of 1957-58, which Cadillac introduced as GM's answer to Ford's Continental Mark II. Ford came out with the Mark II in October, 1955. A production Eldorado Brougham was planned for 1956, but didn't make it until the next year—when the Mark II went out of production.

The other 1955 LaSalle II experimental show car was an extremely short two-seat roadster. Note the chopped-off rear end with fully-open rear fenders! Oval-shaped exhaust pipes project from the rocker panels. The V-6, fiberglass-bodied LaSalle II show cars marked the one and only time Cadillac publicly revived the warmly-remembered LaSalle nameplate. At one time, the LaSalle name was considered for the car that eventually became the 1975 Seville, but the LaSalle, dropped in 1940, was Cadillac's one conspicuous marketing failure and as such the name was rejected. Besides, the 1975 Seville was a premium model and the LaSalle had always been Cadillac's lowest-priced car. There was a definite image problem there.

1955

General Motors Styling reworked the 1953 Cadillac LeMans two-seat sports prototype for the 1955 auto show circuit. While this experimental, fiberglass-bodied car retained its 1954 production-car-style grille, it was given quad headlight units and redesigned rear fenders. The restyled fenders had integral air intakes and modest, pointed tail fins instead of the towering fishtail fins used on the 1953 LeMans show car. The chrome side molding extended the full length of the car. The front license plate reads "1955 Cadillac LeMans".

Besides far-out experimental idea cars like the Eldorado Brougham and LaSalle II, Cadillac's 1955 auto show exhibits included specially-trimmed and equipped versions of standard production body styles. These feature cars were often given colorful show names. This is the 1955 Cadillac Series 62 Coupe de Ville "Celebrity" Two-Door Hardtop. There was also a "St. Moritz" customized Eldorado Convertible and a "Westchester" Sixty Special Sedan. The Celebrity Hardtop was equipped with Cadillac's new "Sabre Spoke" wheels which were standard only on the 1955 Eldorado. Except for its special exterior color and interior trim, this car was very similar to the one on the showroom floor at your friendly neighborhood Cadillac dealer's place of business. The Celebrity name eventually showed up on a 1982 Chevrolet subcompact.

Among the most unusual Cadillacs built this year were six special "Sky View" tour coaches custom-built for the posh Broadmoor Hotel in Colorado Springs, Colo. These special sightseeing cars were built on the 1955 Cadillac commercial chassis by the Hess and Eisenhardt Co. of Cincinnati, a well-known hearse and ambulance manufacturer. The Sky View Cadillacs were designed to carry 11 passengers and the driver. Their roofs were equipped with four large plexiglass panels, so the tourists could enjoy the Rocky Mountain scenery as they rode up and down nearby Pike's Peak. The Sky View fleet was previewed at a special showing at The Broadmoor in June, 1955. This famous resort hotel operated a huge fleet of Cadillac Fleetwood Seventy-Five limousines for its affluent guests for many years. Note the full rear wheel openings and obligatory PUC number painted on the front fender. Technically, this was a bus!

The six special Sky View Cadillac sightseeing cars built by Hess and Eisenhardt were built on the 158-inch wheelbase Series 8680-S commercial chassis, the same one that H & E used for its S & S hearses and ambulances. These super station wagons were finished in Mandan Red with Pecos Beige upper side window trim. Note the angled "C" and "D" pillars. Six of these custom-built, 12-passenger cars were delivered to the Broadmoor Hotel in Colorado Springs, Colo. this year. Several more were built in subsequent years. This is the prototype Sky View as delivered to the Cadillac Engineering Dept. in Detroit for testing and approval.

The Hess and Eisenhardt Co. produced small numbers of custom-built Cadillac station wagons in the mid-1950s. This Cincinnati, Ohio firm had gained valuable station wagon construction experience when it built the fleet of special sightseeing cars for the Broadmoor Hotel in Colorado. Most of H & E's station wagons were conversions of Series 62 four-door sedans. They were available with plain metal sides, like this one, or with simulated wood paneling. Note the ventilator pane built into the rear side window opening.

1955

Here is a custom-built Hess and Eisenhardt "woodie," a distinctive station wagon conversion of a 1955 Cadillac four-door sedan. These station wagons were available— for a price—with or without simulated wood side paneling. Of special interest are the Fleetwood Sixty Special louvers on the lower edge of the rear fenders. The style of chrome rear fender edge molding used indicates that this is a Series 62 Sedan conversion, rather than a Sixty Special. These louvers were available over the parts counter at any Cadillac dealership. The 60-S had a special, wider rear fender edge molding.

The Hess and Eisenhardt Co. built some of the most distinctive flower cars produced by any of the major U.S. funeral car manufacturers. This is the 1955 S & S Cadillac Superline Florentine Flower Car, Style 55420, which sold for $8,081 new. This "Chicago type" flower car, with an open rear well instead of an adjustable stainless steel flower deck, has a simulated convertible coupe roof. Rumors persist that the tops of these cars actually let down, but we've never seen a picture to prove it. The chassis stiffening and engineering required would have been horrendous. At any rate, it's a very authentic-looking soft-top.

The Eureka Co., of Rock Falls, Ill., still used wood-framed bodies in its funeral cars and ambulances. The forward-angled "C" pillar and softly rounded rear roof contour, complete with Cadillac "Florentine curve" and panoramic quarter windows, were distinctive Eureka styling touches. This is a 1955 Eureka-Cadillac Manual Side-Servicing Limousine Funeral Coach which cost $8,629 delivered at the factory. Eureka continued to use the same type of vertical rear fender molding seen on all 1954 Cadillacs and carried over on the 1955 Fleetwood Seventy-Fives.

The long-wheelbase, luxury type Service Car was quickly going the way of the horse-drawn hearse. Very few were being built now, although this type of funeral car was usually still included in most funeral coach manufacturers' catalogs. Most of the service cars being sold now were on less expensive chassis, like Pontiacs. This is a 1955 Superior-Cadillac Service Coach, at $7,685 the lowest-priced model in the Superior Coach Corp.'s 1955 product line. The funeral service car was little more than a glorified panel truck. It was used for first calls, removals, hauling caskets and equipment, and other heavy hauling chores. Note the wreath emblem and ornamental chrome streamers on the blanked-in upper rear body panels.

Most funeral car manufacturers offered their customers a wide choice in funeral coach styling. Superior's limousine style hearses and ambulances were available with or without wraparound quarter windows in the rear. This is a 1955 Superior-Cadillac "Moderne" Limousine Hearse with closed upper rear quarters. Quarter windows were a delete option. Cadillac shipped 1,975 of its 158-inch wheelbase 1955 Series 8680-S commercial chassis to American funeral car and ambulance manufacturers this year. With the demise of the old Henney Motor Co. at the end of the 1954 model year, Superior Coach Corp. emerged as the new industry leader. Formed in 1868, Henney's professional car business died with Packard.

1956

Model year production topped 150,000 for the first time as the Cadillac Motor Car Division chalked up yet another all-time record production year. Two glamorous new models were added to Cadillac's 1956 model lineup, increasing to 10 the number of individual models offered.

The 1956 Cadillacs went into dealer showrooms with modest styling changes. This would be the third and final year for the current series of GM "C" bodies which had originally been introduced for 1954. It was also the ninth, and last, year for Cadillac's famous fishtail fins. Future Cadillacs would continue to set the style for these curious automotive appendages, but they would be of distinctly different types. The breakaway from the knobby fishtails that had made Cadillacs unmistakable everywhere since 1948 had actually begun the previous year when the Eldorado convertible got its own distinctive rear-end styling with bold, new pointed tail fins.

Styling changes on the mildly-revamped 1956 Cadillacs included a modified grille design with a fine crosshatched texture. The Cadillac name appeared in old-style angled script on the upper left-hand corner of the grille on all models. Small golden Cadillac crests adorned the front fenders just forward of the front doors, and model name script nameplates were added to the front fenders on DeVille, Sixty Special and Eldorado models beneath the crests. Cadillac's famed "V" and crest were both wider this year.

Tail fins on the 1956 cars were squared off and capped with a fine chrome bead that extended the full length of the rear fenders. The lower rear fenders of all models had what Cadillac called "slipstream styling" —prominent horizontal fairings that flowed into wide, oval-shaped exhaust ports in the massive new rear bumper. On the Sixty Special this fairing was sheathed in bright chrome. Nine vertical chrome trim pieces ornamented the last foot of this fairing on all models except the Eldorado.

Cadillac customers who really wanted to flaunt it could specify an optional anodized gold grille on the Sixty Special, and gold grille and wheels on the 1956 Eldorados.

Seven of Cadillac's ten 1956 body styles were in the rapidly-proliferating Sixty-Two Series. There was still just one Fleetwood Sixty Special and two Fleetwood Seventy-Fives. Major dimensions, including wheelbases, remained unchanged for the third consecutive year. The two all-new models, both in the 62 Series, included Cadillac's first production four-door hardtop and the first closed-body Eldorado. Designed as a four-door companion car to the very popular Coupe de Ville, the smart new pillarless sedan was called the Sedan de Ville. It's been a staple Cadillac model ever since.

The premium Eldorado convertible was renamed the Eldorado Biarritz this year and was joined by a two-door hardtop coupe called the Eldorado Seville. Both cars took their names from romantic European cities, Biarritz being a fashionable French seaside resort and Seville a fabled old city in Spain.

The spectacular new Seville had the same flamboyant rear-end styling as the Biarritz, which carried over the Eldorado's dagger-tipped 1955 tail fins. The Seville featured the same gracefully-contoured coupe roofline as the lower-priced Coupe de Ville, but the Seville's roof was covered with a special Vicodec fabric material. Interestingly enough, Chrysler offered a "Seville" this year, too: it was a highline DeSoto four-door hardtop. Cadillac used the Seville name through

The 1956 Cadillacs carried over the same basic styling introduced for 1954 for the third and final year. This time, however, there were some significant styling changes. The redesigned grille had a very fine texture, with the Cadillac name in old-style script on the upper left side. Tail fins were squared off and had a fine chrome bead that extended the length of the rear fender. A lower rear fender fairing flowed into large, oval exhaust ports in the massive, new rear bumper. The Series 62 Coupe, Style 6237, remained the "most affordable" Cadillac, with a factory-advertised price of $3,814. Weight was 4,420 pounds and 26,649 Series Sixty-Two Coupes were sold this year.

Front fender model identification appeared on most of the 1956 Cadillacs. All models, from the lowest-priced Series 62 Coupe on up to the Fleetwood Seventy-Five Limousine, had small gold Cadillac crests on their front fenders just ahead of the front doors. DeVille, Eldorado and Sixty Special also had their model names in script below these crests. This is the 1956 Cadillac Series 62 Coupe de Ville Two-Door Hardtop, Style 6237-D, which had a factory suggested price of $4,210. Weight was 4,445 pounds. A total of 24,086 Coupe de Villes were sold this year. The new hood and decklid crests and "V" emblems were wider than previously.

1956

the 1959 model year. It was resurrected on a new, smaller Cadillac 16 years later, in 1975.

The Eldorado Biarritz Convertible and the new Eldorado Seville Hardtop Coupe had the same $6,014 base price. Special Eldorado styling features included new twin-bladed hood ornaments, a ribbed chrome saddle on the upper door sills, "Sabre Spoke" wheels and large rear wheel cutouts without skirts. Customers could specify either aluminum or gold-anodized grille and wheels. Both Eldorados were powered by a special 305-horsepower version of Cadillac's V-8 engine with two four-barrel carburetors.

Power brakes were made standard equipment on all 1956 Cadillacs. Engine displacement changed this year for the first time since the 331 CID V-8 was introduced for 1949. Displacement went up to 365 cubic inches. Rated horsepower went up from 250 to 285, and compression ratio also went up to 9:75-to-1. Cadillac's standard Hydra-Matic automatic transmission was redesigned this year, too, and passenger seat belts were added to the list of available options.

The big auto shows, and especially GM's own Motoramas, were still pulling the people in. Cadillac's 1956 show cars included two Eldorado Broughams, one an actual pre-production prototype of the car that would go into limited production for 1957-58 and the other an experimental Eldorado Brougham Town Car that temporarily revived the classic open-front town car body style of the 1930s. The Brougham Town Car, however, was a fiberglass bodied "dream" car while the companion Brougham four-door hardtop was actually intended for public sale.

Cadillac's 1956 show cars also included four more specially painted-and-trimmed "feature" versions of current production body styles. These baroque show jobs had garish pearlescent metallic exteriors and exotically-upholstered complementing interiors. They included the "Castilian," an Eldorado Seville Hardtop Coupe in an Old Spanish motif; the "Gala," a "Wedding Car" edition of the new Sedan de Ville Four-Door Hardtop; a "Maharani" Fleetwood Sixty Special Sedan with Far Eastern decor, and a Westernized Series 62 Convertible called the "Palomino."

Cadillac built an all-time record 154,577 cars during the hectic 1956 model run, which included its one-millionth postwar car.

The last true Packards were built this year. In its dying gasp, this once-proud marque saw its historic nameplate affixed to a series of poorly-disguised Studebakers for 1957. Ford offered a totally-restyled line of 1956 Lincolns, in addition to the very exclusive Continental Mark II. And Chrysler, for the second model year, offered a limited-production "muscle" hardtop coupe called the Chrysler 300.

The family-oriented Series 62 Four-Door Sedan was no longer Cadillac's most popular body style. The high-styled new Sedan de Ville Four-Door Hardtop deposed the venerable Sixty-Two Sedan in its very first year on the market. The Series 62 Sedan, Style 6219, accounted for 26,222 of Cadillac's sales this year, half its usual level. Base price was $3,903. The lowline Series 62 models—base Coupe, Convertible, and four-door sedan, had no model names below the small Cadillac crests on the front fenders. Power brakes became standard equipment on all Cadillacs this year.

An important new body style was added to the Cadillac model lineup this year. It was a pillarless four-door hardtop sedan which traced its origin back to the 1953 Cadillac Orleans show car. Cadillac's first production four-door hardtop was named the Sedan de Ville. With a factory-advertised price of $4,330 it was a spectacular success in its very first year. A total of 41,732 were sold, finally displacing the Series 62 Sedan as Cadillac's top seller. Style 6239-D, the handsome new Sedan de Ville had the same flowing roofline as Cadillac's other closed cars with the elegant "Florentine" curve at its base, encircling the upper rear deck.

The 1956 Cadillac model lineup again included two Series 62 convertibles, the standard Cadillac Convertible Coupe and a premium Eldorado Biarritz with distinctive rear styling. This is the standard Series 62 Convertible, Style 6267, which weighed 4,665 pounds and had a factory suggested retail price of $4,342. The Detroit plant built 8,300 of these two-door convertibles during the booming 1956 model run. Buyers could choose from 10 leather interior upholstery selections. Engine displacement was increased to 365 cubic inches this year and horsepower was boosted from 250 to 285.

1956

With the addition of a companion hardtop coupe to the premium Eldorado series for 1956, the Eldorado convertible was given a second name—Biarritz. The swank Eldorado soft-top was now known as the Eldorado Biarritz Convertible Coupe, Style 6267-S. The 1956 Eldo retained the distinctive rear end styling introduced on this sporty model the previous year, but sprouted new dual-blade hood ornaments. Horsepower was boosted to a very impressive 305, rivaling Chrysler's hairy 300 "muscle" coupe. Eldorado styling features still included a ribbed chrome saddle sill molding, full rear wheel cutouts without fender skirts and bladed "Sabre Spoke" cast aluminum and steel wheels. A gold-anodized grille and gold wheels were available as no-cost options on 1956 Eldorado models. Both the convertible and new Eldorado Seville hardtop had the same $6,014 base price. Although 2,150 Eldorado Biarritz Convertibles were sold this year, the new Seville Hardtop outsold it by a margin of nearly two-to-one.

The exclusive Cadillac Eldorado's sales appeal was broadened this year with the addition of a two-door hardtop coupe to the Eldorado line. This dashing new model was called the Eldorado Seville. It featured the same flamboyant rear styling as the companion Eldorado Biarritz Convertible but offered Eldorado individuality with the practicality of a closed car. The Seville's roof was covered with a special Vicodec fabric with two lengthwise seams. The new Eldorado Seville's factory suggested retail price of $6,014 was exactly the same as that of the convertible. This exclusive hardtop coupe was apparently just what a lot of upscale car buyers were waiting for. A total of 3,900 Seville Hardtops, Style 6237-SD, were delivered in the 1956 model year compared with only 2,150 Eldorado Biarritz convertibles.

The new 1956 Cadillac Eldorado Seville Two-Door Hardtop Coupe featured the same flamboyant rear-end styling as the companion Eldorado Biarritz Convertible. The pointed rocket-type fins were first used on the 1955 Eldorado convertible. By the 1958 model year they would be extended to every car in the line, from the lowest-priced Series 62 Coupe on up to the hyper-expensive, limited-production Eldorado Brougham. The new Seville's roof was covered with a special Vicodec fabric material with two parallel seams. The Eldos were the only 1956 Cadillacs that did not have rear fender skirts. Both Eldorados were powered by 305-horsepower engines with twin four-barrel carburetors.

The 1956 Cadillac Fleetwood Sixty Special boasted several new styling features. "Slipstream-styled" rear fenders had massive chrome moldings that flowed into large oval exhaust ports built into the massive new rear bumper. The grille had a fine crosshatched texture to it and was available in an optional anodized gold finish. The Cadillac name appeared in an angled script on the upper left side of the revised grille. The "Sixty Special" model name also appeared in script beneath small Cadillac crests on front fenders just ahead of the front doors. New, wider Cadillac "V" emblems and crests were used on the front of the hood on all models. The Fleetwood Sixty Special continued on its own 133-inch wheelbase and was offered in only one body style, a fully-equipped four-door sedan. Factory suggested price was $4,587 and an even 17,000 were delivered this year. The Sixty Special shown has the optional "Sabre Spoke" wheels that were standard only on the Eldorado.

1956

This would be the ninth and last year for Cadillac's famous "fishtail" fins. While these curious automotive appendages would rise to new heights on future Cadillacs, they would be of a distinctly different style than those used since 1948. The break away from the fishtail fin with its towering taillight began with the appearance of long, pointed rocketship-type tailfins on the restyled 1955 Eldorado convertible. For 1956 the classic Cadillac tailfins were flatter and more squared than ever. A fine chrome bead extended up and over the taillight and along the entire length of the rear fender. This is a profile view of the impressive 1956 Cadillac Fleetwood Seventy-Five Eight-Passenger Sedan, Style 7523, which sold for $6,040. A total of 1,095 were built. A formal limousine version with glass division was also available.

Wherever the rich or famous gathered you were sure to find the stately Cadillac Limousine. The impressively long, usually black, Fleetwood Seventy-Five Eight-Passenger Imperial Sedan was the only true production limousine built in America. Chauffeur-driven Fleetwood Seventy-Five Sedans and Imperial Limousines could be found at the front doors of the finest homes, clubs and restaurants, at the airport or theatre, or waiting obediently at the curb in front of the best stores. The longest, heaviest and most expensive car in the 1956 Cadillac line was the Imperial Sedan, or Limousine, Style 7533, which sold for $6,240. All Seventy-Fives had auxiliary seats which folded when not in use into the back of the front seat. A fixed partition with glass division and leather-upholstered chauffeur's compartment were standard in the Limo. The "Imperial" name had denoted Cadillac's most expensive formal sedan for half a century. A total of 955 Fleetwood Imperial Limousines were built this year.

Companion to the pre-production Cadillac Eldorado Brougham Four-Door Hardtop on the 1956 GM Motorama circuit was an experimental Eldorado Brougham Town Car. This fiberglass-bodied idea car, built on a 129.5-inch wheelbase, made its public debut at the opening Motorama at the Waldorf in New York. It revived the formal open-front town car styling of the classic era. The crisply-styled rear roof design appeared on the redesigned 1959 Fleetwood Seventy-Fives. The show car's rear roof was upholstered with polished black landau leather. Basic styling was identical to that of the Eldorado Brougham prototype. The 1956 Cadillac Eldorado Brougham Town Car show job stood only 55.5 inches high and with an overall length of 219.9 inches was two inches shorter than the standard 1956 Series 62 Cadillacs. It is unlikely that this car was ever seriously intended for even limited production, but it sure wowed the crowds.

Cadillac had promised that its limited-production Eldorado Brougham would be available to the public sometime during the 1956 model year. As it turned out, the ultra-expensive Brougham didn't make it into the showrooms until 1957. To whet the public's appetite, however, Cadillac exhibited two Eldorado Broughams at the 1956 GM Motoramas. One was an exotic Eldorado Brougham Town Car, the latest in a series of experimental "dream" cars created to dazzle auto show audiences. The other was a genuine pre-production prototype of the 1957-58 Brougham pillarless four-door hardtop. Known at GM Styling as the XP-48, this car made its public debut at the 1956 Los Angeles Auto Show. It was then flown to Europe for the prestigious Paris Auto Show. On its return to the U.S. it went back to GM Styling in Detroit for minor changes before going on the Motorama tour with the Brougham Town Car. The exclusive Brougham was a hyper-luxurious four-door hardtop with a stainless steel roof and every conceivable gadget.

In addition to the Eldorado Brougham pre-production prototype and the exotic Brougham Town Car, Cadillac created four special "production show cars" for the 1956 auto show circuit. This is the "Gala," also known as "The Wedding Car," a lavishly-appointed Sedan de Ville Four-Door Hardtop finished in a shimmering pearlescent white and silver with a dewey bridal motif. The interior was upholstered in ribbed satin, white pearl leather, and white mouton fur. The front doors were equipped with small umbrellas with rhinestone-encrusted handles. We think you get the idea . . . . .

1956

Another one of the four special "production show cars" executed for the 1956 auto shows was Cadillac's "Palomino," a gussied-up Series 62 Convertible Coupe with a predictably Western flavor. The interior was upholstered in soft natural Palomino hides on the door upper panels, seat backs, and floor mats. Cadillac's accompanying promotional puffery described this feature car as . . . . "A ruggedly-styled convertible with a shimmering beige metallic body that reflects the western flair of wide-open spaces . . ."

The 1956 Cadillac "Castilian" show car was an otherwise standard Eldorado Seville Two-Door Hardtop Coupe finished in a special metallic starlight silver with a white-textured Vicodec roof. The interior styling theme was "Old Spain," with contrasting black and white calfskin seat bolsters. The fourth of these specially-trimmed show cars was the "Maharani," a Fleetwood Sixty Special Four-Door Sedan done in a Far Eastern motif with a rich metallic maroon body and a brilliant gold-colored roof panel.

A prominent funeral car and ambulance builder, the Hess and Eisenhardt Co. of Cincinnati, continued to produce small numbers of station wagon conversions of Cadillac four-door sedans. This is a 1956 model complete with simulated wood side paneling of the same style used on Ford and Mercury wagons at this time. Note how the lower rear molding blends into the rear fender exhaust fairing, which flows into the rear bumper. These exclusive Caddy station wagons offered complementary styling and excellent all-around visibility. Price is unknown, but it must have been equally impressive.

The Hess and Eisenhardt Co. continued to build a few special "Sky View" sightseeing coaches for the exclusive Broadmoor resort hotel in Colorado Springs, Colo. This company had delivered a fleet of six Sky View Cadillacs to the Broadmoor in 1955. This is a 1956 model. These custom-built 12-passenger super station wagons were built on the 158-inch wheelbase Cadillac commercial chassis, the same one Hess and Eisenhardt used for its S & S hearses and ambulances. The Sky View's roof had four large plexiglass panels so tourist passengers could enjoy the mountain scenery as they rode. Exterior color was a deep Mandan Red with Pecos Beige upper body accents.

Cadillac did regular business with five independent funeral car and ambulance manufacturers and two outside coachbuilders. The Derham Body Co. of Rosemont, Pa. did small numbers of formal limousine conversions based on the standard Fleetwood Seventy-Five. The Hess and Eisenhardt Co. of Cincinnati took on many special custom body assignments for individual customers and governments. In 1956, H & E constructed two of these very special "security" cars for the U.S. Secret Service. These sinister-looking four-door convertible sedans preceded, flanked or closely followed the Presidential Limousine (usually a Lincoln) at the Chief Executive's public appearances. Special equipment included side running boards and rear steps on which Secret Service agents assigned to the White House security detail rode in motorcades.

1956

The two 1956 Cadillac "security" cars remained in service well into the 1970s. They were highly visible in nearly every presidential motorcade at home and abroad. One of these rolling security and communications centers followed President John F. Kennedy's 1961 Lincoln parade phaeton (also a Hess and Eisenhardt creation) on the high speed run to Parkland Hospital in Dallas on that fateful day in 1963. On rare occasions the President and First Lady rode in one of these cars. Here, photographed in front of the White House, one of the 1956 security specials is fitted with U.S. and Presidential standards for ceremonial duty. Note the big coaster siren mounted on the left front fender. The Secret Service boys look professionally grim.

The 1956 Cadillac White House security cars were usually seen in public with their tops down. But these big four-door convertible sedans, built on Cadillac long-wheelbase commercial chassis, had full tops which could be installed for use in inclement weather. Note the side running boards and rear steps with special deck lid handles for Secret Service agents. One or both of these very special Cadillacs always remained close to the Presidential Limousine in parades and motorcades. They dualled as rolling surveillance and command posts. This one was photographed in front of the Cadillac main office on Clark Avenue in Detroit before it was shipped to Washington.

The Cadillac Motor Car Division shipped 2,025 of its 1956 Series 8680-S commercial chassis to five U.S. funeral car and ambulance manufacturers this year. These firms were The Eureka Co. of Rock Falls, Ill.; Superior Coach Corp. of Lima, Ohio; Meteor Motor Car Co. of Piqua, Ohio; A. J. Miller Co. of Bellefontaine, Ohio, and Hess & Eisenhardt, builders of S & S professional cars at Cincinnati, Ohio. This is a 1956 Superior-Cadillac End-Loading Landaulet built on the 158-inch wheelbase 1956 commercial chassis. It has a smooth roof. A matte, crinkle-finished roof was optionally available.

The Superior Coach Corp. of Lima, Ohio was now America's largest builder of funeral cars and ambulances. The company's hearses and rescue cars utilized the same basic Superior designed-and-built body shell. This is a most attractive 1956 Superior-Cadillac Limousine Style Ambulance. A standard-headroom model, it has built-in tunnel roof lights front and rear, and panoramic rear quarter windows for extra visibility. Note the special chrome side moldings that serve as a color break for the two-tone exterior paint treatment.

1957

The 1957 model year marked another major advance in product development for the Cadillac Motor Car Division. All of Cadillac's standard models were completely redesigned and restyled for the first time in three years, and the company made a spectacular re-entry into the ultra-luxury passenger car market which it had abandoned with its last sixteen-cylinder cars 17 years earlier.

Cadillac's dramatically-restyled 1957 models were officially introduced at the first New York Automobile Show held since 1940. The dazzling, limited-production Cadillac Eldorado Brougham had its formal public debut at the Los Angeles International Automobile Show.

All of Cadillac's 10 standard production models were built on a new tubular-center "X"-type chassis frame which resulted in a significantly lower vehicle height. Cadillac 1957 sedans were three inches lower than comparable 1956 models. Hoods and rear decks were flatter and lower. Upswept, rubber-tipped front bumper guards framed a wide, low-set cross-hatched grille which gave the 1957 Cadillac an impressively low road stance. Several of the restyled 1957 Cadillac's key styling elements had been taken directly from the Park Avenue show car of 1954. These included hooded headlamps, large front wheel openings, elegantly curved rooflines with notched rear roof pillars and new-style tail fins that sloped forward at the rear. These fins were edged in bright chrome. Brake and backup lights were paired in chrome pods at their base. Long rear fender fairings similar to those on the 1956 cars flowed into the taillight clusters.

Seven of Cadillac's ten 1957 standard production models were Series 62 Body styles. At the lower end of the line were the standard Series 62 Two-Door Hardtop and Four-Door Sedan. The pillared four-door sedan was gone, replaced with a new high-styled four-door hardtop that was virtually identical in appearance to the more expensive Sedan de Ville. But Cadillac marketed this car as a Four-Door Sedan. Next came the more luxurious Coupe de Ville, Sedan de Ville and the unusually handsome Series 62 Convertible.

The very costly Eldorado specialty models—the Biarritz Convertible and Seville Two-Door Hardtop—also got fresh, new styling this year. Once again the Eldorados were given their own, exclusive rear-end styling. The rear fenders and deck were softly rounded. Sharply-pointed tail fins rose out of the rear fenders. The lower rear quarters of the Biarritz and Seville were finished in gleaming chrome. These bright panels swept around the corners of the car forming split rear bumper units that also housed the brake and backup lights.

The 1957 Fleetwood Sixty Special was also a totally-redesigned car. The basic body style had changed overnight from a pillared four-door sedan to a four-door hardtop. Only its 133-inch wheelbase remained unchanged. For the first time in 15 years there were no vertical chrome trim pieces on its flanks. Instead, the entire lower half of the rear fenders were sheathed with a ribbed metallic panel which made this four-door hardtop unmistakable as a Sixty Special from any angle. The Fleetwood Seventy-Fives also got Cadillac's shapely new 1957 styling. Rooflines were basically the same as those used on this exclusive series since 1950, but the door openings now extended into the roof so there was no sacrifice in ease of passenger entry and exit, even though these big cars were noticeably lower than before.

For 1957, the 365 cubic inch Cadillac V-8 engine was uprated to a very impressive 300 horsepower. Compression ratios also went up again, to 10-to-1. A 325-horsepower version of this engine with two four-barrel carburetors was optionally available in the Eldorado Biarritz and Seville. The 325 HP engine was standard in the new Eldorado Brougham. Wheelbases were increased fractionally—half an inch (to 129.5 inches) on Series 62 Cadillacs and to 149.8 inches on the Fleetwood Seventy-Five Sedan and Limousine. Inside, the parking brake handle had disappeared. It was replaced with a foot-operated parking brake which released itself when the car was put in gear. On all but the big Seventy-Fives, air conditioning equipment was relocated from the trunk to a new location under the hood.

As promised two years earlier, Cadillac in 1957 brought its Eldorado Brougham "dream" car to life. Only 400 of these incredibly complex, opulently-equipped, limited-production four-door pillarless hardtops were built in the 1957 model year. The base price was a staggering (for 1957) $13,074 making it the most expensive production Cadillac ever built. Not until the 1975 Seville arrived would a Cadillac carry such a stupendous price tag. The new Eldorado Brougham was priced more than $3,000 higher than Ford's Continental Mark II, which up until the arrival of the Brougham had been America's most exclusive production car.

The new Eldorado Brougham stood only 55.5 inches high. It was built on a 129-inch wheelbase and like the other 1957 Cadillacs utilized a tubular "X"-type frame. Overall length was 216 inches. The Brougham's broad, flat hood and rear deck lid were actually lower than the

All of the standard production 1957 Cadillacs, from the lowest-priced Series 62 Hardtop on up to the Fleetwood Seventy-Five Limousine were completely redesigned. Styling was totally fresh, from gullwing, rubber-tipped front bumperguards to forward-sloped new tail fins. Some of the 1957 Cadillac's key styling elements, including large front wheel cutouts, notched rear roof pillars and redefined tailfins, were taken directly from the 1954 Park Avenue show car. This is the 1957 Cadillac Series 62 Two-Door Hardtop, Style 6237, which at $4,609 was the lowest-priced car in the 1957 model line. Sales of 25,120 Series 62 Hardtops exceeded those of the more expensive Coupe de Ville by 1,307 units. The background is most appropriate: somehow, Cadillacs and oil wells just seem to go together!

front and rear fender lines. The roof was covered in gleaming, brushed stainless steel. Massive stainless steel skirts covered the lower rear fenders. A thin molding swept back from hooded dual headlight units to a louvered vertical molding in the center of the sculptured rear door. A curved cellular grille nestled between projectile-shaped gullwing front bumpers. Slender vertical taillights were incorporated into the trailing edges of the swept-back tail fins. Deep-dish, forged aluminum wheels were equipped with low-profile tires with one-inch wide white sidewalls. Air intakes were built into the tops of the front fenders.

The Eldorado Brougham bordered on technological overkill. Conventional coils and springs were replaced by an extremely complex air suspension system. An anti-dive control built into the front suspension prevented the car's nose from dropping when the brakes were applied. With the ignition on, the engine started automatically when the Hydra-Matic gear selector was moved to "neutral" or "park." The Brougham had no center posts, only a locking mechanism which extended 14 inches above the sill. The center-opening doors automatically locked when the car was put into gear.

As might be expected, the Eldorado Brougham was lavishly equipped. Power brakes, steering, windows (including ventipanes), radio antenna, trunk lid, and door locks were standard. Under the Brougham's broad, flat hood were three "seashell" horns and a "trumpet" horn, which blared simultaneously. Buyers could select from 15 special exterior colors and 45 standard and two special-order interior trim and color choices, including genuine mouton or high-pile Karakul carpeting. The six-way power front seat moved automatically for the convenience of passenger entry and exit, and the driver's seat was equipped with a "favorite position" memory setting. Air conditioning, of course, was also standard.

> Of the 11 models Cadillac took to market for the 1957 model year, seven were Series 62 body styles. All were built on a new 129.5 inch wheelbase, fractionally longer (half an inch) than the 1956 models. Vehicle height was lowered significantly through the use of an all-new "X"-type tubular chassis frame. The venerable Series 62 four-door pillared sedan was replaced this year with a new lowline four-door hardtop. Style 6239, it was still called the Series 62 Sedan, however, and was identical in appearance to the more luxuriously trimmed Sedan de Ville. Factory-delivered price was $4,713. With deliveries of 32,342 cars the Series 62 Sedan regained its title as Cadillac's best-selling car. All Series 62 Cadillacs, except Eldorados, had new twin-blade hood ornaments.

Interior amenities included a front seat cigarette case, tissue dispenser, vanity compact and lipstick, stick cologne, a set of four magnetized gold-finished drinking cups, and vanity mirror. A storage compartment in the rear seat armrest contained a note pad and pencil, portable vanity mirror and a perfume atomizer complete with Arpege, Extrait de Lanvin! In announcing this hyper-luxury car, Cadillac Motor Car Division General Manager James M. Roche asserted that the company had succeeded in building "the finest car possible." The hand-built 1957-58 Eldorado Brougham has already been accorded full classic status and was the last true semi-custom ever built in Cadillac's own plant. In its short production life, the Eldorado Brougham was known as Cadillac's Series Seventy.

There was no General Motors Motorama this year, but Cadillac did create at least one show car for that year's auto show tour. It was "The Director," a specially-equipped 1957 Cadillac Fleetwood Sixty Special that had been converted into a rolling executive suite. The right front seat pivoted 180 degrees to permit a secretary to take dictation from the busy executive who sat at a special desk in the rear. The Director had all the conveniences of the office including dictaphone, telephone and filing space.

Cadillac production dropped slightly this year, from 154,577 cars in 1956 to 146,841 in the 1957 model year. James M. Roche took over as Cadillac's new General Manager on January 1, 1957.

Just as the Eldorado Brougham arrived on the scene, the Ford Motor Company wrapped up its two-year Continental Mark II production program. Only 3,012 of these exclusive $10,000 two-door coupes were built during the 20-month production run which terminated in July, 1957. Across town, Chrysler Corporation introduced a totally-new 1957 Imperial which was selling in record numbers. As the 1957 model year wound down, all eyes were on Ford which was about to take the wraps off its long-awaited Edsel, a new intermediate entry.

> Except for one minor cosmetic detail, it was nearly impossible to tell the 1957 Cadillac Coupe de Ville from the base Series 62 Two-Door Hardtop, and the lowline Series 62 Sedan from the more expensive Sedan de Ville. The lowline Series 62 models, however, had small Cadillac script nameplates on their front fenders, just behind the hooded headlights. Coupe de Ville and Sedan de Ville script nameplates were used in the same locations on the more expensive de Villes. Interior trim and equipment levels were upgraded on de Ville models. This is the 1957 Cadillac Coupe de Ville Two-Door Hardtop, Style 6237-D, which had a factory advertised price of $5,048. Sales totalled 23,815 this year, outselling the companion Sedan de Ville Four-Door Hardtop by only five cars.

1957

Sales of 1957 de Ville two and four-door hardtops were almost evenly split this year, but when the final figures were tallied the two-door Coupe de Ville had outsold the 1957 Sedan de Ville Four-Door Hardtop by just five units. The 1957 Cadillac Sedan de Ville Four-Door Hardtop, Style 6239-D, had a factory-advertised price of $5,188. A total of 23,808 were built. The 1957 Cadillac's totally new front end design featured gullwing front bumper guards with black rubber protective tips that flanked a low, finely-textured cellular grille. This would be the last year for single headlight units on standard production Cadillacs. The new ultra-luxury 1957 Eldorado Brougham was the first production Cadillac with a quad headlight system.

The very words "Cadillac convertible" conjure up delicious mental images of wealth, success and The Good Life. A fire engine-red one was even better and was the suppressed desire of many a red-blooded American youth. The 1957 Cadillac Series Sixty-Two Convertible, with its broad hood and rear deck, low wraparound windshield and big wheel cutouts was an especially racy-looking ragtop. It was a big seller, too. A record 9,000 Series 62 Convertibles were sold in the excellent 1957 model year. The jazzy soft-top was listed as Style 6267. Base price was $5,229 and shipping weight was 4,743 pounds. A wide choice of leather interiors were available.

The distinctive Fleetwood Sixty Special for 1957 was a completely new car. Only its 133-inch wheelbase remained unchanged. Overnight, the Sixty Special's traditional four-door pillared sedan body style changed into a sportier-looking four-door hardtop, although the Sixty Special was still listed as a four-door sedan. Gone at last were the short, vertical chrome louvers that had been a Sixty Special styling cliche since 1942. The totally-restyled 1957 Cadillac Fleetwood Sixty Special Four-Door Sedan was instantly recognizable by the broad, ribbed metal trim panels that covered the lower half of the rear fenders. A horizontal fairing extended from the center of the rear doors to the taillight bezels. Style 6039, the Sixty Special had a factory-delivered price of $5,539. Weight was 4,761 pounds. Sales of this premium model set a new record this year, with an even 24,000 delivered.

Cadillac re-entered the ultra-luxury car market for the first time since the end of the V-16 era in 1940 with its fantastic 1957 Eldorado Brougham. This extremely complex, hyper-expensive, limited production pillarless four-door hardtop was the production version of the 1955 Eldorado Brougham show car and was Cadillac's answer to Ford's Continental Mark II. Standard equipment included air suspension and every imaginable comfort convenience and power assist. Wheelbase was 126 inches and the Brougham was built on an "X"-type chassis frame like all other 1957 Cadillacs. Price was a staggering (for 1957) $13,074. Styling included extensive use of bright stainless steel, including the roof panel. Virtually hand-built, only 400 were produced during the 1957 model year. This car remained in production, with minimal mechanical changes, through the 1958 model year. The 1959-60 Broughams were much less distinctive. The exclusive Eldorado Brougham was powered by a 325-horsepower version of Cadillac's 365 cubic inch V-8 engine. The flat hood and rear decklid were actually lower than the front and rear fenders.

1957

Even the very expensive Cadillac Eldorado Biarritz Convertible and the companion Eldorado Seville Two-Door Hardtop were completely restyled this year. Front-end styling was compatible with the rest of the standard Cadillac line, but the Eldorado again got its own, very distinctive rear appearance. Modestly-proportioned, sharply-pointed tail fins jutted out of softly-rounded rear fenders. The lower rear body behind the skirtless wheel openings was sheathed in bright chrome. These chrome panels extended around the corners of the car and also functioned as split rear bumpers. The standard engine was a 300-horsepower, 365 cubic inch V-8. A 325-horsepower version of this engine with dual four-barrel carburetors was optionally available. Style 6267-S, the 1957 Cadillac Eldorado Biarritz Convertible weighed 4,906 pounds and sold for $6,648. Only 1,800 were built.

A specially-equipped Cadillac convertible instead of the usual very conservative Rolls Royce awaited Queen Elizabeth II when she arrived in Canada on her 1957 Royal Tour. All three major Canadian auto manufacturers—Chrysler Canada Ltd., Ford of Canada and General Motors of Canada—supplied special convertibles for the Queen's use at various points along the Royal Tour route. General Motors of Canada provided this modified 1957 Cadillac Eldorado Biarritz Convertible, certainly a flashy conveyance for a monarch. This car was equipped with a transparent plastic top so Her Majesty could be seen by her subjects even if it rained. This bubbletop could be quickly removed for fairweather motorcades. The royal standard flutters from the right front fender in keeping with strict Buckingham Palace protocol.

For 1957, Cadillac's exclusive Eldorado model offerings were expanded to no less than three different body styles. These included the Eldorado Seville Two-Door Hardtop; the Eldorado Biarritz Convertible, and the all-new, hyper-expensive limited-production Eldorado Brougham. This is the sleekly-styled 1957 Eldorado Seville Two-Door Hardtop, Style 6237-D. Priced at $6,648—the same as the Biarritz Convertible—the Seville Hardtop outsold the soft-top by 300 units. A total of 2,100 were delivered. Like the Biarritz, the 1957 Seville had very flashy rear-end styling with sharply-pointed tail fins and massive chrome trim on the lower rear fenders. These chrome corner pieces also served as split bumpers and housed brake, turn signal, and back-up lights.

Like the rest of the 1957 Cadillac line, the Fleetwood Seventy-Five Sedan and Limousine were completely restyled. Gone was the towering, slab-sided appearance of the previous three years. The 1957 Seventy-Fives looked longer, lower and downright sporty in some aspects. Overall height was noticeably reduced by the use of a new tubular-center "X"-type chassis frame, but wheelbase was increased only a fraction of an inch, to 149.8 inches. This is the impressive 1957 Cadillac Fleetwood Seventy-Five Eight-Passenger Sedan, Style 57-7523, which had a factory-advertised price of $7,348. Production totaled 1,010. Note how the side doors now extended into the roof panel for ease of entry and exit.

1957

With the exception of the ultra-expensive new Eldorado Brougham, the most expensive car in the 1957 Cadillac model lineup was the totally-restyled Fleetwood Seventy-Five Eight-Passenger Imperial Sedan. Style 57-7533, the very impressive Fleetwood Limousine had a factory-advertised base price of $7,586. Only 890 of these formal limousines were built this year. The roofline and rear window follow the same style that distinguished Fleetwood Seventy-Fives since 1950. Round brake and taillights were paired in clusters at the base of new forward-swept tail fins. Vertical exhaust ports were incorporated into the outboard ends of the new rear bumper. Note the heavy chrome moldings around the side windows and the horizontal rear fender fairing that extends from the center of the rear door to the taillights. A power-operated glass division was standard in the big Imperial Sedan.

The Hess and Eisenhardt Co. of Cincinnati was among the most innovative of all U.S. specialty body builders. This company's staple products were S & S funeral cars and ambulances, but the company did a lot of custom body and armored passenger car conversion work. This is one of the most unique hearses ever built by S & S. Only a few were built, in the 1957 and 1958 model years. It is a Convertible Landau Flower Car. Like the stately landaulet limousines of the classic era, the entire roof let down for the open display of floral tributes. With the rear top up it looked like a conventional landau funeral car without carriage bows. This unusual hearse was built on the restyled 1957 Cadillac Series 8680-S commercial chassis.

The author photographed this attractive 1957 Cadillac Fleetwood Seventy-Five Formal Limousine in the courtyard of the Hotel Ritz in Paris in 1965. The uniformed chauffeur unobtrusively awaits the return of his pampered passengers. The car was painted a deep blue with a light tan-coloured fabric-covered roof with closed upper rear quarters. It is not known who did this conversion, but it is similar to the formal sedan conversions of Fleetwood Seventy-Fives still being done by Derham of Rosemont, Pa. A roof luggage rack is provided for extended touring. Note the air conditioning scoops at the base of the roof.

The Eureka Co. of Rock Falls, Ill., built this high-bodied "Eastern Style" flower car on the 1957 Cadillac commercial chassis. The porthole window in the coupe roof is highly unusual. This type of flower car was designed to carry a casket in the compartment below the adjustable stainless steel flower deck. Rear compartment height was 24 inches, four more than the standard model. It took a tall funeral director or attendant to place and arrange the floral tributes on top of that high rear deck in an appropriately dignified manner.

The Cadillac commercial chassis was also redesigned this year. Wheelbase was shortened two inches to 158 inches. Cadillac shipped 2,169 of these special hearse and ambulance chassis to the four major U.S. funeral car builders. The A. J. Miller Co. and Meteor Motor Car Co. merged late in 1956 to form a new company called Miller-Meteor, a division of Wayne Corp. The Superior Coach Corp. restyled all of its professional cars this year with what the company called "Criterion styling". This look, with long, curved rooflines and large glass areas was very compatible with Cadillac's new styling. This is a 1957 Superior-Cadillac Limousine Combination which could be quickly converted for either funeral car or emergency ambulance service.

1958

General Motors observed its 50th Anniversary in 1958, and Cadillac produced its two-millionth car. But the postwar automobile sales boom came to a screeching halt during the 1958 model year as the U.S. economy and the American motor vehicle industry plunged into a serious recession. Most passenger car manufacturers —including Cadillac—suffered sharp drops in new vehicle sales.

The North American motor vehicle industry switched en masse to a new four-headlight illumination system on the 1958-model cars. The Motor Vehicle Manufacturers' Association had decided to adopt this much-improved illumination system two years earlier. Four 5-¾-inch sealed-beam headlights replaced the two seven-inch units used previously. The "quad" headlight system provided much better road illumination under all driving conditions and permitted more precise headlight aiming. Most cars built in the United States and Canada still use this same basic system today.

This would be the second and final year for the first-generation Cadillac Eldorado Brougham ultra-luxury car. Although a new Eldorado Brougham would be offered in 1959 and 1960, the second-generation Broughams were much less distinctive than the 1957-58 cars. Only 304 Series 70 Eldorado Broughams were built during the 1958 model year, for a grand total of just 704 over the two-year production run. The Brougham's very distinctive pillarless four-door hardtop styling with brushed stainless steel roof and rear fender trim was unchanged for 1958. The hyper-luxurious Brougham still included virtually every imaginable comfort and power-assisted convenience, from air suspension to automatic starting and door locks, to a set of magnetized golden tumblers and a perfume atomizer filled with Arpege.

It is probable that General Motors lost money on every Eldorado Brougham that rolled out of the Fleetwood plant door. But, like the classic V-16 Cadillacs of an earlier era, the limited-production, hand-built Brougham left no doubt in the public mind as to who was the king of the luxury car hill and brought new lustre to the Cadillac name.

The only significant change in the 1958 Eldorado Brougham was found under its low, flat front-hinged hood. Horsepower was boosted to an astounding 335, and the standard-equipment 365 cubic inch V-8 engine was now equipped with triple two-barrel carburetors. This extremely powerful engine was also standard in the 1958 Eldorado Biarritz Convertible and companion Eldorado Seville Two-Door Hardtop, and was optionally available in all other 1958 Cadillacs. The Brougham's sophisticated air suspension system was also now available in other Cadillacs as an extra-cost option. This system, which did away with conventional coil and leaf springs in favor of infinitely adjustable air bags at each wheel, included a lifting valve controlled by the driver that hoisted the car an extra five inches to clear steep ramps or driveways.

Electric door locks were optionally available on models equipped with power windows. A new four-link rear suspension system was standard on Cadillacs with conventional coil springs instead of air suspension. All 1958 Cadillacs also had an improved power brake system. Standard-production Cadillacs were powered by a 310-horsepower 365 CID V-8 engine with 10:25-to-1 compression ratio. Wheelbases and major dimensions were unchanged from 1957.

Even though all models had been completely

Despite the fact that the entire line had been completely restyled only the previous year, the 1958 Cadillacs received major styling changes. The most significant of these were quad headlights, a "jewelled" grille design and prominent, sweeping tail fins. These new tail fins were similar in style to those first seen on the 1955 Eldorado. For 1958 they appeared on every model in Cadillac's line, from the lowest-priced Series 62 Coupe on up to the Fleetwood Seventy-Five Sedan and Limousine. The Cadillac name appeared in small block letters at the top of the fins on all but the Sixty Special. This is the 1958 Cadillac Series 62 Two-Door Hardtop, Style 6237. At $4,694 it was the least expensive car in the 1958 lineup. A total of 18,736 were built. Shipping weight was 4,630 pounds.

There were two Series 62 Sedans in the 1958 Cadillac model lineup—a "standard deck" version and a new "extended-deck" model. As in 1957, all Series 62 Sedans were actually four-door hardtops. But they were marketed as sedans. Cadillac's 1958 four-door sedans were equipped with small ventilator panes on their rear doors this year. This is the standard 1958 Cadillac Series 62 Sedan, Style 6239, which sold for $4,801. Production totalled 13,335. Weight was 4,675 pounds. Note the new "jewelled" grille, chrome windsplit molding flowing rearward from the hooded dual headlight units into the doors, and the five chrome "speed streaks" on the lower half of the rear doors.

1958

restyled only the previous year, the 1958 Cadillacs got extensive styling changes. The redesigned grille was wider than ever and consisted of a series of sparkling metallic studs instead of the usual cross-hatched texture. Rubber-tipped front bumper/grille guards were smaller and spaced much farther apart at the outer edges of the grille. A heavy chrome windsplit molding extended back from hooded dual headlight units to the center of the front doors. Full-length rear fender fairings were also chrome-trimmed. Small chrome dummy air intakes were mounted on the front fenders ahead of the wheel openings. Five horizontal chrome "speed streaks" embellished the lower body sides.

All 1958 Cadillacs, from the lowest-priced Series 62 Coupe on up to the Fleetwood Seventy-Five Limousine sported long, pointed tail fins similar in appearance to those first seen on the 1955 Eldorado. The Cadillac name appeared in small block letters at the top of the fins on all models except the Sixty Special. The Fleetwood Sixty Special had massive, ribbed aluminum trim panels covering the lower rear fenders and matching full fender skirts. All four-door hardtops had small ventipanes on their rear doors.

Including the exclusive Eldorado Brougham, there were 12 models in Cadillac's comprehensive 1958 model lineup. Eight were Series 62 body styles. An interesting supplemental body style was added to the line this year, for one year only. It was an "Extended Deck" version of the lowline Series 62 Four-Door Sedan. This model was 8.5 inches longer over-all than the base Sixty-Two Sedan. It was a popular choice among Cadillac's customers accounting for the second-highest sales of any model this year. Only the more expensive Sedan de Ville outsold the Series 62 Extended-Deck Sedan. All 1958 Cadillac "sedans", including the two lowline Series 62 models, the Sedan de Ville and premium Fleetwood Sixty Special, were actually four-door hardtops although Cadillac insisted on calling them sedans.

The Eldorado Biarritz Convertible, Eldorado Seville Two-Door Hardtop and the long-wheelbase Fleetwood Seventy-Five Sedan and Limousine also got Cadillac's fresh 1958 styling. Distinctive "Sabre Spoke" wheels continued as standard equipment on the Eldorado Coupe and Convertible but were available as options on other models.

Again, there was no GM Motorama this year. But Cadillac did create a special car for the regular automobile show circuit. It was a slightly-modified Eldorado Biarritz Convertible with a completely automatic top-raising mechanism.

When lowered, the convertible top was concealed beneath a flush-fitting, three-section metal top panel. This panel was powered by four electric motors and "stored" itself when the top was down. A tiny humidity-sensitive panel was located on the upper rear deck between the top cover panel and the rear deck lid. All it took was a few raindrops on this humidity sensor to trigger the top-raising mechanism, even if the driver was nowhere around. The top latched itself to the windshield header and the side windows, if down, raised themselves! To this day it is not clear if this sensor was ever enough to differentiate between raindrops and/or pranksters with a garden hose or glass of water.

Cadillac built only five of these special Biarritz convertibles which also had four bucket seats and luxurious buffed-leather upholstery.

Cadillac's sales were the lowest in four years, dropping to 121,778 in the recession-plagued 1958 model year. Two decades would pass before the industry would be buffeted by a recession of equal intensity.

Even though the industry was in something of a slump, the antique automobile hobby continued to grow at a vigorous rate. Antique Cadillacs and the custom and semi-custom classics had long been formally recognized by the largest and most discriminating antique auto clubs. But there was no club or organization specifically for Cadillac buffs. In the summer of 1958 however, the Cadillac-LaSalle Club was organized by a small group of Detroit area Cadillac devotees anxious to help one another locate parts and technical information for the restoration and appreciation of antique and vintage Cadillacs and LaSalles. The Cadillac-LaSalle Club has since grown into a worldwide organization with several thousand members and its own newsletter, "The Self-Starter."

Cadillac built its two-millionth car on February 7, 1958. It had taken 47 years to build the first million and less than eight for the second.

After a lingering and ignoble death, the once-distinguished Packard nameplate passed into oblivion joining other once-mighty automotive marques including Pierce-Arrow, Peerless, Winton, Locomobile, and Duesenberg. The Packard name was last used on a series of warmed-over 1958 Studebakers. Within eight years even century-old Studebaker would be gone.

With unprecedented fanfare, the Ford Motor Company introduced its long-awaited Edsel. An upscale intermediate, it bombed. The Edsel, with its controversial horsecollar vertical grille, arrived at precisely the time the market for which it was intended dried up. Fifty-Eight was not a very good year.

> Cadillac expanded its four-door sedan offerings with the addition of an "Extended-Deck" sedan in its 62 Series for 1958. Series 62 Sedans (actually four-door hardtops) used the same 129.5-inch wheelbase "X" frame chassis, but the new Extended-Deck Sedan was 8.5 inches longer than the standard model. The standard Sedan had an over-all length of 216.8 inches: the Style 6239-E Extended-Deck Sedan had an over-all length of 225.3 inches. Factory advertised price was $4,989, or $188 higher than the standard Sedan. The new Extended-Deck Sedan was a hot seller, too. A total of 20,952 were built in the 1958 model year, making it the second most popular body style in the entire line.

1958

The modern office buildings at the new General Motors Technical Center in the Detroit suburb of Warren, Mich. were popular backgrounds in GM product publicity photos. Future models could be photographed against attractive real-world backgrounds, away from prying public eyes. The vast Technical Center was opened in late 1955. This is the 1958 Cadillac Series 62 Coupe de Ville Two-Door Hardtop, Style 6237-D posed at the entrance to one of the Tech Center office buildings. A total of 18,414 were sold, slightly fewer than the 18,736 standard Series 62 Coupes delivered this year. At $5,161 the attractive Coupe de Ville was priced $467 higher than the lowline coupe. Note the bright chrome trim on the angled rear roof pillar and the five chrome "speed streak" trim pieces on the lower half of the rear fender.

The Sedan de Ville Four-Door Hardtop was the best-selling single model in the 1958 Cadillac line. A total of 23,989 of these sleekly-styled four-door hardtop sedans were built this year, only 181 more than the previous year. The U.S. auto industry, however, went into a recession in 1958, so this was no small achievement. The new quad-headlight system introduced on most of the industry's 1958 cars featured four 5-¾-inch headlight units instead of the two seven-inch units used previously. This system resulted in much more effective lighting under all driving conditions and far more precise headlamp aiming. The Cadillac Eldorado Brougham introduced the quad headlamp system in 1957. The 1958 Cadillac Sedan de Ville, Style 6239-D, had a factory-advertised price of $5,407 and weighed 4,855 pounds.

Of the 12 models Cadillac offered its customers in the 1958 model year, eight were Series 62 body styles. All utilized the same 129.5-inch wheelbase "X"-frame chassis. One of the most attractive of all the 1958 cars was Cadillac's sexy Series 62 Convertible Coupe. Those long, sweeping rear fenders capped by jaunty, pointed tail fins looked right at home on this sporty soft-top. The 1958 Cadillacs were slathered in chrome, from their new, jewel-studded grilles to heavy horizontal chrome trim on the front fender windsplits which extended into the doors, and massive fairings that stretched the length of the rear fenders. A total of 7,825 Series 62 Convertibles were sold this year. Style 6267, this Cadillac Convertible sold for $5,364.

American passenger car styling entered a new phase in the late 1950s. It was the Age of Chrome and Tailfins, a low point in taste and one of the more forgettable periods in the industry's history. Bigger was perceived to be better, and stylists applied great gobs of chrome and bright exterior trim to their basic designs. The 1958 Cadillac Fleetwood Sixty Special was a good example of the Age of Excess. The entire lower half of the Sixty Special's rear fenders was plastered with brightwork—a massive anodized aluminum ribbed panel with a bright stainless steel molding. Full fender skirts and broad chrome sill moldings also contributed to this car's very heavy appearance. Bright trim even extended across the rear deck. The rocketship tail fins would be even flashier next year.

The American motor vehicle industry slid into a deep recession in 1958. Most U.S. automakers experienced sharp drops in sales. Cadillac was no exception. Model year production slipped to 121,778 from 146,841 in 1957. Although it was extensively restyled this year and retained its traditional 133-inch wheelbase, the 1958 Fleetwood Sixty Special Four-Door Sedan (actually a four-door hardtop since 1957) suffered worse than most other models. Sales plunged by nearly half, from 24,000 in the 1957 model year to only 12,900 in 1958. Style 6039, the 1958 Fleetwood Sixty Special weighed 4,930 pounds and had a factory-advertised price of $6,117. Note the massive ribbed aluminum trim and full skirts on the rear fenders. The rear doors now had small ventilator panes.

1958

This was the second, and final, year for Cadillac's very exclusive first-generation Eldorado Brougham. Although a second-generation Brougham was offered in 1959 and 1960 it was a much-less-distinctive car than the original 1957-58 version. The Eldorado Brougham got only minor changes for 1958. Horsepower was increased to a whopping 335, and the 365 cubic inch V-8 engine was now equipped with three two-barrel carburetors. This extremely powerful engine was optionally available in all other 1958 Cadillacs. The 1958 Eldorado Brougham retained its trim exterior dimensions: 126-inch wheelbase, 55.5-inch height, and over-all length of 216 inches. Styling features included extensive use of stainless steel on the roof, rear fenders, and broad sill moldings. Price remained at a very high $13,074. Only 304 Style 7059 Broughams were sold in the 1958 model year for a total of 704 over the two-year production run. General Motors lost money on every one, but like the V-16 Cadillac of the 1930s, the limited-production Broughams added new lustre to Cadillac's reputation as the leader in the luxury car field in America.

The 1958 Cadillac Eldorado Seville Two-Door Hardtop and companion Eldorado Biarritz Convertible both got Cadillac's revised front-end styling. The new grille was much wider and consisted of a series of fine, jewellike metallic studs instead of the usual chrome crosshatching or "eggcrate" motif. The rubber-tipped front bumperguards were much smaller and were farther apart at the outer edges of the grille. The big, chrome "Dagmars" had all but disappeared. The 1957 and 1958 Eldorados had no hood ornaments, but chrome blades capped each front fender. Eldorado sales dropped sharply this year. Only 855 Seville Style 6237-D Hardtops were sold along with 815 Eldo convertibles. Both models had identical $7,410 price tags.

Cadillac's distinguished Fleetwood Seventy-Five for many years held the distinction as the only true production limousine built in America. Both Ford and Chrysler commissioned small numbers of long-wheelbase limousine conversions of the Lincoln Continental and Chrysler Imperial over the years, but these were handled by outside coachbuilders. Chrysler at this time was selling a very exclusive Imperial limo with a body built by Ghia in Italy. Fleetwood Seventy-Fives could be found wherever the important or wealthy gathered, at home and abroad. The author photographed this 1958 Fleetwood Seventy-Five Nine-Passenger Sedan in front of Sanborn's, one of the finer department stores in Mexico City. Note the heavy chrome window frames and the rear doors that extend into the roof.

Cadillac's two "standard" Eldorado models also got new front-end styling this year. The rear body was pretty much 1957 carryover, however, except for the addition of 10 vertical chrome trim pieces on the lower rear body between the door and rear wheel openings. These chrome louvers were reminiscent of those used on Sixty Specials from 1942 through 1956. There was a new, narrow "V" emblem on the lower left corner of the rounded rear decklid, and the Eldorado name was spelled out in small block letters beside it. Only 815 Eldorado Biarritz Style 6267-S Convertibles were built this year. The $7,410 price tag was the same for both the convertible and the companion Seville two-door hardtop.

Styling of Cadillac's 1958 Fleetwood Seventy-Five Sedan and Limousine was again fully compatible with the rest of the line. This would be the last year for the same basic roofline and formal rear window style used since 1950. Only 1,532 Fleetwood Seventy-Fives were built this year, including 802 Nine-Passenger Sedans and 730 Imperial Sedans with glass division. The Style 58-7523 Nine-Passenger Sedan had a factory price of $8,310 and the Style 7533 Limousine, $8,525. Here, a uniformed chauffeur holds the rear door of a 1958 Imperial Sedan open for a fur-clad lady on her way down to Sak's Fifth Avenue.

Cadillac came up with only one special show car this year. It was a slightly modified 1958 Eldorado Biarritz Convertible that eliminated all manual effort in raising, lowering, and securing the convertible top. When the top was down it was covered by a flush-fitting metal deck, just like on the original 1952 convertible show car that preceded the production Eldorado. The three-section cover was powered by four electric motors and stored itself when the top was up. A small humidity-sensitive panel on the rear deck automatically raised and locked the top at the first drop of rain. (General Motors made no mention of the possible effects of a wayward garden hose stream.) The windows were also automatically raised. Cadillac built five of the 1958 Biarritz convertibles for exhibition at auto shows. The customized, hand-buffed leather interior featured four bucket seats.

Cadillac shipped 1,915 long-wheelbase Series 8680-S commercial chassis to U.S. funeral car and ambulance manufacturers during the depressed 1958 model year. This professional car chassis was routinely restyled every year along with Cadillac's passenger cars. Many funeral directors operated matched fleets of Cadillac funeral cars, flower cars and limousines. This is a 1958 Superior-Cadillac Limousine Combination which could be used as either a hearse or ambulance. Combinations had practical "airline type" rear window drapes instead of the heavy, formal tasselled draperies used in straight hearses.

In late 1956, the A. J. Miller Co. of Bellefontaine, Ohio and the Meteor Motor Car Co. of Piqua, Ohio merged to form the Miller-Meteor Division of Wayne Corp. Operations were combined in Meteor's Piqua plant. This is a 1958 Miller-Meteor high-headroom rescue ambulance built on the 1958 Cadillac commercial chassis. These heavy rescue units were designed to transport as many as four patients at a time. This model was called the "Volunteer" and had 48 inches of headroom. Note the exclusive Miller-Meteor Safe-T-Vu lights on each corner of the roof and the full-length chrome body side molding.

1958

The U.S. auto industry found itself in a slump in the 1958 model year. It was a year the industry would just as soon forget. But there were some bright spots. General Motors marked its 50th anniversary under the banner "Forward from Fifty," and Cadillac built its two-millionth car. It had taken 47 years to produce the first million and only eight to build the second! Cadillac Motor Car Division General Manager James M. Roche, a future GM Chairman, affixes the "2 Millionth" plate to the front bumper of a 1958 Sedan de Ville while a 52-year Cadillac veteran employee, Joe Malachinski, who went to work for Cadillac in 1903, watches from the front seat of a single-cylinder 1903 Cadillac.

The new Miller-Meteor Division came out with some bold and innovative funeral car styling features in 1957 and 1958. Foremost of these were M-M's "Crestwood" funeral cars with simulated wood paneling on their lower body sides. Miller-Meteor also offered some striking two-tone color combinations that were a far cry from the traditional sombre blacks and greys. This is a 1958 Miller-Meteor Cadillac Limousine Combination finished in white with a contrasting dark blue upper body color and hood. The owner's initial appears on a gold crest on the rear side door just below the window sill. A gold "M/M" emblem was affixed to the rear of each front fender just ahead of the door.

1959

In terms of product styling, the 1959 model year was one of the most memorable in Cadillac's long history. Cadillac had precipitated an industry styling craze when it put daring fishtail fins on its first all-new postwar cars away back in 1948. Tail fins soon became Cadillac's most famous styling feature, but with each successive series of new cars these rear fender appendages grew higher and more flamboyant. By the late 1950s they had reached ludicrous proportions and were of questionable taste.

Just mention the 1959 cars and the average person will still recall Cadillac's spectacular "zap" fins! They remain symbolic of the automotive styling excesses of the Fabulous 'Fifties to this day. Never again would Cadillac's stylists be so daring. The 1960 Cadillacs got more subdued styling with much lower, simpler tail fins.

For the third consecutive model year, Cadillac's new cars were extensively restyled. But few people were ready for what GM's stylists had wrought for the 1959 model year. All 12 of the standard-production 1959 Cadillacs, from the lowest-priced Series 62 Coupe on up to the redesigned Fleetwood Seventy-Five Limousine, sported gigantic tail fins. They reared up out of the fenders just behind the doors, sweeping up in chrome-edged, dagger-tipped blades that were skewered by horizontal pods housing a pair of bullet-shaped taillights. The new grille was a glittering cliff of chrome. And as if one toothy grille wasn't enough there was even a dummy grille across the lower rear deck of most models. A thin, horizontal blade divided the jewelled front grille into upper and lower sections. Parking and turn signal lights were paired in pods at the outer ends of the massive new front bumper. The new rear bumper had huge, chrome outer pods with backup lights recessed in their centers.

The new hood was flat and extremely wide. All models had huge expanses of window glass. Thin, new rooflines were supported by slender pillars.

Two separate rooflines with four or six-window styling were offered on Series 62 and de Ville four-door models. Four-window sedans featured an extremely flat cantilever roof with slight rear visor overhang and a huge, single-piece wraparound rear window. The six-window sedan had a more conventional curved roofline with a broad, flat backlight. All standard models had curved "dogleg" windshield posts which intruded slightly into the front seat area.

The two standard 1959 Cadillac Eldorado models lost their once-distinctive rear body styling. The Eldorado Biarritz Convertible and companion Eldorado Seville Two-Door Hardtop now shared body shells with other 1959 Cadillacs but had unique exterior chrome trim. Except for the Fleetwood Seventy-Fives, all 1959 Cadillacs were built on a new 130-inch wheelbase that was incrementally longer than in 1958. Even the redesigned Fleetwood Sixty Special used this chassis which was three inches shorter than the Sixty Special's traditional 133 inches. All 1959 Cadillacs except for Seventy-Fives were 225 inches long from bumper to bumper and continued on low "X"-type chassis frames.

Cadillac's 1959 engine was bored out to 390 cubic inches. Horsepower increased again to 325, and compression ratio went up to 10:5-to-1. All three Eldorado models—Biarritz, Seville and Brougham—were powered by a blistering 345 HP version of this engine equipped with triple two-barrel carburetors. This high-performance engine was optional in all other models. Air suspension became standard on the 1959 Eldorado Biarritz and Seville and was optionally available on all other Cadillacs. New Freon-filled shock absorbers were used on the re-engineered 1959 suspension.

The 1959 Cadillacs were probably the most radically-styled Cadillacs ever placed into actual production. Their towering, dagger-tipped tail fins with podded, dual-bullet taillights became synonymous with the automotive styling excesses of the 1950s. Cadillac, which had started the tail fin styling craze a decade earlier, carried them to ludicrous heights in the 1959 model year. All of Cadillac's 12 standard 1959 production models sported these monstrous rear fender appendages which were not at all in keeping with Cadillac's quietly conservative former image. This is the 1959 Cadillac Series 62 Two-Door Hardtop, at $4,892 the lowest-priced car in the division's 1959 model line. Production totalled 21,947 making the Style 6237 Coupe the second most popular Cadillac body style this year.

The Six-Window Sedan was a trifle more conservative in styling than the panoramic Four-Window Sedan. A Six-Window Sedan was also offered in the more expensive 1959 Cadillac de Ville series. Style 6229, the 1959 Series 62 Six-Window Sedan had a factory-advertised price of $5,080 — the same as the companion Four-Window Sedan. Production totalled 23,461 making the lowline Series 62 Six-Window Sedan the most popular of all 1959 Cadillac body styles. The Six-Window sedan had a more conventional flat backlight and small, fixed quarter windows. All 1959 Cadillac "sedans" with the exception of the Fleetwood Seventy-Fives were actually pillarless four-door hardtops. Lowline Series 62 Cadillacs had no rear fender series nameplates.

1959

Four of Cadillac's 1959 passenger car lines got new numerical series designations. De Villes, up to now part of the 62 Series, became Series 63 Cadillacs. The Eldorado Biarritz and Seville, also formerly Series 62 models, became a new Series 64. Fleetwood Seventy-Fives became "Series 67" models, although the well-known Seventy-Five name was retained in all advertising.

The restyled Seventy-Fives got a smart, new formal roofline with an elegant crease at the rear. This crisply-tailored roofline distinguished the Seventy-Five Sedan and Limousine through the 1965 model year.

The limited-production Eldorado Brougham for 1959 was a completely new car and was redesignated the Series 69. The second-generation Brougham was larger, less complex, and much less distinctive than the 1957-58 Broughams. The 1959 and 1960 Eldorado Broughams were highly unique in that they weren't even built in this country. The all-new Broughams were hand-built by Pinin Farina 4,000 miles from Detroit in Turin, Italy. Pre-tested Cadillac chassis were crated and shipped to Italy by boat. Completed cars were shipped back to the United States and completed in Cadillac's Fleetwood plant in Detroit before they were released to dealers. The 130-inch wheelbase "X"-type chassis, 390 cubic inch V-8 engine and inner body structure were stock 1959 Cadillac, but all exterior sheet metal and numerous other components were unique to this model. Styling was crisp and clean. Only one body style was offered, a handsomely-styled pillarless four-door hardtop sedan.

Much of the sophisticated gadgetry and lavish interior appointments of the earlier Broughams were gone. But one interesting new technical innovation appeared. As the rear door was opened, the triangular quarter window zipped into a recess for ease of passenger entry and exit. Over-all length was 225 inches, the same as standard-production 1959 Cadillacs. Price continued at a very steep $13,074. The all-new 1959 Eldorado Brougham was introduced at the Chicago Auto Show in January, 1959 four months later than the rest of the 1959 models. Only 99 Eldorado Broughams were built in the 1959 model year making them the rarest of all limited-production postwar Cadillacs.

Cadillac fielded only one experimental show car for the 1959 General Motors Motorama, the first held in three years. It was the XP-74 Cyclone, a truly exotic dream car right out of a science fiction film.

The small 104-inch wheelbase Cyclone had a bubble roof canopy that stowed itself in the rear deck when an open car was desired. Two huge black-tipped nose cones housed a proximity-warning system that alerted the driver to the impending approach of another vehicle or other object. The Cyclone's doors extended out three inches at the touch of a button, then slid back parallel with the car's tubular-shaped body. Huge, sharklike fins jutted out of the rear fenders and smaller fins were built into full-length skeg moldings that ran the length of the body sides. These skeg moldings went into production on the 1961 Cadillacs.

The 1959 Cadillac commercial chassis became the Series 6890. The flamboyantly-styled 1959 Cadillac chassis presented a real challenge to funeral car designers who had to come up with something dignified for this Buck Rogers chassis. Superior Coach Corp. of Lima, Ohio introduced a radically-styled new hearse called the Crown Royale Landaulet. Superior also introduced a 12-passenger super station wagon similar in concept to Hess and Eisenhardt's earlier Broadmoor Sky Views. Superior's new "Caravelle," with tinted glass roof panels, was built on the 1959 Cadillac commercial chassis and utilized the same body shell the company used for its funeral cars and ambulances.

The U.S. economy, and passenger car sales, improved during the 1959 model year but failed to set records. The "Big Three" U.S. auto manufacturers were preoccupied with launching their first compact cars—Chevrolet's Corvair, Ford's Falcon, and Chrysler's Valiant. All three were introduced late in 1959 as 1960 models. A new decade was about to begin.

Cadillac offered its customers a choice of two distinctively different roof styling treatments on its 1959 four-door hardtops. Buyers could have either four or six-window upper body styling at no difference in base vehicle price. This is the 1959 Cadillac Series 62 Four-Window Sedan, Style 6239, which had a factory-advertised price of $5,080. The companion Six-Window Sedan was priced identically. Also available in the more expensive de Ville series, the Four-Window Sedan featured an extremely flat, cantilevered roofline with thin rear roof pillars and a huge, single-piece wraparound rear window. A total of 14,138 of these Series 62 Four-Window sedans were built during the 1959 model year. Shipping weight was 4,770 pounds.

Since the introduction of the Coupe de Ville during the 1949 model year, all Cadillac de Ville models were listed as Series 62 body styles. But in 1959, the de Villes became Series 63s. All Series 62 and 63 models for 1959 were built on the same new 130-inch wheelbase. The only way to tell 1959 de Ville models apart from the less expensive Series 62s was by the small script de Ville nameplates on the rear fenders above the horizontal chrome side molding. Style 6337, the 1959 Cadillac Coupe de Ville had a factory-advertised price of $5,252. Production totalled 21,924 — slightly fewer than the Series 62 Coupe. This profile view shows the 1959 Cadillac's thin roofline and flamboyant rocketship tail fins.

1959

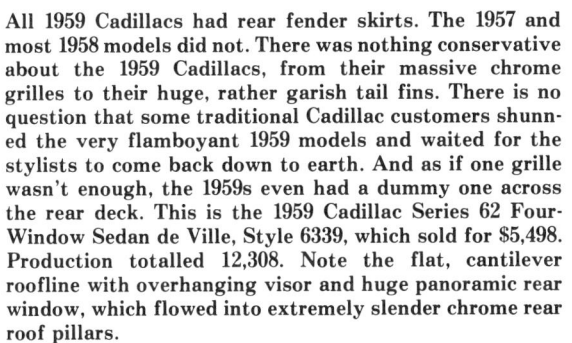

All 1959 Cadillacs had rear fender skirts. The 1957 and most 1958 models did not. There was nothing conservative about the 1959 Cadillacs, from their massive chrome grilles to their huge, rather garish tail fins. There is no question that some traditional Cadillac customers shunned the very flamboyant 1959 models and waited for the stylists to come back down to earth. And as if one grille wasn't enough, the 1959s even had a dummy one across the rear deck. This is the 1959 Cadillac Series 62 Four-Window Sedan de Ville, Style 6339, which sold for $5,498. Production totalled 12,308. Note the flat, cantilever roofline with overhanging visor and huge panoramic rear window, which flowed into extremely slender chrome rear roof pillars.

Cadillac's customers could choose from two distinctly different upper body and roof styling treatments on 1959 four-door sedans. Four and six-window sedans were offered in both the lowline Series 62 and the more expensive Series 63 de Ville Series. This is the 1959 Cadillac Six-Window Sedan de Ville, Style 6329, which had a factory-advertised price of $5,498, the same as the Four-Window Sedan de Ville. The more conservatively-styled Six-Window model was preferred by the majority of de Ville customers this year, with 19,158 Six-Window Sedans delivered compared with 12,308 Four-Window Sedans de Ville. The curved "dogleg" windshield posts were used on most 1959-60 Cadillacs and on the Fleetwood Seventy-Fives through the 1965 model year.

Cadillac's standard convertible coupe remained in the lowline Series 62 for the 1959 model year. Cadillac's notorious 1959 tail fins looked more appropriate on the flashy convertibles than they did on most closed body styles. The taillight fairings built into the fins were painted body color on most 1959 Cadillacs, but were chromed on the Sixty Special. The flamboyant Series 62 Convertible Coupe, Style 6267, was a big seller; a record 11,130 were sold in the 1959 model year. With its top up or down, this was one sexy automobile. Factory-advertised price was $5,455. Eight finely-grained leather interior color combinations were offered. The full-length chrome side moldings tapered to a point just behind the front wheel opening.

Cadillac's 1959 model lineup included an all-new, second-generation Eldorado Brougham. But this Brougham did not come from the Fleetwood plant in Detroit. It was hand-built 4,000 miles away in the Turin, Italy shop of Pinin Farina. Pre-tested Cadillac chassis were crated and shipped to Italy by boat. Completed cars were shipped back to Detroit for final finishing, testing, and shipment to dealers. The 130-inch wheelbase "X"-frame chassis and inner body structure were stock 1959 Cadillac, but no exterior body panels were interchangeable with other models. Only 99 Eldorado Broughams were built during the 1959 model year making this one of the rarest of all limited-edition Cadillacs. Some of the 1959 Brougham's styling features showed up on later Cadillacs. Note the thin, graceful roofline and huge, flat windshield that did away with the "dogleg" pillars which had impeded front-seat entry and exit on all Cadillacs since 1954.

The prestigious Cadillac Fleetwood Sixty-Special Four-Door Sedan was completely redesigned again this year. Wheelbase was reduced three inches to 130 inches, the same as the 1959 Series 62 and 63 models. The new Sixty Special, Style 6029, was given its own distinctive bodyside trim which included a fine chrome molding that extended from the tip of the front fender to the rear bumper, then along the lower body side to the front wheel opening. This molding encircled a prominent dummy air intake built into the rear fender. The Fleetwood name appeared in small block letters on the lower front fenders. The taillight pods faired into the huge tail fins were chromed. Actually a six-window four-door hardtop, the Sixty-Special Sedan was a very heavy-looking car this year. Factory price was $6,233 and 12,250 were built.

The all-new 1959 Cadillac Eldorado Brougham lacked the sophistication and individuality of the first-generation of 1957-58 Broughams. The larger second-generation Broughams had more restrained styling than Cadillac's other 1959 models and some of its styling features were seen on later production cars. The thin, crisp upper body and roofline styling was used on Cadillacs through the mid-1960s, and the low, subdued fins showed up on Cadillac's 1960 cars. The Eldorado's price remained at a stratospheric $13,074 for the third straight year. The Brougham was redesignated the "69" Series this year and was shown as Style 6929. Standard engine in the ultra-luxurious 1959 Eldorado Brougham was a 345-horsepower version of Cadillac's 390 CID V-8 equipped with triple carburetors.

1959

Cadillac juggled some of its series identifications this year. The 1959 Eldorado convertible and hardtop, previously much-glamorized Series 62 models, were redesignated Series 64 cars. The 1959 Eldorado lost its distinctive rear body styling this year but got its own special exterior chrome trim treatment. Broad chrome sill moldings swept up around the ends of the rear fenders and continued in a wide chrome swath along the upper rear fender contour, terminating at the base of the windshield. The Eldorado name appeared in small block letters on the lower edges of the front fenders. This is the 1959 Cadillac Eldorado Biarritz Convertible, Style 6467. Production totalled 1,320. The Biarritz Convertible's hefty $7,401 price tag was identical to that of the companion Eldorado Seville Two-Door Hardtop.

Cadillac's 1959 engine was bored out to 390 cubic inches. Compression inched up to 10:5-to-1 and horsepower was boosted to 325 BHP on standard Cadillacs and to a tremendous 345 on Eldorados. This very powerful engine, with three two-barrel carburetors, was available as an extra-cost option on all other 1959 Cadillacs. The Eldorado name appeared in small block letters on the lower edge of the front fenders on the Biarritz Convertible and Seville Hardtop. It was also spelled out in a narrow chrome band on the lower edge of the decklid. The distinctive Eldo had new deep-dish wheel covers which it shared with the 1959 Sixty Special. Air suspension, Cadillac's automatic Cruise Control, Autronic Eye headlight dimming system, and power door locks were all standard on the prestigious Eldorado.

For the same steep $7,401 base vehicle price, Cadillac customers could choose between the dazzling 1959 Cadillac Eldorado Biarritz Convertible or the more practical Eldorado Seville Two-Door Hardtop. Both models utilized standard 1959 Cadillac body shells but had their own, distinctive exterior chrome trim. Wheelbase of 130 inches was shared with Series 62 and 63 models. Standard equipment on the Eldorado Seville was a color-keyed, fabric-covered roof. Style 6437, the Seville turned in a marginally better sales performance this year, with 975 delivered. There were still three Eldorado models in the Cadillac lineup: Seville, Biarritz, and a brand-new 1959 Eldorado Brougham.

The distinguished Fleetwood Seventy-Five Sedan and Limousine for 1959 got their first complete road-to-roof restyling since 1950. These very impressive cars wore Cadillac's massive, new chrome front end and soaring tailfins very well. The redesigned Seventy-Fives received a crisp, new formal roofline that was continued on this series through the 1965 model year. The 1959 Seventy-Fives also got a new internal series designation — Series 67. The 1959 Cadillac Fleetwood Seventy-Five Nine-Passenger Sedan illustrated was listed as Style 59-6723. Only 710 were built. Factory advertised starting price was $8,750. All 1959 Cadillacs, including the big Seventy-Fives, continued on an "X"-type chassis frame. Wheelbase remained at 149.8 inches.

Like the rest of the 1959 Cadillacs, the Fleetwood Seventy-Five Sedan and Limousine were completely restyled. Even the stately Seventy-Fives sported Cadillac's soaring new tail fins. The 1959 Series Seventy-Fives were redesignated Series 67 models, although the well-known and respected Fleetwood Seventy-Five name was retained in all advertising. This is the Style 59-6733 Fleetwood Seventy-Five Nine-Passenger Limousine, only 690 of which were built this year. Base price was $9,748. The chauffeur's compartment was upholstered in a choice of black, grey, or fawn leather complementing the wool broadcloth and Bedford cord upholstery in the rear passenger compartment. A fixed partition with adjustable glass division was standard.

1959

All three major auto manufacturers provided special one-off limousines for Queen Elizabeth II's 1959 Royal Tour of Canada and the United States. Ford supplied a modified Lincoln Continental convertible, Chrysler a special Imperial-Ghia limousine with open rear, and General Motors of Canada this custom-built 1959 Cadillac Fleetwood Seventy-Five Limousine. The rear roof was cut away in classic landaulet fashion. This car could be paraded with a fully-open rear quarter or equipped with a detachable clear plastic top. The Cadillac Royal Tour Limousine was photographed at the Windsor, Ontario airport during a test-fitting in a Royal Canadian Air Force C-119 Flying Boxcar, which airlifted the Royal Limousines around the country. Note the royal standard on the roof.

What's this? A 1959 Cadillac convertible sedan? That's sure what it looks like, but upon closer examination it's a cut-down Series 62 four-door. The roof has been hacked away just behind the windshield header and at the base of the rear roof pillars. Presumably there is plenty of boiler plate and other reinforcing in the underbody and rear doors. This novel Cadillac parade car showed up at a firemen's parade on the East coast. It was photographed by fire apparatus buff Dan Martin who knows something unusual when he sees it. Cadillac built its last convertible sedan in 1941.

The somewhat bizarre 1959 was simply too far-out for some luxury car buyers. It was not surprising, then, that several independent coachbuilders did their own thing with 1959 Cadillacs for affluent customers who wanted something different. This Raymond Loewy restyle of a 1959 Cadillac sedan was about as far removed from the finny production 1959 as you could get. Styling is unusually clean and beautifully integrated. Taillights are faired into the rear fenders. About the only remaining resemblance to the production 1959 Cadillac is in the roofline and doors. The wide whitewall tires and special Studebaker-type wheel discs further mask this car's basic identity.

Cadillac's stylists succumbed to "fin fever" this year. Fin madness even extended to the division's only 1959 experimental show car. The XP-74 Cadillac Cyclone was right out of a science fiction movie. Built on a 104-inch wheelbase, it was equipped with a power-operated, rear-hinged bubble canopy that raised for entry and exit and stowed in the rear deck when an open car was desired. The doors extended out three inches at the push of a button then slid back along the side of the car. Two huge black-tipped nose cones housed a proximity warning system. A warning light and audible signal warned the driver of the imminent approach of another car or object. The Cyclone also had retractable headlights. Its engine was Cadillac's standard 325 BHP V-8 coupled to a three-speed transmission and final drive unit located behind the two-passenger cockpit. The finned skeg moldings on the lower body showed up on 1961 production Cadillacs.

No, this is not a 1960 Cadillac, but admittedly those low tail fins make it look like one. It is a 1959 Cadillac "Estate Carriage" offered by Hollywood carrossier Peter Stengel. The handcrafted body was built in England. Front-end sheet metal is stock 1959 Cadillac, but the tail fins have been shaved down resulting in very clean, straight fenderlines exactly like those used on production 1960 Cadillacs, right down to the thin, vertical taillights built into the trailing edges of the fins. The rear bumper, however, is pure 1959. This conversion of a four-door sedan cost about $14,000. The "Estate Carriage" was marketed through selected Cadillac dealers. A more streamlined model was offered with a body built in Italy. Note the missing wheel covers. The interior was trimmed in formica, in natural wood pattern.

The Superior Coach Corp. of Lima, Ohio built at least one special sightseeing coach for the Broadmoor Resort Hotel in Colorado Springs, Colo. on the 1959 Cadillac 156-inch wheelbase commercial chassis. The body shell was the same one Superior used for its funeral cars and ambulances. Like the 1955 and '56 Sky Views built for the Broadmoor by competitor Hess and Eisenhardt, the Superior-Cadillac featured large plexiglass roof panels for sightseeing. Superior briefly marketed this coach under the product name of "Caravelle", with nine or 12-passenger seating. A Superior-Pontiac Caravelle was added to the line in 1960. Note the golden "Broadmoor" nameplate on the tail fin. The Caravelle name appeared on a Renault import at about the same time and was later used on a Canada-only version of Chrysler's Dodge Diplomat intermediate.

1959

Funeral car manufacturers must have winced when they got their first working drawings of the Cadillac commercial chassis they would have for the 1959 model year. Creating a dignified funeral car out of such a flamboyantly-styled car was a real challenge. The Superior Coach Corp. pulled off a real styling coup when it introduced its radically-styled Crown Royale Landaulet Funeral Coach. Suddenly, every other hearse made looked stodgy and out of date. The stunning Crown Royale had a forward-sloped "C" pillar capped with a bright chrome band which continued up over the roof giving new definition to the rear roof area. The rear window was cut back at an opposite angle. Landau bows were nearly straight. The 1959 Superior-Cadillac Crown Royale Landaulet set a new standard in American funeral car design. No other hearse introduced since, except perhaps for Superior's 1961 Crown Sovereign Brougham, had such an impact on the industry.

The Superior Coach Corp. offered two series of funeral cars and ambulances in the memorable 1959 model year. Standard models were called Royales and a new premium series was called the Crown Royale. The 1959 Superior funeral coach line included a restyled Royale Coupe de Fleur flower car. Superior introduced this cut-back roofline on its 1957 flower car and used it through the 1964 model year. The Coupe de Fleur had broad, ribbed aluminum panels covering the lower half of the rear fenders. The stainless steel flower deck was hydralically adjustable. Note the modern script nameplates mounted on the front doors.

The 1959 Miller-Meteor Cadillac funeral coaches and ambulances were not nearly as radical in their styling as rival Superior's. Miller-Meteor offered some effective two-tone exterior color combinations that worked well with Cadillac's bold 1959 styling. This Miller-Meteor Cadillac "Futura" Limousine Combination is a good example. The 1959 Cadillac commercial chassis did not have the dummy rear "grille" used on Cadillac passenger cars. Miller-Meteor's products had wide chrome sill moldings. Superior's did not.

Cadillac's special hearse and ambulance commercial chassis also got a new series designation this year. Previously called the 8680-S chassis, it was now the Series 6890. Cadillac delivered 2,102 of these 156-inch wheelbase chassis to the four major U.S. funeral car manufacturers during the 1959 model run. The flashy 1959 Cadillac chassis was a real challenge to the funeral car stylists but it resulted in some spectacular hearses and ambulances. This is a 1959 Superior-Cadillac Crown Royale Limousine Ambulance. Note the angled "C" pillar, up-and-over bright roof molding and aluminum trim panels on the lower rear fenders. The sweeping "zap" fins on the 1959 commercial chassis had chrome taillight fairings, just like the Fleetwood Sixty Special.

1960

Cadillac entered the new decade with restrained styling and lower, much more subdued tail fins. Thirteen 1960 Cadillacs were offered in three standard and one limited-production custom series. This would be the fourth—and last—year for the exclusive, ultra-luxury Cadillac Eldorado Brougham.

Gone were the soaring rocketship tail fins, massive chrome grilles and heavy, very flamboyant styling of the previous model year. The 1960 Cadillacs featured clean, uncluttered new styling with fine detailing and a minimum of chrome trim. Lower, much more restrained tail fins resulted in a smooth, sweeping horizontal body line from the front fenders through the length of the car. Thin taillights were faired into the trailing edges of the new fins providing distinctive side illumination.

A slightly convex, finely-detailed new full-width grille gave the cleanly-styled 1960 Cadillac a wider, lower appearance. There was no grille center bar this year. Very fine horizontal grille bars between rows of jewel-like metallic ornaments were set at a slight angle to provide gleaming highlights in the delicate over-all design. Broad, visor-like chrome housings contained parking and directional signal lamps at each end of the new front bumper. A new, slimmer Cadillac "V" and crest were mounted on the front of the broad hood. Tiny amber directional signal lights were built into the rear face of new front fender crown moldings. A single, finely-tapered chrome molding extended rearward from the front wheel opening to the rear bumper on Series 62, 63, and Seventy-Five models.

New, vertical oval chrome housings contained two circular lights in the ends of the long, flowing rear fenders. The upper lamp was the backup light. The lower lamp functioned as a taillight, brake light and turn signal, operating in unison with the vertical taillight in the fin above. A horizontally-ribbed cove molding spanned the rear deck on Series 62, 63 and Seventy-Five models. On Sixty Special and Eldorado models this molding contained a grille texture similar to that on the front of the car.

The new Fleetwood Sixty Special was an exceptionally "pure" design with almost no bodyside trim. An extended rocker sill molding outlined the rear fender contour. A small cloisonné crest was mounted low on the front fenders. Nine fine vertical louvers finished in body color with chrome edging appeared on the rear fenders. The 1960 Sixty Special also got a grained, fabric-covered roof. The Eldorado Biarritz Convertible and companion Eldorado Seville Two-Door Hardtop had thin, dual chrome side moldings that outlined the soft new rear fender contours. This would be the fifth and final year for the Seville Hardtop. The Seville name would be resurrected 15 years later on a new, smaller Cadillac.

The 1960 Cadillacs also got large, new wheel discs. These discs featured the Cadillac medallion within concentric rings of brushed chrome or optional white or black baked enamel. Sixty Special and Eldorado models had deeply-fluted, spoked wheel covers with three black circles around the center crest.

Wheelbases and major dimensions of the 1960 Cadillacs were unchanged. Engines were also carry-over. Standard models were powered by a 325-horsepower version of Cadillac's 390 cubic inch V-8 with a single four-barrel carburetor. Eldorado models continued to use a 345-HP version of this engine with tri-carb setup. Engineering refinements for 1960 included improved self-adjusting brakes and a vacuum-released parking brake. With the engine running, the foot-operated parking brake was automatically released when the car was put into gear. A vacuum system replaced electric power in the safety door locks. A redesigned, much smaller Guide-Matic automatic headlight dimming system replaced the Autronic Eye. Inside, there was a new trimmer steering wheel with a chrome horn bar instead of the usual horn ring. The new instrument cluster was deeply recessed and had a curved glass facing. Fine leathers and new fabrics were offered in 103 interior trim combinations.

For the fourth consecutive model year, the new Cadillacs received major styling changes. The 1960 Cadillacs were much more restrained in styling than the flamboyant 1959 models. The towering, rocketship tail fins were gone, replaced with lower, much cleaner fins which gave the 1960 Cadillac a long, unbroken body line from the peak of the front fenders to the rear of the car. The lowest-priced car in the 1960 model lineup (and the last priced below $5,000), was the Series Sixty-Two Coupe, Style 6237, which had a factory advertised price of $4,892. This was a very popular body style with 19,978 Series 62 Coupes delivered in the 1960 model year.

The handsomely-styled Series 62 Six-Window Sedan was the top-selling model in Cadillac's 13-car 1960 product lineup. Production totalled 26,824. Like all 1960 Cadillacs, the Six-Window Sedan had a finely-textured, full-width grille and clean, straight body lines. Wheel covers were also new and incorporated concentric rings of brushed chrome. Style 6229, the 1960 Series 62 Six-Window Sedan had a factory-advertised price of $5,080 which it shared with the companion Series 62 Four-Window Sedan. Weight was 4,805 pounds. Over-all length continued at 225 inches for all 1960 models except for Fleetwood Seventy-Fives.

1960

The extensively-restyled 1960 Cadillac Eldorado Brougham was formally introduced at the 52nd Chicago Auto Show in January, 1960. This would be the last year for the remarkable Brougham, which was again built by Pinin Farina in Italy. While the 1960 Eldorado Brougham's grille was generally similar to that used on standard production Cadillacs, its body styling was again unique. The 1960 Brougham's sides were finely sculptured and featured a full-length lower body skeg molding. The very high windshield with two thin vertical pillars and a flat, obliquely-contoured roofline showed up on production 1961 Cadillacs. The softly-sculptured body side and rear fender lines were transferred virtually intact to the 1962 cars.

Again, the Brougham's small quarter windows retracted when the rear door was opened to facilitate entry and exit. Buyers could choose from 15 special Brougham exterior colors and a wide selection of top-grain leather or wool broadcloth interior upholstery. An impressive list of equipment was standard: cruise control, air suspension, air conditioning, power decklid release, power-operated ventipanes, six-way power seat, heater, radio, Guide-Matic headlamp control, E-Z Eye glass, vacuum door locks, and fog lamps. Over-all length of 225 inches was the same as the standard 1960 Cadillacs'. Height was 55 inches.

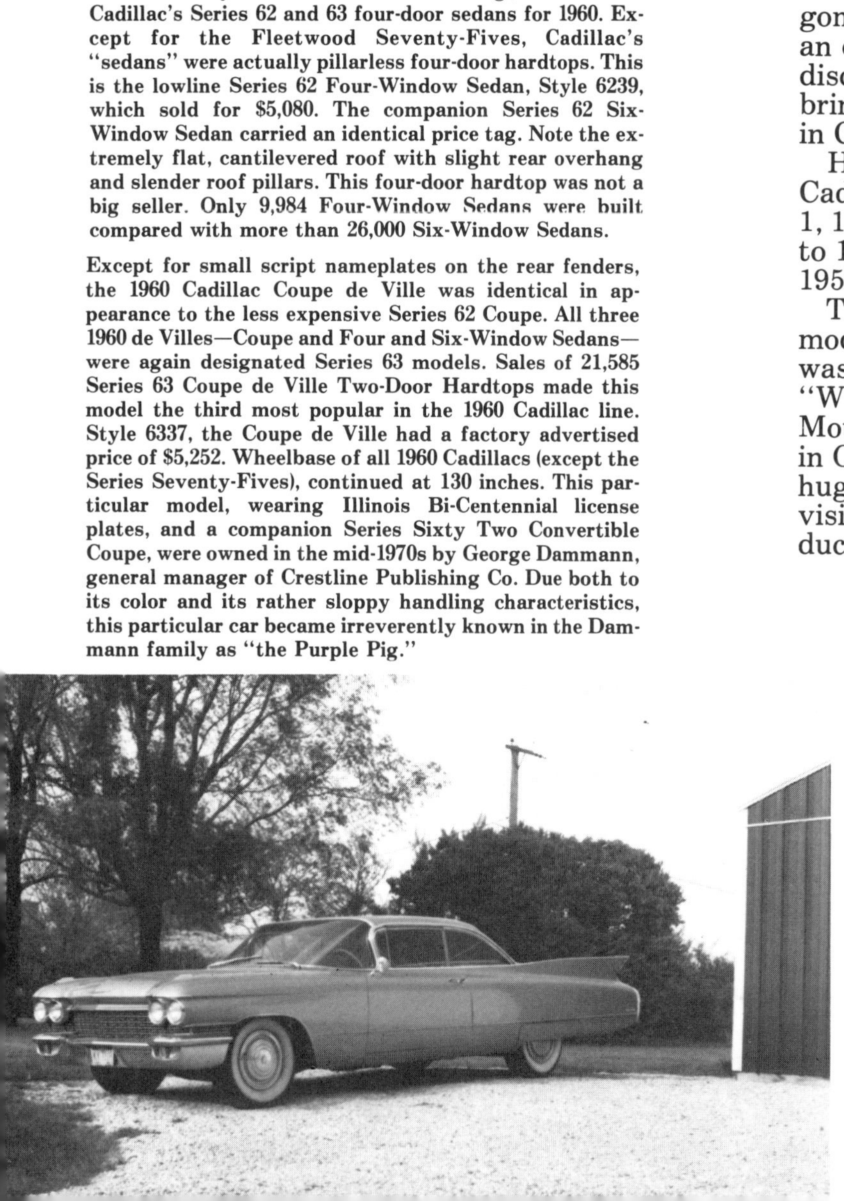

The Eldorado Brougham's chassis, engine and interior body components were again stock Cadillac. Pre-tested chassis were shipped to Italy by boat and completed cars were returned to Detroit for checking and release to dealers. The final Broughams were built in Italy to keep costs down, but it is doubtful if GM ever made a nickel on any of the prestigious Broughams built between 1957 and 1960. The 1960 Brougham was the last true limited-edition "factory" Cadillac ever built. Only 101 were produced in its last year for a total of exactly 200 second-generation 1959-60 Broughams. By comparison, 704 first-generation Broughams had been built in 1957-58 for a grand total of just 904 Broughams in all.

Price remained at a very steep $13,074 for the fourth consecutive year. The ultra-expensive Brougham had served its purpose. The Continental Mark II was long gone and there was simply no longer any need for such an exclusive limited-production car. So it was quietly discontinued at the end of the 1960 model year bringing to an end one of the most intriguing chapters in Cadillac's postwar history.

Harold G. Warner succeeded James M. Roche as Cadillac Motor Car Division General Manager on June 1, 1960. Cadillac's sales dipped very slightly this year to 142,184—exactly 60 fewer cars than in Model Year 1959.

The U.S. auto industry took the wraps off its 1961 models at the 43rd National Automobile Show which was held outside New York for the first time. The "Wheels of Freedom" extravaganza, sponsored by the Motor Vehicle Manufacturers' Association, was held in October, 1960 in Detroit's vast new Cobo Hall. The huge Detroit exhibition, which drew close to 1.5 million visitors, provided a spectacular setting for the introduction of the 1961 Cadillacs.

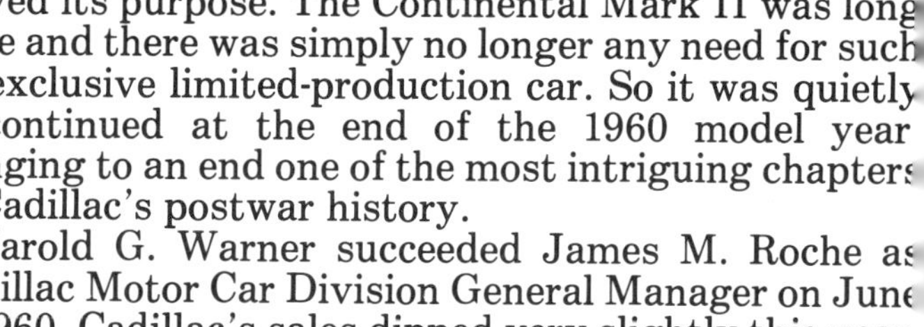

Two distinctly different rooflines were again offered on Cadillac's Series 62 and 63 four-door sedans for 1960. Except for the Fleetwood Seventy-Fives, Cadillac's "sedans" were actually pillarless four-door hardtops. This is the lowline Series 62 Four-Window Sedan, Style 6239, which sold for $5,080. The companion Series 62 Six-Window Sedan carried an identical price tag. Note the extremely flat, cantilevered roof with slight rear overhang and slender roof pillars. This four-door hardtop was not a big seller. Only 9,984 Four-Window Sedans were built compared with more than 26,000 Six-Window Sedans.

Except for small script nameplates on the rear fenders, the 1960 Cadillac Coupe de Ville was identical in appearance to the less expensive Series 62 Coupe. All three 1960 de Villes—Coupe and Four and Six-Window Sedans—were again designated Series 63 models. Sales of 21,585 Series 63 Coupe de Ville Two-Door Hardtops made this model the third most popular in the 1960 Cadillac line. Style 6337, the Coupe de Ville had a factory advertised price of $5,252. Wheelbase of all 1960 Cadillacs (except the Series Seventy-Fives), continued at 130 inches. This particular model, wearing Illinois Bi-Centennial license plates, and a companion Series Sixty Two Convertible Coupe, were owned in the mid-1970s by George Dammann, general manager of Crestline Publishing Co. Due both to its color and its rather sloppy handling characteristics, this particular car became irreverently known in the Dammann family as "the Purple Pig."

The Six-Window Sedan de Ville was far more popular among Cadillac's customers than the companion Four-Window model, even though base vehicle prices were identical. Power windows and front seat were standard equipment on all Series 63 de Villes. The Six-Window Sedan de Ville, Style 6329, retailed for $5,498. Shipping weight was 4,835 pounds. A total of 22,579 Six-Window Sedan de Villes were built during the 1960 model run. The curved "dogleg" windshield posts were used on most 1959 and 1960 Cadillacs and continued on Seventy-Fives through the 1965 model year.

1960

The 1960 Cadillac was a thoroughly restyled car. Gone were the garish rocketship tail fins and heavy chrome accents of 1959. The fresh 1960 look emphasized long, uncluttered body lines and fine, jewel-like details. Brake and back-up lights were housed in vertical oval chrome pods in the ends of the long, low rear fenders. Thin vertical tail lights were built into the trailing edges of the lower, cleaner fins. A horizontally-textured cove molding spanned the rear deck above the redesigned rear bumper. This is the 1960 Cadillac Series 63 Four-Window Sedan de Ville, Style 6339, which listed for $5,498. A six-window Sedan de Ville was also offered at the same price. Only 9,225 Four-Window Sedan de Villes were built this year compared with more than 22,500 Six-Window Sedans.

Sales of the sleek Series Sixty-Two Convertible Coupe set a new record. An even 14,000 of these attractive convertibles were sold during the 1960 model year. This sporty model, Style 6267, wore its clean, new styling especially well. Factory price was $5,455. All 1960 Cadillacs had much more subdued styling than the previous year. Tail fins were lower and cleaner. Full-length chrome body side moldings were finer and were positioned below the center line of the body. The Cadillac name appeared on small rectangular plaques on the front fenders just behind the dual headlights. Whitewalls were becoming narrower now, a trend that would continue well into the sixties.

The Fleetwood Sixty Special Four-Door Sedan (actually a pillarless four-door hardtop) retrieved some of its traditional dignity this year. It shed the ponderous, very busy appearance it had been saddled with since 1958. For the first time, the Sixty Special's roof was covered in a leather-like fabric, heightening the formal effect. Bodyside styling was unusually clean. Broad chrome sill moldings encircled the ends of the rear fenders, and the Sixty Special's distinctive vertical louvers returned for the first time since 1956. Nine of these chrome-accented trim pieces appeared on the ends of the rear fenders. Style 6029, the 1960 Fleetwood Sixty Special sold for $6,233. Production totalled 11,800. Note the small, dark-red cloisonne emblems on the lower front fenders. Wheelbase was again 130 inches and the Sixty Special weighed 4,880 pounds.

The 1960 model year would be the fourth, and last, for Cadillac's exclusive, limited-production Eldorado Brougham. The 1960 Brougham got major styling changes which very accurately predicted the appearance of future Cadillacs. The large, compound curved, windshield with extremely slender pillars and the crisp, thin roofline showed up on the 1961 cars. The full-length body and rear fender sculpturing and lower body skeg moldings were carried over virtually intact on the 1962 Cadillacs. The grille, however, was stock 1960 as were the fluted Eldorado wheel covers. Again, the 1960 Eldorado Brougham was hand-built in Italy for Cadillac by Pinin Farina. But the Pinin Farina name was nowhere to be found. Because the design was strictly GM, Cadillac insisted that only the Fleetwood name appear on the Brougham's door sills. Only 101 Eldorado Broughams were built in the 1960 model year. Price remained at a very steep $13,074 for the fourth straight year.

Only 200 second-generation Cadillac Eldorado Broughams were built in 1959 and 1960. Total Brougham production was just 904 cars between 1957 and 1960. These included 400 in 1957; 304 in 1958; 99 in 1959 and 101 in model year 1960. These were the last true factory-custom Cadillacs ever built. The 1959-60 Broughams were built in Italy, not Detroit. The 130-inch wheelbase "X"-frame chassis and 345-horsepower, 390 CID V-8 engine were stock 1960 Cadillac but no exterior sheet metal was interchangeable with other Cadillac models. Note the stock 1960 rear bumper, small cloisonne emblems on the rear fenders, Eldorado wheel covers and the startling resemblance to 1961 and 1962 production Cadillacs.

1960

As in previous years, Cadillac's premium Eldorado models for 1960 were given some distinctive styling touches. Both the Eldorado Biarritz Convertible and companion Eldorado Seville Two-Door Hardtop Coupe for 1960 had broad chrome sill moldings and fine, dual-beaded chrome moldings that outlined the rear fenders, terminating at the base of the windshield. The two "standard" Eldorados also had finely-textured dummy grilles that spanned their rear decks above the bumper. Style 6467, the 1960 Cadillac Eldorado Biarritz Convertible retained its $7,401 base price for the third straight year. The Seville Hardtop was identically priced. The ritzy Eldo soft-top weighed 5,060 pounds. Only 1,285 were built this year.

This would be the final year for the Cadillac Eldorado Seville. The Seville Two-Door Hardtop had been around since the 1956 model year as a companion car to the premium Eldorado Biarritz Convertible. Both 1960 Eldorados had distinctive styling features including special chrome side trim. The Eldorado name appeared in tiny block letters on the leading edges of the front fenders this year. Also standard were special deep-fluted chrome wheel covers. The Seville hardtop had a textured, vinyl fabric roof covering. Style 6437, the 1960 Eldorado Seville weighed 4,855 pounds and was priced at $7,401 for the third consecutive year. Production totalled 1,075. The Seville name was resurrected on a new, smaller Cadillac in 1975.

The impressive Cadillac Fleetwood Seventy-Five Sedan and Limousine also got Cadillac's fresh, more restrained 1960 styling. This formal sedan body style bore its strong, horizontal body lines very well indeed. As usual there were only two models in the "67" Series—a nine-passenger sedan with two folding auxiliary seats and a formal limousine with glass division. Style 6723, the Nine-Passenger Sedan sold for $9,533 and accounted for 718 sales. The Style 6733 Limousine sold for $9,748 and 832 were built. Both 1960 Fleetwood Seventy-Fives were built on the same 149.8-inch wheelbase and had an over-all length of 244.8 inches. Limousine chauffeur's compartments were upholstered in black, gray, or fawn Florentine leather. Passenger compartments were trimmed in sedate Bedford cords and wool broadcloths.

The Cincinnati, Ohio plant of the Hess and Eisenhardt Co. has turned out some unusual special-order conversions over the years. Anything is available—for a price. Hess and Eisenhardt modified this 1960 Cadillac Fleetwood Seventy-Five Limousine for the du Ponts, scions of the chemical industry fortune. The entire roof was raised several inches so the rear seat passengers could wear silk toppers to and from formal engagements. Note the severely chopped rear roofline. The chauffeur is also isolated from the rear compartment by what appears to be a blind panel. Note also the absence of any body side moldings.

1960

The Hess and Eisenhardt Co. continued to turn out small numbers of Cadillac station wagon conversions. This one, based on a 1960 Cadillac four-door hardtop, was built for Frank Porter, owner of the Central Cadillac Co., a major Cadillac dealer in Cleveland. Styling features included simulated wood side paneling and an extremely flat roofline like that seen on 1959-60 Series 62 and 63 four-window sedans. This very distinctive 1960 Cadillac successfully combined the most desirable features of two body styles—the sporty look of a pillarless four-door hardtop and the utility of a station wagon.

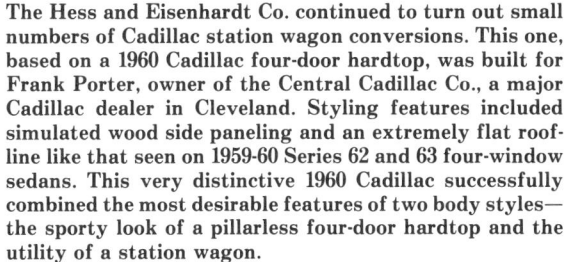

The Eureka Company, of Rock Falls, Ill. was the builder of this 1960 Eureka-Cadillac Combination Landau. This formally-styled coach was designed for use as either a hearse or ambulance, although it looks far more funeral car than ambulance. The red Beacon-Ray light on the roof and "Ambulance" insignia in the rear side windows could be quickly removed when it was to be used for funeral service. Eureka's 1960 Cadillac landaus sported graceful new landau bows. The large chrome coach lamps that distinguished Eureka landaus for many years were gone.

All major funeral car builders also offered special high-headroom ambulances designed for heavy-duty rescue work. These high headroom units were designed to transport as many as four stretcher-patients at a time. This is a 1960 Superior-Cadillac 54-inch Royale Super Rescuer built on the 158-inch Cadillac commercial chassis. The huge roof panel was a single piece of molded fiberglass which kept vehicle weight down.

Cadillac supplied 2,160 special long-wheelbase commercial chassis to the four major U.S. funeral car and ambulance manufacturers during the 1960 model run. The Series 6890 commercial chassis had a 158-inch wheelbase. This is a 1960 Superior-Cadillac Royale Coupe de Fleur flower car. Note the fluted aluminum panel on the lower half of the rear fender. These trim panels were also used on Superior's premium Crown Royale models this year. The coupe-style flower car has an adjustable stainless steel flower deck and a full-length casket compartment. Some funeral directors used this type of coach as a "floral hearse" in lieu of a conventional flower car.

1961

This was another major styling change year for the Cadillac Motor Car Division. All of the 1961 Cadillacs got crisp, new styling with heavily sculptured body sheel metal and finely-detailed grilles and exterior ornamentation. Underneath, there were some significant engineering improvements and refinements.

The redesigned 1961 Cadillac lineup included 12 models in four series on just two wheelbases. Series 62, Series 63 de Ville and Sixty Special models shared a common 129.5-inch wheelbase "X"-frame chassis. The top-line Fleetwood Seventy-Fives were built on a 149.8-inch wheelbase. The hyper-expensive Eldorado Brougham was gone, as was the Seville two-door hardtop, leaving the Biarritz Convertible the only Eldorado left in the line. Even the Biarritz was sharply downgraded. It was now a Series 63 body style, but was not listed as a de Ville.

The 1961 Cadillac was a totally restyled car. Cadillac's fresh, new 1961 styling featured sculptured body sheet metal, a redesigned windshield that did away with the "dog leg" pillars of 1959-60 and thin, extremely flat rooflines. Tail fins were low and sharply pointed. All 1961 Cadillacs except the Seventy-Fives were built on an identical 129.5-inch wheelbase. Lowest priced of the five 1961 Series 62 Cadillacs was the Sixty-Two Coupe, Style 6237, which carried a factory-advertised price of $4,892. This six-passenger coupe weighed 4,560 pounds. Note the huge glass area and slender windshield and roof pillars. New, deep-dish, fluted wheel covers were color-keyed to the body color. Production totalled 16,005 of this model.

Cadillac's all-new 1961 styling featured a wide, finely-textured grille design that was "V"eed" vertically at its center as well as horizontally. Dual headlights were housed in bright chrome bezels that peered out from under the broad new hood. Front fenders were cut away around the headlights. The front bumper was a thin chrome blade. Parking or fog lamps were contained in large round openings in the lower bumper. All 1961 Series 62 Cadillacs carried small Cadillac crests on their lower front fenders. This is the 1961 Cadillac Series 62 Six-Window Sedan, Style 6229, which was priced identically with the companion Four-Window Sedan at $5,080. Shipping weight was 4,680 pounds. It was the second most popular body style in the 1961 Cadillac line, outsold only by the Series 63 Six-Window Sedan de Ville. A total of 26,216 Series 62 Six-Window Sedans were built this year.

A new Short Deck Sedan joined the Series 62 line. This "compact" Cadillac was seven inches shorter than the other standard production 1961 models, with an over-all length of 215 inches compared with 222 inches on the other cars. It was a poor seller, however, with only 3,756 delivered during the 1961 model year. Cadillac's customers evidently were not much interested in a junior edition, at least at this point in the marque's history.

The 1961 Cadillacs were .1 to 1.7 inches higher than the 1960 cars, but their completely redesigned bodies offered improved head and leg room, passenger seating comfort and ease of entry and exit.

All 1961 Cadillacs were powered by the same 325 horsepower, 390 cubic inch V-8 engine used previously. The front suspension was redesigned and a "lifetime" lubrication system eliminated all chassis lubrication fittings, minimizing maintenance requirements. There was a new hydraulic power steering pump under the hood which permitted the use of a new, smaller 16-inch steering wheel in all models.

New options included a new rear window defogger, a new non-slip differential and an anti-smog crankcase ventilation kit that was standard on all Cadillacs sold in California but which was optionally available everywhere else. The 1961 Cadillacs also got a new single exhaust system said to offer twice the life of the dual exhaust system used previously.

Cadillac's fresh, new 1961 styling had been "previewed" on the 1959 and 1960 Eldorado Broughams. The 1961 Cadillac's tall, flat windshield with thin vertical pillars was taken almost intact off the final-generation Brougham. Bodies were crisp-lined and stylishly sculptured with very strong, sweeping horizontal lines and a full-length lower body skeg molding that terminated in an inverted fin. This skeg molding was first seen on the 1959 Cadillac Cyclone show car. Cadillac's traditional tail fins were of modest height and sharply pointed. Vertical taillights were faired into their trailing edges. Rooflines were very thin and flat, again reflecting the styling influence of the earlier limited-production Broughams.

The extensively redesigned 1961 Cadillac line continued to offer four and six-window sedans in both Series 62 and 63. Like the 1959 and 1960 four-window hardtops, the 1961 four-window sedans featured huge wraparound rear windows and cantilevered rooflines. Style 6239, the Series 62 Four-Window Sedan had a factory-delivered price of $5,080. Only 4,700 were built, making the Sixty-Two Four-Window Sedan one of the least popular models in the 1961 line. Note the full-length upper body side sculpturing which was complemented by a lower body skeg molding that terminated in an inverted fin—a styling feature taken from the 1959 Cadillac Cyclone experimental show car.

1961

The redesigned 1961 grille spanned the full width of the car and featured a finely-textured, jewelled look. It was "Veed" both vertically and horizontally at its center. The hood was broad and flat with fine chrome crown moldings atop the front fenders, which were cut away around large chrome dual headlight bezels. Brake and backup lights were paired in horizontal oval chrome pods at the base of the fins. Tail fins and the lower body skeg moldings were edged with fine chrome trim. Body sculpturing extended across the sloped rear deck. New wheel discs were of a deep dish, fluted design and were color-keyed to the car's lower body color.

The 1961 Cadillac Eldorado Biarritz Convertible lost its individuality and was now virtually indistinguishable from the standard Series 62 Convertible, although the Eldo continued to offer many standard luxury features that were extra-cost options on other models and had premium leather interior trim.

The 1961 Fleetwood Sixty Special was given a crisp, new formal roofline with closed upper quarters and a smart crease at the rear. The prestigious Fleetwood Seventy-Fives also got Cadillac's new sculptured styling but carried over the upper body greenhouse of 1959-60, with its high, compound-curved windshield and dogleg pillars. This upperstructure was retained on Seventy-Fives through the 1965 model year.

The 1961 Cadillacs went on sale October 3, 1960. The industry previewed its 1961 cars at the National Automobile Show in Detroit's new Cobo Hall. The U.S. auto industry was buffeted by yet another minor recession. Cadillac built 138,379 cars during the 1961-model production run, slightly fewer than the 142,184 built in 1960.

Cadillac's principal rival, Lincoln, came out with a completely redesigned car for 1961. There were only two body styles compared with Cadillac's dozen. But one was a stunning Lincoln Continental Four-Door Convertible Sedan, the first built by a North American manufacturer in many years. The impressive Continental four-door convertible was offered through the 1967 model year.

The new American-built compact cars were now the rage of the industry and for a time successfully held the imports at bay. Even Cadillac offered a "compact" model this year: it was the Series 62 "Short Deck" Sedan. The 1961 Cadillac Series 62 Short Deck Sedan, Style 6299 had an over-all length of 215 inches—seven inches shorter than the rest of the standard 1961 line. Except for its shorter rear deck and fenders, it was identical in appearance to the Series 62 Six-Window Sedan which had an over-all length of 222 inches. But the Short-Deck Sedan was a poor seller. Only 3,756 were built during the 1961 model production run. Customers were apparently not very interested in a "small" Cadillac.

The convertible body style was nearing the peak of its popularity among American car buyers, and a Cadillac convertible was still the flashiest thing on the road. Sales of Cadillac's Series 62 Convertible Coupe set a new record this year. A total of 15,500 were sold. Style 6267, the 1961 Cadillac Series 62 Convertible carried a $5,455 price tag. Weight was 4,720 pounds. Taillights were faired into the trailing edges of the sharply-pointed fins, and brake and backup lights were housed in horizontal oval chrome pods at the outer ends of the rear bumper. All 1961 Cadillacs had a new single exhaust system. Note the strong sculptured body lines that extended across the sloping rear deck.

The 1961 Series 63 Cadillac de Villes were simply more lavishly-trimmed and appointed versions of the price-leading Series 62 cars. Body styles were generally identical, except that there was no Short Deck Sedan in the 1961 de Ville line. All Series 63 Cadillacs had tiny "de Ville" script nameplates on the leading edges of their front fenders just behind the headlights but no horizontal front fender crests like those used on 1961 Series 62 models. Thlis is the 1961 Cadillac Series 63 Coupe de Ville, Style 6337, which carried a factory-delivered price of $5,252, unchanged from the previous year. Weight was 4,595 pounds. Sales of 20,156 Coupe de Villes fell slightly below those of 1960. Note the thin, flat roofline and new curved windshield pillars which replaced the awkward dogleg posts used in 1959 and 1960.

1961

For some reason, Cadillac's four-window sedans (actually four-door hardtops) were poor sellers. In both the 62 and 63 Series, the identically priced companion six-window sedans outsold them five-to-one. Cadillac's 1961 Four-Window Sedan featured a thin, flat cantilevered roofline and a huge, single piece wraparound rear window similar in design to those used on the 1959 and 1960 four-window sedans. But the '61 models had no rear visor overhang. This is the 1961 Cadillac Four-Window Sedan de Ville, Style 6339, which had a factory-delivered price of $5,498 and weighed 4,715 pounds in stock form. Only 4,847 were sold compared with sales of more than 26,000 Six-Window Sedan de Villes.

Cadillac's 1961 model lineup included four Series 63 models—the Coupe de Ville, four and six-window Sedan de Villes and the downgraded Eldorado Biarritz Convertible. All were built on the same 129.5-inch wheelbase "X"-frame chassis and had an over-all length of 222.0 inches. The Six-Window Sedan de Ville, Style 6329, was the most popular body style in the entire 1961 Cadillac line. Production totalled 26,415—199 more than the lower-priced Series 62 Six-Window Sedan. The Series 63 Six-Window Sedan de Ville carried a factory-delivered price of $5,498—exactly the same as the much less popular Series 63 Four-Window Sedan de Ville. Note the very thin upper body structure with its new windshield design and extremely slender roof pillars.

Pity the poor Eldorado Biarritz! Since 1956 one of the most distinctive of all Cadillacs, the Biarritz lost nearly all of its individuality in the 1961 model year. The 1961 Eldorado Biarritz Convertible was virtually indistinguishable from the less expensive Series 62 Convertible. The only visible differences were tiny Biarritz script nameplates on the front fenders just behind the headlights. The Eldorado Biarritz was repositioned into the 63 Series this year but was not called a de Ville. Style 6367, the 1961 Cadillac Eldorado Biarritz Convertible sold for $6,477—$1,022 higher than the Series 62 soft-top. The Eldo Biarritz had premium interior trim, however, featuring ostrich grain leather with Florentine leather trim in tones of Sandalwood, Nautilus Blue, White, Topaz, Jade, Mauve, Black or Metallic Red. A total of 1,450 Biarritz Convertibles were built this year—165 more than in 1960.

The 1961 Cadillac Fleetwood Sixty Special Sedan retained its distinctive, very formal styling for 1961. The premium Sixty Special got a crisp new roofline this year similar in design to the one on the Fleetwood Seventy-Five Sedan and Limousine. This handsome new roof treatment featured closed upper quarters with a smart crease at the rear. The backlight was framed by a color-keyed molding instead of chrome, heightening the formal effect. The roof was covered with simulated leather. Six fine vertical chrome trim piece ornaments were positioned at the ends of the rear fenders, and again there were cloisonne emblems on the lower front fenders. Sales increased significantly to 15,500. The Style 6039 Fleetwood Sixty Special had a factory-delivered price of $5,233 and weighed 4,770 pounds.

At the top of Cadillac's 12-model 1961 product range were the stately Fleetwood Seventy-Fives—a nine-passenger sedan and a formal limousine. The big Seventy-Fives also got Cadillac's fresh, new 1961 styling, but retained the upper body structure of the 1959-1960 models. This carryover greenhouse included a very high, compound-curved windshield with "dogleg" windshield pillars and the same formal roofline originally introduced for 1959. Both 1961 Fleetwood Seventy-Fives rode on a 149.8-inch wheelbase. Over-all length was 242.3 inches. This is the 1961 Cadillac Fleetwood Seventy-Five Nine-Passenger Sedan, Style 6723, which sold for $9,533. Only 699 of these big Seventy-Five Sedans were built this year. Weight was 5,390 pounds.

1961

With the elimination of the ultra-expensive Eldorado Brougham, the Fleetwood Seventy-Five Limousine once again became the most expensive car in the Cadillac family. The stately 1961 Cadillac Fleetwood Seventy-Five Limousine with standard glass division carried a factory-advertised price of $9,748. A total of 926 were built. Like the Fleetwood Seventy-Five Sedan, the Limo was a nine-passenger car with two folding auxiliary seats. The chauffeur's compartment of the Style 6733 Limousine was trimmed in black, gray, or fawn Florentine leather. The passenger compartment was upholstered in gray or fawn wool broadcloth; or gray, fawn, or dark gray Calais cord with matching broadcloth bolsters and trim. Sumptuous was the word.

One of the Hess and Eisenhardt Co.'s specialties is the armor-plating of standard production passenger cars. The Cincinnati, Ohio firm will fortify anything on wheels, from inconspicuous-looking Fords and Chevies on up to Cadillac limousines—for a price. Hess and Eisenhardt armor-plated this 1961 Cadillac Fleetwood Seventy-Five Limousine for Premier Abdel Karim Kassem of Iraq. Note the much-modified windshield and side window openings, which were altered to accommodate extremely thick bulletproof glass. Such automotive security didn't come cheap: typical conversions cost several times the price of the base car.

The Cadillac Motor Car Division shipped 2,204 Series 6890 commercial chassis to U.S. funeral car and ambulance manufacturers during the 1961 model year. The 156-inch wheelbase hearse and ambulance chassis featured the same new styling as the rest of the line but retained the high, compound-curved windshield and "dogleg" windshield posts used in 1959 and 1960 that were discontinued on all other 1961 models except the Fleetwood Seventy-Fives. This is a 1961 Eureka-Cadillac Side-Servicing Landau Funeral Coach. It is finished in a striking white with white crinkle-finish roof and color-keyed wheel covers. The 1961 Cadillac's sharply-pointed tail fins were edged with chrome, as was the full-length lower skeg molding.

The Hess and Eisenhardt Co. built far fewer flower cars than its rival funeral car builders—Superior, Miller-Meteor, and Eureka. The company's S & S flower cars were true custom creations and were quite rare. Hess and Eisenhardt by this time was producing only a few a year and in some model years didn't even build one. This is a 1961 S & S coupe-style flower car with open flower well. Note the short skeg molding used only on the rear fender. The notched coupe roofline was identical to that being used at the same time by Superior and Miller-Meteor on their flower cars. This impressive professional car was built on the 156-inch wheelbase 1961 Cadillac commercial chassis.

Miller-Meteor of Piqua, Ohio, was the builder of this very efficient-looking Cadillac ambulance. It is a standard-headroom "Sentinel" model with Miller-Meteor's distinctive roof warning lights. The siren was concealed behind the grille. Miller-Meteor used this same body shell for its "Futura" limousine funeral cars and combinations. Miller-Meteor offered a wide range of bold two and three-tone exterior color combinations on its funeral cars and ambulances. The Sentinel Ambulance had 42.5 inches of headroom. Several high-headroom models were also available, including a monstrous "Guardian" rescue unit with a hooded rear door.

1962

The Cadillac Motor Car Division marked its 60th anniversary in 1962 and chalked up its first new sales record in six years.

Cadillac's 1962 cars showed relatively modest styling changes from the previous year, but one new model joined the line raising the division's 1962 model count to 13. Again, there were four series on two wheelbases with three different over-all lengths. Series 62, 63 de Ville and Sixty Special models were all built on the same 129.5-inch wheelbase "X"-frame chassis. Seventy-Fives rode on a 149-inch wheelbase.

The 1962 Cadillac family included two short-deck sedans, one each in the Series 62 and 63 de Ville series. These four-door hardtops were seven inches shorter than the standard sedans in their series. The lowline Series 62 short-deck model was called the Town Sedan. The new Series 63 short-deck sedan was called the Park Avenue Four-Window Sedan de Ville. Only 5,200 short-deck sedans, 2,600 in each series, were sold during the 1962 model year: Cadillac's traditional customers were still showing little interest in "compact" cars. The short-deck sedans had an over-all length of 215 inches compared to 222 inches on Cadillac's standard four-window sedans.

There were five Series 62 Cadillacs for 1962. The line-up included the Sixty-Two Coupe; four and six-window Sixty-Two Sedans; the short deck Town Sedan, and the hot-selling Series 62 Convertible Coupe. The Series 63 line was expanded to five models—Coupe de Ville, four and six-window Sedan de Villes, and Eldorado Biarritz Convertible.

The premium Biarritz Convertible, however, was not listed as a de Ville. The only Eldorado left in the line, the Biarritz got its own exterior trim treatment this year, setting it apart from the nearly identical Series 62 Convertible. Sporty bucket seats were a no-cost option in the Eldorado Biarritz, which also offered a long list of standard luxury equipment and a wide choice of premium leather interiors. Rounding out the 1962 product line were the distinctive Fleetwood Sixty-Special Sedan and the long-wheelbase Fleetwood Seventy-Five Nine-Passenger Sedan and Limousine.

Styling changes on the 1962 Cadillacs included a new flat-faced grille with a fine eggcrate texture and a thin horizontal center bar. The Cadillac name appeared in script in the lower left-hand side of the grille. Parking or fog lamps were contained in large rectangular openings in the lower bumper below the headlights. The sides of the headlight bezels housed new cornering lights, standard on all models, which cast a broad beam of light to illuminate night turns.

The 1962 Cadillac's rear fender sculpturing was identical to that used on the 1960 Eldorado Brougham. Tail fins were slightly lower this year and had no chrome trim on top. Brake, backup, and turn indicator lights were housed in new, vertical rectangular-shaped chrome bumper ends. A series of thin chrome trim pieces spanned the lower rear deck above the rear bumper on most models. This area was occupied by a finely-detailed rear grille on the Eldorado Biarritz and Sixty Special. Redesigned wheel covers had centers that were color-keyed to the body color.

Five of Cadillac's thirteen 1962 models, including both two-door coupes and all four-window sedans, got wide new rear roof pillars. The Sixty Special retained the formal roofline introduced the previous year, but the Sixty Special's distinctive vertical chrome louvers were repositioned from the ends of the rear fenders to a new location at the base of the upper roof quarters.

A heater became standard equipment on all 1962 Cadillacs. But the most important technical advance on the 1962 Cadillacs was a new dual-safety braking system. A split hydraulic master cylinder provided independent pistons and brake fluid reservoirs for front and rear wheel brakes, which assured that at least one pair of brakes functioned in the event of the failure of either.

The 1962 Cadillacs were mildly warmed-over versions of the 1961 cars. Styling and engineering changes were minimal. The redesigned 1962 grille had a flat face and a fine eggcrate texture with a thin horizontal center bar. Two-door coupes and four-window sedans got new, wider rear roof pillars or "sail" panels. For the first time, Cadillac's least-expensive model was priced above $5,000. The Series 62 Coupe for 1962, Style 6247, carried a manufacturers' suggested retail price of $5,025. Standard equipment included Hydra-Matic automatic transmission, power steering and brakes, new cornering lights and, for the first time, a heater. Cadillac built 16,833 of these lowline two-door hardtops during the 1962 model year.

Cadillac's 1962 product line included two Series 62 Four-Window Sedans. While both rode on the same 129.5-inch wheelbase, one was a short-deck model seven inches shorter than the standard four-window sedan. This is the standard 1962 Cadillac Series 62 Four-Window Sedan, Style 6239, which had an over-all length of 222 inches. The companion Series 62 Town Sedan was 215 inches long. The standard Series 62 Four-Window Sedan was the third-best-selling car in Cadillac's 1962 model lineup, with 17,314 built. Base price was $5,213. Styling of both four-window sedans was identical except for the Town Sedan's shorter rear end. This view shows the 1962 Cadillac's revised flat-faced grille. New rectangular parking/fog lamp openings were located in the outer ends of the front bumper below the dual headlights.

1962

All 1962 Cadillacs were powered by the same 390 cubic inch, 325 brake horsepower V-8 engine used for the previous three years. Compression ratio continued at 10:5.

The 60th Anniversary Cadillacs went into dealer showrooms on Sept. 22, 1961. Cadillac built 160,840 cars during the booming 1962 model year, finally topping its previous model year production record that had been set six years earlier, in 1956. Fifty-nine per cent of all Cadillacs now sold were delivered with factory-installed air conditioning.

Cadillac's Cleveland Ordnance plant was reactivated this year and began production of three types of military vehicles for the U.S. Army.

The last National Automobile Show was again held in Detroit, providing a spectacular setting for the introduction of the 1963 Cadillacs.

Cadillac's customers could again choose either four or six-window sedan styling at no difference in price. Up until now, the six-window sedan had always outsold the four-window sedan by a handsome margin, but in the 1962 model year this pattern was reversed. Sales of 17,314 Series 62 Sedans exceeded those of 16,730 six-window sedans. Style 6229, the Series 62 Six-Window Sedan weighed 4,640 pounds. At $5,213 it was identically priced with the Four-Window Sedan and the short deck Series 62 Town Sedan. Note the thin, flat roofline and wide new rear roof "sail" panels used on 1962 two-door coupes and four-window sedans.

There were two short-deck sedans in Cadillac's 13-car 1962 product lineup, one in the lowline 62 Series and another in the more expensive Series 63 de Ville. The Series 62 short-deck sedan was given a name this year: Town Sedan. As in 1961, this four-window sedan was seven inches shorter in over-all length than the standard Series 62 Four-Window Sedan. And again it was a poor seller. Only 2,600 were sold compared with sales of more than 17,000 standard Series 62 Four-Window Sedans. Style 6289, the 1962 Town Sedan had a manufacturers' suggested retail price of $5,213—the same as the longer four-window sedan and the Series 62 Six-Window Sedan. The "Town Sedan" name appeared in small script on the ends of the rear fenders.

The Cadillac Series 62 Convertible kept on setting new sales records. Deliveries of 16,800 Series 62 Convertibles during the 1962 model year easily topped the previous record 15,500 Series 62 soft-tops sold in the 1961 model year. Style 6267, the 1962 Cadillac Series 62 Convertible Coupe had a manufacturers' suggested retail price of $5,588. Leather upholstery was $134.20 extra. Front bucket seats were also optionally available, for another $107.35. This slightly elevated view shows the 1962 Cadillac's new flat-faced grille with horizontal center bar and modestly-refined body side sculpturing, which was very close to that seen on the 1960 Eldorado Brougham. The centers of the new wheel covers were color-keyed to the body color.

With the addition of the new short-deck Park Avenue and the inclusion of the Eldorado Biarritz Convertible, there were now five models in the Series 63 Cadillac family. But only four were called de Villes. The Biarritz was not listed as a de Ville. The lowest-priced Series 63 Cadillac was the ever-popular Coupe de Ville, Style 6347, which carried a 1962 manufacturers' suggested retail price of $5,385. Shipping weight was 4,595 pounds. All 1962 de Ville models had small script series nameplates on the ends of their rear fenders. Cadillac's 1962 two-door coupes and four-window sedans had much wider rear roof pillars or "sail" panels. With deliveries of 25,675 cars, the Coupe de Ville was Cadillac's second-best-selling car behind the six-window Sedan de Ville.

The 1962 Cadillacs received modest styling changes. Besides the new flat-faced grille used on all models, the most visible exterior styling changes on the 1962 cars appeared on two-door coupes and four-window sedans, which got much wider "sail" panels or rear roof pillars. This is the 1962 Cadillac Four-Window Sedan de Ville. Style 6329, which had a manufacturers' suggested retail price of $5,631. The companion Six-Window Sedan de Ville and new Series 63 short-deck Park Avenue Four-Window Sedan were identically priced. The 1962 Cadillac's rear fender sculpturing was taken directly from the 1960 Eldorado Brougham. Production of this four-door hardtop totalled 16,230.

1962

A new short-deck sedan was added to the Series 63 de Ville model lineup this year, raising to 13 the number of individual models available in Cadillac's 1962 model lineup. The Series 63 short-deck sedan was called the Park Avenue. Like the less expensive Series 62 Town Sedan, the Park Avenue was seven inches shorter than the standard de Ville four-window sedan. The Park Avenue was offered only in the 1962 and 1963 model years. Listed as Style 6389, the Park Avenue had a manufacturers' suggested retail price of $5,631—the same as the four and six-window Sedan de Villes. Only 2,600 Park Avenue Sedan de Villes were sold this year. The Park Avenue had an overall length of 215 inches compared with 222 inches on the standard four-window sedan.

The Six-Window Sedan de Ville was the best-selling model in Cadillac's 13-car 1962 product line. A total of 27,348 were delivered. Cadillac's 1962 six-window sedans retained the same flat rooflines and very slender roof pillars used the previous year, while coupes and four-window sedans got wider sail panels. Style 6339, the Six-Window Sedan de Ville had a manufacturers' suggested retail price of $5,631 and weighed 4,675 pounds. Note the fine horizontal trim that spanned the lower rear deck of all 1962 Cadillacs except the Eldorado Biarritz and Sixty Special. Brake and backup lights were combined in new vertical rectangular-shaped bumper ends that capped the rear fenders. Cadillac's 1962 tail fins were lower and did not have chrome edging on top.

The expensive Eldorado Biarritz Convertible got back just a little of its former styling individuality this year. The only way to tell the premium Biarritz convertible from the less expensive Series 62 soft-top was by the fine body side molding that swept from the top of the door down the rear fender sculpturing, then dipped to the tip of the lower body skeg molding. Sporty bucket seats were offered as a no-cost option on this model. Soft leather upholstery in a choice of seven colors plus a Cannes cloth and leather combination were available in the Eldo Biarritz. This luxurious convertible carried a manufacturers' suggested retail price of $6,610. Production totalled 1,450—exactly the same number built the previous year.

Cadillac's—indeed, General Motors'—flagship cars were the distinguished Fleetwood Seventy-Fives. Again, only two Seventy-Fives were offered, a nine-passenger sedan and formal limousine, both on the same 149.8-inch wheelbase and with an over-all length of 242.3 inches. This is the 1962 Cadillac Fleetwood Seventy-Five Nine-Passenger Sedan, Style 6723, which sold for $9,722. Only 696 were built, three fewer than in 1961. The big Seventy-Fives had full-length lower body skeg moldings but no sculpturing on their upper doors. This aristocratic series retained its 1959-60 upper body structure with huge compound-curved windshield and curved dogleg posts.

The 60th Anniversary Cadillac model lineup included the distinctive 1962 Fleetwood Sixty Special Four-Door Sedan, which reflected subtle exterior styling changes from the previous year. In addition to the flat new grille, there was a finely-textured grille panel across the lower rear deck. The Fleetwood name appeared in tiny block letters on the right side of the decklid, and small, red cloisonne emblems graced the lower front fenders for the third and final year. The Sixty Special's traditional vertical louvers were repositioned from the ends of the rear fenders to a new location on the rear roof quarters. Style 6039, the 1962 Fleetwood Sixty Special Four-Door Sedan (actually a four-door hardtop) carried a manufacturers' suggested retail price of $6,366. Weight was 4,710 pounds. Sales of 13,350 Sixty Specials were slightly lower than the previous year.

1962

The most expensive car in Cadillac's comprehensive 1962 model line was the Fleetwood Seventy-Five Limousine. This formal nine-passenger car, with standard glass division, had a base price just under $10,000 ($9,937) and weighed a hefty 5,390 pounds. The chauffeur's compartment was upholstered in black leather, or optional fawn or gray leather. The plush rear passenger compartment was trimmed in luxurious cord or broadcloth. Cadillac built 904 Style 6733 Fleetwood Limousines this year. Note the small, formal rear window and the fine horizontal trim across the lower rear deck above the rear bumper. Air conditioning cost $624 extra. Small A/C vents are visible below the rear window. Only 1962 Series Seventy-Five models had chrome trim on top of their tail fins.

Cadillac sales soared to a new record during the 1962 model year, finally toppling the previous record set six years earlier. But the division marked another milestone, too. The 2,500,000th Cadillac was driven off the assembly line in the Clark Avenue plant in Detroit. It was a 1962 Series Sixty-Two Coupe. Cadillac Motor Car Division General Manager Harold G. Warner, left, and General Sales Manager Fred H. Murray greet the milestone car at the end of the line. Cadillac also observed its 60th anniversary this year.

After the embarrassment of Sputnik and the explosion of the Vanguard rocket on the launching pad in the late 1950s, the United States mounted a massive program to put an American into space during the 1960s. The U.S. space program culminated in Neil Armstrong's spectacular moonwalk in 1969. The first American in space was astronaut John H. Glenn Jr., who was launched into orbit in 1961. Like a true hero, Major Glenn was accorded ticker-tape parades wherever he went. Inevitably, he rode in the rear seat of a Cadillac convertible. Here, Major Glenn waves to the crowd from the tonneau of a 1962 Cadillac Series 62 Convertible in a parade somewhere in middle America. Note the Cadillac script nameplate in the lower left corner of the horizontally-textured grille.

Cadillac continued to supply a special long-wheelbase commercial chassis to major funeral car and ambulance builders. Since 1957, this 156-inch wheelbase hearse chassis had used an "X"-type, tubular-center frame, just like Cadillac's standard passenger cars. Cadillac shipped 2,280 Series 6890 commercial chassis to the four principal U.S. funeral car and ambulance manufacturers this year. This is a 1962 Eureka-Cadillac standard-headroom limousine style ambulance. Note the angled "C" and "D" pillars and the wraparound rear quarter windows on this two-toned emergency car.

The rarest of all custom-built funeral cars were the long, low coupe-style Flower Cars. Well under 100 of these special-purpose cars were built each year. The Superior Coach Corp. of Lima, Ohio built more than any other manufacturer, turning out between 25 and 50 a year. This is a 1962 Superior-Cadillac Royale Coupe de Fleur. Note the short skeg molding which extends only the length of the rear fender. This flower car has an open well rather than an adjustable stainless steel deck. This style of flower car was popular in the Chicago area. It was built on the 1962 Cadillac Series 6890 commercial chassis.

1963

The 1963 Cadillac was a completely restyled, thoroughly re-engineered car. Even its V-8 engine was redesigned for the first time in 15 years. New sales and production records were set for the second straight year.

The Cadillac Motor Car Division went to market with a total of 12 individual models for the 1963 model year—one less than in 1962. The short-deck Series 62 Town Sedan was dropped at the end of the 1962 model year, but there was still a short-deck sedan in the mid-range Series 63 de Ville model lineup. Once more, Cadillac offered its customers four series of cars built on two wheelbases with three over-all lengths.

The 1963 product offerings included four lowline Series 62 Cadillacs: a two-door hardtop coupe, four and six-window Sedans and a sporty convertible. There were five Series 63 models: the stylish Coupe de Ville (the best-selling car in the entire 1963 line); the short-deck Park Avenue Four-Window Sedan de Ville; standard-length four and six-window Sedan de Villes, and the premium Eldorado Biarritz Convertible. While listed as a Series 63 body style since 1961, the Eldorado Biarritz was never referred to as a de Ville. Rounding out the line were the distinctive Cadillac Fleetwood Sixty Special Four-Door Sedan (since 1957 actually a four-door hardtop) and the aristocratic Fleetwood Seventy-Five Sedan and Limousine.

All 1963 Cadillacs except the top-line Fleetwood Seventy-Fives were built on the same 129.5-inch wheelbase "X"-frame chassis. All except the Seventy-Fives and the short-deck Park Avenue had a bumper-to-bumper length of 223 inches—one inch longer than the previous year.

With an over-all length of 215 inches, the Park Avenue was eight inches shorter than the rest of Cadillac's standard 1963 models. This would be the third and final year Cadillac would offer a short-deck model. The first had been introduced for the 1961 model year. There were two in the 1962 line and just one for 1963. Combined sales over the three model years totalled only 10,531 short-deck sedans. Cadillac's customers just weren't interested in a compact Caddy—yet.

Fleetwood Seventy-Fives rode on a long 149.7-inch wheelbase and had an over-all length of 243.3 inches.

The 1963 Cadillac's crisp, new styling featured broad, flat surfaces with none of the sheet-metal sculpturing used in the previous two model years. Tail fins were lower than ever. The 1963 Cadillac's new front-end ensemble was faintly reminiscent of the 1959 models but was more restrained. The flat-faced, crosshatched grille had a thin horizontal center bar and a Cadillac script nameplate in its lower left corner. The chamfered hood projected out slightly over the grille. Viewed in profile, both ends of the 1963 cars were starkly vertical. There were smart, new rooflines on all models (except Seventy-Fives) complemented by a redesigned windshield with straight side pillars. Coupe rooflines had been shortened by as much as seven inches which resulted in much longer-looking rear decks, yet with no sacrifice in interior room. Rear roof pillars were thicker on most models with slightly smaller rear windows. Series 62 and 63 customers could still choose four or six-window sedan styling at no difference in price.

Narrow, vertical chrome rear bumper ends contained brake, backup, parking and directional signal lights. There were also thin vertical taillights in the ends of the bladelike tail fins. Body sides were high and flat. On all but the Sixty Special and Eldorado Biarritz, a

Cadillac's fresh 1963 styling did away with the sculptured, bevelled look of the previous two years. The 1963 cars had crisp, clean styling that emphasized broad, flat surfaces. In profile, the front and rear ends were absolutely vertical. This is the 1963 Cadillac Series 62 Four-Window Sedan, Style 6239, which sold for $5,214. Weight was 4,595 pounds. The division produced 16,980 of this model. Cadillac's customers could still choose either four or six-window sedan styling at no difference in price. The short-deck Series 62 Town Sedan was discontinued this year leaving this model the only four-window sedan in the lowline Sixty-Two series.

Cadillac's 1963 models were completely redesigned. Front-end styling was faintly reminiscent of 1959 but much more subdued. Tail fins were lower than ever, gradually receding into the rear fenders. Body lines were crisp and flat. New, shorter rooflines emphasized the length of the rear deck. This is the 1963 Cadillac Series 62 Coupe, one of four Series 62 models. Wheelbase continued at 129.5 inches but over-all length was increased one inch to 223 inches. Style 6257, the Sixty-Two Coupe was the lowest-priced car in the 1963 model line with a manufacturers' suggested retail price of $5,036. Weight was 4,505 pounds. A total of 16,786 Sixty-Two Coupes were built this year.

1963

heavy chrome molding extended the full length of the car, from grille to rear bumper at the body center line. The Biarritz and Sixty Special had no body side trim. These highline models had broad chrome rocker sill and lower rear fender moldings instead. Model names were spelled out in small block letters on the lower edges of the front fenders, and there were distinctive new chrome laurel wreath and crest emblems on the ends of the rear fenders.

A significant new styling option was available on two 1963 Cadillacs. A cross-grained vinyl fabric roof covering was offered as an extra-cost option on the 1963 Coupe de Ville and Fleetwood Sixty Special. Vinyl roofs were becoming the industry's latest styling fad. They would soon be available on every size and type of car made in America, from compacts to luxury cars. Other new Cadillac options for 1963 included a six-position adjustable tilt steering wheel, AM-FM radio and an integrated heater and air conditioning control panel.

But the most important engineering change in the 1963 Cadillac lay under its restyled hood. Cadillac's V-8 engine had been redesigned for the first time since the milestone short-stroke, high-compression 331 cubic inch engine had been introduced away back in 1949. Its new cast iron cylinder block was shorter and narrower and had a light-alloy timing gear case cover. Many of its internal components were also new. The redesigned 325 HP engine was more than 50 pounds lighter than in 1962 and was even lighter than the original 1949 engine. Bore and stroke (4.0 by 3.8 inches) were unchanged, as was its 390 cubic inch displacement and 10:5-to-1 compression ratio. Power-to-weight ratio, however, was now higher than ever and a new "true-center" driveline featured a re-engineered propeller shaft with two constant velocity joints and a single universal that delivered a remarkably smooth flow of power under all road and load conditions. The epic 1949 V-8 engine design, which ranked with the first V-type, eight-cylinder engine of 1915 and the impressive 12 and 16-cylinder engines of the 1930s, had proved extremely adaptable. It had powered Cadillacs through 14 model years with few changes aside from increased displacement and compression.

The biggest new product news of the year, however, was generated by another GM division. It was the 1963 Buick Riviera, a sleek new two-door "personal luxury" hardtop coupe that invaded a market that had been dominated up to now by Ford's Thunderbird. The beautifully-styled 1963 Riviera was one of GM styling chief William J. Mitchell's personal favorites, along with the original 1938 Cadillac Sixty Special. At one time, the car that became the Riviera had been considered for Cadillac, possibly as a LaSalle II, but it was ultimately assigned to the Buick Motor Division. Cadillac would wait another four years before making its own spectacular entry into the increasingly-important luxury-personal car market.

Cadillac production again set a new record—163,174 cars. The division announced a major expansion program that added more than 471,000 square feet of floor space along Clark Avenue. Included were a new engineering building, a sales courtesy and showroom building, modifications to the Administration Building, and expanded manufacturing and assembly floor space.

The U.S. auto industry's 1963 cars were introduced at what was to be the last National Automobile Show ever held. This lavish auto show was again held in Detroit, in the early fall of 1962.

Cadillac's customers could again choose between four and six-window sedan styling at no difference in cost in both the lowline 62 Series and the more expensive Series 63 de Villes. The four-window sedans were the top sellers in each. Cadillac's fresh 1963 styling included a redesigned windshield and new rooflines on most models. This is the 1963 Cadillac Series 62 Six-Window Sedan, Style 6229, of which 12,929 were built. Note the large glass area and thin rear roof pillars. Series 62 four- and six-window sedans were priced identically at $5,214. The Six-Window Sedan weighed 4,610 pounds.

Sales of the sporty Series 62 Convertible continued to set records. A total of 17,600 Series 62 Convertible Coupes were delivered during the 1963 model year, topping the previous record 16,800 sold the previous year. Manufacturers' suggested retail price of the Sixty-Two Convertible was $5,590 and shipping weight was 4,545 pounds. A heavy chrome side molding extended the entire length of all 1963 Cadillacs (except the Eldorado Biarritz Convertible and Fleetwood Sixty Special), protecting the car's high, flat sides. Note the very narrow whitewalls, which were coming into vogue. In the background, a group of Cub Scouts are about to look over GM's Firebird I experimental turbine car on a visit to the General Motors Technical Center in Warren, Mich.

1963

Fifteen years after it had been introduced as a 1949 model, the stylish Coupe de Ville remained one of Cadillac's most popular models. Sales of 31,749 Series 63 Coupe de Ville Two-Door Hardtops during the 1963 model year made it the best-selling car in the line and second only to the record 33,000 sold in Model Year 1955. A cross-grained, vinyl fabric-covered roof was available as a new extra-cost option on the 1963 Coupe de Ville and undoubtedly heightened this sporty coupe's appeal. Style 6357, the Coupe de Ville weighed 4,520 pounds and had a manufacturers' suggested retail price of $5,386. The 1963 Coupe de Ville also sported a smart new roofline and an extremely long rear deck. Note the narrow vertical chrome rear bumper ends which housed brake, parking, turn signal and back-up lights. The 1963 Cadillac's low tail fins also contained thin vertical taillights.

There was only one short-deck model in Cadillac's 1963 product lineup. It was the Park Avenue, a four-window sedan in the Series 63 de Ville. The lowline Series 62 short deck Town Sedan had been dropped at the end of the 1962 model year. This would be the last year Cadillac would offer a short-deck model in its standard line. Only 1,575 Series 63 Park Avenue Four-Window Sedan de Villes were built during the 1963 model production run. The Park Avenue, which took its name from the 1954 show car, was offered only in 1962 and 1963. Except for its eight-inch shorter rear end, the "compact" Park Avenue was identical in styling to the standard 1963 Four-Window Sedan de Ville. Price $5,633 and the Park Avenue weighed 4,590 pounds.

Most of Cadillac's extensively-restyled 1963 cars got new windshields and rooflines. Series 62 and 63 (de Ville) six-window sedans had thin rooflines with large glass areas and triangular quarter windows. This is the Six-Window Sedan de Ville, Style 6329. Cadillac sold 15,146 of these spacious-looking four-door hardtops. The Six-Window Sedan de Ville weighed 4,650 pounds and was priced identically with the companion four-window Sedan de Ville and short deck Park Avenue Sedan de Ville, at $5,633. All 1963 Cadillacs had Cadillac script nameplates in the lower left-hand corner of their finely-textured, crosshatched grilles. Attractive new wheel covers appeared on all 1963 Cadillacs.

Little by little, the premium Cadillac Eldorado Biarritz Convertible was regaining its former styling individuality. Still a Series 63 body style, the Biarritz, however, was never called a de Ville. The 1963 Eldorado Biarritz Convertible had the same high, flat flanks as the companion Fleetwood Sixty Special with no chrome side moldings down the center. Like the Sixty Special, it had wide chrome rocker sill moldings and a distinctive chrome laurel wreath and crest on the ends of its rear fenders. The magic Eldorado name was spelled out in small block letters on the lower edges of the front fenders—just like in 1959 and 1960. Style 6367, the Eldo commanded an impressive base price of $6,608. Weight was 4,640 pounds. The 1963-model production run included 1,828 Eldorado Biarritz Convertibles—378 more than the previous year.

The Four-Window Sedan de Ville was the second-most-popular single body style in Cadillac's 12-car 1963 product line. A total of 30,579 of these attractively-styled four-door hardtops were sold this year, second only to the Coupe de Ville Two-Door Hardtop. The Four-Window Sedan de Ville, Style 6339, carried a manufacturers' suggested retail price of $5,633—exactly the same as the companion Six-Window Sedan and the short deck Park Avenue. Weight was 4,605 pounds. Note the finely-textured vertical grille with thin center bar and the chamfered hood which extended out slightly. Cadillac's crisply-styled 1963 front-end ensemble was faintly reminiscent of the 1959 model but much finer in detail. The hood, front fenders and rear deck were longer this year giving the 1963 Cadillac an unmistakable new profile.

The 1963 Fleetwood Sixty Special Four-Door Sedan retained the very distinctive formal styling it had borne since the 1960 model year. Like the companion Eldorado Biarritz Convertible, the Sixty Special had no mid-body chrome side trim. Special exterior styling touches included broad chrome sill moldings and an elegant new laurel wreath and crest emblem on the ends of the rear fenders. The Fleetwood name was spelled out in tiny block letters on the lower edges of the front fenders. The cloisonne emblem was gone. Five fine vertical louvers continued on the upper rear roof quarters. The 1963 Fleetwood Sixty Special weighed 4,690 pounds and carried a manufacturers' suggested retail price of $6,300. A total of 14,000 were sold this year.

Vinyl-covered roofs were the U.S. auto industry's latest styling craze. Simulated leather-covered roofs lent a look of sporty elegance to any car, from compacts on up to luxury cars. They were especially popular on two-door hardtops. It was not surprising then, when Cadillac came out with vinyl roof options for two of its restyled 1963 models—the hot-selling Coupe de Ville and the luxurious Fleetwood Sixty Special Four-Door Sedan. Cadillac's cross-grained vinyl fabric roof covering was an $85 extra on the Coupe de Ville and cost $125 on the Sixty Special. This is a 1963 Fleetwood Sixty Special equipped with the smart new vinyl roof option. Soon they would be offered on most models. Vinyl roofs made possible a whole, new world of two-tone color combinations—a light lower body color with a complementing darker vinyl roof color.

The most expensive of Cadillac's twelve 1963 models was the stately Fleetwood Seventy-Five Limousine, a formal nine-passenger car with glass division. This model was intended to be chauffeur-driven, and most of course, were. Except for the carryover 1959-60 windshield and greenhouse, the Fleetwood Seventy-Fives were compatible in styling with the rest of the 1963 Cadillacs. Style 6733, the Fleetwood Seventy-Five Limousine weighed 5,300 pounds and had a manufacturers' suggested retail price of $9,939. A total of 795 of these impressive limousines were built this year. The Fleetwood Limousine traditionally outsold the slightly less expensive Nine-Passenger Sedan, which had no division between the front and rear passenger compartments.

Since the end of the Second World War, the Derham Body Co. in the Philadelphia suburb of Rosemont, Pa. had produced small numbers of formal sedan conversions of standard Cadillac Fleetwood Seventy-Five sedans and limousines. But in the mid-1960s, Cadillac began to produce some of these formal limousines in its own plant on a special-order basis. Special styling features included closed rear quarters and a thickly-padded top with a small "frenched" rear window. Most also had small decorative landau irons. This is a 1963 Cadillac Fleetwood Seventy-Five Formal Limousine photographed at the 1972 Hershey Flea Market. The thermos bottle on the rear deck was not standard equipment.

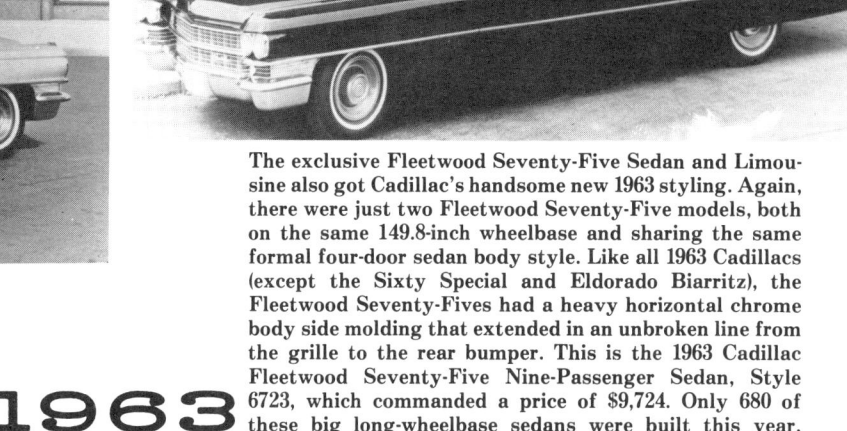

1963

The exclusive Fleetwood Seventy-Five Sedan and Limousine also got Cadillac's handsome new 1963 styling. Again, there were just two Fleetwood Seventy-Five models, both on the same 149.8-inch wheelbase and sharing the same formal four-door sedan body style. Like all 1963 Cadillacs (except the Sixty Special and Eldorado Biarritz), the Fleetwood Seventy-Fives had a heavy horizontal chrome body side molding that extended in an unbroken line from the grille to the rear bumper. This is the 1963 Cadillac Fleetwood Seventy-Five Nine-Passenger Sedan, Style 6723, which commanded a price of $9,724. Only 680 of these big long-wheelbase sedans were built this year, many for funeral and commercial limousine livery service. Standard weight was a hefty 5,200 pounds.

Cadillac's 1963 commercial chassis was compatible in styling with the redesigned 1963 Cadillac passenger cars. Cadillac shipped 2,527 Series 6890 commercial chassis to the four major U.S. funeral car and ambulance manufacturers during the 1963 model year. This is a 1963 Superior-Cadillac Crown Royale Landaulet Funeral Coach. Note the new, curved rear quarter windows. The license plate was mounted in the center of the rear door. Superior's 1963 models had wide chrome rocker sill moldings and a small, ribbed trim panel just ahead of the rear fender skirts. Note how the rear door extends into the roof to provide the maximum rear door opening.

1964

Cadillac production and sales set records for the third consecutive year. While exterior styling changes were modest, all 1964 Cadillacs were powered by an improved 429 cubic inch engine.

The Cadillac Motor Car Division's 1964 product offerings included 11 body styles on two wheelbases. Customers could choose from 21 exterior color choices and no less than 129 interior trim selections. There were now only three series—the lowline Series 62, the mid-range de Villes and the premium Cadillac Fleetwoods.

The short-deck Park Avenue Sedan was discontinued at the end of the 1963 model year. Also gone was the Biarritz nameplate which had distinguished Cadillac's most expensive convertible since the 1956 model year. Previously a Series 63 (de Ville) body style, the premium convertible was upgraded for 1964 as a companion car to the Fleetwood Sixty Special Sedan and was now called the Fleetwood Eldorado, even though it kept its internal Style 6367 model designation. Similarly, Cadillac's standard convertible, which had long been a Series 62 model, was upgraded for 1964 into the more expensive de Ville family.

All 1964 Cadillacs except the Fleetwood Seventy-Fives were built on the same 129.5-inch wheelbase "X"-frame chassis and had an over-all length of 223.5 inches. The low-volume Seventy-Fives continued on a 149.7-inch wheelbase with a bumper-to-bumper length of an impressive 243.8 inches.

The 1964 Cadillac catalog showed only three Series 62 models, a base two-door hardtop coupe and four and six-window sedans. There were four Series 63 cars; the ever-popular Coupe de Ville, four and six-window Sedan de Villes and the upgraded de Ville convertible. The exclusive Cadillac Fleetwood series for 1964 encompassed four models—the Sixty Special Four-Door Sedan, Fleetwood Eldorado Convertible and the long-wheelbase Fleetwood Seventy-Five Sedan and Limousine.

Styling changes for 1964 were subtle, but effective. The revised grille design, with a strong horizontal theme, featured a wide center divider painted the same color as the body. Slightly reminiscent of Cadillac's 1961 grille, it was "V'eed" horizontally and vertically. Cadillac's famous tail fins receded even farther into the rear fenders and were lower than ever. Vertical rear bumper ends which capped the long rear fenders were bright chrome chevrons which, with a full-length body side molding and fine horizontal detailing on the front fender cornering lights, made the 1964 cars look longer and lower. The Eldorado Convertible had no rear fender skirts for the first time since 1958.

Cadillac now offered a handsome new landau roof option for its prestigious Fleetwood Seventy-Five Limousine. Similar in style to those produced in small numbers over the years by Derham, the new formal roof option featured closed upper rear quarters spanned by ornamental landau irons. The roof was thickly padded. This new top was made entirely in Cadillac's own plant.

Fleetwood Seventy-Five nine-passenger sedan and limousine bodies were built in GM's old Plant 21 on Detroit's near-east side within sight of the General Motors Building. Production was only about 12 bodies

Although the 1964 Cadillacs got only modest styling changes, the appearance refinements were very effective. The 1964 cars carried over their basic 1963 styling but redesigned grilles and rear ends made them instantly recognizable. The revised 1964 grille thrusted out at its center and had a wide divider painted the same color as the body. Upper and lower sections featured Cadillac's familiar "eggcrate" texture, and there was a Cadillac script nameplate in the lower left corner of the grille for the third straight year. Tail fins were lower than ever. All 1964 Cadillacs got this new grille. The car shown is a 1964 Cadillac De Ville Convertible. If you look closely you can see the optional Guide-Matic headlight control sensor built into the top of the left front fender above the headlights.

Cadillac's cheapest (or, in the words of the division's ad agency writers, "most affordable") 1964 model was the Series 62 Two-Door Hardtop Coupe. This would be the last year for the Sixty-Two Series. Only three Series 62 body styles were offered this year, the Coupe and four and six-window Sedans. The 1964 Cadillac Series 62 Two-Door Hardtop had a factory suggested retail price of $5,026. Weight was 4,475 pounds. Cadillac's record 1964 production included 12,166 Sixty-Two Coupes. Cadillac's 1964 engine was bored out to 429 cubic inches and produced a very potent 340 horsepower. This body style was listed as Style 6257.

1964

a day. The painted, trimmed bodies were hauled in closed trailers to the Clark Avenue plant on the other side of town and fed onto the body drop line according to a very tight schedule. All other Cadillac bodies came from the huge Fisher-Fleetwood plant on West Fort St., a few miles west of the main Cadillac plant and office complex.

The most important engineering changes for 1964 were again found under the hood. Displacement of Cadillac's V-8 engine went up to 429 cubic inches this year, from 390. Rated at 340 horsepower, this was the most powerful engine yet offered in standard production Cadillacs. Bore and stroke also went up, to 4.13 and 4.0 inches respectively. Compression remained at 10:5-to-1. Cadillac's 1964 power trains included two different automatic transmissions. Series 62 and Seventy-Five models were equipped with GM's standard Hydra-Matic, but 1964 de Villes and the Fleetwood Sixty Special and Eldorado got a new Turbo Hydra-Matic torque converter transmission as standard equipment.

Vinyl roofs were now available on three models—the Coupe de Ville, Four-Window Sedan de Ville and Sixty Special. New comfort and convenience options for 1964 included a completely automatic, thermostatically-controlled "Comfort Control" heating and air conditioning system. Theoretically, the owner could set the Comfort Control to the desired setting upon taking delivery of the car and never touch it again as long as he owned it.

Factory-installed air conditioning was now being ordered in 75 per cent of all new Cadillacs. The division had built its 500,000th air-conditioned car the previous year, the first auto manufacturer to reach this level. Factory air was a $474 extra-cost option on standard Cadillacs and cost $624 on Seventy-Fives.

Also new this year was a "Twilight Sentinel" headlamp control that automatically switched the lights on at dusk and off at sunrise. A built-in delay kept the headlights on for up to 90 seconds after the driver left the car. The electronic eye unit was gone from the top of the instrument panel. A new Guide-Matic headlamp control was cleverly concealed in the top of the left front fender just above the dual headlamps. Inside, the instrument panel was redesigned and seat belts became standard equipment.

The American automobile industry showcased its products at the 1964-65 New York World's Fair. Cadillac's exhibit in the huge General Motors Futura Pavilion included the division's first experimental show car since 1959. It was the Cadillac Florentine, a customized version of a standard two-door hardtop. The Florentine featured closed upper rear quarters with small vertical quarter windows that retracted into the sail panels. The top was covered with elegant vinyl suede. High-back bucket seats were trimmed in embroidered leather with a Michaelangelo motif. Full skirtless rear wheel openings were similar to those on the 1964 Eldorado Convertible.

Cadillac built a record 165,959 cars during the prosperous 1964 model year. Before the calendar year was out the three-millionth car would come off the Cadillac Motor Car Division's busy assembly line in Detroit.

Ford, which had abandoned the long-wheelbase limousine market to Packard and Cadillac following the Second World War, offered an Executive Limousine version of its 1964 Lincoln Continental. Available through its Lincoln-Mercury dealers, this four-door "stretch" job with fixed center section was built for Ford by Lehmann-Peterson, Inc. of Chicago. Chrysler Corporation had earlier made incursions into the limited-production limousine market with a semi-custom Imperial limousine built for Chrysler by Ghia in Italy.

Cadillac customers could again choose either four or six-window sedan styling at no difference in base vehicle price. The lowline Sixty-Two series included four and six-window hardtops which carried identical $5,214 manufacturers' suggested retail prices. This is the 1964 Cadillac Series 62 Four-Window Sedan, Style 6239, the most popular car in the division's lowest-priced series with deliveries of 13,670 cars. Weight was 4,550 pounds. Series 62 Cadillacs had no rear fender nameplates. Wheelbase continued at 129.5 inches with an over-all length of 223.5 inches.

Six-window sedans were offered in Cadillac's 62 and 63 series from 1959 through the 1964 model year. The six-window sedan's styling featured a very large glass area, thin, flat roofline and slender rear roof pillars. Triangular quarter windows were also a standard element in this design. Style 6229, the Series 62 Six-Window Sedan was the least popular body style in its three-car series. A total of 9,243 were sold in the 1964 model year compared with deliveries of more than 13,600 four-window Sixty-Two Sedans and 12,000 Series 62 Coupes. Series 62 and Seventy-Five models for 1964 were equipped with GM's standard Hydra-Matic transmission. De Villes and Fleetwood Sixty Special and Eldorado models got a new Turbo Hydra-Matic torque converter transmission this year. Note the redesigned "chevron" bumper ends and extremely low 1964 tail fins.

1964

Cadillac's extremely popular Series 63 de Ville line included four body styles for 1964. Cadillac's biggest seller the previous year, the handsome Coupe de Ville Two-Door Hardtop slipped to second place in the 1964 model year. But its sales performance was spectacular nonetheless. Deliveries of 38,195 Cadillac Coupe de Villes finally topped the previous record 33,300 sold away back in 1955 and even exceeded 1963's respectable 31,749 by a wide margin. But when all the numbers were in, the Coupe de Ville had been displaced by the Four-Window Sedan de Ville as Cadillac's best-selling single model. Style 6357, the 1964 Cadillac Coupe de Ville had a manufacturers' suggested retail price of $5,386 and weighed 4,495 pounds. The padded vinyl roof was a $91.25 option.

The best-selling car in the 1964 Cadillac line was the Series 63 Four-Window Sedan de Ville. A very impressive 39,674 of these four-door hardtops were delivered during the big 1964 model year. A padded vinyl roof was available as a new option on this model and cost an extra $140. Style 6339, the Four-Window Sedan de Ville was priced identically with the companion Six-Window Sedan de Ville at $5,663. Weight was 4,575 pounds. The 1964 Cadillac's revised front-end styling had a strong horizontal theme with a painted divider between angled upper and lower grille sections. Fine horizontal trim also covered the standard cornering lights which wrapped around the front fenders and incorporated parking and turn signal lamps.

The 1964 model year would be the last for Cadillac's six-window sedans, which were available in both 62 and 63 series. The six-window sedans had distinctive upper body structures with large glass areas and slender rear roof pillars. Style 6329, the 1964 Cadillac Six-Window Sedan de Ville was identically priced with the companion Four-Window Sedan de Ville at $5,633 but didn't sell nearly as well. A total of 14,627 Six-Window Sedan de Villes were built during the 1964 model year compared with sales of more than 39,000 Four-Window Sedans. All Series 63 models had small de Ville script nameplates on the ends of their rear fenders above the full-length chrome body side molding. Note the front fender windsplit which continued into the door. Front fender windsplits were a basic Cadillac styling element from 1941 through this year.

Cadillac juggled its convertible model offerings this year. The division's "standard" convertible was upgraded from the lowline Series 62 into the more expensive Series 63 De Ville family, while the Eldorado convertible—a Series 63 body style since 1961—was bumped up into the prestigious Fleetwood series. But confusing matters was the fact that the new De Ville Convertible retained its earlier model number, Style 6267. Whatever it was officially called, the standard Cadillac convertible continued to set new sales records. An even 17,900 were sold this year topping the previous year's record by 300 units. Factory retail price of this glamorous soft-top was $5,590. Weight was 4,545 pounds.

The impressive Fleetwood Sixty Special Four-Door Sedan continued as the flagship of Cadillac's standard model line. Loaded with luxury features available only as extra-cost options on other models, it was truly the "Cadillac of Cadillacs." The 1964 Fleetwood Sixty Special got the marque's revised grille design and new rear bumper ends but otherwise carried over its distinctive 1963 styling. One cosmetic change was the relocation of the Fleetwood laurel-wreath-and-crest emblem from the ends of the rear fenders to a new position on the upper rear roof quarters, in place of the vertical louvers used in 1963. The Fleetwood name appeared in small block letters on the lower front fenders. The Sixty Special and the companion Fleetwood Eldorado Convertible had no bodyside trim except for broad chrome rocker sill and lower rear fender moldings. A total of 14,500 Style 6039 Fleetwood Sixty Specials were built this year. Starting price was an appropriately expensive $6,366.

1964

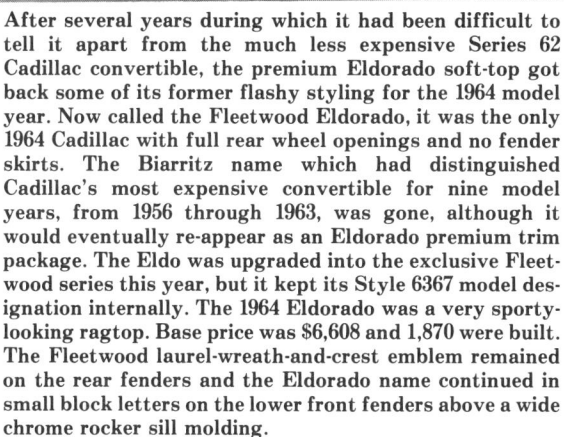

After several years during which it had been difficult to tell it apart from the much less expensive Series 62 Cadillac convertible, the premium Eldorado soft-top got back some of its former flashy styling for the 1964 model year. Now called the Fleetwood Eldorado, it was the only 1964 Cadillac with full rear wheel openings and no fender skirts. The Biarritz name which had distinguished Cadillac's most expensive convertible for nine model years, from 1956 through 1963, was gone, although it would eventually re-appear as an Eldorado premium trim package. The Eldo was upgraded into the exclusive Fleetwood series this year, but it kept its Style 6367 model designation internally. The 1964 Eldorado was a very sporty-looking ragtop. Base price was $6,608 and 1,870 were built. The Fleetwood laurel-wreath-and-crest emblem remained on the rear fenders and the Eldorado name continued in small block letters on the lower front fenders above a wide chrome rocker sill molding.

The builder of this customized Cadillac limousine is not known for certain, but it may be a Derham. The author found this interesting car, which was wearing District of Columbia plates, on a side street in New Orleans in 1978. The rear roof has been substantially altered to provide extra headroom in the rear seat area. This car appears to be similar in design to the one done by Hess and Eisenhardt in 1960. The roof is leather-covered. Note the large backlight (for a formal limousine) and the angular, chopped-off rear roofline. The bashed-in door and accessory rear bumper guards indicate that it operated in an urban traffic environment. This could be either a 1964 or 1965 model, as 1964 styling was carried over to the following year on the Seventy-Five series cars, which were not restyled until 1966.

The stately 1964 Fleetwood Seventy-Five Nine-Passenger Sedan and Limousine wore Cadillac's modified 1964 styling very well, even though the upper body structure dated back to 1959. Again there were two Fleetwood Seventy-Fives which shared the same formal four-door sedan body shell and 149.7-inch wheelbase chassis. The big Seventy-Fives regained their earlier "75" model designations for one, last time. The Fleetwood Seventy-Five Nine-Passenger Sedan, Style 7523, for 1964 had a factory-suggested retail price of $9,724. Only 617 of these long-wheelbase sedans were built. Note the new wheel covers and very narrow white sidewalls. The Seventy-Five Sedan tipped the scales at 5,215 pounds.

Ever since it had ended custom and semi-production body production prior to the Second World War, Cadillac had met its infrequent requirements for special formal sedan versions of its standard Fleetwood Seventy-Five limousines through the independent Derham Body Co. in the Philadelphia suburb of Rosemont, Pa. Derham turned out small batches of formal sedan conversions of Cadillac limos, with closed rear quarters and thickly-padded roofs with tiny rear windows for maximum privacy. But in 1964, Cadillac introduced its own landau roof option for the Fleetwood Seventy-Five Nine-Passenger Limousine. This elegant formal limousine had a richly-grained padded roof and closed upper quarters spanned by ornamental landau irons. The new Fleetwood Seventy-Five Landau Limousine was one of the most impressive Cadillac semi-customs built in the postwar era. The standard 1964 Cadillac Fleetwood Seventy-Five Limousine with glass division was priced just below $10,000 at $9,939. The hand-fitted landau top added several thousand dollars to the price. A total of 808 Style 7533 Fleetwood Limousines were built in the 1964 model year, but it is not known how many were sold with the landau roof option.

1964

American automobile manufacturers used the 1964-65 New York World's Fair as a showcase for their products. All of GM's five automotive divisions, including Cadillac, had large exhibits in the General Motors Futurama Pavilion at Flushing Meadows. In addition to a selection of current models, Cadillac exhibited its first new show car since 1959 at the fair in 1964. It was the Cadillac Florentine, a customized two-door hardtop. Styling features included a formal roofline with narrow quarter windows that retracted into the rear roof pillars, a vinyl suede top covering, and highback bucket seats upholstered in embroidered leather. Note the wire wheels, full rear wheel openings, handleless doors, and total absence of body side trim.

After 93 years, The Eureka Company of Rock Falls, Ill. closed its doors. The old multi-story plant was shut down at the end of the 1964 model year leaving only three firms, all within a 75-mile radius in Ohio, producing funeral cars and ambulances on the Cadillac commercial chassis. Eureka had had a long and proud history in the hearse and ambulance industry. The company's hand-crafted hearses, ambulances, combinations and flower and service cars had attained an enviable reputation for quality and dependability. This is a 1964 Eureka-Cadillac Landau Funeral Coach, one of the last Eureka models offered. It was built on the Series 6890 Cadillac 156-inch wheelbase commercial chassis. The Eureka name was resurrected by a Canadian funeral car and ambulance builder in 1980.

The Hess and Eisenhardt Co. of Cincinnati built the highly-respected S & S (Sayers and Scovill) line of funeral cars and ambulances on the Cadillac commercial chassis. Cadillac shipped 2,639 Series 6890 hearse and ambulance chassis to the four principal U.S. funeral coach and ambulance manufacturers still in business in 1964. This is a 1964 S & S Park Row Limousine Combination Funeral Car and Ambulance. Note the leather-covered roof and "airline" style rear window drapes, which could be closed for privacy. The Cadillac commercial chassis was powered by the same 429 cubic inch, 340-horsepower V-8 engine used in the division's passenger cars.

Now in its last year in business, The Eureka Company of Rock Falls, Ill. was the builder of this handsomely-styled standard-headroom emergency ambulance. Note the angled "C" and "D" pillars, a Eureka styling feature since 1947. The triangular rear quarter windows were a delete option on Eureka coaches. The rear fender tail fin profile has been altered slightly just behind the rear door. All Eureka models had broad chrome rocker sill moldings. The roof has built-in flashing red lights above the windshield header.

1965

The 1965 Cadillacs were the division's most thoroughly redesigned and re-engineered cars in nearly a decade. The major 1965 model changeover was comparable in scope to the milestone 1948 and 1957 restyles. Everything but the engine, itself redesigned only two years earlier, was brand-new.

Cadillac's dramatically restyled 1965 models featured clean, bevelled body lines with three distinct planes. And after 15 years, Cadillac's traditional tail fins were gone. There was just the slightest hint of a fin in the bladed peaks of the long, straight rear fenders. At the 1965 Cadillac National Press Preview held in Detroit on Sept. 15, 1964, Cadillac Motor Car Division General Manager Harold G. Warner left the decision as to whether the 1965 Cadillacs had fins or not entirely up to the individual interpretation of the automotive writers present. The consensus was, however, that Cadillac had cleverly — but definitely — done away with its most famous styling element.

A broad, finely cross-hatched new grille nestled between protruding front fenders that had vertically-stacked headlights. Parking lamps were recessed in a massive new front bumper which had integral side cornering lamps. The totally-restyled bodies had long, uncluttered lines and were unusually handsome in their simplicity. All upper body structures were also redesigned and featured innovative new curved side glass. Even the rear end was pleasingly symmetrical. Concave reveals encircled the side window openings on closed body models.

New vertical rear bumper ends with rectangular centers contained tail, stop, directional, and back-up lights.

The most significant engineering change in the redesigned 1965 cars was the switch to an all-new perimeter type chassis frame which replaced the "X"-type chassis used under all Cadillacs since 1957. The new box-section perimeter frame provided greater torsional rigidity with impressive improvements in ride, handling, smoothness and quietness of operation. A new engine mounting system was used on this new chassis and the engine was repositioned six inches further forward. A redesigned single-piece propeller shaft utilized the division's proven constant velocity joints. All 1965 Cadillacs were powered by the same 429 cubic inch, 340-horsepower V-8 engine used in the 1964 cars, coupled to an improved Turbo Hydra-Matic transmission that became standard in all models.

Other important engineering improvements included a new four-link rear suspension system and modified front suspension; increased front and rear track; cross-flow radiator, and a sound-deadening new exhaust system. Cadillac boasted that its 1965 cars were its quietest ever. Optional white sidewall tires were also smaller with a wider tread. A unique triple-stripe whitewall design was used for the first time.

An automatic levelling device that maintained normal vehicle height regardless of passenger or trunk load was standard on the 1965 Fleetwood Sixty Special and Eldorado convertible but was optionally available on all other models except Fleetwood Seventy-Fives.

New comfort and convenience options this year included an infinitely adjustable tilt-and-telescope steering wheel and redesigned power door lock system.

Although there were still 11 models in Cadillac's 1965 product lineup, the division extensively revised its numerical model designation codes. The long familiar four-digit designators, which had indicated the series in the first two digits, was replaced with a new five-numeral code. Under this rationalization, Fleetwood Seventy-Five models became "69" series cars and all others were "68"s. The 62 Series, which had identified Cadillac's lowest-priced models since 1952, was gone, replaced by a noble new name —

The 1965 Cadillacs reflected the division's most sweeping product design change since 1957. Everything but the engine was new. Cadillac also replaced its old four-digit model code with a more complex five-digit code this year. Even the Series 62 name was gone, replaced with a new name that would identify Cadillac's least expensive cars for the next 12 model years—Calais. This is the very handsome 1965 Cadillac Calais Coupe, Model 68257, at $5,059 the lowest-priced car in the 1965 model lineup. A total of 12,515 of these two-door hardtop coupes were built this year. Shipping weight was 4,435 pounds. Note the broad, finely-textured grille nestled between extended front fenders with new vertically-stacked headlights.

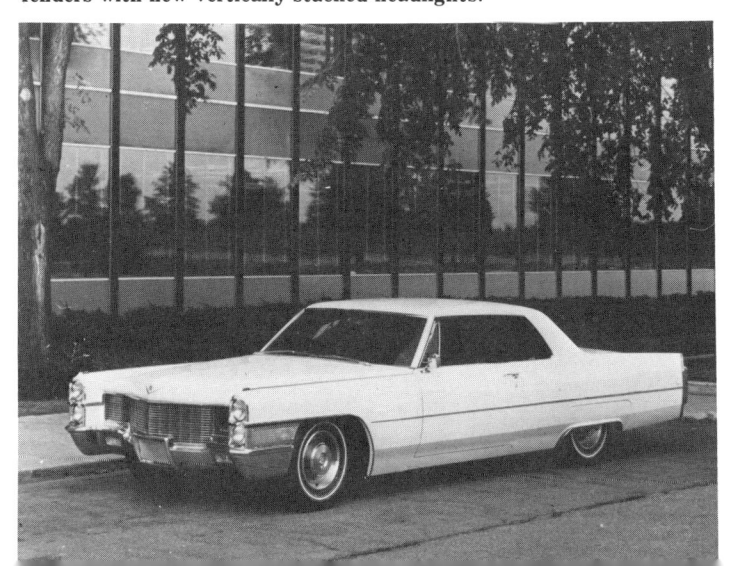

Cadillac's attractive new 1965 styling included clean, bevelled body lines with a strong horizontal theme from headlights to rear bumper. After 15 years, Cadillac's traditional rear fender tailfins were gone: there was only the slightest hint of a fin in the bladelike peaks of the long, straight rear fenders. Introducing the 1965 Cadillacs to the automotive press in Detroit, Cadillac General Manager Harold G. Warner left it entirely up to the auto writers to decide for themselves whether the new car had fins or not. This is a rear view of the 1965 Calais Coupe. Note the extremely narrow triple-ring whitewalls and vertical rear bumper ends which incorporated taillights, brake, backup, and turn indicator lights.

1965

Calais. The Calais name was used for the next 12 model years, through the 1976 line. It later re-appeared on an Oldsmobile Cutlass model in the early 1980s.

Each of Cadillac's three series got one new model this year. A new four-door sedan with thin "B" pillars replaced the six-window sedans in the Calais and de Ville series. There was also a pillared sedan in the Sixty Special line. This marked the return of a true pillared four-door sedan to the Cadillac family for the first time since the 1956 model year. Although the "sedan" body style designation had been carried forward without interruption, these were actually pillarless four-door hardtops from 1957 through 1964.

The new lowline Cadillac Calais series consisted of a two-door coupe, four-door hardtop, and a pillared four-door sedan.

There were four de Villes: the very popular Coupe de Ville, Hardtop Sedan de Ville (top selling car in 1965), the pillared Sedan de Ville, and the sporty de Ville Convertible. Calais and de Ville models and the Fleetwood Eldorado Convertible were all built on the same new 129.5-inch wheelbase chassis and had an over-all length of 224 inches, half an inch longer than the previous year. The premium Fleetwood Sixty Special Four-Door Sedan regained its exclusive 133-inch wheelbase for the first time since 1958. Over-all length was 227.5 inches, or three-and-a-half inches longer than the other cars in the line.

A new formal roof option was available for the Sixty Special this year. This elegant, grained top covering featured a twin beading around its edges and there was a "Brougham" script next to the Fleetwood wreath and crest on the rear roof quarters. When ordered with this popular new option, the Sixty Special became the Fleetwood Brougham. The following year the luxurious Brougham became a separate model rather than just an options package. Padded vinyl roofs were now available on six Cadillacs — Coupe de Ville, both Sedan de Villes, the Sixty Special, and the two Fleetwood Seventy-Fives.

In an unusual move, the Fleetwood Seventy-Fives carried over their 1964 exterior styling. They did not receive a complete restyle until the 1966 model year. There was a precedent for this strategy: the 1947 Seventy-Fives were carried through the 1949 model year while the rest of the line was totally restyled for 1948 and 1949. In each case, extra time was required to engineer all-new "D" bodies for these low-volume specialty models.

The 1965 Seventy-Fives also kept their tubular center "X"-type chassis frames and high upper body structures which dated back to 1959. There were again just two models, a nine-passenger sedan and limousine on a long 149.5-inch wheelbase. An elegant formal landau roof option was optionally available for the stately Fleetwood Limousine. The only way one could tell a 1965 Seventy-Five from a 1964 model was by the three-ring original equipment whitewall tires on the 1965 cars.

Cadillac's three-millionth car, a 1965 Fleetwood Brougham, was driven off the Clark Avenue assembly line on November 4, 1964. The 500,000 square-foot facility expansion program undertaken two years earlier was completed on schedule and included a new Engineering Center and an additional 200,000 square feet of production floor space. And 83 per cent of all Cadillacs now being sold were ordered with factory installed air conditioning.

Production of 181,435 Cadillacs during the booming 1965 model year established another all-time record for the fourth consecutive year.

Carl A. Rasmussen was appointed Chief Engineer of GM's Cadillac Motor Division on March 1, 1965, succeeding Charles F. Arnold who had held this important post for the previous 15 years.

The Series 62 name had identified Cadillac's least-expensive cars ever since the 61 Series was dropped at the end of the 1951 model year. The 62 Series actually went all the way back to the 1941 model year when Cadillac juggled its lowest-priced models after the demise of the LaSalle. The 62 Series disappeared at the end of the 1964 model year, replaced for 1965 by a new Cadillac Calais series. There were three Calais body styles—a two-door coupe, four-door sedan, and this nicely styled pillarless four-door hardtop, Model 68239. The 1965 Cadillac Calais Hardtop Sedan weighed 4,500 pounds and had a manufacturers' suggested retail price of $5,247. A total of 13,975 of these lowline four-door hardtops were built during the record 1965 model run.

Cadillac's new lowline Calais series included two four-door models, a pillarless four-door hardtop and a companion four-door sedan. Only a thin "B" pillar on the sedan set them apart. In fact, with its side glass raised the Calais Hardtop Sedan was nearly indistinguishable from the Calais Sedan. This is the 1965 Cadillac Calais Sedan, Model 68269 which at $5,247 was priced identically with the Calais four-door hardtop. The Calais Sedan was the least-popular model in the series with only 7,721 delivered during the record 1965 model year. Note how the front fender cornering lights have been integrated into the bumper design.

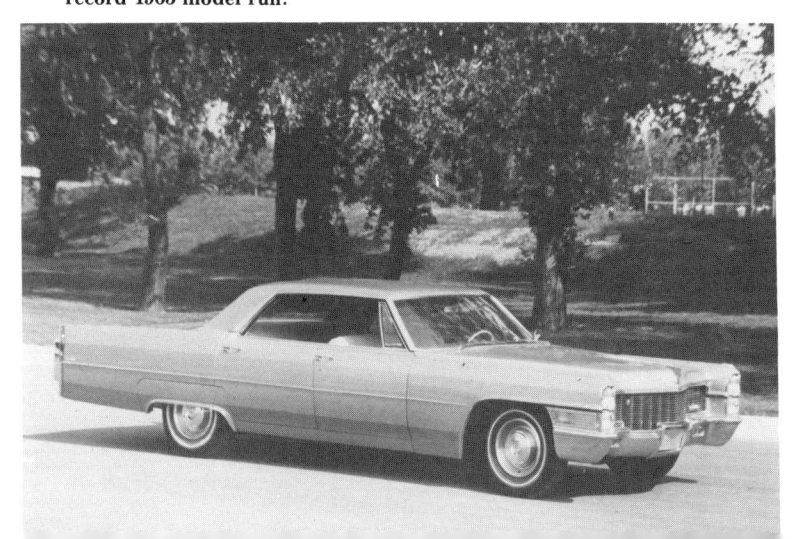

1965

All of Cadillac's "standard" 1965 models—Calais, de Ville and the Fleetwood Eldorado Convertible—were built on a new 129.5-inch wheelbase perimeter-frame chassis and had an over-all length of 224 inches. Except for the Fleetwood Seventy-Fives, which carried over their 1964 styling, all of the 1965 Cadillacs got fresh, new styling. This attractive new styling featured strong, chiselled body lines, expansive, flat surfaces and softly-rounded upper body structures. Calais and de Ville models had thin horizontal body side moldings and tiny script nameplates on their finless rear fenders. The redesigned 1965 models also featured curved side glass. This is the 1965 Cadillac Calais Sedan.

Cadillac sales rocketed to another new all-time record this year. The top-selling model in the 11-car 1965 product lineup was the four-door Hardtop Sedan de Ville. A record 45,535 of these pillarless four-door hardtops were delivered during the 1965 model year. Under Cadillac's revised numerical model designations, the Hardtop Sedan de Ville was listed as Model 68339. Manufacturer's suggested retail price was $5,666 in base form. Note the chrome window sill moldings which set de Ville models apart from the less-expensive Cadillac Calais series.

Cadillac offered "pillared" four-door sedan body styles in its 1965 Calais, de Ville, and Fleetwood Sixty Special series for the first time since 1956. Although the "sedan" body style designation had been carried forward without interruption, these cars were in fact four-door hardtops. The new thin-pillared four-door sedans replaced the six-window sedans offered in the 62 and de Ville series between 1959 and 1964. This is the 1965 Cadillac Sedan de Ville, Model 68369, which had a manufacturer's suggested retail price of $5,666—the same as the companion de Ville four-door hardtop. Shipping weight was 4,555 pounds and an even 15,000 were built.

Although sales of the handsomely-restyled 1965 Cadillac Coupe de Ville set yet another record for this popular body style, it took second place to the companion four-door Hardtop Sedan de Ville. An impressive 43,345 Coupe de Villes were sold during the booming 1965 model year. Model 68357, the Coupe de Ville carried a manufacturer's suggested retail price of $5,469. A padded vinyl roof was a $121 extra-cost option on this model. All four 1965 de Ville models had small script series nameplates on the ends of their rear fenders just above the chrome side molding. Note the license plate location in the center of the clean, symmetrical rear end.

The convertible reached the peak of its popularity among American car buyers this year. The U.S. auto industry built an all-time record 507,000 convertibles in 1965. Of these, 21,325 were Cadillacs, including 19,200 de Ville convertibles and 2,125 Fleetwood Eldorado soft-tops. This is the dashing 1965 Cadillac de Ville Convertible, Model 68367, which sold for $5,639. De Ville convertible sales also topped out this year. A total of 11 interior color combinations were offered in the de Ville Convertible along with five convertible top colors—white, black, blue, green and sandalwood. Bucket seats with a center console were a $184 extra-cost option.

1965

Although externally similar to the companion Fleetwood Sixty Special, the redesigned 1965 Cadillac Fleetwood Eldorado Convertible was 3.5 inches shorter. The distinctive Fleetwood Eldorado got its fender skirts back this year. Like the Sixty Special, the Eldo had no body side trim but featured wide chrome rocker sill moldings and rear fender extension trim capped with a fine chrome molding. The Fleetwood name was spelled out in small block letters on the lower edge of the front fenders. A choice of bench or bucket seats was offered at no charge. Eight premium interior trim combinations were offered in lush, perforated leathers. Model 68467, the Fleetwood Eldorado, weighed 4,660 pounds. Starting price was a steep $6,738. Production of this luxurious convertible remained low. Only 2,125 were delivered this year.

The prestigious Cadillac Fleetwood Sixty Special returned to its traditional 133-inch wheelbase this year for the first time since 1958. Like all standard Cadillacs, the 1965 Sixty Special was built on a new perimeter-type chassis frame instead of the "X"-frame used from 1957 through 1964. The completely redesigned and restyled 1965 Fleetwood Sixty Special was also a true pillared sedan for the first time in eight years. Although always listed as the Sixty Special Sedan, it was actually a four-door hardtop from 1957 through 1964. The 1965 Sixty Special, Model 68069, had a factory suggested retail price of $6,479. Weight was 4,670 pounds and over-all length was 227.5 inches, 3.5 inches longer than Calais and de Ville models. The record-breaking 1965 model production run included 18,100 Fleetwood Sixty Specials, including both the standard sedan shown here and the even more luxurious Fleetwood Brougham.

The all-new 1965 Cadillac Fleetwood Sixty Special featured the same distinctive exterior styling as the slightly shorter companion Fleetwood Eldorado Convertible. There was no bodyside trim on the Sixty Special's broad, bevelled flanks. A fine chrome detail molding extended the length of the lower body above wide chrome rocker sill and rear fender extension moldings. The Fleetwood name appeared in small block letters on the lower front fenders and on the right-hand side of the rear deck lid. The Fleetwood laurel and crest emblem continued on the upper rear roof quarters. Note the slender "B" pillar and new curved side window glass.

An important new option broadened Cadillac's Fleetwood Sixty Special offerings this year. In addition to the standard Sixty Special Sedan, Cadillac buyers could opt for a new formal roof option called the Fleetwood Brougham. On the exclusive Brougham, the entire roof was padded and covered with grained vinyl fabric set off by a twin-beaded molding that bordered the side windows and lower roof edges. There was also a thin band of body color between the window frame and the dual molding which imparted new distinction to the upper body. A "Brougham" script nameplate was added to the upper rear quarter, next to the Fleetwood wreath and crest. The Brougham option added $194 to the price of the car. It proved an extremely popular option and the Fleetwood Brougham became a separate model the following year.

1965

Here is a "special" Fleetwood Sixty Special Sedan with the new Brougham package and a custom touch—a small, formal rear window. This rear roof treatment was similar to the one offered on the Fleetwood Seventy-Five Landau Limousine. The standard Sixty Special and Fleetwood Brougham featured a large conventional backlight. This may have been a special order job. The 1965 Cadillac Sixty Special featured bold geometric body contours and was an especially clean and handsome design.

Not all of Cadillac's 1965 cars got fresh styling. The limited-production Fleetwood Seventy-Five Sedan and Limousine carried over their 1964 exterior appearance. An all-new Seventy-Five was not introduced until 1966. There was a precedent for this unusual move. In the late 1940s, Cadillac had carried over its 1947 Seventy-Fives through the 1949 model year, while the 1948 and 1949 cars got totally new styling. The 1965 Fleetwood Seventy-Fives were described in a separate catalog, although this illustration of the Model 69723 Fleetwood Seventy-Five Sedan appeared with the other Fleetwood models in the deluxe sales catalog. The 1965 Seventy-Fives retained the same high windshield with "dogleg" pillars and roof profile that had originally been introduced for 1959. Only 455 Fleetwood Seventy-Five Nine-Passenger Sedans were built this year. Suggested retail price was $9,746.

Cadillac's standard de Ville model offerings for 1965 included a Coupe de Ville, Sedan de Ville, and Hardtop Sedan de Ville—but no station wagon. A New Jersey firm, however, offered a custom-built "Wagon de Ville" based on the 1965 Cadillac de Ville four-door hardtop. Standard equipment included a power-operated sun roof and a rooftop luggage carrier. Designed to seat eight or nine passengers, the Wagon de Ville sold for about $14,950 through authorized Cadillac dealers. It was a product of Cadillac Wagons Ltd. of Linden, N.J. Note the turbine wheel covers and unique rear roof canopy.

The only way to tell a 1965 Cadillac Fleetwood Seventy-Five from a 1964 model was by the three-ring white sidewall tires supplied as original equipment on 1965 models. The 1965 Fleetwood Seventy-Five Sedan and Limousine carried over their now strangely-dated 1964 styling for one extra year. The 1965 Cadillac Fleetwood Seventy-Five Limousine, a nine-passenger car with standard glass division, was the heaviest and most expensive car in the 1965 line. Its base price was $9,960 and shipping weight was 5,260 pounds. This is a 1965 Fleetwood Limousine equipped with the formal Landau roof option, which featured a padded, vinyl-covered roof, small "opera" rear window, and closed upper rear quarters spanned by ornamental landau bows. Only 795 Fleetwood Limousines were produced this year.

Except for its engine, everything about the 1965 Cadillac was new. The most important engineering change was the replacement of the "X"-frame chassis used from 1957 through 1964 with an all-new box-section perimeter-chassis frame that ran under the outer edges of the body shell. This new chassis provided a lower center of gravity and greater torsional rigidity. An improved body mounting system made the 1965 Cadillacs the quietest cars yet. All 1965 Cadillacs also got GM's new Turbo Hydra-Matic automatic transmission which had been introduced on some 1964 models. Standard engine in all 1965 Cadillacs was the division's one-year-old 429 cubic inch, 340 horsepower V-8. The new chassis was also equipped with a new one-piece drive shaft.

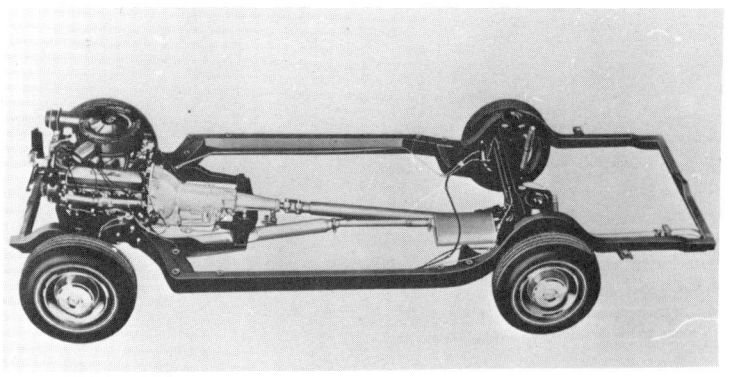

1965

Cadillac's long-wheelbase commercial chassis was also redesigned this year. Wheelbase continued at 156 inches, but the new Cadillac hearse and ambulance chassis featured the same new box-section perimeter-type frame used under the division's standard passenger cars. This is a 1965 Superior-Cadillac Crown Sovereign Landaulet Funeral Coach, a premium side-loading hearse with power-operated casket table. Note the ribbed lower rear fender applique and the broad chrome "C" pillar that encircled the roof. This was Superior's most prestigious model—the ultimate way to go.

The Hess and Eisenhardt Co. of Cincinnati, Ohio, builder of S & S funeral cars and ambulances, built this long, very low coupe style flower car. It utilized Cadillac's 156-inch wheelbase commercial chassis and the upper body structure of a Cadillac two-door coupe. It has an open well instead of the conventional flower deck. The flower well and boot were made of brushed stainless steel. The roof was vinyl-covered. Hess and Eisenhardt also offered a deck-type flower car with a higher coupe roof compatible with the styling of its S & S funeral coaches.

Miller-Meteor of Piqua, Ohio, offered three series of professional ambulances on the 1965 Cadillac commercial chassis. Two headroom heights—42 and 48 inches—were available. This is the high-headroom Classic 48-Inch model which was designed to transport as many as four litter patients. Note the extremely large rear door, built-in rear bumper step, and the five roof lights. Finish was in a very striking white and bright red. Chrome ambulance crosses adorn the rear roof quarters.

Superior Coach Corp. of Lima, Ohio redesigned its Coupe de Fleur flower car this year for the first time since 1957. Gone was the low, rearward-notched coupe roofline. Superior's new flower car "greenhouse" utilized the same cowl assembly and high windshield used on the company's standard hearses and ambulances. The new sloped roofline included triangular quarter windows. This same roofline was used on Superior's flower cars through the 1976 model year. Cadillac supplied 2,669 long-wheelbase commercial chassis to the three principal U.S. funeral car and ambulance manufacturers during the 1965 model year.

1966

Cadillac production and sales set new records for the fifth straight year. The 1966 Cadillacs received only minor exterior styling changes, but one new model was added and the prestigious Fleetwood Seventy-Fives were totally redesigned for the first time in seven years.

The division's model offerings were increased to 12 for 1966 with the addition of the ultra-luxurious Fleetwood Brougham. Introduced a year earlier as a padded vinyl roof option for the 1965 Fleetwood Sixty Special, the Brougham was upgraded to a separate model this year. For the first time in many years, Cadillac's upscale customers had a choice of two Sixty Specials—the standard well-equipped Fleetwood Sixty Special Four-Door Sedan and the even more lavishly equipped and appointed new Brougham. Addition of the 1966 Fleetwood Brougham marked the return of this distinguished model name to the Cadillac family for the first time since 1960. Both models continued on their own exclusive 133-inch wheelbase.

As in 1965, the Fleetwood Brougham featured an elegant grained vinyl roof. Inside, there were such luxurious appointments as genuine walnut trim, top-grade leather or rich embroidered cloth upholstery, two lighted tables that dropped down out of the front seat backs, carpeted foot hassocks for rear seat passengers, and adjustable individual reading lamps.

Styling changes on the 1966 Cadillacs were very minor. The grille treatment was revised with a horizontal center bar and a coarser cross-hatched grille texture. New rectangular parking lights were built into the outer ends of the center divider and the Cadillac script nameplate remained on the lower left corner of the grille. The stacked headlight bezels were color-keyed to the body color and redesigned cornering lights were incorporated into the sides of the front fenders. The new cornering lamps had a vertical lens texture with tiny Cadillac crests on the leading edge.

Massive chrome rear bumper ends contained two thin red vertical taillights which were surrounded by white backup light lenses. All 1966 Cadillacs also got new wheel covers with six thin-bladed spokes radiating out from large, flat centers.

All 1966 Cadillacs were equipped with a new variable ratio power steering system that reduced steering effort considerably. The suspension system was also refined for greater vehicle stability and smoothness. Padded vinyl roofs were optionally available on the Coupe de Ville, both de Ville sedans and the all-new Fleetwood Seventy-Fives. Air conditioning was now being ordered on 88 per cent of all Cadillacs sold, and automatic Climate Control was made standard equipment on the Seventy-Fives.

The most intriguing new option for 1966 was an electric seat warmer. Available as an extra-cost option ($78.95) on the front seats of all Cadillacs, this cold-weather comfort convenience was also offered for the rear seats of the redesigned Fleetwood Seventy-Five. A new Hazard·Warning Flasher system automatically caused all front and rear turn signal lights to blink simultaneously for driving or roadside emergencies.

The regal Cadillac Fleetwood Seventy-Five Sedan and Limousine got their first complete restyle since 1959. These big formal sedans—favorites of wealthy individuals, bankers, funeral directors, and diplomats— were once again compatible in styling with the rest of Cadillac's other cars. The completely re-engineered Fleetwood "D" body featured long, bevelled body contours and a softly-rounded upperstructure with broad sail panels. The handsomely redesigned Seventy-Fives looked much lower than previous models. The doors extended into the roof for ease of entry and exit. The totally-new 1966 Seventy-Fives were built on the same box section, perimeter-type chassis frame introduced on all other Cadillac models the previous year. There were still only two models, a roomy Nine-Passenger Sedan and the formal Fleetwood Limousine with standard division glass between the chauffeur's and rear passenger compartments.

The 1966 Cadillacs showed only minor styling changes from the previous year. All 1966 Cadillacs, including the totally-redesigned Fleetwood Seventy-Fives, got this revised grille design which featured a horizontal center bar and new rectangular parking lamps mounted at the outer edges of the more coarsely-textured eggcrate grille. The stacked headlight bezels were now painted the same color as the body. Overall, there was less exterior chrome this year. The Cadillac script nameplate remained on the lower left corner of the redesigned grille. This is the 1966 Cadillac Calais Four-Door Sedan.

The Calais Coupe remained the "most affordable"—or lowest-priced—model in the 12-car 1966 Cadillac product lineup. The Calais Coupe's price was actually reduced this year, from $5,059 in 1965 to $4,986 in 1966, a drop of $73. Cadillac built 11,080 of these lowline two-door hardtops, Model 68257, this year. Shipping weight was 4,390 pounds. All three Cadillac Calais models rode on the same 129.5-inch wheelbase. Note the concave sheet-metal sculpturing around the side window openings. The 1966 Cadillacs showed only cosmetic styling changes from the previous year.

1966

Full vinyl roofs were available on both models and an elegant Landau Roof option with closed rear quarters and decorative carriage bows was offered on the Fleetwood Limousine. These impressive new limousine bodies were produced for the next five model years with annual facelifts and exterior "skin" changes.

This would be the last year for the rear-wheel-drive Fleetwood Eldorado Convertible. A front-wheel-drive Eldo soft top would eventually rejoin the Cadillac product line-up, but not until the 1971 model year. The Eldorado Convertible had been introduced in 1953 and had featured its own distinctive exterior styling since 1955.

Three series—Calais, de Ville and Fleetwood—continued on three separate wheelbases. The three Calais models, four de Villes and the Eldorado Convertible were built on the same 129.5-inch wheelbase. The Sixty Special and Fleetwood Brougham had their own 133-inch chassis, and the new Seventy-Fives were built on a 149.8-inch wheelbase. The top-selling car in the Cadillac line this year was again the Hardtop Sedan de Ville. More than 60,000 were sold, followed closely by the Coupe de Ville with sales that topped 50,000 for the first time. Sales of the de Ville Convertible reached their peak this year. Cadillac production hit an all-time record 196,675 cars and the division was eagerly anticipating the day when sales would top the 200,000 mark.

There were some rapid personnel changes in the General Manager's office at the Cadillac Motor Car Division this year. On April 1, Kenneth N. Scott succeeded Harold G. Warner as Cadillac's chief executive, but exactly five months later Mr. Scott moved on to a new position in General Motors and Calvin J. Werner was named General Manager of the division.

Cadillac now used a sophisticated $250,000 road simulator in its elaborate quality control engineering process. The division's 429 cubic inch, 340 horsepower engine and powertrain were unchanged from 1965.

The biggest product news of the year was Oldsmobile's new front-wheel-drive Toronado, a high-styled personal luxury car introduced in the fall of 1965 as a 1966 model. More than 50,000 of these sleek two-door fastback coupes were sold in the car's first year on the market. Unknown to many, Cadillac also had a front-wheel-drive personal car in the works.

After 114 years, Studebaker dropped out of the automobile business midway through the 1966 model year. What was left of this once-proud automotive entity had shifted its miniscule production from South Bend, Indiana, to Canada three years earlier. Formed in 1852, Studebaker had ceased to be a factor in the U.S. luxury car market since it stopped building its highly-regarded President series in the 1930s.

The most popular body style in Cadillac's lowest-priced series for 1966 was the Calais Hardtop Sedan. This pillarless four-door hardtop, Model 68239, had a factory advertised price of $5,171. A total of 13,025 were built during the record 1966 production run. Note the revised grille treatment and vertically-textured cornering lights, which were no longer part of the front bumper. The Calais Hardtop Sedan weighed 4,465 pounds in standard trim. A new Climate Control automatic air conditioning/heater system was a $484 extra on most 1966 Cadillacs.

Sales of the Calais Sedan dropped sharply this year, from a lacklustre 7,721 in 1965 to only 4,575 in the 1966 model year. Cadillac's low-end buyers overwhelmingly preferred the sportier appearance of the companion Calais Hardtop Sedan. Both Calais four-door models were identically priced at $5,171. The 1966 Cadillacs showed only minor styling changes this year. Vertical rear bumper ends were new, as were the six-spoked wheel covers. Listed as Model 68269, the Calais Sedan weighed in at 4,460 pounds.

Sales of Cadillac's extremely popular Coupe de Ville passed the 50,000 mark during the 1966 model year for the first time, but continued to trail those of the four-door Hardtop Sedan de Ville. De Ville closed models had chrome trim in the concave reveal around the side window openings, including the window sills. The handsome 1966 Cadillac Coupe de Ville had a manufacturer's suggested retail price of $5,339 and was listed as Model 68357. A padded vinyl roof was a $121 extra-cost option. Weight was 4,460 pounds. When 1966 model production ended in the summer of that year, a record 50,580 Coupe de Villes had rolled off the Clark Avenue plant assembly lines.

The four-door Hardtop Sedan de Ville was Cadillac's top-selling single model during the prosperous 1966 model year. An astonishing 60,550 Sedan de Ville four-door hardtops were sold, outselling the perennially popular Coupe de Ville by nearly 10,000 units! A padded vinyl roof was optionally available on both the Sedan de Ville and Hardtop Sedan de Ville for an extra $136. The 1966 Cadillac's modestly-refined styling included new vertical rear bumper ends with two thin red lenses, surrounded by white backup lights. Note the gently-sloping rear deck and large rear window.

1966

The four-door pillared sedans were the slowest sellers in both the Calais and de Ville series this year. But the four-door Sedan de Ville handsomely outsold the Calais: 1966 model year production included 11,860 Sedan de Villes compared with just 4,575 Calais Sedans. Style 68369, the Sedan de Ville was priced identically with the far more popular companion Hardtop Sedan de Ville. Note the standard chrome wheel opening lip moldings and de Ville script nameplates on the rear fenders.

Sales of the sporty Cadillac de Ville Convertible in 1966 equalled the all-time record for this hot-selling model set only the previous year. Exactly 19,200 de Ville Convertibles were built in both 1965 and 1966—the highest ever. Manufacturer's suggested retail price was $,5,555. The Model 68367 two-door, six-passenger soft-top weighed 4,445 pounds. Leather upholstery was standard with perforated seat and seat back inserts. A bench-type seat was standard but bucket seats were optionally available for an extra $184. Eight interior color choices were offered.

For the first time in many years, two Fleetwood Sixty Special four-door sedans were available to discriminating Cadillac buyers. The 1966 product line included a standard Fleetwood Sixty Special Sedan and an even more luxurious new Fleetwood Brougham. This is the standard Sixty Special Sedan, Model 68069, which sold for $6,378. Only 5,445 Sixty Special Sedans were sold this year, compared with deliveries of 13,630 Broughams. Within a few years the "plain-Jane" Sixty Special sedan would be dropped. Note the revised 1966 exterior trim treatment, consisting of a full-length chrome molding along the bevelled lower body sides. Wheel covers were also new and featured six thin "spokes." Narrow three-ring whitewalls continued as extra-cost $56 options on all standard models.

This would be the last year for the luxurious Fleetwood Eldorado Convertible, at least for a while. A front-wheel-drive Eldorado convertible rejoined the Cadillac family in 1971. The 1966 Fleetwood Eldorado Convertible, Model 68467, was a soft-top companion car to the Fleetwood Sixty Special four-door sedan but was three-and-a-half inches shorter. Wheelbase was 129.5 inches, the same as the lower-priced de Ville Convertible. Sixty Specials rode on an exclusive 133-inch wheelbase. Manufacturer's suggested retail price of the 1966 Fleetwood Eldorado was $6,631. Bucket seats were again offered as a no-cost option. Eight sumptuous leather interior color choices were available. Triple-stripe whitewall tires were standard.

An extremely popular options package for the Sixty Special the previous year, the Fleetwood Brougham became a separate, 12th Cadillac model in 1966. The opulently-equipped Brougham was a premium version of Cadillac's premier four-door sedan. The Fleetwood Brougham had a padded vinyl roof with beaded edge molding and Brougham script on its upper rear roof quarters. Interior appointments included rich embroidered cloth or top-grade leather upholstery, illuminated tables that dropped down out of the front seat backs, carpeted footrests, and adjustable reading lamps for rear seat passengers. Genuine walnut trim was also standard. The elegant Fleetwood Brougham had a manufacturer's suggested retail price of $6,695—$317 higher than the standard Sixty Special Sedan. The plush Brougham was very popular with Cadillac's upscale customers; 13,630 were sold in 1966 compared with only 5,445 standard Sixty Specials. The new Fleetwood Brougham was designated Model 68169.

The exclusive Fleetwood Seventy-Five Sedan and Limousine for 1966 got their first complete redesign and restyle in seven years. The 1965 version had carried over the Seventy-Five's staid 1964 styling, but the 1966 model was totally new. Styling was now fully compatible with the rest of the Cadillac family. The 1966 Fleetwood Seventy-Five Sedan and Limo featured long, bevelled body lines and softly-rounded upper body contours. All four doors extended into the roof panel for ease of entry and exit. Seating was provided for nine passengers. This basic "D" body was produced for five years through 1970, with annual grille and "skin" changes. Note the chrome edge molding around all windows.

1966

The totally-restyled 1966 Cadillac Fleetwood Seventy-Five was an unusually handsome car, especially for a long-wheelbase formal body style. Its general appearance was that of a greatly-elongated Sedan de Ville. The all-new Seventy-Five Sedan and Limousine were built on a 149.8-inch wheelbase and rode on a new perimeter-type chassis frame identical to the one introduced on all 1965 Cadillacs except the carryover Seventy-Fives. The distinctive Fleetwood wreath and crest embellished the front of the hood and the rear decklid. Note the bright window reveals, broad sail panels and small backlight. Automatic Climate Control air conditioning and Level Control became standard equipment on the new Fleetwood Seventy-Fives.

Vinyl roofs were optionally available on both redesigned 1966 Cadillac Fleetwood Seventy-Fives, the nine-passenger sedan and limousine. A formal Landau Roof option was again offered on the Limousine. This elegant exterior decor package included a full padded roof, closed rear quarters, and a tiny "opera" rear window for utmost privacy. Stylish landau irons were also part of this package, although some formal limos were ordered without them. Fleetwood Seventy-Five prices passed the five-figure mark for the first time. The Nine-Passenger Sedan, Model 69723, sold for $10,312 and the Style 69733 Limousine had a base price of $10,521. Sales increased significantly this year because of the fresh, new styling. Deliveries included 980 Sedans and 1,037 Limousines.

The Superior Coach Corp. of Lima, Ohio, built more special-purpose flower cars than all other funeral car manufacturers combined. Rivals Miller-Meteor and Hess and Eisenhardt (S & S) produced only a few a year. Superior built 30 of these Model 609 Coupe de Fleur flower cars on the 1966 Cadillac commercial chassis. The stainless steel flower deck was hydraulically adjustable to accommodate various floral arrangements. There was a full-length casket compartment underneath. Factory-delivered price was $11,683.

Not all hearses were painted funeral black. While grays, silvers and even whites had always been popular alternates among style-conscious funeral directors, the trend was now toward more daring two-tone combinations. This is a 1966 Superior-Cadillac Crown Sovereign Limousine Funeral Coach with a special two-tone exterior, with chrome color break molding on the upper body sides. Superior's premium Crown Royale and Crown Sovereign models had a wide, chrome-sheathed "C" pillar which swept up and over the roof. The rear roof area had a crinkle finish. The 1966 Cadillac commercial chassis featured the same perimeter-type chassis frame used under all other Cadillacs and retained its 156-inch wheelbase.

1967

Cadillac formally entered the burgeoning luxury-personal car market this year with a spectacular new model called the Fleetwood Eldorado. Offered in only one body style, a very distinctive two-door hardtop coupe, the totally-new 1967 Cadillac Fleetwood Eldorado was the division's first production front-wheel-drive car.

Planning for the new Eldorado actually began in 1959 as the finishing touches were being made to the last of the exclusive Eldorado Broughams for the 1960 model year. Rival GM passenger car division, Buick, had beaten the more conservative Cadillac into the luxury-personal car market with the 1963 Riviera. Oldsmobile was the first GM division to offer a front-wheel-drive car, the 1966 Toronado. But the stunningly styled 1967 Fleetwood Eldorado by Cadillac was the first to combine front-wheel-drive, variable ratio steering, and automatic level control as complementary standard equipment.

The new Fleetwood Eldorado was built on its own, separate assembly line in the former foundry in Cadillac's sprawling Clark Avenue plant in Detroit. The Eldorado shared its two-door "E" body shell with the Riviera and Toronado. These bodies were built in Fisher Body's Euclid, Ohio, plant and were trucked to the various divisions' vehicle assembly plants. The high-styled new Eldorado replaced the Fleetwood Eldorado Convertible in Cadillac's premier car series leaving the de Ville convertible the only soft-top in the Cadillac Motor Car Division's product line.

The 1967 Cadillac Eldorado was built on its own, exclusive 120-inch wheelbase which was nine-and-a-half inches shorter than Cadillac's standard 1967 models. Its over-all length was 221 inches, three inches shorter than the Coupe de Ville. Height was 53.3 inches, about an inch lower than the de Ville. The stylish new Eldorado's unmistakable profile included an extremely long hood, formal coupe roofline, and a short rear deck. Full wheel openings and large, slotted wheel covers gave the car a powerful, sporty flair.

Up front, the Eldorado sported a broad cross-hatched grille. Dual headlight units were concealed for the first time behind grille-textured doors which dropped down when the headlights were switched on. Rectangular parking lamps were incorporated into the front bumper. The Eldorado coupe had no front ventilator panes, resulting in an exceptionally tidy side appearance. Small vertical quarter windows retracted into the rear roof structure. The beautifully-styled Eldorado had very long, smoothly contoured front fenders and short, bobbed rear fenders that terminated in thin, vertical blades that divided the taillights. Back-up lights were mounted in the rear bumper. One of the Eldorado's most distinctive styling features was its creased, V-shaped rear window.

The first-generation FWD Eldorado continued in production for four model years and was produced through 1970 with only minor exterior styling refinements. All are sought by special-interest car collectors today, especially the milestone 1967 model.

The 1967 Eldorado was powered by the same 429 cubic inch, 340 horsepower V-8 engine used in all other 1967 Cadillacs but was modified for front-wheel-drive. The Turbo Hydra-Matic automatic transmission was positioned alongside the engine and was directly connected to the differential which drove the front axles and wheels. A key component was an innovative flexible, lightweight drive chain developed in conjunction with Borg-Warner. The Eldorado's chassis frame, drive train and suspension were also unique. The suspension design used torsion bars up front with single leaf springs, a drop-center axle and four shock absorbers in the rear.

A spectacular new model joined Cadillac's product lineup this year. It was the Fleetwood Eldorado, an all-new "luxury personal" car that replaced the Fleetwood Eldorado Convertible. The new Eldorado was Cadillac's first front-wheel-drive car. It was offered in just one body style, a distinctively-styled two-door hardtop coupe. The classically-styled 1967 Fleetwood Eldorado featured the long hoodline, short rear deck profile that was steadily gaining favor among U.S. auto manufacturers. With a suggested retail price of $6,277 the new Eldo was the least-expensive car in the premium Fleetwood series. It was also by far the most popular with first-year deliveries of 17,930 cars. The new Eldorado was listed as Model 69347, and weighed an even 4,500 pounds. Note the sporty-looking slotted wheels.

The all-new Cadillac Fleetwood Eldorado Coupe was built on its own assembly line in the division's Clark Avenue plant in Detroit. It also had its own exclusive 120-inch wheelbase, 9.5 inches shorter than Cadillac's standard 1967 models, and was three inches shorter over-all and an inch lower than the Coupe de Ville. The Eldorado was powered by the same 340-horsepower, 429 cubic inch V-8 engine used in all other Cadillacs but modified for front-wheel-drive. Exterior engine changes included a new oil pan, exhaust manifolds, accessory belt drive, accessory mounting arrangement, and a unique engine mounting system. Frame and suspension design were also unique to this model. The high-styled 1967 Fleetwood Eldorado with its long hood, short deck and creased rear window is already a bona fide special interest car.

1967

The all-new Fleetwood Eldorado was introduced along with the rest of the 1967 Cadillacs on October 6, 1966. Base price was $6,277—lowest in the premium Fleetwood series. Nearly 18,000 were sold in its first year on the market. Production was kept to one shift to ensure a high level of product quality—and limited availability.

While the spotlight was on the all-new Fleetwood Eldorado, Cadillac's other 1967 models were extensively changed, too. All models got fresh, new styling. Front fenders and broad, cross-hatched grilles "leaned" forward, and a powerful sculptured design line swept from the tops of the front fenders down to the rear bumper. The 1967 Cadillacs sported raised haunches with more than just a hint of tail fins in the redesigned rear fenders. Coupes were given a smart, new formal roofline taken directly from the Florentine show car exhibited at the 1964 New York World's Fair. Four-door hardtops also received stylish new roof structures. Revised rear bumpers with massive vertical chrome fender caps contained brake, parking, directional signal and taillights and provided a strong Cadillac rear identity.

Again, there were a total of 12 models. The 1967 lineup included three lowline Calais body styles, four mid-range de Villes, and five Fleetwood models including the new Eldorado, a standard Sixty Special, the premium Fleetwood Brougham, and the big Seventy-Five Sedan and Limousine. Although they had been completely redesigned and restyled only the previous year, the Seventy-Five's also got Cadillac's fresh 1967 exterior styling.

The 1967 Cadillacs rode on four different wheelbases. Calais and de Ville models were built on a 129.5-inch perimeter frame chassis. The new Eldorado rode on its own 120-inch stub-frame chassis. Sixty Special and Fleetwood Broughams shared a 133-inch wheelbase and the Seventy-Fives continued on a 149.8-inch wheelbase. Prices ranged from just over $5,000 for the Calais Coupe on up to more than $10,500 for the stately Fleetwood Limousine.

Sparked by consumer advocate Ralph Nader's earlier attack on GM's Chevrolet Corvair, the U.S. auto industry was now preoccupied with vehicle safety. In addition to the energy-absorbing steering column now standard on all GM cars, the 1967 Cadillacs were equipped with an energy-absorbing steering wheel that deflected under severe impact and cushioned the driver against a padded surface within the circumference of the wheel.

Upper instrument panel surfaces were fully padded and controls were padded, repositioned, and recessed for added passenger safety. Other safety improvements included flexible vinyl coat hooks and an interior rear-view mirror with padded rim designed to break away on impact. A red warning light on the dash indicated any malfunction in the dual power braking system.

In addition to 16 standard exterior color choices, Cadillac's customers could now choose from five rich new "Firemist" premium colors. These lustrous color options—Olympic Bronze, Crystal, Ember, Atlantis Blue, and Tropical Green—were a $132 extra.

Cadillac production and sales shattered records for the sixth straight year. Production reached an important new plateau: the division produced 200,000 cars in a single model year for the first time. Annual production had doubled since 1950.

The 1967 Cadillacs were extensively restyled. Prominent styling features included a forward-leaning front end, long sculptured body lines and redefined rear fenders that had more than just a hint of tail fins in them. Coupes got smart new rooflines that were inspired by the Florentine show car created for the 1964 New York World's Fair. The lowest-priced in the 12-car 1967 Cadillac line was again the Calais Coupe, Model 68247, which was priced just over $5,000 at $5,040. Shipping weight was 4,447 pounds. Cadillac's record-breaking 1967 production included 9,085 of these lowline two-door hardtops.

The Cadillac Calais entered its third year as the division's least expensive series. Again, there were three body styles on a 129.5-inch wheelbase. The most popular of these was the Calais Hardtop Sedan, a pillarless four-door hardtop. Model 68249, the Calais Hardtop Sedan had a manufacturer's suggested retail price of $5,215. Production totalled 9,880 cars. Shipping weight was 4,495 pounds. Color choices included 16 standard and five optional "Firemist" exterior finishes. Note the forward-thrusted front-end design with stacked dual headlight units and new roofline.

This would be the third and final year for the slow-selling Calais Sedan. Cadillac's customers clearly preferred sportier hardtop sedan styling. Only 2,865 Calais Sedans were built during the record 1967-model production run and this lowline pillared four-door sedan body style, Model 68269, was not offered in 1968. The Calais Sedan was identically priced with the companion Calais Hardtop Sedan at $5,215. At 4,499 pounds it was only four pounds heavier. Note the revised rear-end styling with raised haunches, split "grilles" bracketing the license plate, and new vertical chrome bumper ends.

1967

The 1967 Cadillac was a thoroughly restyled car. The 1967 cars were given a powerful frontal appearance with forward-thrusted grille and front fenders. The new full-width "eggcrate" grille was flanked by dual stacked headlight units for the third consecutive year. Rectangular parking lamps were built into the outer edges of the grille. Redesigned bodies featured chiselled sculpturing and a strong hint of tail fins. This is the 1967 Cadillac Coupe de Ville, the second-best selling car in the line this year. A total of 52,905 Coupe de Villes were delivered. Model 68347, the Coupe de Ville had a manufacturer's suggested retail price of $5,392. Power windows and front seat were standard and a choice of 19 leather or cloth interiors were available.

Cadillac's two-door coupes—Calais and Coupe de Ville—got a smart new formal roofline this year. This elegant new coupe styling was inspired by the Cadillac Florentine show car created for the 1964 New York World's Fair. Just like on the show car, the quarter window glass retracted rearward into the sail panel. The restyled 1967 Cadillacs had a strong character line that swept down from the tops of the sculptured front fenders to the rear bumper. This is an aft view of the 1967 Coupe de Ville, second best-selling car in the line. The padded vinyl roof was a $132 extra-cost option. Most Coupe de Ville buyers sprang for it.

The four-door Hardtop Sedan de Ville was again the top-selling car in Cadillac's 12-model product line. Nearly 60,000 of these handsomely-styled pillarless four-door hardtops were sold in the record-setting 1967 model year. Actual production totalled 59,902 cars. Model 68349, the Hardtop Sedan de Ville carried a manufacturer's suggested retail price of $5,625—exactly the same as the companion Sedan de Ville. The padded vinyl roof cost $137 extra and was available on both de Ville four-doors. Note the slim new roofline and body side sculpturing which swept down from the front fender peaks to the rear bumper.

With the replacement of the Fleetwood Eldorado Convertible by the new front-wheel-drive Eldorado Coupe, the de Ville convertible was the only soft-top left in the Cadillac line. After two all-time record model years, sales of Cadillac convertibles began to decline. Within five years the de Ville Convertible would be gone altogether. A total of 18,200 de Ville Convertibles were built during the 1967 model year. Manufacturer's suggested retail price was $5,608 and the Model 68367 convertible coupe weighed 4,479 pounds. Note the soft crease in the lower body sheetmetal and the raised haunches that appeared on all 1967 Cadillacs.

By far the least popular of the four models in the 1967 Cadillac de Ville series was the pillared four-door sedan. Only 8,800 four-door Sedan de Villes were built this year, compared with sales of nearly 60,000 Hardtop Sedan de Villes. Both cars were identically priced at $5,625. Model 68369, the Sedan de Ville weighed 4,534 pounds. All 1967 Cadillacs continued to use curved side glass. The front fender cornering lights had horizontally-textured lenses. Note the powerfully forward-thrusted front-end design and softly rounded roof contours.

1967

The luxurious Cadillac Fleetwood Brougham entered its second year as a separate model. A less-expensive standard Fleetwood Sixty Special Four-Door Sedan was still offered but was outsold four-to-one by the lavishly equipped and appointed Brougham. The 1967 Fleetwood Brougham got Cadillac's new sculptured styling but retained its distinctive Sixty Special exterior styling features, including clean, uncluttered body sides and massive chrome rocker sill and rear fender extension moldings. The Fleetwood name again appeared in tiny block letters on the front fenders, and Brougham script and the Fleetwood wreath and crest emblem ornamented the rear roof quarters. Model 68169, the Brougham weighed 4,715 pounds and carried a suggested retail price of $6,739. Cadillac's record 1967 model run included 12,750 Fleetwood Broughams.

Cadillac buyers still had a choice of two Fleetwood Sixty Specials—a standard Sixty Special Four-Door Sedan and a premium version called the Fleetwood Brougham. The "base" Sixty Special, Model 68069, carried a manufacturer's suggested retail price of $6,423—$316 less than the more opulently equipped and trimmed Brougham. Both models were built on the same exclusive 133-inch wheelbase. Most of Cadillac's customers opted for the more expensive Brougham. A total of 12,750 Fleetwood Broughams were sold in the 1967 model year compared with deliveries of 3,550 Sixty Specials. By 1971 the "plain vanilla" model was gone and the Fleetwood Brougham stood alone in its premium price class.

Although they had been completely redesigned only the previous year, the exclusive Cadillac Fleetwood Seventy-Five Sedan and Limousine were restyled for 1967. Exterior styling of these formal, long-wheelbase models was again compatible with the rest of the cars in the line. The Fleetwood Seventy-Five Nine-Passenger Sedan, Model 69723, was still favored by funeral directors and commercial limousine operators as well as many corporate customers. Wheelbase continued at 149.8 inches. Only 835 Seventy-Five Sedans were built this year. Manufacturer's suggested retail price was $10,360. This Seventy-Five Sedan has the optional padded vinyl roof. The Fleetwood name appeared on a small rectangular nameplate on the rear roof panel.

The heaviest and most expensive car in Cadillac's 12-car 1967 model lineup was again the regal Fleetwood Seventy-Five Limousine. This impressive nine-passenger car with two folding auxiliary seats came with a standard fixed partition with adjustable division glass and leather-upholstered chauffeur's compartment. The big Seventy-Five Limousine weighed 5,436 pounds in standard trim and its base price was an equally impressive $10,570. Production totalled 965 this year. The Fleetwood Limousine shown was photographed in front of Cadillac's customer reception and new car delivery lounge across the street from the factory on Clark Avenue in Detroit.

A small number of Cadillac Fleetwood Seventy-Five Sedans and Limousines were being built each year with the distinctive Landau roof option. This elegant decor package, which added more than $2,000 to the price of the car, featured a full padded roof, closed upper roof quarters spanned by decorative landau irons, and a small rear window for maximum privacy. Note also the fine chrome window edge and roof moldings and the Fleetwood wreath and crest on the rear fenders. This Fleetwood Seventy-Five Landau Limousine was photographed at the 1967 Detroit Auto Show. Visible in the background is its competition from Dearborn—a Lincoln Continental Executive Limousine built for Ford by Lehmann-Peterson of Chicago.

1967

Cadillac's new Fleetwood Eldorado had barely arrived when it was picked up for conversion by several customizing houses, including Barris' Kustom City in North Hollywood, California. Barris offered an "Eldorado del Caballero," which was built for Barris by Universal Coach at shops in both Los Angeles and Detroit. Available through Cadillac dealers, the del Caballero package cost anywhere from $2,500 to $4,500 above the cost of the base car depending on options chosen. The front section of the roof retracted into the sedanca-type rear landau roof. Note the front fender parking lamps which did not show up on production Eldorados until the 1968 model year. Sterling silver owner's nameplates were mounted on both front fenders as well as on the instrument panel, and gold wreath hood and deck lid ornaments were standard.

Miller-Meteor's "Landau Traditional" funeral coach was unchanged in its basic upper body styling from 1963 through the 1976 model year, when the full-sized Cadillac commercial chassis went out of production to be succeeded by a new downsized hearse and ambulance chassis for 1977. Lower sheet-metal, of course, was alway compatible with current Cadillac passenger car styling. Miller-Meteor's professional car bodies featured flat, angular lines. Landaus had triangular chrome shields behind the rear side windows and nearly straight landau bows. This is a 1967 Miller-Meteor Cadillac End-Loading Landau Funeral Coach with crinkle roof finish. Roof warning lights were becoming popular on hearses in larger cities and were no longer limited to ambulances and other emergency vehicles.

Combination coaches, which could be used for either ambulance or funeral service, were alway very popular in Southern and Southwestern states. This is a 1967 Miller-Meteor Cadillac Classic Limousine Combination finished in a sparkling neutral white. Note the crisp, angular styling of this dual-purpose professional car. Miller-Meteor, located in Piqua, Ohio, rivalled slightly larger Superior Coach Corp. of Lima, Ohio, for supremacy in the U.S. funeral car and ambulance industry. Cadillac's Series 69890 commercial chassis featured a 156-inch wheelbase and a perimeter-type chassis frame. Cadillac shipped 2,333 to the three major U.S. funeral car and ambulance manufacturers this year.

The Superior Coach Corp. of Lima, Ohio, built only 20 deck-type funeral flower cars on the 1967 Cadillac 156-inch commercial chassis. Model S-608, Superior's Royale Coupe de Fleur had a factory price of $11,869. Superior's flower cars used the same front doors and special windshield as the company's hearses and ambulances. Triangular quarter windows and a conventional coupe backlight afforded the driver excellent all-around visibility—unless the rear deck was banked with flowers. Even then, the stately flower car led the funeral procession at a dignified pace.

Industry leader Superior Coach delivered 1,100 funeral cars, ambulances and flower cars on the 1967 Cadillac commercial chassis. Of these, only 300 were Cadillac ambulances. This is a 1967 Superior-Cadillac "Royale Rescuer" with standard 42-inch headroom. Note the triple Beacon-Ray roof lights and Federal "Q" siren mounted on the roof. A rear compartment ventilator is visible on the rear roof. Twin spotlights mounted on the windshield pillars were popular on professional type ambulances. Styling of the Cadillac commercial chassis remained compatible with that of the division's passenger cars. It was not uncommon for larger funeral directors to operate sizeable fleets of matching Cadillac hearses, limousines, flower cars, and an ambulance or two.

1968

The 1968 Cadillacs were given a modest facelifting but generally carried over their crisp 1967 exterior styling. Cadillac's big news this year, however, was a brand-new 472 cubic inch V-8 engine.

Standard in all 1968 Cadillacs, including the front-wheel-drive Fleetwood Eldorado, the new engine's displacement and 525 foot-pound torque rating were the highest of any production passenger car engine in the world at the time. Horsepower was rated at 375 and compression ratio was 10.5-to-1. Bore and stroke were 4.30 inches and 4.06 inches respectively. This efficient, new luxury car powerplant was eventually bored out to a whopping 500 cubic inches. With annual improvements and refinements, this same basic engine powered Cadillacs for the next nine model years, through 1976.

All but four minor components of the milestone 1968 engine were new. The engine was the first in the industry designed with optional air conditioning compressor mountings as integral parts of the engine itself. The new engine also featured an internal distribution system for the air injector reactor, heart of Cadillac's pollution control system. Another unique feature of the 472 engine was a metal temperature monitoring device located in the engine head. This device activated a red warning light on the dash and sounded a buzzer if the engine overheated for any reason. The new V-8 was subjected to more than two million miles of testing before the first one was installed in a production Cadillac, and despite its impressive displacement the new engine was designed with a slightly smaller, simplified configuration with fewer parts.

At Cadillac's 1968-model press preview in Detroit, Chief Engineer Carl A. Rasmussen noted that the engine's large displacement was required to provide adequate output to power the many accessories customers were now ordering on their cars—air conditioning (now in 96 per cent of all Cadillacs), power steering and brakes, windows, seats and other equipment, with an acceptable degree of over-all operating performance.

The engine was described as the "fourth-generation Cadillac V-8 since Henry M. Leland introduced the company's first in its 1915 cars. The others were the inherently-balanced engine of 1923; the all-new high compression, overhead-valve 1949 engine, and the totally redesigned 429 cubic inch V-8 introduced in the division's 1964 cars.

All 1968 Cadillacs got new, longer hoods. On standard models (Calais, de Villes, Fleetwood Sixty Special and Brougham, and Seventy-Fives) the redesigned hood was 6.5 inches longer than in 1967. The Eldorado's new hood was 4.5 inches longer. The new hood extended all the way to the base of the windshield, concealing the windshield wipers below its upward-curved end. This made for a tidier appearance although the new "hidden" wipers proved to be something of a leaf-and-snow catcher. All standard models also got a restyled, sculptured decklid. The new decklid, with a center windsplit, had a raised top surface that provided increased trunk space.

The 1968 Cadillac's revised grille design featured a distinct center section capped with a round-shouldered chrome header bar. The grille had a finely-textured horizontal theme. The entire frontal assembly "leaned" forward, as in 1967. A diagonal sculpture line continued from the tops of the front fenders down the side of the car to the rear bumper. The most visible exterior change on the Eldorado were clear-lensed vertical parking and turn signal lights built into the

All eleven 1968 Cadillacs, including the lowline Calais series, were powered by a new 472 cubic inch V-8 engine. The Calais series was reduced to just two models this year—a two-door coupe and a four-door hardtop. The four-door pillared sedan was discontinued at the end of the 1967 model year. The least expensive Cadillac offered for 1968 was the Calais Coupe, Model 68247, which carried a manufacturer's suggested retail price of $5,284. Cadillac's 1968 production run included 8,165 of these price leader two-door hardtop coupes. Weight was 4,570 pounds. Exterior styling was identical to that of the more expensive Coupe de Ville.

The 1968 Cadillacs went into dealer showrooms with relatively modest exterior styling changes. All models got new, longer hoods that extended to the base of the windshield, concealing the windshield wipers. All but the Eldorado also had a new, sculptured rear decklid that provided more luggage space. And, the 1968 Cadillacs were powered by a brand-new 472 cubic-inch engine. The redesigned grille featured a distinct center section. The new grille header had rounded "shoulders" and the grille itself consisted of a series of thin horizontal bars. Rectangular parking and turn signal lights continued in the outer ends of the grille. As in 1967, the whole frontal ensemble "leaned" forward. This is a 1968 Coupe de Ville. Note the Cadillac script nameplate on the left side of the grille and the air intake slots at the rear of the restyled hood.

1968

leading edges of the front fenders. Side marker lights were adopted by most of the U.S. auto industry this year. The 1968 Cadillacs had small red triangular side marker lights in the sides of the chrome rear bumper ends. On the front of the car, amber side marker lights were located in new, finely-detailed cornering lights. On the Eldorado, the rear marker was a cleverly disguised Fleetwood wreath and crest emblem on the rear fenders.

Power windows were now standard in all Cadillacs, as was a new 15-plate battery with heavier cables. Front wheel disc brakes were standard on Eldorado and optionally available on all other models. Inside, there was a new instrument panel with additional safety padding.

Cadillac's 1968 model lineup had one less model this year. The slow-selling Calais four-door pillared sedan was dropped at the end of the 1967 model year. The 1968 line included a total of 11 body styles in three series—Calais, De Ville and Fleetwood. All 1968 Cadillacs (except Eldorado) were seven-tenths of an inch longer. Wheelbases were unchanged. Over-all length increased to 224.7 inches on the Calais and de Ville; 228.2 inches on the Sixty Special and Brougham, and 245.2 inches on the big Seventy-Fives. The Eldorado continued at 221 inches over-all.

Cadillac chalked up its seventh consecutive record production year, building just over 230,000 cars (230,003) during the 1968 model run. Cadillac's three-millionth postwar car came off the line in Detroit on June 10, 1968. It was a gold 1968 de Ville convertible. This year marked the first time that Cadillacs were built outside Detroit. The General Motors Assembly Division plant in Linden, N.J., began production of de Ville two and four-models for the 1968 model year. Except for small numbers of knocked-down Cadillacs and LaSalles assembled by General Motors of Canada Ltd. in the late 1920s and early 1930s, all Cadillacs had been built in the Motor City, most of them on Clark Avenue.

The Eldorado got some vigorous new competition this year. In April, 1968 Ford entered the luxury-personal market with its all-new Continental Mark III. Like the Eldorado, the Mark III featured long hoodline, short rear deck styling. The Mark was five inches shorter and two inches lower than standard Lincolns and was built on a 117.2 inch wheelbase. It was offered only as a high-styled two-door hardtop coupe and was a rear-wheel-drive car powered by Lincoln's 460 cubic inch V-8 engine.

Cadillac's least expensive series was reduced to only two body styles this year. The slow-selling Calais Sedan was dropped from the line at the end of the 1967 model year. Both 1968 Cadillac Calais models, a two-door hardtop coupe and this pillarless four-door hardtop, continued on a 129.5 inch wheelbase and had an over-all length of 224.7 inches. Style 68249, the 1968 Calais Hardtop Sedan accounted for 10,025 sales this year. Calais script nameplates continued on the rear fenders of both models, above the thin full-length side molding.

The Coupe de Ville two-door hardtop remained one of the most popular of all Cadillac models. Again, the Coupe de Ville was the second best-selling car in the entire Cadillac line. Sales increased by more than 10,000 units this year to a new high of 63,935, but the Coupe de Ville was still outsold by the four-door Hardtop Sedan de Ville. Model 68347, the 1968 Cadillac Coupe de Ville had a base price of $5,520 and weighed 4,595 pounds. The padded vinyl roof continued to be an extremely popular extra-cost option on this model and was also available on the companion Sedan de Ville and Hardtop Sedan de Ville.

The pillarless four-door Hardtop Sedan de Ville was again far and away the most popular single body style in the 11-car 1968 Cadillac product lineup. A record 72,662 were delivered in the 1968 model year, nearly 13,000 more than in 1967. At $5,754 the Hardtop Sedan de Ville was priced identically with the much less popular pillared Sedan de Ville. The 1968 Cadillac Hardtop Sedan de Ville was listed as Model 68349 and weighed 4,675 pounds. All four 1968 Cadillac de Ville models continued on a 129.5-inch wheelbase but were seven-tenths of an inch longer, at 224.7 inches.

Except for the premium Sixty Specials, sales of pillared four-door sedans now accounted for only a tiny proportion of Cadillac's annual passenger car sales. The lowline Calais Sedan was dropped at the end of the 1967 model year leaving the Sedan de Ville the only pillared four-door body style in the standard product line. Only 9,850 Sedan de Villes were sold in the 1968 model year compared with retail deliveries of nearly 64,000 Hardtop Sedan de Villes. Model 68369, the Sedan de Ville was identically priced with the far more popular four-door hardtop at $5,754.

All of Cadillac's standard 1968 models—Calais, de Villes, Sixty Special, Fleetwood brougham, and Fleetwood Seventy-Fives—got a new sculptured decklid with a raised surface that provided increased trunk space. The split "grilles" that flanked the rear license plate opening on the 1967 cars was replaced by a new horizontal trim treatment that spanned the rear of the car. New red side marker lights were located in the sides of the vertical rear bumper ends. Note the windsplit down the center of the redesigned decklid and the familiar Cadillac "V" and crest ornamentation. All Calais and de Ville models also had a Cadillac script nameplate on the lower right-hand side of the decklid. This is the Hardtop Sedan de Ville.

"Youthful in appearance . . . agile in performance," is how the 1968 sales catalog described the sporty Cadillac de Ville Convertible. Sales of the de Ville Convertible declined very slightly this year, from 18,200 cars in 1967 to 18,025 in the 1968 model year. The de Ville Convertible remained the only soft-top in Cadillac's product line. A choice of 11 leather interiors and five fabric top colors were available to Cadillac convertible buyers this year. Manufacturer's suggested retail price was $5,736. The Model 68367 de Ville Convertible weighed an even 4,600 pounds in stock form.

Posed with its top raised is one of the four 1968 de Ville Convertibles owned over the years by George Dammann, general manager of Crestline Publishing Co. This particular model, white with a black interior and top, is shown on a cold spring day sharing yard space with a modified 1968 Ford Torino convertible behind the horse barn on the Dammann's former home near Glen Ellyn, Ill. The 1968 de Ville Convertible was a particular favorite of George, who considers it one of the finest designed Cadillac convertibles ever built. In addition to this white example, George also owned a silver with black top and interior, a turquoise with white top and interior, and a brown with beige top and seats. The silver example was the most extensively used, registering well over 100,000 fast mid-western miles before being retired. All four cars wound up in Sweden, having been purchased by Clarence Engborg, Crestline's distributor for northern Europe.

The front-wheel drive Fleetwood Eldorado went into its second model year with significant engineering and styling changes. The Eldorado's hood was lengthened 4.5 inches and "concealed" the windshield wipers. The Eldorado also got Cadillac's all-new 472 cubic inch V-8 engine. The most noticeable styling changes were new vertical parking and directional signal lights built into the leading edges of the long front fenders, and round side marker lights on the rear fenders. The red rear marker lights were cleverly disguised Fleetwood laurel wreath and crest emblems. Eldorado sales increased from 17,930 in 1967 to 25,528 in the 1968 model year. Note the creased V-shaped rear window and sloped, bladed vertical taillights.

1968

The distinctive Fleetwood Eldorado continued in only one body style, a classically-proportioned, high-styled two-door hardtop coupe. The Eldorado's broad, cross-hatched grille with concealed dual headlamp units was unchanged for 1968, but parking and turn signal lights were relocated from the front bumper to the leading edges of the front fenders. Front wheel disc brakes were standard on the Eldorado and optional at extra cost on all other 1968 Cadillacs. The front-wheel-drive Fleetwood Eldorado, Model 69347, weighed 4,580 pounds and went into the showrooms with a suggested retail price of $6,574. Wheelbase and over-all length remained at 120 and 221 inches respectively. Eight cloth and 12 leather interior color and trim selections were available for Cadillac's entry in the growing luxury/personal car market.

Cadillac's stylists created a special Eldorado town car for the 1968 auto shows. Called the Eldorado Biarritz Towne Coupe, it was patterned after the classic open-front town cars of the 1930s. The Eldorado Biarritz Towne Coupe had no roof over the front seats and its landau-type rear roof had no quarter windows. The front end featured a finely-textured, brushed aluminum grille and frosted parking light lenses in the leading edges of the front fenders. The exterior was finished with 20 coats of gold-flecked, dark olive Firefrost enamel. The interior was trimmed in posh antique gold velours with mouton carpeting and carved antique walnut accents. This show car's golden theme emphasized the Eldorado name, taken from the legendary city of gold sought by the Spanish conquistadors.

A Detroit Cadillac dealer and a local coachbuilder collaborated to create this impressive long-wheelbase Fleetwood Eldorado Limousine. Built by the Universal Coach Co. of Detroit in collaboration with Klett Cadillac, this one-off custom was constructed on a very long 168-inch wheelbase and measured a truly impressive 259 inches from bumper to bumper. Retail price was shown as $32,500. Interior appointments included a glass division, bucket seats that faced the rear seat, sliding sun roof and massive chrome grille header. Note the center-opening doors and landau rear roof treatment with chrome edge molding and carriage bows.

Cadillac's front-wheel-drive Fleetwood Eldorado quickly became a popular "base" car for customizers and custom body shops. Barris Kustom Industries in Los Angeles designed and built this intriguing "Casa de Eldorado Town and Estate Wagon" conversion of a 1968 Fleetwood Eldorado Coupe. The upper rear body structure included a double roll-bar cage. Note the simulated sidemount spare tire bulges atop the front fenders. Standard equipment included "his and her" storage compartments in the rear fender panels. Optional equipment included a sliding sun roof, television set, and cocktail bar.

Cadillac's 1968 model lineup again included two premium sedans—the standard Fleetwood Sixty Special and the more luxurious Fleetwood Brougham. The Fleetwood Brougham was immediately recognizable by its standard padded, vinyl-covered roof. The highline Brougham outsold the less expensive Sixty Special by a margin of nearly five to one. Both models continued on their exclusive 133-inch wheelbase. Styling changes included a new, longer hood with "hidden" windshield wipers and a restyled rear deck. This is the 1968 Cadillac Fleetwood Brougham, Model 68169, which weighed 4,805 pounds and carried a manufacturer's suggested retail price of $6,867. A total of 15,300 Fleetwood Broughams were built this year.

1968

With an over-all length of more than 19 feet, the Fleetwood Sixty Special Four-Door Sedan and more expensive Fleetwood Brougham were second only to the Fleetwood Seventy-Five sedan and limousine in sheer length. The Sixty Special Sedan was recognizable by its painted metal roof. Sales of the Sixty Special ran far below those of the more luxurious Fleetwood Brougham. Only 3,300 Fleetwood Sixty Specials were sold in the 1968 model year, compared with deliveries of 15,300 Broughams. On the Sixty Special, the Fleetwood wreath and crest appeared on the ends of the rear fenders. On the premium Brougham, they were on the roof quarter panels. Model 68069, the Sixty Special at $6,552 was priced only $315 lower than the much more popular Brougham.

At the top of Cadillac's 1968 product offerings were the two Fleetwood Seventy-Fives—a nine-passenger sedan and formal limousine. These prestigious models continued on their own, exclusive 149.8-inch wheelbase with an overall length of 245 inches. This is the 1968 Cadillac Fleetwood Seventy-Five Sedan, Model 69723, which sold for $10,598. Weight was 5,300 pounds. The Clark Avenue plant in Detroit built an even 1,800 Fleetwood Seventy-Fives during the 1968 model run, 805 of which were Seventy-Five Sedans. This one was photographed in front of Detroit's new Pontchartrain Hotel. The padded vinyl roof was an extra-cost option.

A Pittsburg, Pa., Cadillac dealer had this one-off 1968 Cadillac Station Wagon designed and built for his own personal use. The base car was a Calais Four-Door Hardtop. The conversion was done over a period of six weeks by National Coach Distributors of Knightstown, Ind. for W. H. Hufstader, president of Hufstader Cadillac. The nine-passenger wagon was equipped with rear seats that folded down to form a flat bed. Instead of a conventional tailgate, Mr. Hufstader's exclusive Caddy wagon featured a full rear door. The conversion cost more than $6,000 on top of the price of the car.

The most expensive car in Cadillac's 1968 model lineup was again the stately Fleetwood Seventy-Five Limousine. The long-wheelbase Fleetwood Limo came with a standard glass partition and leather-upholstered chauffeur's compartment. Customers could choose from 21 exterior colors including five optional Firemist shades. This 1968 Cadillac Fleetwood Seventy-Five Nine-Passenger Limousine is finished in white with a contrasting black padded vinyl landau roof. The optional landau roof package included closed upper rear quarters, ornamental landau irons, and a small formal rear window. Model 69733, the Fleetwood Limousine weighed 5,385 pounds and had a starting price of $10,736. A total of 995 were built this year.

Here, in one sweeping view, are all eleven of Cadillac's 1968 models. This advertising photograph was set up with meticulous care somewhere on the Western desert. Timing was critical to take advantage of the lengthening shadows for added effect. All 11 cars were finished in the same sparkling white. Starting from the foreground are: De Ville Convertible; Fleetwood Eldorado Coupe; Coupe de Ville; Fleetwood Brougham; Fleetwood Seventy-Five Limousine; Hardtop Sedan de Ville; Sixty Special; De Ville Sedan; Calais Coupe; Calais Hardtop Sedan; and the Fleetwood Seventy-Five Sedan. Now that's a lensful!

1968

The biggest change in the 1968 Cadillacs lay under their new longer hoods. It was an all-new 472 cubic inch V-8 engine. Rated at 375 horsepower, this engine put out 525 foot-pounds of torque—highest of any production passenger car engine in the world at the time. Bore was 4.30 inches and stroke 4.06 inches. This new engine featured a unique metal temperature monitoring device which sounded a buzzer and activated a red warning light on the dash if it overheated for any reason. The new engine was also the first designed specifically for optional air conditioning.

Cadillac's special 156-inch commercial chassis was still the basis for most of the funeral cars and ambulances built in the United States. Cadillac shipped 2,413 hearse and ambulance chassis during the 1968 model year. Nearly half of these went to the Superior Coach Corporation of Lima, Ohio. This is a 1968 Superior-Cadillac Sovereign End-Servicing Landaulet Funeral Coach finished in striking white with a matching white crinkle-finished roof. Cadillac's 1968 commercial chassis also got the division's new 472 cubic inch engine.

The prestigious, soft-riding Cadillac-chassised ambulance, taken off standard passenger car tooling, was nearing its prime. Within a decade it would be completely replaced by more practical and roomier van-type ambulances. This is a 1968 Superior-Cadillac Super Rescuer with 51-inch headroom, a fully-equipped professional emergency vehicle that could transport as many as four stretcher patients. The entire roof was a huge, single piece of fiberglass which kept weight down. Note the roof mounted coaster siren and bullet-type red warning lights.

Full-sized funeral flower cars had become very expensive. Built on the long-wheelbase Cadillac commercial chassis, standard deck-type flower cars now cost $15,000 or more. An enterprising Indiana auto leasing firm, McClain Sales and Leasing of Anderson, Ind., developed an attractive flower car conversion of a standard Cadillac coupe. These flower cars featured a stainless steel-lined flower well and boot. This is a 1968 McClain-Cadillac. Note the chopped-off roofline and black sidewall tires.

There were now only three companies left building funeral cars and ambulances on the Cadillac commercial chassis. All were located within 75 miles of each other in Ohio. The largest was Superior Coach Corp. in Lima. Next came Miller-Meteor, a division of Wayne Works, in Piqua. The smallest of this triumvirate was The Hess and Eisenhardt Co. of Cincinnati, builders of S & S professional cars. This is a 1968 S & S Cadillac Parkway Ambulance. Note the unique roof corner lights. Hess and Eisenhardt hearses and ambulance were always significantly more expensive than the competition, a premium some funeral directors were quite willing to pay for the added prestige of owning an S & S.

1969

Cadillac built its four-millionth car during the final months of the 1969 model year, and the division got a new general manager. All of the 1969 Cadillacs, including the front-wheel-drive Fleetwood Eldorado, went into dealer showrooms in the fall of 1968 with extensive exterior styling changes.

The lunging frontal appearance of 1967-68 was gone. The restyled 1969 Cadillacs had a crisp, new squared-off profile that was reminiscent of the 1965-66 cars. All standard models got new grilles, redesigned rear ends and in most cases, handsome new rooflines. The biggest change up front was a return to horizontally-mounted dual headlight units for the first time since 1964. The attractive new rectangular grille consisted of a series of thin horizontal and vertical bars with fine fins recessed in each eggcrate opening, creating what Cadillac's stylists called a "floating fin" effect. Parking and turn signal lamps wrapped around the bladelike front fenders and joined the cornering lamps. A new front bumper framed the lower portion of the grille. The redesigned hood was 2.5 inches longer and the cowl vent louvers at the rear of the hood were gone.

The most visible exterior change on Cadillac's extensively redesigned 1969 bodies was the elimination of vent panes on the front doors of all standard models. Coupled with improved interior ventilation and air conditioning systems, this change enhanced both appearance and visibility. New vertical taillights were framed in body color paint and incorporated brake, backup and directional signal lights as well as functioning as side marker lights. New sculptured rear decks blended into the redesigned rear bumper. The decklid windsplit extended into a new, creased V-shaped backlight like the one on the Eldorado. Th highline Fleetwood Sixty Special and Brougham als got elegant new formal rooflines which made thes models look longer and more impressive than ever.

After seven consecutive years, the familiar Cadilla script nameplate no longer appeared on the marque' grille.

The front-wheel-drive Eldorado was given its firs real facelifting since it was introduced two year earlier. The concealed headlamp system with it vacuum-operated drop-down doors was gone, replace with fixed-position, dual headlights that flanked finely-textured new center grille. This new grille wa divided into upper and lower sections by a protectiv chrome bar that extended the full width of the car Cornering lights were made of a new amber refle material. The Eldorado's new vinyl "halo" roof wa bordered by a thin chrome edge molding and there wa a paint separation around the outer edge of the roo New wheel covers were also exclusive to this model. new back-up light was built into the fuel filler door o the rear of the car. Inside, the 1969 Cadillac Eldorad had a completely new "control center" instrumen panel. A soft, padded divider separated the instrumen and control panel from the rest of the dash.

Head restraints were now standard on front sea backs of all models. Cadillac's new, canted seats ha higher backs and a "dual-comfort" 60/40 front bench type seat was standard in the Fleetwood Brougha and optionally available in Sixty Special and all de Vil models except the pillared sedan. Doors were reinforc ed with heavy longitudinal bars. An improved energ absorbing steering column was equipped with an ant theft steering, ignition and shift lock system

The 1969 Cadillacs received extensive styling changes. Grille and rear end treatments were all new on Calais, de Ville, Fleetwood Sixty Special, and Brougham and Fleetwood Seventy-Five models. All standard cars got new rooflines. Redesigned grilles featured horizontally-mounted headlights for the first time since 1964. Parking, directional signal and cornering lights were integrated and extended around bladelike new front fenders. The lowest-priced car in the 1969 Cadillac line was again the Calais Coupe, Model 68247, which had a manufacturer's suggested retail price of $5,466. The Calais Coupe weighed 4,555 pounds, and 8,165 of these lowline two-door hardtops were sold this year.

Cadillac's least-expensive 1969 series included only two body styles—a Calais two-door hardtop coupe and this four-door hardtop. The Calais Hardtop Sedan, Model 68249, was the more popular of the two with 10,025 sold. Suggested retail price was $5,642. The Calais Hardtop Sedan weighed 4,630 pounds. Note the new vertical taillamps which wrapped around the ends of the long tailfins and housed brake and back-up lights and directional signals as well as serving as side-marker lights. New two and four-door rooflines, with creased backlights, appeared on all 1969 Calais and de Ville models.

1969

Cadillac's new standard seat belt system was equipped with automatic adjusters and "mini" buckles.

All 1969 Cadillacs were powered by the same 472 cubic inch, 375-horsepower V-8 engine announced the previous year. Redesigned frames under standard models were 7.5 inches longer and had new body mounts. The 1969 engine cooling system featured the industry's first "closed" radiator design with a translucent reservoir which permitted visual checking of coolant fluid levels.

Cadillac's 1969 product offerings again included a total of 11 models in three series on four different wheelbases. Major dimensions were unchanged. Nineteen of the 21 exterior color choices were new this year and Cadillac's customers could choose from no less than 205 interior trim combinations including 103 in cloth, 100 in leather and two vinyls. The 1969 model lineup included the lowline Calais coupe and four-door hardtop; four de Villes—the top-selling Hardtop Sedan de Ville, Coupe de Ville, the division's only remaining convertible, and a thin-pillared four-door Sedan de Ville. There were five premium Fleetwood body styles including the Eldorado luxury-personal coupe, Sixty Special, the even more luxurious Fleetwood Brougham and the long-wheelbase Seventy-Five sedan and formal limousine.

The Cadillac Motor Car Division produced its four-millionth car since 1902 on June 19, 1969. It was a light-blue de Ville hardtop. Waiting at the end of the Clark Avenue assembly line as the milestone car was driven off was its proud owner, Lee Mannes of the Detroit suburb of St. Clair Shores. On display at the end of the line was the division's own 1903 single-cylinder Model "A" Cadillac. It had taken 47 years for Cadillac to reach the one-million plateau, nine years to build the second million cars, six for the third and just five years to produce the four-millionth.

For the first time in six years, Cadillac's model year sales failed to set a new record. The division built 223,237 cars during the 1969 model run. George R. Elges succeeded Calvin J. Werner as Cadillac General Manager on July 1, 1969. Sister GM division Pontiac entered the booming luxury-personal car market this year with a high-styled 1969 model called the Grand Prix, and Chevrolet was ready to go with a similar entry called the Monte Carlo.

One of the biggest changes on all standard series 1969 Cadillacs was the elimination of ventilator panes on the front doors. This feature had been introduced two years earlier on the 1967 Fleetwood Eldorado. Removal of the vent panes gave all models a cleaner, more open appearance and improved visibility. The rear deck windsplit extended to the rear window, just like on the Eldorado. All 1969 Cadillac Calais and de Ville models also got smart new rooflines. Script nameplates continued on the ends of rear fenders on Calais and de Villes. This is the 1969 Coupe de Ville with padded vinyl roof option.

Cadillac's top seller was once again the Hardtop Sedan de Ville. A record 72,958 of these pillarless four-door hardtops were built during the 1969 model run. Manufacturer's suggested retail price was $5,936. The Hardtop Sedan de Ville weighed 4,660 pounds and was listed as Model 68349. The 1969 Cadillac's revised grille design featured horizontally-mounted dual headlight units that flanked a grille composed of fine horizontal and vertical bars. Fins recessed in each rectangular opening created a "floating fin" effect. The new hood was also longer, and the cowl vent louvers at the rear of the hood were eliminated this year.

Sales of the Cadillac Coupe de Ville continued to set new records, but still trailed the companion Hardtop Sedan de Ville by several thousand units. A record 65,755 Coupe de Ville two-door hardtops were delivered during the 1969 model year. Base price of this extremely popular model was $5,703 and standard weight was 4,595 pounds. The Coupe de Ville was listed in the 1969 Cadillac data books as Model 68347. The vinyl padded roof remained a popular extra-cost option. The new front bumper framed the lower grille and extended to the front wheel openings.

1969

The 1969 Cadillacs featured strong, vertical styling for the first time since the 1965-66 models. New chisel-faced front fenders and redesigned taillights were vertical in profile. Most standard models also sported crisp, new rooflines. Elimination of front door vent windows gave the extensively-restyled 1969 cars a cleaner, more open look. The softly-sculptured diagonal sweep line that extended from the tops of the front fenders to the base of the new taillights was similar to that on the 1967-68 models. All standard models except the Sixty Special and Brougham had a fine chrome full-length molding mounted at the body center line. Here's the top-selling Hardtop Sedan de Ville in profile.

The mid-range Cadillac de Ville series for 1969 offered a choice of four body styles, all on the same 129.5-inch wheelbase. These included the extremely popular Hardtop Sedan de Ville, Coupe de Ville, Cadillac's only convertible, and this pillared four-door sedan. The Sedan de Ville was the least popular car in the series, with only 7,890 built this year. Note the softly-rounded roof contours. Model 68369, the Cadillac Sedan de Ville weighed 4,640 pounds and at $5,936 was identically priced with the far more popular Hardtop Sedan de Ville. All 1969 Cadillacs rode on redesigned chassis frames that were 7.5 inches longer and had new body mounts.

The most glamorous car in the Cadillac family was still the flashy de Ville Convertible. Cadillac's soft-top sales continued to taper off. A total of 16,445 de Ville Convertibles, Model 68367, were built during the 1969 model run compared with more than 18,000 the previous year. More and more buyers were opting for closed cars with air conditioning and exotic stereo sound systems instead. The de Ville Convertible looked even racier without front door vent panes, which were eliminated on all 1969 Cadillacs. Manufacturer's suggested retail price was $5,887. Cadillac's new, sculptured hood was 2.5 inches longer this year.

The front-wheel-drive Fleetwood Eldorado went into its third model year with a major front-end restyle. The redesigned grille featured fixed dual headlamp units instead of the concealed headlight system used in 1967 and 1968. The Eldorado's new grille was finely textured and had upper and lower sections separated by a chrome blade that extended the full width of the car. The 1969 Eldo also sported new wheel covers. Model 69347, the Fleetwood Eldorado Coupe carried a base price of $6,693 and 23,333 were sold—slightly fewer than the 24,528 sold the previous model year. The 1969 Cadillac Eldorado's new padded vinyl roof featured a fine chrome edge molding and a paint separation around the outer edge. A new back-up light, providing improved illumination, was built into the fuel filler door on the rear of the Eldo.

1969

Cadillac gave auto show visitors a glimpse into the near-future when it exhibited a special Fleetwood Eldorado Coupe with an electrically-powered retractable sun roof at the New York International Automobile Show in March of that year. This flip-top feature was not offered on production Cadillacs until the following year. Within a few years, sun roofs and T-tops—coupled with air conditioning and impending federal rollover production standards—killed off the American convertible. The Eldorado show car's sun roof panel had a padded, vinyl covering. Photographed on the roof of Cadillac's engineering building in Detroit, the Eldorado show car wears Michigan manufacturer's license plates.

The late 1960s and early 1970s saw the emergence of a whole, new automotive aftermarket industry—the sale of customizing conversion packages for popular luxury cars. The most popular accessories were massive chrome grille enclosures and padded, landau-style roofs, but individuality was limited only by the purchaser's imagination—and pocketbook. This is a fairly mild custom package on a 1969 Cadillac Fleetwood Eldorado Coupe. This car has a bolt-on custom grille, Thunderbird-style headlamp doors, a vinyl-covered padded roof and door sill trim, and chrome landau bows. Wheel covers are also non-stock.

The Fleetwood Sixty Special Sedan and more luxurious Fleetwood Brougham were also extensively redesigned this year. The distinctive Fleetwoods got elegant new formal rooflines that accented the perimeter of the broad roof panel. Window openings were large and rectangular. Ventilator panes were gone from the front doors. The Sixty Special and Brougham retained their exclusive 133-inch wheelbase and distinctive exterior styling features, including broad chrome rocker sill and rear fender extension moldings and lack of bodyside trim. The Fleetwood name continued in tiny block letters on the lower front fenders. The Fleetwood Brougham's standard padded roof was edged with a fine chrome "halo" molding.

The Fleetwood Brougham was among the most impressive of all 1969 Cadillacs. This premium four-door sedan was thoroughly restyled, with a new formal roofline and "halo" padded vinyl top. The Brougham and less expensive Fleetwood Sixty Special Sedan retained their distinctive, smaller rear windows. Both continued on a 133-inch wheelbase. Cadillac built 17,300 Fleetwood Broughams during the 1969 model year. The Brougham's price crept past the $7,000 mark this year to $7,092. Model 68169, the Fleetwood Brougham weighed 4,770 pounds. Vinyl roofs were available in six color choices and the Brougham continued to offer a long list of standard equipment, including automatic level control and rear seat footrests.

1969

Sales of Cadillac's Fleetwood Sixty Special Sedan continued to run far below those of the more expensive companion Fleetwood Brougham. The Sixty Special was easily identified by its bare metal top: the Brougham came with a standard padded vinyl roof. Model 68069, the Fleetwood Sixty Special was price-positioned between the Fleetwood Eldorado luxury-personal car and the Brougham, at $6,761. Only 2,545 Sixty Specials were delivered this year compared with sales of 17,300 Broughams. The next year would be its last. Cadillac's customers were more than willing to fork over an extra $331 to move up into a Brougham. All Fleetwood models carried the distinctive laurel wreath and crest on the front of their hoods and on decklids.

Cadillac's longest, heaviest, most expensive—and exclusive—1969 models were the big Fleetwood Seventy-Five. As had been the case since the end of the Second World War, there were only two models in this limited production series, a nine-passenger sedan and a companion formal limousine. Cadillac delivered 880 Seventy-Five Sedans and 1,156 Fleetwood Limousines during the 1969 model year. Model 69723, the nine-passenger sedan weighed 5,430 pounds and sold for $10,823. The Model 69733 Fleetwood Seventy-Five Limousine, with standard glass division, weighed 5,555 pounds, and at $10,961, was the most expensive car in the 11-model 1969 Cadillac line. Two folding auxiliary seats, automatic climate control air conditioning, and level control were standard on both Seventy-Fives. Wheelbase continued at 149.8 inches and over-all length was an impressive 245.3 inches.

Cadillac customers who insisted on privacy and formal elegance could still specify the stately Landau Roof option for the Fleetwood Seventy-Five Sedan or Limousine. This exclusive upper body styling treatment, which had been built in-house at Cadillac since 1964, included closed upper rear roof quarters, a tiny rear window and full padded vinyl top covering. This top could be had with or without ornamental carriage bows. Most landau customers ordered the landau irons. This is a 1969 Fleetwood Seventy-Five Landau Limousine, exhibited at the 1969 Detroit Auto Show. The highly reflective neon ceiling lights are the bane of every photographer who ever tried to shoot a car in Detroit's vast Cobo Hall.

The Cadillac Motor Car Division built its four-millionth car on June 19, 1969. The light blue Coupe de Ville was delivered to Lee Mannes, of St. Clair Shores, Mich., who was waiting for it with his two teen-aged sons as it came off the line. The trio was seated on Cadillac's 1903 single-cylinder Model "A" Cadillac as their new car came to the end of the Clark Avenue plant assembly line. It took Cadillac 47 years to build its first million cars, nine years for the second million, six years for the third and only five years for the fourth million. Cadillac was also marking its postwar millionth-car rolloffs, which confused the milestone car picture somewhat.

1969

The 1969 Cadillac-chassised funeral cars and ambulances were formally introduced at the 1968 National Funeral Directors Association Convention in Detroit in October, 1968. Three manufacturers—Superior, Miller-Meteor and Hess and Eisenhardt (S & S) introduced their Cadillac hearses and ambulances at this show. Superior's 1969 landau models got a new style landau bow. This is the 1969 Superior-Cadillac Crown Sovereign Landaulet, the most expensive model in Superior's 1969 funeral coach line. Note the sharply-angled "C" pillar with broad up-and-over roof band and the chrome-tipped beading above the rear door.

McClain Sales and Leasing of Anderson, Ind., found an enthusiastic market for its moderately-priced Cadillac flower car conversions. This open well-type flower car was based on a standard 129.5-inch wheelbase 1969 Cadillac Calais Coupe. The roof was chopped off and a rear-raked "B" pillar added. The stainless steel rear boot covered the area formerly occupied by the rear deck lid. These flower cars were available for considerably less than the price of a long-wheelbase flower car built on the Cadillac commercial chassis by one of the major funeral car manufacturers.

Cadillac shipped 2,550 long-wheelbase commercial chassis to U.S. funeral car and ambulance manufacturers during the 1969 model year. The Series 69890 (technically a Fleetwood series unit because of its "69" numerical prefix) hearse and ambulance chassis had a 156-inch wheelbase. As delivered to Superior, Miller-Meteor and Hess & Eisenhardt, this chassis included only the frame, front-end sheet metal and cowl. Professional car bodies were designed and built for this chassis by the three specialty firms. This is a 1969 Superior-Cadillac Sovereign Limousine Combination. Note the rear compartment air conditioning inlet scoop on the rear fender. A combination coach was intended for either ambulance or funeral service.

The Superior Coach Corp. of Lima, Ohio, built only 24 Coupe de Fleur deck-type flower cars on the 1969 Cadillac commercial chassis. While the lower body sheet-metal was all new, the upper body structure was unchanged since 1965. The Coupe de Fleur featured a full-length, stainless-steel lined casket compartment below the adjustable flower deck. Some funeral homes operated these vehicles as "floral hearses", in lieu of a conventional funeral coach. This metallic silver Superior-Cadillac flower car has owner nameplates mounted on the front and rear door.

Miller-Meteor's 1969 professional car line included two ambulances built on the 156-inch wheelbase Cadillac commercial chassis. This is the standard Miller-Meteor Cadillac "Classic 42" Ambulance with 42-inch rear compartment headroom. A raised-roof 48-inch version was also available. The Classic 42 was designed to carry as many as four litter patients. Note the "Ful-Vu" 360-degree warning lights on each corner of the roof, a Miller-Meteor exclusive.

1970

Cadillac's big news this year was a 500 cubic inch, 400-horsepower engine for the front-wheel-drive Fleetwood Eldorado—the largest production passenger car engine in the world. Cadillac Motor Car Division production also established a new model year record.

The 8.2 litre V-8 engine produced a remarkable 550 foot-pounds of torque and was exclusive to the Eldorado. All of Cadillac's other 1970 models were powered by the same 472 cubic inch, 375-horsepower engine introduced two years earlier. The Eldorado's new 500 cubic inch engine got its larger displacement from a slightly longer piston stroke rather than from any increase in cylinder bore. The 8.2 litre engine's bore and stroke were 4.3 and 4.304 inches respectively. This huge-displacement engine eventually found its way under the hoods of all standard Cadillacs before it was phased out of production at the end of the 1976 model year.

All of the 1970 Cadillacs got relatively minor exterior styling changes. Standard models— Calais, de Villes, Fleetwood Sixty Special, Brougham and Seventy-Fives —sported a revised grille design with 13 prominent vertical bars in a framed, rectangular grille opening with fine horizontal cross-hatching. Dual headlamp units in bright chrome bezels were edged with body color giving new definition to the front of the car. Parking and cornering lights swept around the bladelike front fenders and had new horizontal chrome trim. Redesigned taillights consisted of narrow V's above and below the rear bumper. Reflective vertical side marker lights were mounted on the upper ends of the rear fenders.

While basic styling and major dimensions were unchanged this year, there were some significant detail and ornamentation changes. Small vertical wing emblems appeared on the leading edges of the front fenders on all de Ville and Calais models. These were inspired by the raised-wing emblems that adorned the hoods of Cadillac's fondly-remembered 1941 models.

Cadillac's world-famous "V" emblem, an extremely successful identity hallmark since 1946, disappeared from beneath the Cadillac crest on the hoods and deck lids of the Calais and de Villes. Premium Fleetwood models bore the distinguished laurel wreath and crest. All 1970 Cadillacs except the Eldorado also got bright new wheel discs highlighted by fine radial fins. The Fleetwood Sixty Special Sedan and companion Fleetwood Brougham got full-length body side moldings for the first time. These moldings had color-keyed vinyl inserts to protect the car's broad flanks in close parking situations.

Cadillac's basic styling at this time followed firm two year cycles. The 1965-66, 1967-68 and 1969-70 cars were nearly identical in appearance except for subtle but highly effective exterior detail changes. New colors also had a way of making the cars look different. Of Cadillac's 21 exterior color choices for 1970, including 14 standard shades and five optional Firemist hues, 15 were new.

The first-generation Fleetwood Eldorado luxury personal coupe body got minor cosmetic changes as it entered its fourth and final year. The grille had a stronger center section theme. The left side of the revised grille carried an Eldorado script nameplate below which was a bright metallic plaque that read "8.2 Litre," drawing attention to the car's new 500 cubic-inch engine. From the rear, the Eldorado got restyled, much thinner vertical taillights. A power operated sun roof was added to the Fleetwood Eldorado's list of extra-cost options during the 1970 model year. This fresh-air feature had been previewed on a special Eldorado show car the previous year.

The 1970 Cadillacs were given minimal styling changes. All standard models got a revised front-end treatment which featured a series of 13 vertical bars in the cross-hatched, rectangular grille. Wrap-around parking and cornering lights had horizontal chrome trim. Winged crests on the leading edges of the front fenders replaced the traditional Cadillac "V" on the hoods and decklids of Calais and de Ville models. Taillights and wheel covers were also new. The lowest-priced car in Cadillac's 1970 product line was again the Calais Coupe, Model 68247, which sold for $5,637. Of the more than 238,000 Cadillacs sold this year, fewer than 10,000 were in the lowline Calais series. Only 4,724 were Calais two-door hardtop Coupes.

All standard 1970 Cadillacs—Calais, de Ville, Fleetwood Sixty Special, Brougham, and Seventy-Fives—got redesigned taillights. The revised taillamps had narrow V-shaped vertical lenses that housed taillights, stoplights and directional signals. Reflective vertical side markers were mounted on the ends of the rear fenders. Back-up lights flanked the license plate opening in the rear bumper. Cadillac's famed "V" emblem disappeared from below the crest on the hood and decklids of Calais and de Ville models. The best-selling of the two lowline Calais models offered in 1970 was the Hardtop Sedan, Model 68249. A total of 5,187 of these four-door hardtops were sold this year. Base price was $5,813, or $176 higher than the companion Calais Coupe.

1970

Cadillac's new vertical wing emblems also appeared on the clear parking light lenses built into the leading edges of the Eldorado's front fenders.

A new "hidden" radio antenna consisting of two tiny black wires embedded in the windshield glass replaced the familiar fender-mounted mast antenna. While the built-in aerial improved appearance and deterred vandals, its performance proved disappointing and fender-mounted, retractable radio antennas were soon back.

Exclusive to Cadillac this year was a new signal-seeking radio tuner that sought out stereo stations. All 1970 Cadillacs rode on bias-ply, glass-belted radial tires. Important mechanical changes included the first all-new rear axles in 30 years on standard models. The new axle was quieter, more durable and boasted the highest torque capacity of any car in the industry at the time. It was easier to service, too, and provided a lower driveline. Another Cadillac first was a totally-new, single-piece ductile iron steering knuckle. A new feature on the anti-theft steering column lock prevented the driver from accidentally reaching the accessory position. Cadillac's customers could choose from 167 cloth, leather, and vinyl interior trim combinations.

The first Cadillacs of the Seventies went into dealer showrooms September 18, 1969. George R. Elges, Cadillac's new general manager, noted that there were again 11 body styles in three series on four wheelbases.

Three Cadillac body styles came to the end of the line this year. The glamorous de Ville convertible, which traced its ancestry back to the first Series 62 convertible coupe in 1940, and the slow-selling pillared Sedan de Ville, were dropped at the end of the 1970 model year. This was also the last year that two Sixty Specials would be offered. The "base" Fleetwood Sixty Special Sedan was dropped at the end of the 1970 model run. Its place in Cadillac's premium four-door sedan market had been all but taken over by the more luxurious Brougham. A single model called the Fleetwood Sixty Special Brougham would be offered starting in 1971.

Cadillac passenger car production set a new record for the first time in two years. A total of 238,745 Cadillacs were built during the 1970-model production run. This total included the output of both Cadillac's "home" plant in Detroit and the GMAD assembly plant at Linden, N.J.

Cadillac styling during this period followed two-year cycles. The 1965-66, 1967-68, 1969-70 models differed only in detailing and subtle exterior trim modifications. This was especially true of the 1970 cars. As minor as some of these changes were, they were extremely effective. New colors also made the cars look different from year to year. Of the 21 exterior color choices offered for 1970, including 14 regular and seven premium Firemist shades, 15 were new. Principal styling refinements on the 1970 Cadillacs included new cornering and taillights, a revised grille texture, and bright new wheel covers highlighted by fine radial fins. Ornamentation was also changed slightly. This is the 1970 Cadillac Coupe de Ville.

The four-door Hardtop Sedan de Ville again topped Cadillac's sales charts as the division's best selling model. Sales of 83,274 of these pillarless four-door hardtops set another new record and once more exceeded those of the companion Coupe de Ville by a fair margin. Model 68349, the 1970 Cadillac Hardtop Sedan weighed 4,725 pounds and had a manufacturer's suggested retail price of $6,118. Most were delivered with the optional padded vinyl roof. Note the "color break" around the outer edges of the roof. Cadillac introduced this "halo" roof decor on the 1969 Fleetwood Eldorado and Brougham.

By far the most popular of all 1970 Cadillac body styles were the de Ville two and four-door hardtops. Of the 238,000 Cadillacs sold during the 1970 model year, nearly 160,000 were Coupe and Hardtop Sedan de Villes. Sales of the perennially popular Coupe de Ville set another record this year but once more trailed those of the companion four-door hardtop. Production of 181,719 de Villes included a record 76,043 Coupe de Villes. Model 68347, the 1970 Cadillac Coupe de Ville had a manufacturer's suggested retail price of $5,884. The vinyl roof was still a very popular option on this model. All standard 1970 Cadillacs got new wheel covers this year.

1970

The 1970 model year would be the last for the de Ville pillared four-door sedan. The Sedan de Ville joined the lowline Calais Sedan on Cadillac's list of discontinued models. Only 7,230 Sedan de Villes were built during the 1970 production run compared with more than 83,000 Hardtop Sedan de Villes, although the pillared sedan's $6,118 price was identical. Weight was 4,690 pounds. Cadillac's revised 1970 grille design featured 13 thin vertical bars in the rectangular grille opening, which was flanked by dual headlamp units mounted in bright chrome bezels. The new winged emblems on the leading edges of the front fender blades were reminiscent of the raised-wing emblems on the fondly-remembered 1941 Cadillacs.

This would be the last year for the de Ville convertible. In 1971, its place would be taken by an all-new Fleetwood Eldorado Convertible. The de Ville Convertible had been offered since Model Year 1965 as the successor to the Series 62 Convertible, which in turn went all the way back to 1940. Cadillac's convertible sales had been shrinking annually. From a record 19,200 in 1965 and 1966, retail deliveries had tapered off to 16,445 in 1969, and declined further to 15,172 this year. Model 68367, the 1970 Cadillac de Ville Convertible had a sticker price of $6,068. This pristine example was owned by Crestline publisher George H. Dammann, a confessed convertible devotee who still mourns the soft-top's passing. A rear-window defogger was available on the convertible for the first time this year.

The 1970 model year would be the last in which a convertible was offered in Cadillac's standard passenger car line. In 1971, the only soft-top in Cadillac's product line-up would be an all-new premium Fleetwood Eldorado. This classic two-door convertible body style dated all the way back to the 1940 model year, to a time when the Series 62 included both a convertible coupe and a four-door convertible sedan. The glamorous Series 62 Convertible was upgraded to a de Ville in 1965. This rear view of George Dammann's favorite shows the last de Ville convertible's broad rear deck with windsplit and new ornamentation. Ten leather interiors were available and doors and instrument panels were trimmed with simulated Oriental Tamo hardwoods. This particular model was finished in white with black leather interior and black top, matching a similarly finished Ford X-L owned by the Dammanns.

This would be the fourth and final year for the first-generation front-wheel-drive Fleetwood Eldorado. A larger, all-new Eldorado series which included a coupe and convertible, was introduced the following year, for 1971. Eldorado sales in 1970 slipped to their lowest level since this distinctive luxury-personal car was introduced in 1967: 20,568 were delivered in the 1970 model year. Model 69347, the 1970 Fleetwood Eldorado Coupe had a manufacturer's suggested retail price of $6,903 and weighed 4,830 pounds. The Eldorado got new, thinner vertical taillights and a full-length body side molding between its large wheel cutouts.

The 1970 Fleetwood Eldorado's big story lay under its long, broad hood. It was a new 500 cubic inch displacement, 400-horsepower V-8 engine—the largest production passenger car engine in the world at the time. This huge engine developed an incredible 550 foot-pounds of torque and eventually found its way under the hoods of all full-sized Cadillac models before it was phased out of production at the end of the 1976 model year. A small plaque that read "8.2 Litre" was mounted on the left side of the 1970 Eldorado's modestly-changed grille. This would be the fourth and final year for the first-generation Eldorado body, which had originally been introduced for 1967. Other exterior changes this year included the Eldorado name in small block letters on the lower edges of the front fenders and a protective rub-strip molding along the center of the body. The halo-type vinyl roof first appeared on the Eldorado the previous year.

1970

A power operated sun roof was added to the Fleetwood Eldorado's list of optional equipment during the 1970 model year. An Eldorado show car with a sun roof had been exhibited in Cadillac's display at several major auto shows the previous year. The sun roof—which brought back memories of the poorly-received "Sunshine Roof" offered on some Cadillacs in the late 1930s and early forties—was a metal panel over the front seat that retracted into the rear roof at the touch of a button. The sun roof was vinyl-covered and color-keyed to match the Eldo's standard "halo" padded vinyl roof. Sun roofs soon became very popular on many premium U.S. cars and were responsible in some measure for the demise of the convertible. The sun roof provided some of the wind-in-the-hair exhilaration of a ragtop, but the foul-weather comfort of a closed car was available at the flick of a switch. (Still, it was a pretty poor substitute for the true rag-top—G.H.D.)

Cadillac's most prestigious production cars, the long-wheelbase Fleetwood Seventy-Five Sedan and Limousine, remained fully compatible in styling with the rest of the cars in the line. The 1970 Seventy-Fives got the division's revised front and rear appearance and bright, new wheel covers. Separate automatic Climate Control air conditioning systems for front and rear compartments and automatic level control were standard on both models, as was nine-passenger seating with two folding auxiliary seats. Just 876 Fleetwood Seventy-Five Sedans, Model 69723, were built in the 1970 model year. Prices started at more than $11,000, but optional equipment and the inevitable taxes quickly pushed the Seventy-Five Sedan's bottom-line well beyond its $11,039 starting point.

The prestigious Fleetwood Brougham and less expensive Sixty Special Sedan got what amounted to a major styling change this year. For the first time, full-length body side moldings adorned the broad flanks of these premium four-door sedans. Sixty Specials had always been distinguished by their distinctive exterior styling which had long eschewed the use of any body side trim. The new rub moldings had raised vinyl centers to protect the sides of the car in close parking situations. But the Brougham and Sixty Special retained their exclusive 133-inch wheelbase, chrome rocker sill, and rear fender extension moldings, and lower front fender Fleetwood lettering. Model 68169, the 1970 Cadillac Fleetwood Brougham sold for $7,284. Weight was 4,835 pounds and 16,913 were built.

This would be the last year Cadillac's customers would be able to choose from two Fleetwood Sixty Special models. The "base" Fleetwood Sixty Special Sedan was quietly discontinued at the end of the 1970 model year. Its niche at the top end of Cadillac's model offerings had been all but taken over by the more luxuriously-appointed Fleetwood Brougham. Only 1,738 bare-roof Sixty Special Sedans were built in the 1970 model year compared with sales of nearly 17,000 Broughams. Henceforth, only one model would be offered in this premium series: it would be called the Fleetwood Sixty Special Brougham. The Brougham had been introduced as a premium options package for 1965. The following year it became a separate model that eventually displaced the "base" Sixty Special. The Model 69069 Fleetwood Sixty Special Four-Door Sedan had a suggested retail price of $6,953.

The stately Cadillac Fleetwood Seventy-Five Limousine remained the only true production line limousine built in America. This would be the fourth and final year for this upper body styling on the Seventy-Five. The nine-passenger Fleetwood Limousine came with a standard fixed partition with adjustable glass division and a leather-upholstered chauffeur's compartment. Five rear compartment trim selections were offered, including three Divan cloths, one Decordo cloth, and a Dumbarton cloth with leather. Two separate air conditioning systems and automatic level control were standard. Options included a formal landau roof or the padded vinyl roof shown. A total of 1,240 Model 69733 Fleetwood Nine-Passenger Limousines were sold this year. Base price was $11,178.

1970

Lehmann-Peterson Inc. of Chicago was well-known for its Executive Limousine stretches of standard four door Lincoln Continentals, but this firm also did a few special-order Cadillacs too. One of these L-P customs was a four-door convertible sedan based on the 1970 Cadillac de Ville convertible. Cadillac's last production convertible sedan had been offered in 1941. Lincoln marketed a four-door convertible sedan between 1961 and 1967. It is believed this was the only 1970 Cadillac convertible sedan built. We wonder if it's still around . . .

What could be more exclusive than a Cadillac station wagon? A Cadillac Fleetwood Sixty Special Station Wagon, that's what! The builder of this exotic conversion isn't known for sure, but the accessories indicate that it may have been a product of the Custom Craft Division of the Automobile Specialty Corp. in the Detroit suburb of Southgate, Mich. The author photographed this car at Greenfield Village in Dearborn, Mich. Note the massive chrome grille surround complete with classic Cadillac "goddess" hood ornament replica. The Fleetwood Wagon was towing an enclosed classic car trailer.

In 1969, the former Superior Coach Corp. of Lima, Ohio, became a division of a Toledo-based conglomerate, Sheller-Globe. The name was changed to Superior Division. Superior continued to market a line of funeral cars and ambulances built on the Cadillac commercial chassis and a lower-priced line of Pontiacs built on a spliced Bonneville chassis. This is a 1970 Superior-Cadillac Crown Royale Limousine Funeral Coach. Superior built 1,156 hearses and ambulances on the Series 69890 Cadillac commercial chassis this year. Note the lower rear fender applique, which distinguished Superior's more expensive models between 1959 and 1970. This is an air-conditioned combination coach.

Superior continued to turn out far more funeral flower cars than all of its competitors combined. Miller-Meteor, located a few miles down I-75 in Piqua, Ohio, produced its last Cadillac flower cars this year. Superior built a total of 23 of these Coupe de Fleur deck-type flower cars on the 1970 Cadillac commercial chassis. Factory price was $13,588. Superior used this coupe style upper body structure with large triangular quarter windows between 1965 and 1970. The company's flower cars were completely restyled the following year. The coupe roof had a crinkle finish. Note the aluminum rear fender applique.

The Hess and Eisenhardt Co. of Cincinnati continued to offer a full line of funeral cars and ambulances built on the 156-inch wheelbase Cadillac commercial chassis. This company's most expensive standard production model was the S & S Professional High Body Ambulance, a fully-equipped emergency room on wheels designed for heavy-duty rescue service. The high-headroom Professional High Body Ambulance offered 54-inch headroom in the rear compartment. The company's other standard ambulances were the 50-inch Kensington, also a raised-roof body style, and the standard headroom 42½-inch S & S Parkway. This is the 1970 S & S Cadillac Professional High Body Ambulance in action at a highway accident scene.

1971

For the first time in a decade, every one of Cadillac's cars from the lowest-priced Calais on up to the flagship Fleetwood Seventy-Fives were completely redesigned and restyled in a single model year. The front-wheel-drive Fleetwood Eldorado series was expanded with the addition of a glamorous new convertible.

Cadillac's totally-new 1971 product lineup included nine models, three fewer than the previous year. Gone were the de Ville convertible and pillared four-door sedan and the lowline Fleetwood Sixty Special Sedan. The revised 1971 model lineup included a Calais Coupe and Hardtop Sedan; only two de Villes—the extremely popular Coupe de Ville and top-selling Hardtop Sedan de Ville; an all-new Fleetwood Sixty Special Brougham four-door sedan; a Fleetwood Eldorado Coupe and Convertible, and the regal Fleetwood Seventy-Five Nine-Passenger Sedan and Limousine.

Calais and de Ville models were built on a new 130-inch wheelbase, incrementally longer than before. The Fleetwood Sixty Special retained its exclusive 133-inch wheelbase. Seventy-Fives rode on a new 151.5-inch wheelbase 1.7 inches longer than previously, and the totally-new Eldorado's wheelbase was increased 6.3 inches, from 120 to 126.3 inches. Over-all lengths were only slightly longer for all models this year—225.8 inches on Calais and de Ville (225 inches in 1970): 228.8 inches on the Sixty Special (228.5 inches); 221.6 inches on Eldorado (221 inches), and 247.3 inches on Seventy-Fives (245.3 inches in 1970).

All 1971 Cadillacs got redesigned, stronger front-end chassis structures and reinforced safety bumpers. Powertrains were generally unchanged. All standard models were powered by the 472 cubic inch V-8 engine introduced three years earlier, while Fleetwood Eldorados continued to use the 500 cubic inch version of this engine introduced the previous year. Compression ratios, however, were reduced from 10:5-to-1 to 8.5-to-1 to operate more efficiently on new no-lead and low-lead gasolines. The redesigned Eldorado got a coil spring rear suspension system.

Cadillac's sleekly-styled new "C" and "D" bodies featured softly-rounded tubular contours with modest sheet-metal sculpturing on hoods and decklids. Up front, a rectangular center grille with vertically-textured crosshatching and two massive built-in bumper guards was flanked by widely-spaced dual headlamps recessed in bright chrome bezels. Small vertical winged emblems like those used on the front fenders of some 1970 models were positioned between the headlights, emphasizing vehicle width. There was a prominent elliptical fairing in the rear fender sheet-metal. Side marker lights consisted of two thin red slits that straddled a fine full-length body side molding. New crowned hoods and decklids had full-length windsplits on their top surfaces. Window sills swept up at the rear to meet new rooflines. Windshield pillars were thinner. Color-keyed "light monitors" that informed the driver that head, parking, and turn signal lamps were operating properly were mounted on the tips of the new front fenders.

De Villes got bright chrome rocker sill moldings and script-type model nameplates on their front fenders. New style cornering lights were built into the sides of the front fenders and had vertically-textured chrome trim.

The 1971 Fleetwood Sixty Special Brougham retained its traditional styling individuality. The elegantly-styled new Sixty Special featured a very distinctive upper body structure with strong formal styling accents. Flush-mounted side window frames had rounded corners and were edged with thin chrome moldings—just like the ones on the original Sixty Special away back in 1938. Small opera lamps were mounted on the closed rear roof quarters. The standard padded vinyl roof extended down to the beltline between the doors, which were separated by a thick "B" post. The Fleetwood name continued in small block letters on the front fenders and the redesigned hood and decklid bore the Fleetwood laurel wreath and crest. Interiors were lavishly upholstered and appointed in the finest Sixty Special tradition.

But the most spectacular new Cadillacs were the Fleetwood Eldorados, which got their first complete restyling since this luxury-personal car was introduced for the 1967 model year. A dazzling new convertible body style was added to the line. The all-new 1971 Eldorado retained the long-hood, short deck profile of

From the lowest-priced Calais on up to the flagship Fleetwood Seventy-Fives, all of the 1971 Cadillacs were totally restyled. Completely redesigned bodies on all standard series cars got fresh, new "tubular" contours with modest sculpturing. Cadillac's new grille design featured a crosshatched center grille flanked by widely-spaced dual headlamp units. Small vertical "wing" emblems like those used on the front fenders of some 1970 models were located between the bright rectangular headlight bezels. The redesigned grille also had a pair of massive vertical bumper guards. Note the broad, crowned hood with full-length windsplit and "hidden" windshield wipers. This is the 1971 Cadillac Calais four-door hardtop.

Fewer than 7,000 Cadillac Calais models were sold during the 1971 model year, compared with retail deliveries of more than 135,000 de Villes. Cadillac's lowest-priced series again included two body styles, a two-door coupe and four-door hardtop. Sales were almost evenly split, with deliveries of 3,360 Calais Coupes and 3,569 Calais Hardtop Sedans. Model 68247, the Calais Coupe had a manufacturer's suggested retail price of $5,899, making it the lowest-priced car in Cadillac's nine-car 1971 model lineup and the only one still below $6,000. The Calais name appeared in small script on the lower right-hand side of the redesigned decklid.

1971

the first-generation 1967-70 Eldos. Styling, however, was much more massive with heavily-sculptured body and fender lines. A vertically-textured, rectangular grille set in a bright chrome frame was situated low between dual headlamps. Parking and turn signal lights were incorporated into the leading edges of chiselled, squared-off front fenders.

The distinctive new rear deck projected slightly and was attractively bevelled. Slots were provided on top of the decklid for Cadillac's new flow-through ventilation system. The Eldorado Coupe had tall, rectangular "coach" windows in its elegant new upper body structure. Both models had vertical simulated air intake moldings on the leading edges of their rear fenders. A spring-mounted, stand-up wreath-and-crest ornament was mounted at the front of the long, sculptured hood.

The flamboyant new Fleetwood Eldorado Convertible (first Cadillac to bear this name since 1966) was equipped with a unique inward-retracting "Hideaway" convertible top. The top disappeared into a large well behind a full-width, three-passenger rear seat. When down, the top boot was nearly flush with the rear deck.

The stately Fleetwood Seventy-Five nine-passenger sedan and limousine also got Cadillac's fresh, new 1971 styling. Seventy-Five upper body styling was similar to that of the new Sixty Special with rounded upper door frames and formal opera lights. Doors again extended into the roof panel for ease of entry and exit. Dual-Comfort Control air conditioning and automatic level control were standard in both 1971 Fleetwood Seventy-Fives. A formal landau model with closed rear roof quarters was still available.

New curved instrument panels were used in all models. An electrically-operated sun roof was available as an extra-cost option on the Fleetwood Eldorado Coupe, Sixty Special Brougham and both the de Ville models.

Also on Cadillac's long list of extra-cost options was a new "Track Master" computerized rear wheel anti-skid control braking system.

Just as 1971 model production was shifting into high gear, General Motors was hit by a nationwide strike. The three-month work stoppage was one of the longest in recent GM history and sharply curtailed availability of the corporation's 1971 models until early in the new year. The strike by the United Auto Workers Union began virtually on announcement day in late September and was not settled until mid-December. Consequently, Cadillac production fell to its lowest level in six years. The division built 188,537 cars in the strike-shortened 1971 model run.

Cadillac's totally-restyled 1971 "C" and "D" bodies featured softly rounded tubular contours with elliptical sculpturing on the ends of the rear fenders. New vertical rear bumper ends were reminiscent of those used on the 1964 Cadillacs. All standard models had fine full-length chrome side moldings that extended from the front wheel openings to the rear bumper. Side marker lights were two thin red slits above and below the body side molding on the rear fenders. Rooflines and wheel covers were also new. This is the 1971 Cadillac Calais Hardtop Sedan, Model 68249, which sold for $6,075. Of the two lowline Calais body styles offered, the four-door hardtop was the more popular with 3,569 delivered.

A three-month strike sharply curtailed availability of all of GM's restyled 1971 models in the last quarter of 1970. The 1971 cars had just been introduced when the strike began. Production was not resumed until very late in the year, and dealers did not begin to receive new cars in quantity until early in the new year. Consequently, retail deliveries of all 1971 Cadillacs fell well below levels of the previous model year. A total of 66,081 Coupe de Ville two-door hardtops were built in the 1971 model year, nearly 10,000 fewer than in 1970. The very handsomely-styled 1971 Cadillac Coupe de Ville, Model 68347, sold for $6,264. It was again outsold by the four-door Hardtop Sedan de Ville.

Cadillac's 1971 product lineup included only two de Villes. The pillared four-door Sedan de Ville and de Ville convertible were discontinued at the end of the 1970 model year. Both Calais and de Ville models rode on a new 130-inch wheelbase. De Ville models got chrome rocker sill and rear fender extension moldings this year. The Hardtop Sedan de Ville retained its pre-eminent position as the most popular single luxury car in the world. Despite a strike-shortened 1971 model run, 69,345 Model 68349 Hardtop Sedan de Villes were built. Starting price was $6,498. Upper body styling of all standard Cadillac models included new rooflines with large glass areas. Padded vinyl roofs remained among the most popular of all de Ville extra-cost options. Model names appeared in script on the front fenders of both 1971 de Villes.

1971

The most changed of all the 1971 Cadillacs was the front-wheel-drive Fleetwood Eldorado. Completely redesigned for the first time since it was introduced for 1967, the premium Eldorado was now available in two body styles—a two-door coupe and what would be Cadillac's last production convertible. The all-new, second-generation Eldorado convertible replaced the de Ville soft-top as the only convertible in Cadillac's product lineup. At a hefty $7,751, the new Model 69367 Fleetwood Eldorado Convertible was priced only $12 lower than the Fleetwood Sixty Special Brougham. An even 6,800 Eldorado Convertibles were built during the 1971 model year.

The all-new 1971 Cadillac Fleetwood Eldorado Convertible was the only luxury convertible still in production in America. Ford and Chrysler had earlier discontinued production of their Lincoln and Imperial soft-tops. This same basic car, with annual styling revisions, was produced for six model years, through 1976 when it went out with a real splash. The companion Eldorado Coupe continued in production for another two years, through 1978. The totally-restyled 1971 Fleetwood Eldorado featured long, sculptured body lines and a protruding, bevelled rear deck. The inward-folding "Hideaway Top" retracted completely into a well behind the full-width, three-passenger rear seat. Soft, pliable Sierra grain leather upholstery and automatic level control were standard. This particular model is still owned by George Dammann, General Manager of Crestline Publishing Co.

The front-wheel drive Fleetwood Eldorado for 1971 was completely restyled for the first time since Cadillac's personal-luxury car had been introduced for the 1967 model year. The redesigned Eldorado Coupe retained the distinctive long hood, short rear deck profile of the first-generation Eldo. Wheelbase was increased from 120 to 126.3 inches but over-all length increased only incrementally, to 221.6 inches. The Eldorado Coupe's new upper body styling featured narrow "opera" windows in the formal rear roof quarter. A total of 20,568 Model 69347 Eldorado Coupes were built in the strike-shortened 1971 model year. The Eldorado Coupe's base price went up $480, to $7,383.

The totally redesigned and restyled 1971 Cadillac Fleetwood Eldorado featured a heavily-sculptured rear end. Both Eldorado body styles, coupe and convertible, got a stand-up Fleetwood laurel wreath-and-crest hood ornament. Eldorado styling highlights included sculptured fender lines and a vertical simulated air intake molding on the leading edge of the rear fenders. Coupes had narrow vertical "opera" windows and large backlights. A thin chrome body side molding protected the Eldorado's broad flanks and both coupe and convertible models had heavy chrome rocker sill moldings. The Eldorado Coupe far outsold the glamorous new Eldorado convertible. More than 20,000 coupes were built compared with just 6,800 convertibles.

1971

Cadillac's premium Fleetwood Sixty Special four-door sedan underwent extensive changes this year. Exterior styling was totally new. There was only one Sixty Special for 1971. The slow selling "base" Fleetwood Sixty Special Sedan was dropped at the end of the 1970 model year and the far more popular Fleetwood Brougham was upgraded to a new single model called the Fleetwood Sixty Special Brougham. The Sixty Special Brougham styling features included softly rounded side window frames that were nearly flush with the roof and formal opera lamps on the rear roof quarters just behind the rear doors. Wheel covers were also new. A total of 15,200 Fleetwood Sixty Special Broughams, Model 68169, were built during the abbreviated 1971 model run.

The all-new 1971 Cadillac Fleetwood Sixty Special Brougham retained its exclusive 133-inch wheelbase and distinctive exterior styling touches. Note the thick pillar between the front and rear doors. The 1971 Sixty Special also got a new upper body structure with rounded side window frames. The standard padded vinyl roof extended to the beltline between the doors. The thin, chrome-edged side window frames were similar to those used on the very first Sixty Special away back in 1938. Other Sixty Special styling features included the Fleetwood name in small block letters on the front fenders and decklid and the Fleetwood laurel wreath and crest emblem on the front of the redesigned hood and on the rear deck. The formal backlight had no chrome molding.

The flagship cars of the Cadillac fleet, the stately Fleetwood Seventy-Five Sedan and Limousine were also completely restyled this year. These were extremely impressive formal cars with their sleekly-redesigned long-wheelbase "D" bodies. Their new upper body styling was similar to that of the all-new 1971 Fleetwood Sixty Specials, with rounded upper door frames edged with thin chrome moldings. Small opera lamps were mounted on the upper rear doors. All four doors again extended into the roof for ease of entry and exit. Only 1,600 Fleetwood Seventy-Fives were built this year, including 752 Model 69723 Nine-Passenger Sedans and 848 Model 69733 Limousines. The Seventy-Five Sedan listed for $11,869 while the formal limo, with standard leather chauffeur's compartment, movable glass division, dual air and level control, carried a starting price of $12,008.

Bodies for the limited-production Fleetwood Seventy-Five sedan and limousine were built on their own, separate assembly line in GM's old Plant 21 a few blocks east of the General Motors Building in Detroit's New Center area. Only a dozen or so of these special "D" bodies were built each working day. They were trucked across town in closed vans to the main Cadillac passenger car assembly plant on Clark Avenue where they were mounted on the special 151.5-inch wheelbase Seventy-Five chassis. The 1971 chassis was 1.7 inches longer than previously. Much hand work still went into these formal bodies. Here, trimmers hand-fit a padded vinyl roof. A landau model with closed rear roof quarters was still optionally available.

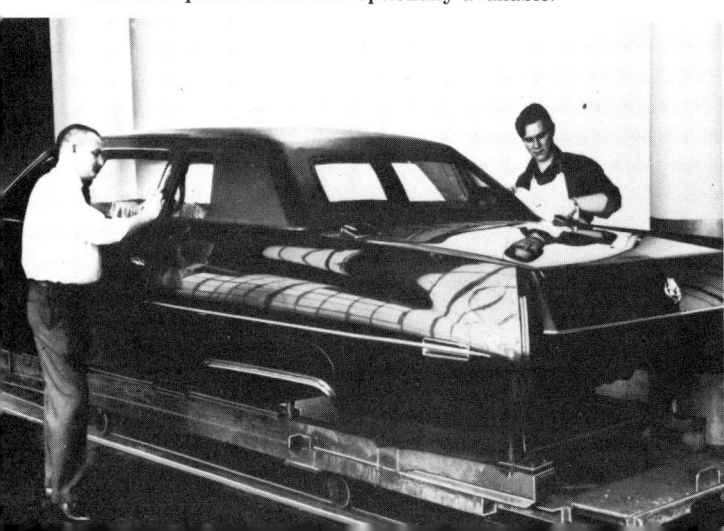

Every year, it seemed, some conversion shop came out with another custom-built Cadillac station wagon. Production of these cobbled-up wagons seldom exceeded more than half a dozen annually, at prices that virtually guaranteed exclusivity. R. S. Harper of Fraser, Mich., designed and built this Cadillac Estate Brougham conversion of the Calais or de Ville four-door hardtop. Standard equipment included a power-operated rear window, padded vinyl roof, dual-action rear door and automatic level control. Note the roof luggage rack and four-door hardtop styling.

1971

Here is another Cadillac station wagon conversion, this one based on a Fleetwood Sixty Special Brougham. Called the Fleetwood Brougham Astro Estate Wagon, it was designed and built by the Custom Craft Division of the Automobile Specialty Corp. of Southgate, Mich. The entire rear roof and retractable two-piece tailgate from a Chevrolet or Pontiac station wagon was grafted onto the rear end of this car. This company also offered a similar conversion of the new Eldorado Coupe called the El Dorado Astrella Station Wagon. Note the roof luggage rack and the opera lights, which have been retained on the "C" pillars.

Cadillac's special long-wheelbase hearse and ambulance chassis was also redesigned this year. Wheelbase was increased 1.5 inches to 157-½. Just over 2,000 Series 69890 Cadillac commercial chassis were shipped to Superior, Miller-Meteor, and Hess and Eisenhardt during the strike-shortened 1971 model run. Commercial chassis styling was again fully compatible with all of the other standard 1971 Cadillacs. This is a 1971 Superior-Cadillac Sovereign Series Side Servicing Landaulet Funeral Coach with three-way casket table. Superior's redesigned professional car bodies featured crisp, straight-lined styling.

The Miller-Meteor Division of Wayne Corp., introduced its premium Citation Landau Funeral Coach on Cadillac commercial chassis for the 1966 model year. Miller-Meteor's most expensive funeral car, the Citation was designed to compete with Superior's top-line Crown Sovereign Landaulet and was offered in side-servicing, end-loading or combination hearse/ambulance versions through 1979, when Miller-Meteor went out of business. The rear roof of the Citation was finished in a choice of deluxe padded vinyl or crinkle or parchment paint finishes and was outlined with bright chrome trim. This is the 1971 M/M Cadillac Citation End-Loading Hearse.

Industry leader Superior redesigned its funeral coach and ambulance bodies this year. Superior's fresh styling was crisp and angular with strong, straight lines. This is a 1971 Superior-Cadillac standard-headroom ambulance on the redesigned Series 69890 commercial chassis. Note the shortened second side windows and closed upper rear quarters. Rear doors on most ambulances were hinged on the "B" pillar for safety reasons. A total of 2,014 Cadillac hearse and ambulance chassis were built this year.

The Hess and Eisenhardt Co. of Cincinnati, manufacturer of S & S funeral cars and ambulances, continued to turn out very few flower cars. These were available only on a special-order basis and production by this time rarely exceeded one or two a year, usually for favored customers. This impressive open-deck coupe style flower car was built on the new 1971 Cadillac commercial chassis. Unlike Superior and Miller-Meteor, H & E preferred low, standard-headroom coupe rooflines on its flower cars with conventional passenger car windshield glass. This made the unit look dramatically lower and longer.

1972

The Cadillac Motor Car Division marked its 70th Anniversary this year and established another model year sales record. Retail deliveries of new Cadillacs also passed the quarter-million level for the first time in any model year.

The 70th Anniversary 1972 Cadillacs showed only minor exterior styling changes from the previous year, continuing Cadillac's traditional two-year styling cycle. Wheelbases, major dimensions and power trains were unchanged from 1971.

The 1972 Cadillacs featured improved, impact-resisting front and rear bumper systems designed to prevent vehicle damage in typical low-speed collisions. The front bumper was designed to withstand five mile-an-hour impacts without damage to the bumper, lights, or front-end sheet metal. A thin rub strip was built into the full width of the bumper face to further reduce the possibility of vehicle damage.

The 70th Anniversary Cadillacs got only a minor face-lifting from the previous model year. The revised grille now had a horizontal theme, enclosed in a bright chrome frame. New vertical parking lamps, with wing emblems etched in their lenses, were repositioned from the front bumper to new locations between widely-spaced square headlight bezels, Cadillac's famous "V" emblem returned after a two year absence, below the Cadillac crest on the front of the bevelled hood and the rear decklid on Calais and de Ville models. The lowest-priced car in Cadillac's 1972 product lineup was the Calais Coupe, Model 68247, which was still priced under $6,000 at $5,719. An even 3,900 were sold.

Sales of Cadillac's two de Ville models soared to new record highs this year. Combined sales of the perennially popular Coupe de Ville and Hardtop Sedan de Ville totalled nearly 195,000 cars. Fewer than 8,000 Calais models were sold in comparison. The handsome Coupe de Ville got only cosmetic changes this year. Grille treatment was revised and Coupe de Ville script nameplates were repositioned from the front fenders to the rear roof quarters. Padded, vinyl roofs were almost standard on this model now, and a power-operated sun roof was optionally available. Model 68347, the Coupe de Ville accounted for 95,280 sales this year. Suggested retail price was $6,116. Dual-ring white sidewall tires and vaned wheel covers were 1971 carryovers.

Only 7,775 lowline Cadillac Calais coupes and four-door hardtops were built during the record-breaking 1972 model year. Sales were almost evenly divided between the coupe and four-door. Retail deliveries included 3,875 of these Calais Hardtop Sedans, Model 68249, and 3,900 Calais Coupes. The two-door Coupe, in fact, outsold the Hardtop Sedan by exactly 25 units. Manufacturer's suggested retail price of the 1972 Cadillac Calais Hardtop Sedan was $5,886. Note the return of Cadillac's famous "V" emblem to the rear decklid for the first time since model year 1969. Back-up lights were built into the rear bumper.

Cadillac's two-year styling cycle continued. The 70th Anniversary 1972 Cadillacs were nearly identical to the 1971s, except for minor grille and detail changes. The next major change didn't come until 1974, and the same basic bodies were built from 1971 through the 1976 model year with annual styling revisions. From some angles, it was hard to tell a 1972 Cadillac from a 1971. This is the 1972 Coupe de Ville. Giveaways are the "V" emblem on the decklid and the Coupe de Ville script on the upper roof quarter. Note the padded vinyl "halo" roof with color break around the side windows. The Cadillac script appeared on the right side of the decklid.

1972

Cadillac's 1972 product lineup again included nine models in three series on four different wheelbases. The lowest-priced cars were the Calais Coupe and Calais Hardtop Sedan, both of which were priced below $6,000 this year. The two de Villes—Coupe de Ville and companion Hardtop Sedan de Ville—accounted for more than 72 per cent of all Cadillac's 1972 model year sales. There was again just one Fleetwood Sixty Special, an opulently-appointed and equipped four-door sedan on its own, exclusive 133-inch wheelbase. Sales of the high-styled Fleetwood Eldorado Coupe and companion Convertible set a new record for this front-wheel-drive luxury-personal car series, now in its sixth year, and the flagships of the Cadillac fleet were still the stately Fleetwood Seventy-Five Sedan and Limousine.

All standard Cadillacs got a modest front-end facelifting. The revised 1972 grille featured a strong horizontal bar theme enclosed in a bright chrome upper frame.

Parking and directional signal lamps were repositioned from the front bumper to new locations between widely-spaced square headlamp bezels. The new vertical parking lights had Cadillac's raised-wing emblem etched on their lenses. Cadillac's famous "V" emblem returned after a two-year absence to the front of the hood and rear decklids on Calais and de Ville models, below the Cadillac crest. Fleetwood models continued to bear distinctive laurel wreath-and-crest identity.

The 1972 Fleetwood Eldorado got a revised grille texture with a vertical bar theme. This refined grille was enclosed in a heavier outer frame with the Cadillac name in script on the left side of the header. Eldorado script nameplates appeared above the cornering lamps on the sides of the front fenders, and "8.2 Litre" was spelled out in small block letters on the ends of the front fenders, below a thin chrome body side molding.

An elegant "Custom Cabriolet" roof option was introduced during the 1972 model year for the Eldorado Coupe. Exclusive to this model, the Custom Cabriolet package featured a distinctive rear roof treatment covered with a rich, padded Elk grain vinyl. A chrome strap bordered by vinyl welts highlighted the leading edge of the cabriolet roof. Other details included a rolled edge around the backlight, a French seam that shaped the padded area around the rear window and Fleetwood wreath-and-crest emblems on the broad sail panels, behind the vertical coach windows. The elegant Custom Cabriolet roof option was available in a choice of seven colors, with or without an electrically-operated sun roof.

Beginning in 1971, Cadillac expressed its engine horsepower output in both gross and SAE net horsepower. The 472 cubic inch V-8 engine used in all standard 1971 and 1972 Cadillacs, therefore, was shown as developing 345 gross horsepower at 4,400 RPM, and 220 net SAE horsepower. The 500 cubic-inch Eldorado engine was rated at 365 gross and 235 net horsepower.

One of the most unusual customer new car orders in Cadillac's long history came directly to the factory from the White House in May, 1972. President Richard M. Nixon, on three days' notice, wanted a Fleetwood Eldorado Coupe to take with him on a state visit to Russia as a gift from the American people to Soviet Communist Party Secretary Leonid Brezhnev, who was reputed to be something of a car buff. Cadillac manufacturing executives scrambled to find a Sable Black Eldorado coupe body with red pinstripe on the assembly line. Finally, after much searching, a suitable Eldorado body was located on the receiving dock, minutes after it had arrived from Fisher Body. The car was literally walked all the way down the assembly line. After testing on Detroit freeways, the car was trucked to Selfridge Air Force Base near Detroit and loaded aboard a military transport for the long journey overseas. The car was shipped with a selection of popular stereo tapes, spare parts and service manuals. Brezhnev was reportedly delighted with the ultimate symbol of Western capitalism and decadence.

Now the oldest automobile manufacturer left in the City of Detroit, Cadillac capped its 70th Anniversary by building a record 267,787 cars.

The last car of the 1972 model year was driven off the Clark Avenue assembly line on July 7. The previous model year record was the 238,745 Cadillacs built in the 1970 model year. Nineteen Seventy-Two also marked the first time Cadillac produced more than a quarter of a million cars in a single model year, although the division had built 266,789 Cadillacs in Calendar Year 1969, which included both 1969 and 1970 models.

The Cadillac Hardtop Sedan de Ville was far and away the world's most popular single luxury car. Sales of this model came very close to topping the 100,000 mark. A total of 99,531 Model 68349 Hardtop Sedan de Villes were sold during the record 1972 model year, establishing another new high for this body style. With 95,280 sales, the companion Coupe de Ville was close behind. The attractively-styled Hardtop Sedan de Ville combined the open-air spaciousness of a sporty hardtop with the closed-car comfort and practicality of a four-door sedan. The 1972 Cadillac's revised grille design had a series of thin horizontal bars in a bright chrome frame. Note the "light monitors" on the tips of the front fenders. These informed the driver of correct operation of head, parking and turn signal lights and also warned when the washer fluid level was low.

1972

The impressive Cadillac Fleetwood Sixty Special Brougham entered its second year as a single model, opulently appointed and equipped premium four-door sedan. "This is Cadillac in the grand manner" read the 1972 ads. The Sixty Special got minimal styling changes this year. Basically, these consisted of a new horizontal grille texture and repositioned vertical parking lights between the dual headlamp units. A wide range of interior upholstery color choices were available, including nine Sierra grain leathers, four Matador cloths or a combination of Matador cloth and leather; a new Minuet fabric in three colors, or a lush, new Medici crushed velour. Standard Sixty Special Brougham equipment included individual rear seat reading lamps, Dual Comfort front seats, and automatic level control.

The 1972 Fleetwood Sixty Special Brougham retained its exclusive 133-inch wheelbase and distinctive formal upper body styling. Note the fine chrome moldings around all window openings and the small opera light on the roof quarter panel. The Sixty Special also had broad chrome rocker sill and rear fender extension moldings. The Fleetwood name continued in small block letters on the front fenders below full-length chrome body side moldings. Model 68169, the impressive Sixty Special weighed 4,858 pounds and had a suggested retail price of $7,585. Sales increased to 20,750 this year.

Like the rest of the 1972 Cadillacs, the front-wheel-drive Fleetwood Eldorado got only cosmetic exterior styling changes. The grille had a new vertical texture. The Cadillac name appeared in script on the left side of a new, slightly heavier grille header and there were new Eldorado script nameplates above the cornering lights on the sides of the front fenders. The words and numerals "8.2 Litre" appeared in tiny block letters below the thin, chrome body side molding on the ends of the front fenders. Eldorado sales rocketed to a new all time record this year. More than 40,000 were sold, including 32,099 of these Fleetwood Eldorado Coupes. President Richard M. Nixon presented an Eldorado Coupe like this one to Soviet Communist Party Secretary Leonid Brezhnev on his trip to Russia in May, 1972.

During the record 1972 model year, Cadillac added a luxurious new "Custom Cabriolet" roof option for the Fleetwood Eldorado Coupe. Exclusive to the Eldorado Coupe, the Custom Cabriolet package included an electric sun roof with a distinctive new elk-grain padded vinyl roof treatment that covered the rear half of the upper body structure. Detailing included a "halo" molding around the entire roof, a rolled perimeter around the rear window and a tailored French seam that shaped the padded area. A bright chrome strap bordered the leading edge of this elegant cabriolet roof.

The most glamorous of all nine 1972 Cadillac body styles was the flashy Fleetwood Eldorado Convertible, the only luxury soft-top still being built in North America. Model 69367, the Eldorado Convertible commanded an appropriately impressive price of $7,511. A new two-piece "hard" convertible top boot could be quickly and easily snapped into place when the top was lowered, and customers could choose from eight soft Sierra grain leather interior colors. As attractive as the sporty convertible was, most Eldorado customers preferred closed cars. Only 7,975 of the 40,070 Fleetwood Eldorados sold in the 1972 model year were convertibles. Both 1972 Eldorado models got restyled wheel covers with concentric outer rings.

1972

The aristocratic Fleetwood Seventy-Fives bordered on the awesome in their intimidating appearance and overwhelming proportions. Captains of commerce, Government leaders, and VIPs of every persuasion were whisked from place to place in hushed comfort in these impressive nine-passenger cars. The Seventy-Fives were still the only long-wheelbase formal automobiles designed and production-line-built as limousines in North America. The Cadillac Motor Car Division of General Motors built 1,915 Fleetwood Seventy-Fives during the 1972 model year, of which 955 were Model 69723 Seventy-Five Sedans. The big Seventy-Five Sedan, which came with folding auxiliary seats for three passengers, had a base price of $11,695 and weighed 5,515 pounds.

Long black Cadillac limousines and funeral directors were synonymous in America. Most funeral directors owned at least one of these long-wheelbase Fleetwood Seventy-Five sedans or limousines which usually followed a matching Cadillac chassised hearse in funeral processions. Cadillac at this time was doing some cooperative advertising with the three major U.S. funeral car and ambulance manufacturers—Superior, Miller-Meteor, and Hess & Eisenhardt. In this funeral service trade publication ad, a 1972 Fleetwood Seventy-Five Limousine poses with a Miller-Meteor Cadillac Landau Traditional. Only 960 Model 69733 Fleetwood Seventy-Five Limousines with formal glass division were built during the 1972 model year. Its $11,827 base price was reduced slightly from the previous year.

Customizing shops continued to turn out small numbers of Cadillac station wagon conversions. The builder of this estate wagon conversion of a Cadillac Fleetwood Sixty Special Brougham four-door sedan is not known, but Maloney Standard Coach Builders in the Chicago area was doing conversions of this type at the time. The roof has been removed aft of the rear doors and a General Motors "C"-body station wagon roof added. The rear roof is vinyl-covered and is equipped with a conventional station wagon roof luggage rack. Note the Fleetwood wreath-and-crest emblem on the "C" pillar. This photo turned up in Cadillac's photographic archives and contained no information.

Cadillac shipped nearly 2,500 special long-wheelbase commercial chassis to U.S. funeral car and ambulance manufacturers during the 1972 model year. The 157-½-inch hearse and ambulance chassis should not be confused with the Fleetwood Seventy-Five limousine chassis, which had a 151.5-inch wheelbase. The commercial chassis came with front-end sheet metal only. The rest of the body from the cowl back was built by three specialized body builders utilizing Cadillac outer door panels and rear quarter panels. This is a 1972 Miller-Meteor high headroom ambulance built on the Cadillac commercial chassis. These heavy rescue units were favored by fire departments and professional rescue squads.

Maloney Standard Coach Builders, a Chicago area shop best known for its Lincoln Continental Executive Limousine stretches, now offered a wide range of customizing accessory packages for standard Cadillacs. This is the "El Doral" Cabriolet roof option as fitted to a Cadillac Coupe de Ville. The vinyl padded rear landau roof covering extended to the door sills and the leading edge of the roof was trimmed with a bright chrome molding. The grille was modified with a massive chrome frame.

1973

Cadillac—and the U.S. automobile industry—shattered all previous production and sales records this year. The Cadillac Motor Car Division produced its five-millionth car before the memorable 1973 model year was out, and the division's sales passed the epochal 300,000 mark for the first time.

The 1973 Cadillacs were given moderate exterior styling changes, their most extensive since all-new bodies had been introduced two years earlier for the 1971 model year. The most interesting, and practical engineering innovation on the 1973 Cadillacs were new energy-absorbing bumpers and grilles designed to prevent vehicle damage in low-speed impacts, the kind commonly encountered in parking situations.

Cadillac's most interesting engineering innovation this year was an advanced impact-absorbing front and rear bumper system designed to prevent vehicle damage in low-speed impacts such as those commonly encountered during parking maneuvers. The 1973 Cadillac grille was attached directly to the front bumper, which in turn was mounted to the car's frame by a pair of energy-absorbing snubbers. The entire grille, and bumper, retracted several inches when impacted, without damaging body sheetmetal. The rear bumper was also equipped with GM's new energy-absorbing mounts. Bumper rub strips further reduced the possibility of damage. This is the 1973 Cadillac's redesigned front end with its new, wider grille and new rectangular parking lamps mounted between widely-spaced dual headlight units. Note also the new chamfered hood and massive vertical bumper guards.

The lowest-priced car in Cadillac's 1973 model lineup and the only one still priced below $6,000 was the Calais Hardtop Coupe. Model 68247, this lowline two-door hardtop carried a manufacturer's suggested retail price of $5,866. The Calais Coupe was the more popular of the two Calais body styles offered this year. The Calais and more expensive de Villes continued on a 130-inch wheelbase. De Villes had chrome rocker sill moldings: Calais models did not. Six cloth and two vinyl interiors were available in the "most affordable" 1973 Cadillacs.

Grilles on all 1973 Cadillacs, including the front-wheel-drive Eldorado, were attached directly to the front bumper. When the bumper was struck at low speeds, the entire grille retracted inward several inches preventing damage to the grille and front-end sheet metal. Rear bumpers were also equipped with GM's improved impact-cushioning system. Pliable, color-keyed urethane fillets occupied "crush" space provided between the ends of the rear fenders and vertical rear bumper ends. Cadillac had introduced five-MPH bumper protection systems on its 1972 cars, but the 1973 system marked a significant advance in damageability protection. Rubber impact strips were also used across the full width of front and rear bumpers and on the faces of the vertical front bumper guards that flanked the redesigned grille.

The 1973 Cadillac's restyled front end featured a wider, vertically-textured center grille with a fixed lower section below the bumper. Large new rectangular parking and directional signal lamps were located between the dual headlight units. The redesigned hood had a flat, chamfered face with the Cadillac crest and "V" at its center. On the rear of all standard models, back-up lights were repositioned to the top of the new rear bumper. All standard models also got bright new chrome wheel covers.

The Fleetwood Eldorado got a bold new "eggcrate" grille and restyled, chamfered hood. New parking and turn signal lights with vertically-textured lenses wrapped around the leading edges of the front fenders, which also bore Cadillac's now-familiar raised-wing emblems. The simulated air intake rear fender moldings of the previous two years were gone, replaced by a heavy full-length chrome body side molding. Lighted, circular Fleetwood wreath and crest side marker lights, last used in 1970, returned to the ends of the rear fenders. The rear end of the 1973 Eldorado was extensively redesigned. The projecting rear deck of 1971-72 was replaced by a more conventional bevelled rear deck and straight bumper. The Eldo's taillights were also redone but remained in the trailing edges of the car's sculptured rear fenders.

For the first time in 34 years, there was no Sixty Special, as such, in the premium Fleetwood model

The four-door Hardtop Sedan was the less popular of the two Cadillac Calais body styles offered in the record-breaking 1973 model year. Only 8,000 lowline Calais models were sold this year, compared with deliveries of more than 216,000 de Villes. Model 68249, the Calais Sedan sold for $6,038. A total of 3,798 were built. The 1973 Cadillacs were given extensive styling changes, including new hoods, grilles and wheel covers. Cadillac script nameplates appeared only on the front fenders of Calais models this year.

1973

lineup. This car was now advertised as simply the Fleetwood Brougham. The first Cadillac Sixty Special was introduced in 1938. This distinguished model name had identified the division's most exclusive standard four-door sedans ever since.

During the 1973 model year, an ultra-luxurious interior decor package was added to the list of extra-cost options available for the Brougham. Called the "Brougham d'Elegance", it was the first of a series of premium special edition packages that would eventually be available to Brougham customers who desired even more than the aristocratic Fleetwood four-door sedan's conventional standard of luxury. The Brougham d'Elegance option included lavish interior trim and appointments. An adjustable reading lamp for the right front seat passenger became standard, in addition to the two already provided for rear seat passengers. New Fleetwood Brougham options also included a lap robe and pillow, deep-pile carpeting and shirred elastic pockets on the backs of the front seats. The Brougham's standard padded vinyl roof was available in a choice of eight colors and 19 of the 21 rich exterior color choices were new this year.

The optional padded vinyl roof available on the 1973 Coupe de Ville and Hardtop Sedan de Ville was of a new center seam design that permitted the use of a wider sun roof panel (optional) on the four-door hardtop. Other new Cadillac options for 1973 included a left-hand outside rear view mirror with a thermometer built into its base; a lighted vanity mirror on the right-hand sun visor, and steel-belted radial ply tires for all models except Seventy-Fives. Also new this year was a "theft deterrent system" that sounded the horn and flashed the headlights.

Engineering refinements included an exhaust gas recirculation (EGR) system that reduced emissions of oxides of nitrogen. Control of exhaust emissions had replaced vehicle safety as the industry's principal federally-mandated goal.

Cadillac's 1973 product lineup again included nine models; a lowline Calais coupe and four-door hardtop; the perennially popular Coupe de Ville and Hardtop Sedan de Ville, both of which topped the 100,000 mark in model year sales for the first time; the Fleetwood Brougham (formerly sixty special) on its own exclusive 133-inch wheelbase; the front-wheel-drive Fleetwood Eldorado coupe and convertible, sales of which hit a record 50,000-plus in the 1973 model year, and the flagship Fleetwood Seventy-Five nine-passenger sedan and limousine.

A Cotillion White Fleetwood Eldorado Convertible paced the 57th Indianapolis 500-Mile Race on May 28, 1973, driven by veteran Indy driver Jim Rathmann. This marked the fifth time the "500" was paced by a Cadillac Motor Car Division product and only the second time that a front-wheel-drive car had led the pack. The only other FWD Indy 500 pace car was a 1930 Cord which had been driven by E. L. Cord himself. LaSalles had been selected as pace cars in 1927, 1934, and 1937, and a V-12 Cadillac driven by "Big Bill" Rader paced the pack in 1931.

The 1973 model year was another one for the record books. Cadillac built a record 304,839 cars in the fantastic 1973 model year, which brought back pleasant memories of record-breaking 1955. A blue Hardtop Sedan de Ville which came down the assembly line June 27, 1973 was the five-millionth Cadillac produced since 1902. Robert D. Lund was named General Manager of the division effective January 1, 1973. The entire North American motor vehicle industry looked forward to another all-time model year record as 1974 production began that fall. No one foresaw the energy crunch that lurked around the very next corner.

After years of playing second-fiddle in the division's sales race, the Cadillac Coupe de Ville finally outsold the companion Hardtop Sedan de Ville this year. An incredible 112,849 Coupe de Ville two-door hardtops were built during the 1973 model year, setting yet another sales record for this extremely popular body style, and passing the 100,000 mark for the first time. The Coupe de Ville thus became the world's most popular single luxury car. Model 68347, the Coupe de Ville had a retail delivered price of $6,268 exclusive of local taxes, shipping, and licensing charges. The "halo" vinyl roof remained among the most popular of all de Ville options. Note the bumper rub strips and bright new wheel covers.

Sales of the four-door Hardtop Sedan de Ville slipped into second place behind the Coupe de Ville during the record 1973 model year. But the perennially-popular Sedan de Ville's sales performance was outstanding nonetheless. Sales topped the 100,000 mark for the first time and set a new record of 103,394. Model 68349, the 1973 Cadillac Hardtop Sedan de Ville sold for an even $6,500. Optional padded vinyl roofs on both 1973 de Villes featured a new center seam, which permitted the use of a wider sunroof panel (also an extra-cost option) on the Sedan de Ville. Interior trim included "soft pillow" door panels with pull straps, over a rosewood vinyl base. Also new this year was a Cadillac clock with Roman numerals.

The rear fenders of all 1973 Cadillacs (except the long-wheelbase commercial chassis) were equipped with pliable, color-keyed extensions as part of Cadillac's new energy-absorbing bumper system. The ends of the fenders were trimmed with a fine chrome bead that followed the vertical "V" profile of the rear bumper ends. Rear bumpers were also redesigned this year and equipped with a full-width rub strip. Back-up lights were relocated in the upper surfaces of the bumper and were better protected. This was the third model year for these basic body shells which had been introduced for 1971. With annual revisions, they were produced for another three years, through 1976. This is the 1973 Cadillac Hardtop Sedan de Ville.

1973

The luxurious Cadillac Fleetwood Sixty Special Brougham received the same moderate exterior styling changes as the rest of Cadillac's standard models this year, including a redesigned front end, flat-faced hood, color-keyed rear fender extensions and attractive new chrome wheel covers. The 1973 model year also saw the first of a series of ultra-luxurious interior decor options for this premium four-door sedan. A "Brougham d'Elegance" option was added during the year, and an adjustable reading lamp for the front seat passenger became standard along with the two lamps for the rear-seat passengers. New Sixty Special options also included a lap robe and pillow, deep-pile shag carpeting and shirred elastic pockets on the backs of the front seats. Model 68169, the Fleetwood Sixty Special Sedan sold for $7,765. Sales hit a record 24,800 finally topping the previous record 24,000 Sixty Specials delivered in model year 1957.

The front-wheel-drive Fleetwood Eldorado got substantial exterior styling changes this year. The Eldorado coupe and convertible had bold, new "eggcrate" grilles, more massive front bumpers with integral rub strips, and redesigned hoods and decklids. Parking and directional signal lamps wrapped around the leading edges of the front fenders. The Eldorado's grille was attached to the energy-absorbing front bumper and retracted inward several inches for improved damageability protection in low-speed impacts. Sales of the Fleetwood Eldorado Coupe, Model 69347, rocketed to a new record of 42,136 this year. Base price was $7,360. Cadillac's raised-wing emblem, a favorite with the division's stylists since 1970, adorned the leading edges of the 1973 Eldo's chisel-tipped front fenders.

The Fleetwood Eldorado underwent its most extensive restyling in several years. Both Eldorados got the same impact-absorbing front and rear bumper system used on all other 1973 Cadillacs, including a bumper-mounted grille that retracted on impact. The Eldorado's rear end was thoroughly restyled this year. Gone was the prominent center-section trunk of 1971-72. The new straight bumper had a full-width protective rub strip. Vertical tail-lamps were built into the ends of the rear fenders. Lighted wreath-and-crest circular side marker lights, last seen in 1970, returned to the ends of the rear fenders. The vertical rear fender moldings were also gone, replaced with a new full-length chrome body side molding. Both coupe and convertible retained their heavy chrome sill moldings.

Fleetwood Eldorado sales hit a record 51,451 in the extremely prosperous 1973 model year. Of these, 9,315 were Eldorado Convertibles. Model 69367, the front-wheel-drive Fleetwood Eldorado Convertible sold for $7,681. The Eldorado's 20th anniversary passed almost unnoticed. It is interesting to note that despite two decades of normal price increases and inflation, the original 1953 Eldorado Convertible was priced $390 higher than the 1973 version. A white 1973 Cadillac Fleetwood Eldorado Convertible like this one, driven by Jim Rathmann, paced the 57th Indianapolis 500-Mile Race on May 28th, the fifth Cadillac Motor Car Division product and only the second front-wheel-drive car ever to do so.

1973

The most impressive, and expensive, cars in Cadillac's nine-car 1973 product lineup were again the imposing Fleetwood Seventy-Fives. These long-wheelbase specialty models—longtime favorites of bankers, diplomats and funeral directors—got the same noticeable exterior styling changes as the rest of the division's other 1973 cars, including redesigned grilles, hoods, wheel covers, and improved impact-absorbing front and rear bumper systems. Only 2,060 Fleetwood Seventy-Fives were built this year. This total included 1,017 Model 69723 Nine-Passenger Sedans and 1,043 Model 69733 Limousines with formal glass division. The Seventy-Five Sedan sold for $11,948 and the Limousine carried a base price of $12,080 making the latter the most expensive car in the 1973 Cadillac line. Note the front bumper rub strips. The pliable, color-keyed rear fender extensions were used this year only.

Best known for its S & S funeral cars and ambulances, the Hess and Eisenhardt Co. of Cincinnati also has a very low-profile specialty car division. This division's specialty is the design and construction of custom-built "security" vehicles and the armor-plating of conventional passenger cars and limousines. The specialty car division has its own highly-secured "plant" within the main H & E plant where these conversions are done, often under code names for customers unknown to all but one or two senior company executives. Bulletproofing a car requires stripping the interior to the bare shell and the installation of "boilerplate" in key areas, and bullet and projectile-resistant glass in all window openings. This is an armored 1973 Cadillac Fleetwood Seventy-Five Limousine done for an unnamed foreign government. The only indication that this is no ordinary limo are the extra-heavy side window frames required to hold the thick bulletproof glass.

The 1973 Cadillac commercial chassis did not have the new energy-absorbing rear bumper system found on the division's standard passenger cars. There was no soft plastic fillet between the end of the rear fender and the vertical bumper ends. The commercial chassis grille, however, was attached to the front bumper and retracted several inches, just like on the passenger cars. Cadillac shipped 2,212 Series 69890 commercial chassis this year. This is a 1973 Superior-Cadillac Flower Car. Superior produced this style of deck-type flower car between 1971 and 1976. Note the padded, vinyl-coverd roof and upper body side trim and slender chrome landau irons.

For many years, the Hess and Eisenhardt Co.'s best known product was its S & S "Victoria" Landau Funeral Coach, a formally-styled hearse this company introduced in 1938. The distinctive Victoria was produced from 1938 through the 1981 model year, when H & E sold its S & S funeral car division to Superior Coaches of Lima, Ohio. Superior continues to market the S & S Victoria as a premium model. Here is a trio of 1973 S & S Victoria Landaus built on a 157-½-inch wheelbase Cadillac commercial chassis. Note the full formal window draperies, thickly-padded roofs and "V"-contoured rear door.

In the larger cities, especially in the "inner-city" areas, many Cadillac dealers were enhancing their profit margins by marketing a variety of "customizing" packages for standard Cadillac models. These options ranged from relatively modest bolt-on grille surrounds, like this one, to bizarre custom restyles with heavily-padded cabriolet roofs and door uppers, phoney continental extensions, fake hood and rear deck straps and garish interiors. This three-piece brass grille cap, richly nickel and chrome-plated, was offered by Maloney Coachbuilders, of Chicago. It is seen fitted to a 1973 Fleetwood Eldorado Coupe.

Cadillac continued to produce about 2,500 special hearse and ambulance chassis annually. These were sold to three specialized funeral car and ambulance manufacturers, all located within 75 miles of one another in Ohio. Superior Coach was still the largest, closely followed by Miller-Meteor. Hess and Eisenhardt, builders of S & S funeral cars and ambulances, was the smallest. This is a 1973 Superior-Cadillac Crown Sovereign Limousine Funeral Coach built on the Cadillac 157-½-inch commercial chassis. The Crown Sovereign was Superior's most expensive series. Limousine-style funeral cars by now had become fairly rare. Most funeral directors preferred the more formal landau styling.

1974

Sales of 1974-model passenger cars in the United States were unexpectedly affected by political turmoil on the other side of the world. Late in 1973, the oil-producing nations in the Middle East embargoed the flow of Arab oil to the U.S. Faced with an impending fuel shortage and the prospect of gasoline rationing, a panicky public scrambled for smaller, more fuel-efficient cars. The market for large cars was temporarily depressed. But by the spring of 1974, when it was obvious that we weren't about to run out of gasoline after all, the customers resumed buying the same kinds of cars they had always bought—big ones, including Cadillacs.

Coming off an all-time record model year, automakers were looking for more of the same as they took the wraps off their 1974 cars in the fall of 1973. All of the 1974 Cadillacs got extensive styling changes inside and out. There were some significant engineering innovations and improvements, too.

All of Cadillac's standard models got bold, new eggcrate grilles. Massive new cornering, parking, and directional signal lamps wrapped around the corners of the front fenders. Dual headlight units were mounted closer together. Two-door coupes got fixed rectangular quarter windows. Restyled rear fenders had a lower profile and were equipped with flexible, color-keyed urethane extensions. Redesigned vertical chrome bumper ends with built-in taillights telescoped into the ends of the rear fenders to cushion minor impacts.

Front and rear bumpers were equipped with new white-on-gray full width rub strips. The improved bumpers with Delco energy-absorbing mounts added 2.2 inches to over-all length.

New Cadillac options for 1974 included the controversial "air bag" passive restraint system; a high energy electronic ignition system, and a space-saver spare tire for greater luggage compartment capacity. A new pulsating-action windshield wiper system with "intermittent" position for driving in light rain was also added to the optional equipment list. New steel belted radial tires built to GM specifications were optional on all models. Low-pressure, bias-belted fiber glass tubeless tires were standard. Padded vinyl roofs were now optionally available on the lowline Cadillac Calais series.

All 1974 Cadillacs featured a completely redesigned curved instrument panel with a built-in digital clock. Key warning lights were grouped in a narrow band across the top of the new panel. Principal operating instruments were grouped in a chrome housing directly above the steering column. All 1974 Cadillacs also got attractive new wheel covers.

Several new "special edition" packages were offered this year. De Ville buyers could upgrade their cars with "d'Elegance" and "Cabriolet" options. Premium d'Elegance interior and exterior decor packages were available on both Coupe de Ville and Sedan de Ville. The Cabriolet roof package was available only on the Coupe de Ville. The "d'Elegance" interior appointments included velour upholstery, deluxe door pads, storage pockets in the backs of the front seats and deep-pile carpeting and floor mats.

The 1974 Cadillacs showed extensive styling changes from the previous year. All standard models got new eggcrate grilles. Massive new cornering, parking and directional signal lights wrapped around the corners of the front fenders. Redesigned rear quarter panels and rear bumpers increased over-all length by 2.2 inches. Two-door coupes got new fixed quarter windows. The lowest-priced cars in the 1974 Cadillac product lineup were the two Calais models, a two-door coupe and four-door hardtop. Only 6,883 Calais models were built this year, including 4,559 Model 68247 Calais Coupes and 2,324 Model 68249 Calais Hardtop Sedans. Padded vinyl roofs were now optionally available on both lowline Calais body styles. The Calais Coupe sold for $5,997 and the Calais Hardtop Sedan for $6,169. Note the thin chrome header molding above the grille opening.

All nine 1974 Cadillacs got extensive styling changes inside and out. Grilles and rear bumpers were redesigned and inside there was a totally-new curved instrument panel with a built-in digital clock. Wheel covers were also restyled. Fixed rectangular quarter windows were used on both two-door coupes, the lowline Calais Coupe and the more expensive Coupe de Ville. New vertical rear bumper ends, which also housed the taillights, telescoped into the rear fenders to absorb minor impacts. There were also color-keyed flexible urethane extensions between the ends of the redesigned rear fenders and bumper ends. Calais and de Ville models had script series nameplates on their rear fenders. This is a profile view of the handsome 1974 Cadillac Coupe de Ville.

1974

The Cabriolet option featured a landau type rear roof bordered on its leading edge by a bright chrome strip. The ultimate Coupe de Ville combined the d'Elegance and Cabriolet packages with the optional power-operated sun roof. The d'Elegance package included a standup see-through hood ornament in place of the usual "V" and crest on the front of the hood and special tape stripe accents on the hood, bodysides and decklid.

The ultra-luxurious "Brougham d'Elegance" option first offered on the 1973 Fleetwood Brougham (formerly Sixty Special) was continued. An incredibly opulent "Fleetwood Talisman" option was added for 1974. (A talisman is something that produces "magical or miraculous effects"). The Fleetwood Talisman option provided truly sumptuous seating for just four people in deeply cushioned armchair style seats. Running between the two-plus-two seats were upholstered consoles with an illuminated writing set in front and a vanity in the rear. There was rich Medici crushed velour everywhere. Other Talisman touches included a reclining front passenger seat, assist straps and special wheel covers. This was certainly the most lavishly-appointed production Cadillac seen since the classic 1930s.

The Talisman's special turbine wheel discs were also part of the Brougham d'Elegance package and were optionally available on the standard Fleetwood Brougham and Seventy-Fives, and without the wreath emblems, on de Villes and Calais.

A power-operated antenna was now standard with all Cadillac radios. The air conditioning system was also redesigned. Fifteen of the 21 exterior color choices were new this year.

The front-wheel-drive Fleetwood Eldorado coupe and convertible got a new fine-mesh, vertically-textured grille capped with a bright brushed chrome header. The Eldorado's rear end was also redesigned and had the same improved rear bumper system as all other models. The dashing Eldorado Convertible remained the only luxury soft-top still being built in the U.S. A rear stabilizer bar was used on both 1974 Eldorados for improved road handling.

Standard models were powered by essentially the same 472 cubic inch V-8 engine introduced for 1968. The Eldorado used the same 500 cubic inch version of this engine introduced for 1970. Compression ratios, however, were lowered again to 8:25-to-1, to accommodate new lower-octane fuels.

Cadillac's 1974 model lineup again included nine cars: Calais Coupe and Hardtop Sedan; Coupe de Ville and Hardtop Sedan de Ville; Fleetwood Brougham Four-Door Sedan; Fleetwood Eldorado Coupe and Convertible, and the stately Fleetwood Seventy-Five Nine-Passenger Sedan and Limousine. Standard models rode on a 130-inch wheelbase. The Fleetwood Brougham continued on its exclusive 133-inch wheelbase chassis, and Seventy-Fives rode on a 151.5-inch wheelbase. Eldorados were built on a 126.3-inch wheelbase. Over-all length this year was 230.7 inches for Calais and de Villes; 233.7 inches for the Brougham; 224.1 inches for Eldorados and 252.2 inches for the big Seventy-Fives.

Cadillac built 242,330 cars including 40,412 Eldorados during the 1974 production run. Despite the sales drop caused by the Arab oil scare, Cadillac actually improved its share of the U.S. luxury car market from 30.3 per cent in 1973 to 34.5 per cent in the 1974 model year.

All of Cadillac's standard 1974 models (Calais, de Villes, Fleetwood Brougham and Seventy-Fives) got bold, new eggcrate grilles. Dual headlamp units were also mounted closer together. The Coupe de Ville was by far the most popular single model in Cadillac's nine-car 1974 product lineup and the only one that topped 100,000 sales this year. A total of 112,201 Model 68347 Coupe de Villes were delivered during the 1974 model year, 648 fewer than the record 112,849 sold the previous year. Factory retail price was $6,399. Coupe de Ville buyers could also choose from two new "special editions"—a Coupe de Ville Cabriolet and Coupe de Ville d'Elegance. The ultimate Coupe de Ville was the d'Elegance combined with the Cabriolet roof option, with or without a power-operated sun roof.

Sedan de Ville customers could upgrade their cars by specifying the new "d'Elegance" interior and exterior decor option. Interior appointments included rich velour and leather upholstery, deep-pile carpeting, deluxe door pads and storage pockets in the backs of the front seats. Special exterior touches included a standup see-through hood ornament, "d'Elegance" script on the sail panels, and accent striping on the hood, body sides and rear deck. Sedan de Ville sales plunged by more than 40,000 this year, to 60,419. Model 68349, the Hardtop Sedan de Ville sold for $6,631. This is the standard 1974 Cadillac Hardtop Sedan de Ville. New white-on-gray impact strips were used on front and rear bumpers of all 1974 Cadillacs.

1974

Two ultra-luxurious "special edition" packages were available for the 1974 Fleetwood Brougham. The "Brougham d'Elegance" option introduced for 1973 was continued. New this year was a positively decadent "Talisman" option. The Fleetwood Talisman's lavishly trimmed and appointed interior provided seating for just four people in obscene comfort in deep armchair-type seats. Medici crushed velour upholstery was used throughout. Consoles between the two-plus-two seats contained an illuminated writing set and vanity. Other Talisman luxuries included a reclining front passenger seat, assist straps, seat-back pockets and deep-pile carpeting and floor mats. Special exterior touches included special wheel discs, a standup hood ornament, and "Fleetwood Talisman" script on the rear roof quarters. The standard Model 68169 Fleetwood Brougham shown sold for $7,896 and weighed 5,143 pounds. A total of 18,250 were built during the 1974-model production run.

An old, established builder of airport limousine "stretches" introduced a premium six-door Cadillac limousine conversion this year. Armbruster/Stageway Inc. of Fort Smith, Ark. had been doing cut-and-splice stretches of standard passenger cars since 1928. The company supplemented its Pontiac, Buick, and Chrysler six-door limousine offerings, intended primarily for airport limousine operators, hotels, and funeral directors, with an impressive six-door limousine conversion based on the 1974 Cadillac Fleetwood Brougham. The car was literally cut in half and an additional set of doors was inserted in the middle. This 1974 Armbruster/Stageway Cadillac Fleetwood Brougham Six-Door Limousine was one of three operated by the Tulsa-Whisenhunt Funeral Home in Tulsa, Okla.

The Cadillac Motor Car Division built only 1,900 Fleetwood Seventy-Five nine-passenger sedans and limousines during the 1974 model year, which was adversely affected by the first Middle East oil scare. Sales included 895 Model 69723 Fleetwood Seventy-Five Sedans and 1,005 Model 69733 Fleetwood Limousines with formal glass division. Both 1974 Fleetwood Seventy-Fives got Cadillac's redesigned front and rear ends. The Seventy-Five Sedan sold for $12,344. At $12,478 the long-wheelbase Fleetwood Limousine was the most expensive car in the 1974 line. A landau roof option was still available for both models, and Seventy-Five customers could specify the same lush Medici crushed velour upholstery offered in the shorter Fleetwood Brougham.

The front-wheel-drive Fleetwood Eldorado luxury-personal car also got numerous styling changes this year. Among these was a finely-detailed, vertically textured grille and redesigned rear fenders. The Cadillac name appeared in script on the left side of a new brushed chrome grille header. Rear fenders were equipped with bumper/taillight units that telescoped to absorb low-speed impacts. Inside, there was an all-new, curved instrument panel. Wheel covers were new this year, too. This is the 1974 Fleetwood Eldorado Coupe equipped with the "Custom Cabriolet" roof option. The Custom Cabriolet roof cost an extra $385, but when a power-operated sun roof was added the Custom Cabriolet option package price shot up to $1,005 over and above the base price of the car.

The Fleetwood Eldorado Coupe outsold the companion Eldorado Convertible by a margin of four-to-one. Of the 40,412 front-wheel-drive Eldorados built during the 1974 model year, 32,812 were two-door coupes. The Model 69347 Fleetwood Eldorado Coupe carried a starting price of $7,491. The 1974 Eldorado got a new fine-mesh grille with a broad brushed chrome header bar. Accent striping was also bolder this year. Rear stabilizer bars were used on all 1974 Eldorados for improved handling, and the compression ratio of the 500 cubic inch V-8 engine exclusive to the Eldo was lowered again, to 8:25-to-1 to accommodate new lower-octane gasolines. This Fleetwood Eldorado Coupe looks right at home in front of an elegant, columned residence.

1974

The dashing Fleetwood Eldorado Convertible remained the only luxury soft-top still in production in the United States. The Eldorado Convertible retained its unique inward-retracting convertible top mechanism and full width, three-passenger rear seat. Both 1974 Eldorados—Coupe and Convertible—were powered by essentially the same 500 cubic inch V-8 engine that had been introduced for 1970. An even 7,600 Eldorado Convertibles were sold this year compared with deliveries of well over 32,000 Eldorado Coupes. Model 69367, the luxurious Fleetwood Eldorado Convertible carried a base price of $7,812, less than $100 below that of the premium Fleetwood Brougham four-door sedan. Nine Sierra grain leather interior trim choices were offered for the sporty Eldorado Convertible this year.

About the only thing more exclusive than a custom-built Cadillac station wagon would have to be a genuine Caddy pickup truck—if such a classy conveyance could even politely be referred to as a truck in the general sense. The Hillcrest Motor Co. of Beverly Hills, Calif. (where else?) was doing some interesting quality conversions at this time. This very attractively-styled pickup was created from a standard 1974 Cadillac Coupe de Ville... just the thing for toting hay bales around the country place, or for dazzling the other guests at fashionable parties. We doubt that this elegant "cowboy Cadillac" was ever used for hauling rocks or cleaning out the stables.

A number of aftermarket specialty firms offered a wide assortment of custom conversion packages for standard production Cadillac models. One of the largest of these was the Custom Craft Division of the Automobile Specialty Corp., a component of Heinz Prechter's American Sunroof Corp. in the Detroit suburb of Southgate, Mich. This is Custom Craft's "El Deora" custom appearance option for the 1974 Cadillac Coupe de Ville. Custom styling touches included a thickly-padded cabriolet roof with two horizontal quarter windows on each side and a simulated spare tire cover on the rear decklid. The continental-style "spare" was also vinyl-covered to complement the rear roof.

Here's a different kind of Cadillac "stretch" job. Custom Craft, a division of the Automobile Specialty Corp. of Southgate, Mich., grafted a couple of extra feet of rear fender onto a standard Fleetwood Eldorado Coupe. Part of Custom Craft's "El Dorado" option was a functional continental spare tire mounted between the rear bumper and decklid. This package also included a cabriolet or landau-style rear roof treatment with ornamental chrome carriage bows. Custom Craft also offered Cadillac Eldorado and Fleetwood Brougham station wagon conversions.

The builder of this impressive Cadillac Fleetwood Eldorado station wagon is not known for sure, but it was likely the product of one of several prosperous conversion shops that had sprung up in Southern California. The basis for this estate wagon conversion was a standard Fleetwood Eldorado Coupe. The design features a built-up rear roof and huge rear liftgate with integral continental spare tire cover. The area over the front seat is equipped with a sun roof. Note the chrome roof luggage rack.

Cadillac continued to turn out about 2,265 long-wheelbase hearse and ambulance chassis each model year. These special commercial chassis were supplied to three U.S. funeral car and ambulance manufacturers who designed and built their own bodies for this 157.5-inch wheelbase chassis. Cadillac provided only the extra-long perimeter frame chassis, running gear and front-end sheet metal. Rear fenders and door panels were shipped separately. The largest company still in this highly specialized field was the Superior Division of Sheller-Globe, in Lima, Ohio. This is a 1974 Superior Cadillac "Sovereign" Side-Servicing Landau Funeral Coach. Note the crinkle-finished roof and chrome landau bows.

1975

Cadillac's big news this year was the division's first truly "small" car, the compact Cadillac Seville which was introduced late in the 1975 model year.

The Cadillac Motor Car Division's product planners were keenly aware of the increasingly-larger share of the U.S. luxury car market that was going to expensive foreign cars like the Mercedes-Benz. By the early seventies, import penetration of Cadillac's traditional market had reached the point where a totally new kind of product - a significantly smaller one but with a very high level of luxury - was not only viable but essential if the division was to maintain leadership in its chosen segment of the North American industry. With the go-ahead from GM management, Cadillac's long-rumored small car project shifted into high gear in 1973. The totally-new "international-sized" Cadillac Seville was formally announced in April, 1975 and went into dealer showrooms May 1.

The "1975½" Cadillac Seville was available in only one body style, a conservatively-styled five-passenger, four-door notchback sedan. Built on a 114.3-inch wheelbase and with an over-all length of 204 inches, the Seville was 27 inches shorter and eight inches narrower than the 1975 Sedan de Ville. It also weighed half a ton less. The Seville's new "K" body was developed off the same basic GM "X"-car platform shared at the time by the compact Chevrolet Nova, Pontiac Ventura, Oldsmobile Omega and Buick Apollo.

All exterior sheet-metal, however, including the distinctive upper body structure, was exclusive to the new, small Cadillac. The Seville was powered by a 350 cubic-inch (5.7 litre) V-8 engine purchased from Oldsmobile but built to Cadillac specifications with a standard Bendix electronic fuel injection system. This 180 BHP engine, with 8.0-to-1 compression ratio, was coupled to a standard Turbo Hydra-Matic automatic transmission. With a manufacturer's suggested retail price of $12,479 the Seville was the most expensive production Cadillac offered since the exclusive Eldorado Broughams of 1957-60.

Cadillac's new, small car revived a well-remembered division model name used on Eldorado hardtop coupes between 1956 and 1960. At one point, Cadillac's marketing staff had seriously considered naming their new baby the LaSalle, but the LaSalle was one of the Cadillac's few conspicuous marketing failures and had always been the division's lowest-priced car, while the new K-body program was targeted at the top end of the price scale.

The Seville went into production in March, 1975 on the former Eldorado assembly line in Cadillac's home plant in Detroit. Eldorado production was integrated with the standard models on the main Cadillac assembly line. The Seville's restrained styling featured an unmistakable Cadillac eggcrate grille flanked by dual rectangular headlamp units with wraparound parking and cornering lights; crisp, straight body lines, generous glass area and large, round wheel openings.

The Seville was an exceptionally well-equipped car. Standard equipment included a padded vinyl roof, air conditioning, automatic level control, power seats, windows and door locks, 15-inch steel-belted radial ply whitewall tires, tilt-and-telescope steering wheel, AM/FM stereo radio and a choice of supple leather or richly-tailored cloth upholstery. The Seville was an immediate sales success. More than 16,000 were built before the end of the 1975 model year.

Cadillac's other 1975 cars received numerous styling and engineering changes. The most noticeable of these was the appearance of four rectangular headlamp units on all models, the most basic front-end styling change since dual headlights were adopted in 1958. The new rectangular headlights resulted in a lower, wider, more unified frontal appearance. Cadillac's standard models got new finely-detailed grilles surrounded by bright

After years of rumor and speculation, Cadillac took the wraps off its long-awaited new "small" car in April, 1975. It was the Seville, the smallest and most expensive standard production Cadillac since the ultra-expensive Eldorado Broughams of 1957-58. The totally-new Seville was defined by its builder as "international" in size and was Cadillac's carefully-considered answer to the Mercedes-Benz, Jaguars, and BMWs which were accounting for increasingly larger shares of the U.S. luxury car market. The Cadillac Seville utilized GM's new "K" body, built off the same basic X-car platform used for the Chevrolet Nova, Pontiac Ventura, Oldsmobile Omega, and Buick Apollo. It was powered by a 350 cubic inch V-8 engine purchased from Oldsmobile but modified by Cadillac with electronic fuel injection. The Seville was offered in only one body style, a conservatively-styled four-door notchback sedan. The base price was a very expensive $12,479. It was an immediate success with 16,355 built through the balance of the 1975 model year.

The totally-new Cadillac Seville was 27 inches shorter and eight inches narrower than the 1975 Sedan de Ville. Wheelbase was 114.3 inches. Over-all length was 204.0 inches, width was 71.8 inches and height was 54.7 inches. The Seville weighed 4,345 pounds; nearly 1,000 pounds less than the de Ville. Styling highlights included an unmistakable Cadillac cross-hatched grille, four rectangular headlamps, large wheel cutouts, and a conservative notched roofline. Luxurious cloth and leather interior choices were offered and the Seville's long list of standard equipment included softly-tinkling chimes that asked—not ordered—driver and passengers to buckle up. With minor annual styling changes, the "first-generation" Seville was produced through the 1979 model year.

1975

chrome frames. New fixed triangular quarter windows were incorporated into the rear roof sail panels on Calais and the de Ville four-door hardtops. Redesigned "lamp monitors" were mounted on the ends of the front fenders. The front-wheel-drive Fleetwood Eldorado also got extensive appearance changes this year. For the first time since 1970 there were no rear fender skirts, which gave the Eldorado coupe and convertible a sportier appearance. The revised grille design had a bold, vertical texture. The coupe got larger fixed quarter windows that dipped down to the rear fenderline. New rectangular cornering lights were built into the sides of the front fenders and parking lights were repositioned in the lower portion of the front bumper.

All 1975 Cadillacs except the Seville were powered by the 500 cubic-inch V-8 engine previously used only in the Eldorado. This would be the largest engine ever used in a production Cadillac. The 1975 engine was equipped with GM's high energy electronic ignition and an improved Quadrajet carburetor with electric choke. Lower axle ratios were used on all models (except limousines) to further improve fuel economy. An electronic fuel injection system was optionally available. One of the most important engineering changes on the 1975 cars was not readily visible: General Motors equipped virtually all of its 1975 models with catalytic converters to further reduce vehicle emissions.

A new Cadillac option this year was an "Astroroof" a power-operated tinted glass panel that let the sun shine in. An adjustable sun shade could be pulled forward to shut the daylight out. A conventional metal sun roof was also still available on most models. Another new option was an illuminated vehicle entry system actuated by pressing the front door handle button. this illuminated the lock cylinder and turned on all interior lights for 20 seconds. Luxurious "special edition" options were again offered. De Ville buyers could choose a "Cabriolet" package for the Coupe de Ville and "d'Elegance" options were available for both Coupe and Sedan de Ville. Fleetwood Brougham customers could again specify the posh "Fleetwood Talisman" four-seater package or the popular "Brougham d'Elegance" upgrade option. Eldorado Coupe buyers could opt for the "Custom Cabriolet" roof option.

Electronic fuel injection was optionally available for the 500 cubic inch V-8 engine beginning in March, 1975.

With the addition of the new Seville, Cadillac's model offerings for 1975 increased to 10 in four series. The lineup included lowline Calais Coupe and Hardtop Sedan; Coupe de Ville and Hardtop Sedan de Ville; Fleetwood Brougham and Fleetwood Eldorado Coupe and Convertible; Fleetwood Seventy-Five Nine-Passenger Sedan and Limousine, and the compact Seville four-door sedan.

During the 1975 model year, the U.S. auto industry was buffeted by the worst economic recession since dismal 1958. Sales, which had just begun to pick up after the 1973-74 oil scare, went into a slump again. Cadillac's 1975 model year production totalled 264,731 cars - an improvement, to be sure, over the 242,330 Cadillacs built in 1974 but well below 1973's record 304,839.

Chrysler Corporation discontinued its flagship Imperial at the end of the 1975 model year, abandoning the luxury passenger car market to Cadillac and Lincoln for the next five years. The Imperial had been Chrysler's finest car for 49 years, since 1926. But it would be back.

All 1975 Cadillacs got new, rectangular headlamps — the most dramatic basic change in front-end styling since dual headlights were introduced across the board in the 1958 model year. All standard models also sported finely-detailed new crosshatched grilles set in bright chrome frames. Calais and de Ville four-door models now had triangular quarter windows in their sail panels. The lowest-priced cars in the 1975 Cadillac product lineup were the Calais Coupe and companion Calais Hardtop Sedan. Only 8,300 Cadillac Calais were built during the 1975 model year, including 5,800 Model 68247 Coupes and 2,500 Hardtop Sedans. The lowline Coupe sold for $8,197, and the Model 68249 Hardtop Sedan for $8,390. Front fender lamp monitors, a high-energy electronic ignition, and power door locks were standard on the least-expensive Cadillacs. Padded vinyl roofs were optional on both body styles.

Two "special edition" options packages were again available to Cadillac de Ville customers this year. Luxurious "d'Elegance" options were offered for both Coupe and Sedan de Ville. A "Cabriolet" roof option was available only on the Coupe. When the "d'Elegance" option was specified, the Cadillac "V" emblem and crest were deleted from the front of the hood and replaced by a see-through standup hood ornament in the shape of the Cadillac crest. Accent tape stripes on the hood, body sides and decklid were also part of the d'Elegance package. This well-equipped 1975 Cadillac Coupe de Ville, with both the d'Elegance and Cabriolet roof options, was owned by Bob Withey of Glen Ellyn, Ill. It is shown here at Bob's Truck Rental Agency with Bob on the left and George Dammann, Crestline General Manager, on the right.

1975

The Coupe de Ville retained its pre-eminent position as by far the most popular of all 10 (including the new Seville) Cadillac models, and consequently, the most-preferred single luxury automobile in the world. Although sales slipped by nearly 2,000 units this year, the Coupe de Ville was still the only Cadillac model to top the 100,000 mark. A total of 110,218 were built during the 1975 model year. Model 68347, the Coupe de Ville carried a manufacturer's suggested retail price of $8,613. New options for 1975 included an illuminated entry system actuated by pressing the front door handle, which turned the interior lights on for 20 seconds. The key cylinder was also illuminated. The 1975 Cadillac Coupe de Ville shown is equipped with the "Cabriolet" roof option as well as the Coupe de Ville d'Elegance package.

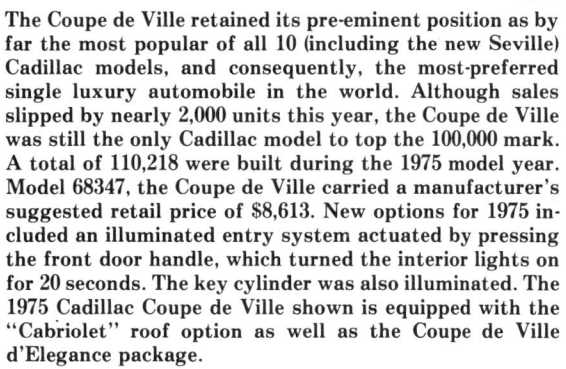

The two de Ville body styles accounted for two-thirds of all of the Cadillac Motor Car Division's sales. Of the 242,330 cars built during the 1975 model year, 172,620 bore the "de Ville" script on their rear fenders. The four-door Hardtop Sedan de Ville accounted for 63,352 sales this year. Factory price was $8,814. Both the lowline Calais and more expensive de Ville four-door hardtops got triangular quarter windows in their rear roof sail panels this year. The Cadillac "V" and crest on the hood of this 1975 Cadillac Sedan de Ville indicate that it is a standard model and is not equipped with the Sedan de Ville d'Elegance upgrade package. The Sedan de Ville was listed as Model 68349.

The 1975 Cadillacs received some significant exterior styling changes. The most noticeable was a switch to rectangular headlamp units which flanked a new, finely-textured grille. Both standard series four door hardtops — the Calais Hardtop Sedan and the more expensive Hardtop Sedan de Ville — got fixed quarter windows in their rear roof pillars. Buick and Oldsmobile four-door hardtops also got the new "plug" quarter windows. The standup hood ornament indicates that this 1975 Cadillac Sedan de Ville four-door hardtop is equipped with the luxurious "d'Elegance" option. All 1975 Cadillacs were equipped with GM's new catalytic converter emission control system which required the use of unleaded fuel only.

The distinctive Cadillac Fleetwood Brougham, no longer officially described as a Sixty Special, went into its fifth consecutive model year with essentially cosmetic exterior styling changes. All 1975 Cadillacs (except the new Seville) were powered by the same 500 cubic-inch engine V-8 previously used only in the Eldorado. The 1975 Fleetwood Brougham got the same front-end restyle seen on all standard Cadillacs this year, with new rectangular headlamp units that gave the front end a lower, broader, more unified appearance. Whitewall tire stripes were getting wider. Fleetwood Brougham sales increased slightly this year, to 18,755 and the Model 68169 Brougham's base price finally climbed into the five-figure bracket, starting at $10,427. Two special editions — Fleetwood Talisman and Brougham d'Elegance — were again available.

The front-wheel-drive Fleetwood Eldorado got a number of substantial styling changes this year. Fender skirts disappeared. Wraparound cornering, parking and turn signal lights of the previous two years were replaced by large, rectangular cornering lights mounted in the sides of the front fenders. Parking and directional signal lights were relocated in the outer ends of the lower front bumper. The grille had a bold vertical texture. The narrow rectangular coach windows used since 1971 gave way to large quarter windows that extended down to the rear fenderline. The Fleetwood wreath-and-crest appeared on the rear roof quarters. The two-door coupe was by far the more popular of the two 1975 Cadillac Eldorado body styles offered for 1975, accounting for 35,802 of the 44,752 Eldorados built this year. Model 69347, the Fleetwood Eldorado Coupe sold for $9,948.

1975

The convertible body style had all but vanished from the U.S. auto industry. The Fleetwood Eldorado Convertible was the last true convertible still in production. Even its days were numbered, as Cadillac had already served notice that the 1976 model would be the last. And it was, at least until 1982 when Chrysler resurrected the ragtop. Speculators drove prices of 1976 Eldorado convertibles to ridiculous heights after the last one was produced in April of that year, but the fact remains that the 1975 model is more rare. Only 8,950 Eldo Convertibles were built during the 1975 model year, compared with 14,000 "last" 1976 models. Model 69367, the 1975 Fleetwood Eldorado Convertible sold for $10,367 marking the Eldorado's ascension into the five-figure bracket. New options for 1975 included electronic fuel injection and a reclining front passenger seat with six-way power adjuster.

The regal Fleetwood Seventy-Five Sedan and Limousine also got Cadillac's attractive new front-end styling, with dual rectangular headlamp units and a finely-detailed rectangular grille set in a bright chrome frame. Only 1,671 Seventy-Fives were built in the depressed 1975 model year, including 876 Model 69723 Seventy-Five Nine-Passenger Sedans and 795 Model 69733 Fleetwood Limousines. Prices shot up to over $14,000 this year— $14,231 for the Seventy-Five Sedan and $14,570 for the Limo—finally eclipsing the $13,074 of the legendary 1957-60 Eldorado Broughams. The flagships of the General Motors fleet, the prestigious Seventy-Fives continued to offer two automatic Climate Control air conditioning systems, automatic level control and folding auxiliary seats for three extra passengers as standard equipment. A glass division and leather chauffeur's compartment were standard in the Limousine, and full-vinyl roofs and a landau top continued on the options lists.

Cruise ships sailing into Nassau's Harbour are met by hordes of Cadillac Fleetwood Seventy-Five sedans and limousines of varying vintages, which are available to take tourists on whirlwind tours of this popular Bahamian resort island. The owner-operators of these ritzy taxicabs go to great lengths to "personalize" their cars. This 1975 Fleetwood Seventy-Five has the full front-end treatment, complete with massive chrome grille cap, "goddess" hood ornament replica and huge "Superfly" headlight covers popularized by the ace black detective of American movie fame. Superfly headlights were more commonly found on Eldorados and Fleetwood Broughams in the big U.S. cities.

Armbruster/Stageway, Inc. of Fort Smith, Ark., a well-known builder of "stretch" limousine conversions for airport limousine operators and funeral directors, enhanced its product line with the introduction this year of an ultra-luxurious "Silverhawk Executive Limousine". The Silverhawk was a much-elongated Cadillac Fleetwood Brougham. A standard Fleetwood Brougham was literally cut in two and a fixed center section was inserted between the front and rear doors. For the funeral and livery trade, Armbruster/Stageway also offered a companion six-door Fleetwood Brougham Limousine conversion. The impressively-long Silverhawk Executive Limousine was also available on Buick, Pontiac, or Chrysler chassis.

Hess and Eisenhardt, a prominent funeral car and ambulance builder located in Cincinnati, has done many "security" limousine conversions for Governments and wealthy individuals over the years. This is an armor-plated Cadillac Fleetwood Limousine done for an unidentified client in 1975. Identities of customers and details of such modifications are closely-guarded secrets for security reasons. The only indication that this car is anything special is the extra-heavy upper door frames, required to hold thick bulletproof glass.

1975

No sooner had the new Cadillac Seville hit the market than the conversion builders did their own thing—or things—with it. Heinz C. Prechter's Automobile Specialty Corp. in the Detroit suburb of Southgate, Mich., marketed this "Charisma" two-door coupe conversion of a 1975 Seville. Design features included a redesigned grille, raised hood, formal rear window, and curved "rear lounge". Custom two-tone paint and premium wire wheels gave the "Charisma" added custom appearance. Base price of this conversion, sold through Cadillac's dealer organization, was $19,750. A padded cabriolet roof was also available for the charismatic Charisma.

Traditional Coach Works of Los Angeles designed and built this high-styled station wagon conversion of a 1975 Cadillac Fleetwood Brougham. Officially called the Mirage Sports Wagon, its intended buyer was..."the discriminating sportsman, rancher or gentleman farmer." Notable styling features included a sun roof over the front seat, a broad, forward-angled "C" pillar that swept up and over the roof, and a sharply-raked rear window/liftgate with integral wind deflector. Traditional Coach Works was one of a number of specialty car conversion shops now flourishing in the L.A. area.

Traditional Coach Works Ltd. of Los Angeles, also offered a Cadillac pickup conversion. This elegant luxury utility vehicle was also called a "Mirage". It was based on a standard 130-inch wheelbase Cadillac Calais Coupe or Coupe de Ville chassis. Of special interest is the full quarter window and padded, vinyl-covered upper body. Wire wheels and circular Fleetwood wreath-and-crest side marker lights on the rear fenders contribute to the custom look. The backlight is recessed between upswept rear roof panels for an attractive canopy coupe appearance.

The sun was slowly setting on the traditional full-sized Cadillac hearse and ambulance. The funeral car industry was already working on new bodies for an all-new, "downsized" commercial chassis still two years down the road. The 157.5-inch wheelbase Series 69890 Cadillac commercial chassis was built through the 1976 model year. The funeral car and ambulance business was negatively affected by the same economic downturn that depressed 1975 car sales. Only 1,328 Cadillac commercial chassis were shipped this year. This is a 1975 Superior-Cadillac Coupe de Fleur flower car sold to the Dyer Metropolitan Funeral Chapel in Tulsa, Okla.

Coach, or opera, windows had become a styling cliche throughout the U.S. auto industry. Formal opera windows could be found on "personal" cars from the Chevrolet Monte Carlo on up to the Cadillac Eldorado. It was not surprising, therefore, that the style-conscious funeral car industry picked up this styling gimmick and applied it to the most formal cars of all—hearses. Superior again led the way. For 1975, Superior introduced a new Sovereign Regal Landaulet built on the Cadillac commercial chassis. The Sovereign Regal featured a narrow triangular "coach" window in its upper rear body quarters, just behind the second side door and small, vertical opera lights. The 1975 Cadillac commercial chassis also got Cadillac's handsomely redesigned front end with dual rectangular headlamp units.

1976

Cadillac has been credited with numerous industry "firsts" over the years, but 1976 is remembered as a year of "lasts". The division attracted national attention when it built the last production convertible in America. The 1976 model year was the 12th, and last, for the lowline Calais series, and it was also the last year for the traditional "big" Cadillac. Next year's cars, already being prepared for production, would be significantly smaller and lighter.

Wednesday, April 21, 1976 was a memorable day in Cadillac's long history. On that sunny spring morning, the last convertible built by a major U.S. auto manufacturer came off the assembly line in Cadillac's Clark Ave. plant in Detroit. A white Fleetwood Eldorado Convertible, it was driven off by Cadillac Motor Car Division General Manager Edward C. Kennard and Manufacturing Manager Bud Brawner. Detroit Mayor Coleman Young and several Cadillac assembly workers rode as passengars. The occasion was a major news event and hundreds of newsmen and guests were waiting as "The Last of the Breed" came down the final line. It was a sad day and truly the end of an era. Cadillac had offered its first true convertible, a Type 53, some 60 years earlier, in 1916. Convertible production peaked in the early 1960s, then rapidly declined with the corresponding increase in popularity of air conditioning, stereo sound systems, vinyl roof hardtop styling, and "sun roof" options. Cadillac had served notice that the 1976 Eldorado Convertible would be its last. The Eldo soft-top had been the only luxury convertible still in production in the U.S. since 1971.

The factory was besieged with orders for the milestone "last" car. The problem was resolved by scheduling production of 200 identical "last" convertibles, which were quickly snapped up by dealers and collectors. The actual last one was kept by Cadillac for its own historical collection. The 200 "last" Eldorado convertibles were painted white with white convertible tops, white-painted wheel covers and matching white leather upholstery. All bore special dash plaques. Cadillac built an even 14,000 Eldorado convertibles during the 1976 model year — 60 per cent more than it had produced in 1975. Speculators soon drove 1976 Eldorado Convertible prices into the stratosphere, but they eventually came down with a resounding crash. Within months, conversion builders were chopping the tops off Eldorado coupes and standard Cadillacs and selling Cadillac convertibles of their own. Six years later, Chrysler Corporation resurrected the convertible body style with a premium soft-top version of its 1982 Chrysler LeBaron.

Cadillac went to market with 10 models in four series for 1976. These were grouped into four size categories: the "family-sized" Calais Coupe, Hardtop Sedan, Coupe de Ville and Hardtop Sedan de Ville; "personal-sized" Fleetwood Eldorado Coupe and Convertible; the new "international-sized" Seville, and the "executive-sized" Fleetwood Brougham, Fleetwood Nine-Passengar Sedan and Fleetwood Limousine. Major dimensions were unchanged from 1975. The 1976 Cadillacs got only minor cosmetic styling changes. All full-sized models got new cross-hatched grille textures. Cornering lamps on Calais, de Ville, Fleetwood Brougham, and the Fleetwood Nine Passenger Sedan and Limousine had new horizontal chrome trim.

Calais and de Ville coupes had a new vinyl roof treatment with the top molding a continuation of the door belt molding. Vinyl roofs were of a new elk-grain material while Sevilles and Fleetwood nine-passenger models had cross-grain padded vinyl roofs. Exterior opera lights like those on the Fleetwood Brougham and limousines were made part of the Coupe de Ville d'Elegance package and were optionally available on

The 1976 model year was the 12th, and last, for the lowline Cadillac Calais. Only 6,200 Calais coupes and four-door hardtops were sold this year compared with retail deliveries of more than 182,000 more expensive de Villes. Starting in 1977, the de Ville became Cadillac's "most affordable", or lowest-priced, series. The 1976 Cadillacs carried over their 1975 styling with very minor cosmetic appearance changes. These included a new grille texture and horizontal chrome trim on the wraparound cornering, parking and directional signal lamps in the front fenders. This is the 1976 Calais Coupe, Model 68247, which sold for $8,629. Only 4,500 were built.

A mere 1,700 Calais Hardtop Sedans were built during the record 1976 model year, making this lowline body style by far the rarest of all standard 1976 Cadillac body styles, including the much-vaunted "last" Fleetwood Eldorado Convertible. The Calais, Cadillac's lowest-priced series since 1965, was discontinued at the end of the 1976 model year. The de Ville took its place at the bottom of Cadillac's model lineup in Model Year 1977. Model 68249, the Calais Hardtop Sedan sold for $8,825 making it the second least-expensive model in the 1976 line, behind the companion Calais Coupe. Script series nameplates again appeared on the rear fenders of both Calais body styles.

1976

Calais and de Ville coupes. The distinctive Fleetwood Brougham's exterior signature now appeared in styled block letters in matching car color. Styling changes on the front-wheel-drive Fleetwood Eldorado included a Cadillac script nameplate on the left side of the hood above the grille header. Taillights were continuous red strips in large bezel frames. New Eldorado wheel discs had black-painted centers, and the 1976 Eldorado was equipped with standard four-wheel disc brakes. Optional wheel covers for all 1976 Cadillacs included simulated wire wheels and turbine-vaned discs.

The highly-successful Seville went into its first full model year with minimal changes. Exterior color choices were broadened to the full Cadillac spectrum of 15 standard and six optional Firemist colors, and sun roofs and the "Astroroof" tinted glass roof were added to the list of Seville options. Cadillac's first "small" car had been introduced as a 1975-½ model only the previous May. More than 60,000 Sevilles were sold through the end of the 1976 model year.

Electronic fuel injection was again available as an extra-cost option on all models except Seville, on which it was standard.

This would be the last year for the immense 500 cubic inch (8.2 litre) V-8 engine, the largest ever used in a production Cadillac. The compact Seville was powered by a 350 cubic inch (5.7 litre) V-8.

New Cadillac options for 1976 included an improved automatic door locking system that locked all doors when the transmission shift lever was moved into any forward drive position; a power-operated passenger seat back recliner, and a "weather band" feature for AM/FM stereo radios. Weather information in cities with weather transmitters was available by simply pressing a button.

The Cadillac Motor Car Division set a new production record for the first time in three years. A record 309,139 Cadillacs rolled off the division's assembly lines by the end of the 1976 model run, finally eclipsing the previous record 304,839 cars built in model year 1973.

The 1977-model changeover would be one of the most extensive in many years. A whole, new generation of smaller, lighter, more fuel efficient Cadillacs were on the way.

Cadillac again rightfully advertised its ubiquitous Coupe de Ville as "America's favorite luxury car." Sales spurted to a new record 114,482 Coupe de Villes during the 1976 model year, surpassing the previous record 112,849 built in model year 1973. Model 68347, the Coupe de Ville now carried a $9,067 price tag. The distinctive "Cabriolet" roof option was available for the third consecutive year, and the Coupe de Ville d'Elegance luxury package now included an opera light on the rear roof quarter behind the fixed quarter window. The opera light was optionally available on the Calais Coupe and Coupe de Ville. This 1976 Cadillac Coupe de Ville is equipped with both the Cabriolet and d'Elegance options, as well as the "Astroroof" introduced the previous year. These options pushed the price well over $12,000.

Sales of Sedan de Ville four-door hardtops continued to run well below those of the companion Coupe de Ville. A total of 67,677 Hardtop Sedan de Villes were built during the 1976-model production run, compared with sales of more than 114,000 Coupe de Villes. Model 68349, the Hardtop Sedan de Ville carried a base price of $9,265. A d'Elegance luxury decor option was again available. Standard Calais and de Ville models bore the traditional Cadillac "V" and crest on the front of the hood. When the d'Elegance option was specified, these were deleted and a see-through, standup hood ornament was mounted at the front of the chrome windsplit that ran down the center of the car's broad hood.

The majestic Cadillac Fleetwood Brougham Four-Door sedan, still on its exclusive 133-inch wheelbase, went into its sixth and final model year with the same basic body originally introduced for 1971. Fleetwood Brougham sales increased substantially, from 18,755 in 1975 to an even 24,500 in 1976 — 300 less than the record 24,800 built in 1973. Price of admission to the exclusive Fleetwood Brougham class was now $10,935. Two lavish decor options were again available, including the opulent "Fleetwood Talisman" four-seater package and the popular "Brougham d'Elegance." New premium wheel discs with black-painted centers and bearing the Fleetwood wreath and crest were optionally available, as well as wire wheel discs and chrome turbine-vaned wheel covers.

1976

Extensively restyled only the previous year, the front-wheel-drive Fleetwood Eldorado received only minor exterior styling changes for 1976. A Cadillac script nameplate appeared on the left side of the hood and four-wheel disc brakes became standard equipment. New wheel covers with black-painted centers also became standard and were optional on other 1976 Cadillacs. New taillights were continuous red slots above the rear bumper on each side of the licence plate opening. Manufacturer's suggested retail price for the Model 69347 Fleetwood Eldorado Coupe was $10,586. Sales held up well with 35,184 Eldorado Coupes delivered during the 1976 model year. This body style was continued through 1978. The companion Eldorado Convertible was dropped at the end of the 1976 model year.

Cadillac had served notice that it would not offer a convertible body style after the 1976 model year. The Fleetwood Eldorado Convertible, in fact, for the past few years had been the last production convertible built in America. Cadillac marketed its 1976 Eldorado Convertible as "The Last of the Breed". This, of course, drew a lot of attention. By the time the last Eldorado Convertible rolled off the Clark Ave. assembly line in Detroit in April, 1976, the division had built an even 14,000 Eldorado Convertibles for 1976, a 60-per cent increase over the 8,950 produced for 1975. All 1976 Eldo soft-tops became instant "collector's items" and the speculators drove prices into the stratosphere. Manufacturer's suggested retail price of the Model 69367 Eldorado Convertible was $11,049 but cars were being sold for well over twice that and asking prices were often at the 3-X mark. Once the hype receded, so did prices — drastically. The 1976 Fleetwood Eldorado's rear end featured revised, linear taillights.

Even before the first 1976 model was built, Cadillac sensed that there would be extraordinary demand for the "last" Cadillac convertible. From the beginning, the division decided to keep the actual last car for its own historical collection. But the division decided to end convertible production with a real splash. The final 200 Fleetwood Eldorado Convertibles produced would be identical replicas of the actual "last" car. All were finished in sparkling white with white convertible tops and wheel covers, and white leather upholstery with red piping and matching red carpeting and instrument panel. All 199 were immediately snapped up by dealers and collectors. This is the true last Cadillac convertible — complete with special "LAST" Michigan bicentennial licence plates — photographed in Dieppe Park in Detroit's neighboring Canadian city of Windsor, Ontario.

It was the automotive media event to top all such events; the end of an era; a time for nostalgia, and even to shed a tear or two. The last production American convertible (for awhile) was a white Cadillac Fleetwood Eldorado Convertible that came off the Clark Ave. assembly line in Detroit on the morning of Wednesday, April 21, 1976. It was the last of 200 identical "last" Eldorado convertibles. The drive-off was a gala news event that drew national attention. Bearing special Michigan bicentennial licence plate "LAST", the milestone soft-top was driven off the line by Cadillac Motor Division General Manager Edward C. Kennard and Manufacturing Manager Bud Brawner, with Detroit Mayor Coleman A. Young riding as a passenger. This historic car was kept by Cadillac and is displayed alongside the division's 1903 single-cylinder Model "A" and classic 1931 V-16 Phaeton.

Cadillac's first small car, the Seville, was carried over into 1976 with few changes. The "international-sized" Seville had been introduced as a "1975-½" model only the previous May. It was again available in only one body style, a luxuriously-equipped four-door notchback sedan. Price continued at $12,479. The Cadillac Seville, powered by a 350 cubic inch (5.7 litre) V-8 engine with standard electronic fuel injection, was a instant sales success. Production totalled 43,772 this year and Cadillac dealers gleefully took in many Mercedes-Benzes on trade. Standard equipment included a cross-grain padded vinyl roof, fuel monitoring system, air conditioning, stereo radio, level control, and nearly every other comfort and convenience item. The Seville's 114.3 inch wheel base was 15 inches shorter than the Calais and de Villes.

1976

The 1976 model year was the last for the extremely long (151.5-inch wheelbase, 252.3-inch over-all length) and heavy (5,800 pounds) last-generation "big" Cadillac limousines. The 1977 models were considerably smaller and lighter. This body style was produced for six model years, from 1971 through 1976. Cadillac built 1,815 Fleetwood Seventy-Five Nine-Passenger Sedans and Limousines during the 1976-model production run. This total included 981 Model 69723 Seventy-Five Nine-Passenger Sedans and 834 Fleetwood Seventy-Five Limousines with standard glass division. Prices were $14,889 and $15,239 respectively. Like all other 1976 standard series Cadillacs, the big Seventy-Fives got a new grille texture and revised cornering lamps with horizontal chrome trim.

Hess and Eisenhardt, a Cincinnati specialty car builder, continued to turn out highly-specialized armored and "security" car conversions of standard Cadillac limousines. Specifications and customer identity were closely-guarded secrets. Clients' names were known only to one or two senior executives, and the conversions were done under mysterious code names. This is an H & E "security" car conversion of a Fleetwood Seventy-Five Limousine. The entire rear roof has been removed and replaced with a convertible top. Note the running boards, handrails and rear bumper steps for security personnel. This impressive vehicle was commissioned for an unnamed government.

This is how the custom-built Hess and Eisenhardt "security car" looked with its top up. The entire rear roof could be folded back for parade appearances. Running boards and handrails around the upper front doors were provided for secret service agents, and rear steps and grab handles accommodated two more bodyguards on the rear of the car. The grille has also been modified. Armor-plating and security equipment details are closely-guarded secrets. This Cincinnati firm has done numerous special car conversions for foreign governments.

Several independent firms were now offering various "stretch" conversions of standard Cadillac nine-passenger limousines. Moloney Coachbuilders, then in the Chicago suburb of Rolling Meadows, Ill. introduced this extra-long wheelbase "Embassy Coach Limousine." Based on the standard Fleetwood Limo, the car was cut in half and a 12-inch body extension was added between the front and rear doors. Wheelbase was stretched from 151.5 inches to 162.5 inches. Interior decor and equipment was limited only by the purchaser's imagination and budget. Price started at $36,000. Note the landau bows on the upper rear roof panels and the Fleetwood laurel wreath and crest on the extra-wide "B" pillar.

Perhaps the most exclusive Cadillac Built in 1976 was this one-off 10-passenger "stretch" done by Wisco Corp. of Ferndale, Mich. for King Khalid Ibn Abdul Aziz of Saudi Arabia. Dollie Cole, wife of former General Motors Corp. President Edward N. Cole, was involved in designing the car — which was based on a concept advanced by Tamwin Industries of Washington, D.C. A standard Cadillac Nine-Passenger Limousine was literally cut in two and stretched four feet. Weight was increased by some 1,800 pounds. King Khalid commissioned this huge Cadillac custom limousine as his official car. More than 3,000 manhours went into the conversion.

1976

With the demise of the convertible, fresh-air freaks still wanted some way to get that open-air feeling in their cars. A promising alternative to the soft-top was the T-top, which featured large removeable roof panels above the front seat. These detachable panels were stored in the trunk when not needed and could be either solid metal or tinted glass. A center roof "spine" preserved body structural integrity. T-Bar roofs became quite popular at this time and were standard on Chevrolet Corvette Coupes. This Fleetwood Eldorado T-Bar Coupe was done by the Phaeton Coach Corp. of Dallas, Texas, a well-known conversion builder. Barris Kustoms of Los Angeles did some similar conversions for the Hillcrest Motor Co. of Beverly Hills.

Traditional Coach Works Ltd. of Los Angeles was successfully merchandising exclusive station wagon and pickup conversions of standard Cadillacs. These semi-customs were marketed under the name of "Mirage." The Cadillac Mirage pickup was a nicely-styled utility car conversion of the 130-inch wheel base Calais Coupe or Coupe de Ville. Somehow, the word "truck" just doesn't suit such a high-toned vehicle.

The day of the luxurious, long-wheelbase Cadillac ambulance was fast drawing to a close. Very few were built after Cadillac downsized its special commercial chassis for 1977, and Superior turned out the last in 1979. California ambulance buff Tom Parkinson photographed this fully-loaded 1976 Miller-Meteor high-headroom ambulance on 1976 Cadillac commercial chassis. Note the roof-mounted Federal "Q" coaster siren and four huge roof beacons. Maximum vehicle illumination was a must in emergency situations on California's teeming freeways.

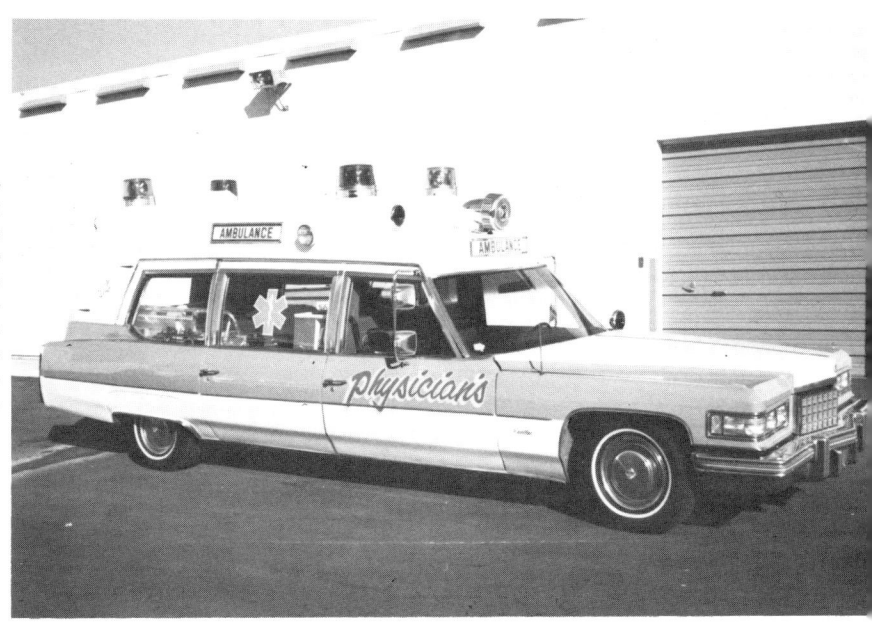

This would be the last year for the traditional "big" Cadillac hearse. The Cadillac commercial chassis was downsized for 1977, and somehow the industry was never again the same. Cadillac routinely built about 2,200 Series 69890 hearse and ambulance chassis which were supplied to the three principal U.S. funeral coach and ambulance manufacturers — Superior, Miller-Meteor, and Hess & Eisenhardt (S & S). This is an S & S "Victoria" Landau Funeral Coach by Hess and Eisenhardt on the 157.5-inch commercial chassis. Note the location of the parking and backup lights in the special rear bumper supplied with this chassis.

Until the arrival of the compact Seville, a key element in Cadillac's success was the product's impressive length. Biggest was still equated with best. Leo Weiser, operator of a New York City driving school, dreamed of owning the longest Cadillac of them all. Unsatisfied with the exterior dimensions of the standard long-wheel base Cadillac Limousine, Weiser set out to build his own long, long Caddy. Weiser had his limo stretched five feet, eight inches and rushed to have it listed in the Guiness Book of World Records. Alas for Weiser, a Santa Barbara, Calif. customizer came up with a tandem-axle 1967 Cadillac stretched that measured 32 feet between bumpers — five feet longer than Weiser's creation!

1977

The 1977 model year was one of the most momentous in modern Cadillac Motor Car Division history. The division proudly marked its 75th anniversary. Production and sales soared to new records. But most importantly, every one of Cadillac's standard cars was totally redesigned and restyled for a changing automotive world. The smaller, lighter, more fuel-efficient 1977 Cadillacs were introduced as... "The Next Generation of the Luxury Car."

Every one of GM's full-sized 1977 cars, from the lowest-priced Chevrolet on up to the long-wheelbase Cadillac limousine, were "downsized" in a massive crash program to make them more energy-efficient. General Motors led the industry in this unprecedented transition which would continue in successive waves until every one of the corporation's cars were similarly downsized. It was the start of a billion-dollar product revolution required to meet tough impending Government fuel economy regulations. Each of GM's five passenger car divisions was assigned the new full-sized "C"-body, which it styled and equipped to meet the division's own specific requirements.

The all-new 1977 Cadillacs were eight to 12 inches shorter and nearly 1,000 pounds lighter than the 1976 models they replaced. Average weight reduction totalled more than 950 pounds per car. This dramatic weight saving was accomplished through the use of advanced computer technology and lightweight materials. The new bodies were 3.5 inches narrower, but with virtually no sacrifice in interior room. Their trim, new exterior dimensions made the 1977 cars easier to handle and park.

All standard 1977 Cadillacs, including the carryover Eldorado, were powered by a brand-new 425 cubic-inch (7.0 litre) V-8 engine. The compact Seville, also carried over with only minor styling changes, continued to use the 350 cubic-inch (5.7 litre) V-8 with standard electronic fuel injection. Four-wheel disc brakes were now standard on the Fleetwood Brougham and Seville as well as the Eldorado.

Cadillac's model lineup was thoroughly shuffled this year. Only seven body styles were offered compared with 10 the previous year. After 12 years, the lowline Calais was gone. The extremely popular Coupe de Ville and Sedan de Ville replaced the Calais as the division's lowest-priced series. The redesigned Fleetwood Brougham now shared the same wheelbase as the lesser de Villes. The Eldorado convertible was gone. For the first time in many years there was no Fleetwood Seventy-Five at the top of Cadillac's model line. The totally-new long-wheelbase formal sedans would henceforth be known as simply the Fleetwood Eight Passenger Limousine and Seven-Passenger Formal Limousine. (The "formal" indicated that this model was equipped with glass division). Cadillac had downplayed the Series Seventy-Five name for the previous two years and in 1977 dropped it altogether. This distinguished series, introduced in 1936, at one time identified a complete range of body styles from sporty coupes and convertible sedans on up to stately open front town cars, but since 1941 was reserved for the exclusive senior series formal sedans and limousines.

The 1977 Cadillac Coupe de Ville, Sedan de Ville and Fleetwood Brougham rode on a new 121.5-inch wheelbase chassis, down from 130 inches the previous year. De Villes and Brougham had an over-all length of 221.2 inches, 9.5 and 12.5 inches shorter than comparable 1976 models. Limousines were built on a new 144.5-inch wheelbase, seven inches shorter than previously. Over-all length of the new Cadillac Limousine was 244.2 inches, eight inches shorter than before.

Despite their road to roof and bumper-to-bumper redesign, Cadillac's new 1977 models retained their unmistakable Cadillac styling identity. The front end featured a low, wide eggcrate grille with horizontal theme, flanked by widely-spaced dual rectangular headlamps. Full rear wheel openings on the new bodies resulted in a lighter side appearance. De Villes had

> Cadillac announced its totally-redesigned 1977 standard models as "The Next Generation of the Luxury Car". Every one of GM's full-sized cars, from the lowest-priced Chevrolet on up to the Cadillac Limousine, were completely redesigned and restyled in one massive, billion-dollar swoop. All emerged shorter, lighter and significantly more fuel-efficient than the vehicles they replaced. General Motors led the industry in this "downsizing", which was necessary to meet tough U.S. federal fuel economy standards. The 1977 Cadillacs were eight to 12 inches shorter and an average 1,000 to 950 pounds lighter than comparable 1976 models. All (except the Seville) were powered by a brand-new 425 cubic inch V-8 engine. Although GM's new "C" body was shared by all five passenger car divisions, each was free to give it distinctive divisional styling identity. This is the front-end of the 1977 Cadillac Sedan de Ville.

> Just as the Series 62 Cadillac bumped the lowline Series 61 at the end of the 1951 model year, the de Ville displaced the Calais as Cadillac's lowest-priced series when the 1977 Cadillacs went into the showrooms September 23, 1976. After 12 years, the slow-selling Calais was dropped at the end of the 1976 model year. The totally-redesigned and restyled Coupe de Ville retained its reputation as the world's favorite luxury car, and chalked up a new sales record in the process. A record 138,750 were built in the 1977 model year. Base price was now up to $9,654. D'Elegance and Cabriolet decor options were carried over for the all-new 1977 models. The Coupe de Ville was listed as Style 6CD47.

1977

ramed door glass with body-colored chrome accents (all chrome on Broughams) and the Fleetwood Brougham got distinctive, tapered center door pillars and a custom-trimmed backlight. Broughams also had opera lights on their rear roof quarters and wide chrome rocker sill moldings. Bodies, hoods and rear decks were crisply sculptured and rear ends had new, forward-angled vertical taillights.

The front-wheel drive Fleetwood Eldorado Coupe's revised grille and headlamps were unified into a horizontal design capped with a bright brushed-chrome molding. The Eldorado name was spelled out in block letters on the face of the hood and there were new rectangular side marker lights on the rear fenders. The Eldorado's revised rear styling included new vertical taillights in the bumper ends. A premium "Custom Eldorado Biarritz" option introduced late in the 1976 model year joined the Eldorado line for 1977.

The Custom Biarritz option revived an honored Cadillac model name that had not been used since 1964. The Biarritz package featured a thickly-padded cabriolet roof with formal rear quarter windows and backlight and standard opera lamps; color-keyed wheel discs, heavy brushed-chrome side moldings that swept from the front of the hood and encircled the base of the roof, and pillow-type 50/50 split front seats upholstered in standard leather.

The 1977 Cadillac Seville, now in its third model year, got a new, vertically-textured grille. A bare, painted metal roof option was available for the first time in addition to the standard padded vinyl roof.

Cadillac "special editions" for 1977 again included luxurious "d'Elegance" options for the Coupe and Sedan de Ville and Fleetwood Brougham. The four-seat "Fleetwood Talisman" option for the Brougham was discontinued. "Astroroof" and sun roof options were available on all models except limousines. New padded "Tuxedo grain" vinyl roofs were featured for all models except the Eldorado, which continued to use an elk-grain vinyl roof. In all, 16 vinyl roof color choices were available along with 21 exterior colors, 18 of them new.

New instrument panels featured a "center control area" which provided both driver and passenger access to radio, air conditioning and accessory switches. All models also got new two-spoke steering wheels. Automatic level control was standard on all models except de Villes, on which it was available as an extra-cost option.

Expanded radio options included an AM/FM stereo pushbutton radio with 23-channel CB. The new stereo radio option featured digital display of time, date, and elapsed trip time.

Even the Cadillac commercial chassis—the special long-wheelbase chassis-cowl unit built for funeral car and ambulance manufacturers—was significantly downsized this year. Wheelbase was reduced from 157.5 to 144.5 inches. All three major funeral car builders—Superior, Miller-Meteor, and Hess and Eisenhardt—had to design totally-new bodies for this shorter and lighter chassis.

Two years after the introduction of the compact Seville, Ford responded with a "baby" Lincoln. The 1977-½ Lincoln Versailles was introduced late in the 1977 model year. The Versailles was offered in only one body style, a formal four-door sedan built off Ford's compact Granada/Monarch platform. Standard engine was Ford's 351 cubic-inch V-8. The stubby Versailles' rear deck featured Lincoln's traditional simulated continental spare tire. Chrysler introduced a new upscale intermediate series called the Chrysler LeBaron.

Cadillac marked three-quarters of a century as the oldest, continuous automobile manufacturer in Detroit by setting another new production record—358,487 cars. The division also built its six millionth car, a Seville destined for a customer in California. It had taken only 3-½ years to build the six-millionth car compared with 47 years to produce the first million.

Although the popular de Ville series became Cadillac's least-expensive cars this year, replacing the lowline Calais, prices increased significantly. The four-door Sedan de Ville now carried a base sticker price of $9,864—without options. Sedan de Ville sales increased sharply this year, from 67,677 in 1976 to 95,421 by the end of the 1977-model production run. Standard equipment included Comfort Control air conditioning, power windows and door locks, six-way power seat adjuster, and match-mounted wheels and tires. A "Sedan de Ville d'Elegance" upgrade option was again available at extra cost. The Sedan de Ville was listed as Model 6CD69.

Like the rest of the 1977 Cadillacs, the premium Fleetwood Brougham underwent a complete transformation this year. Gone was the exclusive 133-inch wheelbase reserved for this four-door sedan from 1965 through 1976. The redesigned Brougham now shared the same 121.5-inch wheelbase as the lesser de Villes. The Fleetwood Brougham, however, retained its traditional high level of standard luxury. Four-wheel disc brakes became standard this year. The lavish "Brougham d'Elegance" option was still available but the four-seat Talisman was no more. Despite the downsizing, Cadillac buyers liked what Cadillac did to the Brougham: sales rose to a record 28,000. The Fleetwood Brougham Four-Door Sedan carried a base sticker price of $11,546 and was listed as Model 6CB69.

1977

The "downsized" 1977 Cadillac Fleetwood Brougham was a full 12.9 inches shorter over-all than the 1976 version. Wheelbase was shortened from 133 to 121.5 inches. At a glance, it was hard to tell the premium Fleetwood Brougham from the standard Sedan de Ville. Tip-offs were the distinctive tapered center side pillars between the windows and a smaller backlight, or rear window. Other styling features exclusive to the Brougham were broad chrome rocker sill moldings, color-keyed wheel discs, and small coach lamps on the upper rear roof quarters. The Fleetwood name appeared in block letters on the front fenders and rear decklid. All standard 1977 Cadillacs got new vertically-integrated taillights and sculptured decklids.

For the first time in nearly 40 years, there was no Fleetwood Seventy-Five in Cadillac's model lineup. The division had downplayed the Seventy-Five series name in most of its advertising for the previous two years. In 1977, this distinguished name was dropped altogether. The two special long-wheelbase sedans at the very top of GM's product offerings were now simply called the Fleetwood Eight-Passenger Limousine and Formal Seven-Passenger Limousine. Despite the downsizing, the completely-redesigned 1977 Cadillac Limousines sold very well. A total of 2,614 were built during the 1977-model production run, including 1,582 Model 6DF23 Eight-Passenger Limousines and 1,032 Model 6DF33 Formal Limousines with glass division. Prices were $18,193 and $18,858 respectively. Note the ventilator pane at the front of the rear door and the coach light on the door pillar ahead of the fixed quarter window.

The front-wheel-drive Fleetwood Eldorado went into its seventh model year with the same basic body that had originally been introduced for 1971. Compared with Cadillac's all-new, downsized standard models, the carry-over 1977 Eldo was huge. Styling changes were minimal. The revised grille got a fine, new texture and the Eldorado name was spelled out in block letters above the brushed chrome grille header. New rectangular side marker lights appeared on the rear fenders. The Eldorado Convertible was gone but a new premium Eldorado "Custom Biarritz" coupe option was added for 1977. Base price was $11,187 for the standard 6EL57 Fleetwood Eldorado Coupe. Sales increased to a record 47,344.

Late in the 1976 model year, Cadillac introduced a new "special edition" option package for the Eldorado Coupe. It was called the Eldorado Custom Biarritz, reviving an honored Eldorado name last used in 1964. The Custom Biarritz became a premium Eldorado Coupe option for 1977 and it proved very popular with Eldorado customers. The Biarritz package featured a thickly-padded cabriolet rear roof treatment with "Frenched" rear window, coach lamps, and special "Biarritz" nameplates on the upper roof quarters behind the fixed quarter windows. A heavy spear-tipped chrome molding swept from the front of the hood, thickening at the rear fender kickup, and continued around the base of the roof. Inside were contoured, pillow-style seats upholstered in standard Sierra grain leather. A sun roof or Astroroof were also available at extra cost.

The "international-sized" Cadillac Seville entered its third model year with only minor changes. The revised grille design got a new vertical texture, and a bare, painted metal roof was available for the first time. The deluxe chrome wire wheel discs on the car shown were extra-cost options. Also available were optional turbine-vaned wheel covers. Four-wheel disc brakes became standard equipment on the 1977 Cadillac Seville. Base price increased to $13,359. The very well-equipped Seville was a big seller. A total of 45,060 were sold during the 1977 model year. The Seville was listed as Model 6KS69.

1977

Cadillac's first "small" car, the expensive Seville, was now in its third model year. For 1977, Seville customers could choose either the standard, color-keyed padded vinyl roof or a bare, painted metal roof. Four-wheel disc brakes became standard equipment this year. The rear-wheel-drive Seville continued to use a 350 cubic inch (5.7 litre) V-8 engine with standard electronic fuel injection. The most visible styling change was a new, vertically-textured grille. Standard equipment included air conditioning, automatic level control, and variable-ratio power steering. The few Seville options available included a power trunk pull-down that automatically closed and locked the trunk, and power-operated Astroroof or sun roof.

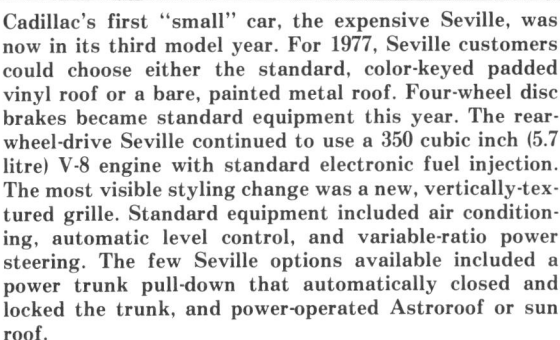

Along with its cars, Cadillac also downsized its special hearse and ambulance chassis this year. This was one of the most traumatic basic changes in the highly-specialized funeral car and ambulance industry in many years. Wheelbase of the 1977 Cadillac commercial chassis shrank from 157.5 inches to 144.3 inches. Funeral coach manufacturers did not begin to receive this new chassis until fairly late in the year. Prices also went up significantly. Consequently, funeral car and ambulance industry production dropped sharply this year. This is the 1977 Superior-Cadillac Sovereign Landau Funeral Coach. All three manufacturers—Superior, Miller-Meteor, and Hess & Eisenhardt—came up with totally-redesigned professional car bodies to fit this new chassis.

Very few ambulances were built on the Cadillac commercial chassis after the new, shorter hearse and ambulance chassis was introduced for 1977. The passenger car-based ambulance was being quickly replaced by increasingly more sophisticated van conversions and modular-type units. Only Superior and Miller-Meteor offered ambulances built on Cadillac chassis this year. This is Miller-Meteor's top-line high-headroom "Lifeliner", with full limousine styling. Cadillac shipped only 1,299 commercial chassis this year—half as many as it had produced in 1976. The industry would never again be the same.

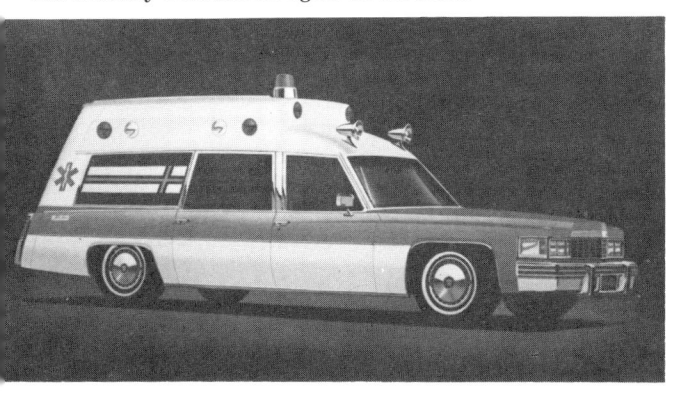

The Cadillac Motor Car Division created this "1977-½" Seville show car for its large exhibit at the 1977 New York International Automobile Show. Special styling touches included a striking two-tone exterior paint treatment and attractive color-keyed full wheel discs with a series of concentric rings. The roof was painted rather than covered with vinyl trim—a styling option that was gaining favor with Seville customers.

With the "downsizing" of the former nine-passenger Fleetwood Seventy-Five Limousine, several conversion builders developed "stretches" of the new, smaller Cadillacs to fill a new niche in the marketplace. Moloney Coachbuilders, of the Chicago suburb of Schaumberg, Ill., offered this 40-inch stretch of the 1977 Cadillac Fleetwood Brougham. The cost of the conversion was $14,900 over and above the price of the car. This conversion included two rearward-facing companion seats, an oil-rubbed walnut console cabinet and separate air conditioning system for the rear compartment.

1977

Another major conversion builder is the Phaeton Coach Corp. of Dallas, Texas. Phaeton Coach offered this six-window "stretch" limousine conversion of the 1977 Cadillac Fleetwood Brougham. The car was literally cut in two and a fixed center section inserted between the doors. Other major conversion limousine builders in business at this time included Moloney Coachbuilders in the Chicago area, and American Custom Coachworks in Los Angeles.

Cadillac had built the North American auto industry's last production convertible in 1976. That milestone car had barely rolled out the door when several conversion builders rushed in to fill the void. One of the first was Convertibles, Inc., located in Lima, Ohio. This small shop offerd a full convertible conversion of a standard Fleetwood Eldorado Coupe for $6,500 over and above the price of the car. This then, is a real, honest-to-goodness 1977 Fleetwood Eldorado Convertible, a body style no longer found in the Cadillac Motor Car Division's catalogs.

One of the largest and most successful of all conversion builders in the U.S. is the American Custom Coachworks of Beverly Hills, Calif. American Custom Coachworks retails its products only through authorized automobile dealers. This L.A. firm offered a full convertible conversion of the new, downsized 1977 Cadillac Coupe de Ville. Such conversions retailed for close to $37,000. Ah, the price of exclusivity. But, in that faddish part of the country, automotive individuality is important, and after all, it's only money.

Some well-heeled automobile buyers, unable to find exactly what they wanted among the various automakers' standard offerings, could satisfy their needs by going to a conversion shop. Not everyone, it seems, wants a long-wheelbase limousine. Sometimes, only a few extra inches will do. In the Chicago area, Moloney Coachbuilders did this "EM" stretch of a Cadillac Seville. The car was cut in half and a four-inch extension added between the doors—just enough to increase rear-seat legroom to the desired length. Wonder what that worked out to in dollars-per-inch?

One of the most bizarre creations ever seen on a Cadillac chassis was this incredible vehicle that was exhibited at the 1978 Paris Auto Salon. Called the "TAG Function Car" (?), it was based on a 1977 Cadillac Fleetwood Eldorado. About all that remained of the original car was the front-wheel drive running gear and front-end forward of the doors. The raised rear body contained a fully-equipped office with posh seats along the right-hand side. Note the tandem rear axles and cast wheels. This breathtaking beast was built in Switzerland. What more can we say?

Here is another 1977 Cadillac conversion by Moloney Coachbuilders. It's a subtle "stretch" of a standard Sedan de Ville. All the extra length is in the rear, between the rear door and wheels, and in the rear roof area. Custom bodywork like this is very expensive, but those with the means never seem to balk at a little thing like price when it comes to getting what they really want, or need.

1978

Cadillac's highly-successful "downsized" cars received only minor styling changes for 1978, but there were some important new options and engineering refinements. General Motors' new Diesel V-8 passenger car engine became optionally available in the Cadillac Seville midway through the 1978 model year, and a premium "Seville Elegante" was added to the line.

The Seville Diesel option was announced at the Chicago Auto Show in February, 1978 along with a sophisticated electronic trip computer. The 5.7-litre Diesel engine was sourced from Oldsmobile, which had introduced it as an option in its full-sized 88 and 98 models at the start of the 1978 model year. Cadillac Motor Car Division General Manager Edward C. Kennard predicted that about 1,500 Seville Diesels would be built during the balance of the 1978 model run. The new Diesel V-8 delivered remarkable fuel economy—nearly 50 per cent better average MPG than the gasoline version of the 350 CID engine. Because it was fired by compression rather than spark ignition, the V-8 Diesel needed no points or spark plugs. It also did not require a catalytic converter. Within a few years, Diesel engines would be available in all Cadillacs.

Cadillac's product lineup was essentially unchanged for 1978. There were again seven standard models in four series: Coupe and Sedan de Ville, Fleetwood Brougham Four-Door Sedan, Fleetwood Eldorado Coupe, the compact Seville, and two Fleetwood Limousines. No less than four new "special edition" Cadillacs were also offered this year. These included the ultra-luxurious Seville Elegante; a premium Eldorado Custom Biarritz Classic Coupe, and sporty new simulated-convertible versions of the Coupe and Sedan de Ville, called Cadillac Phaetons. Luxurious "d'Elegance" upgrade options were continued for the Coupe and Sedan de Ville and Fleetwood Brougham, as well as "Cabriolet" options on both Coupe de Ville and Eldorado. Exterior opera lamps were part of the Cabriolet roof option on Coupe de Ville and Eldorado.

The sporty new Cadillac Phaetons were fitted with very realistic-looking imitation convertible tops. These stylish roofs were complete in detailing right down to welts, stitching and convertible-type rear windows. The Phaeton package also included wire wheel covers, special bodyside accent stripes, leather seats and a leather-wrapped steering wheel. Three exterior color combinations were offered.

Cadillac customers could choose from a total of 21 exterior color choices, 17 new and 15 exclusive to Cadillac. There were also 16 vinyl roof color selections and seven new interior color choices along with leather and vinyl trim and three new cloths.

De Villes, the Fleetwood Brougham and Limousines got a bolder eggcrate grille design. A Cadillac script nameplate appeared on the left side of the hood face above the chrome grille header. There was also a new solid standup hood ornament. New chrome bumper ends housed vertical taillights and featured built-in side marker lights. The front-wheel-drive Fleetwood Eldorado Coupe also got a revised grille texture. This would be the eighth and final year for the second-generation Eldorado body which had been originally introduced for 1971. There were no less than four Eldorado Coupe variations this year: the standard Coupe, Eldorado Custom Cabriolet, the very popular Eldorado Custom Biarritz which was continued from 1977, and a top-line, limited-edition Eldorado Custom Biarritz Classic. The Biarritz Classic sported a distinctive two-tone exterior paint treatment and other custom touches.

The 1978 Cadillac Seville got restyled taillights with chrome ornamentation and insignia. The paint accent stripe extended across the decklid for a wider rear

The 1978 Cadillacs showed only minor exterior styling changes from the extremely successful "downsized" 1977 models. The 1978 grille got a bolder eggcrate texture and a Cadillac script nameplate on the left side of the hood face above the chrome grille header. There was also a new solid standup hood ornament. The Coupe de Ville was again America's favorite luxury car with 117,750 built during the 1978-model production run. At $10,399 it was the lowest-priced of Cadillac's seven body styles. Premium "d'Elegance" upgrade options were again available for both coupe and sedan in the division's lowest-priced series, as well as a Cabriolet option exclusive to the coupe. The 1978 Cadillac Coupe de Ville shown is equipped with the Cabriolet roof and new optional chrome wire wheels. Note the "V" and crest on the rear roof quarter. Opera lamps were part of the Coupe de Ville d'Elegance package.

During the 1978 model year, Cadillac added a sporty simulated convertible top option for both body styles in its popular de Ville series. Called the "Cadillac Phaeton", this high-styled appearance package included a fixed dummy convertible top, special accent body striping, chrome wire wheels and "Phaeton" nameplates on the rear fenders. Inside, there were dual-comfort, leather-upholstered seats and a leather-trimmed steering wheel. Color choices included Cotillion White with a dark blue top, Platinum with black top, and Arizona Beige with a dark brown top. Note the new chrome rear bumper ends with built-in side marker lights.

1978

appearance. Customers could choose a painted metal or fully-padded vinyl roof at no extra cost and opera lights were now available on most Sevilles. The very distinctive new Seville Elegante was offered in a combination of black and platinum or two shades of brown. Authentic chrome wire wheels, special exterior moldings and a painted metal top were standard along with a leather-trimmed interior with front seat center console. The Seville's standard 350 cubic inch, electronically-fuel-injected V-8 engine was equipped with a new, 31-pound lighter intake manifold. Seville sales hi a record 53,000 cars, including more than 5,000 Ele gantes, during the 1978 model year.

The new "Trip Computer" option was exclusive t the Seville. This highly-sophisticated on-boar computer displayed 11 separate items of informatio from average and actual fuel economy to estimate time of arrival. A digital speedometer and fuel readou gauge were part of the package. Heart of the syster was a tiny micro-processor. The Cadillac Tri Computer was announced simultaneously with th Seville Diesel engine option at the 1978 Chicago Aut Show. A new electronic level control system wa standard on Seville, Brougham, and Eldorado, an optional on de Villes.

All standard 1978 Cadillacs, including the Eldorade were powered by the 425 cubic inch (7.0 litre) V- engine introduced the previous year. Most 1978 Flee wood Broughams, as well as Sedan de Villes with th electronic fuel injection option and Sedan de Ville destined for sale in California, were equipped with a aluminum hood that was 45 pounds lighter than th conventional steel hood. The lightweight hood kep these models within the 4,500-pound EPA weigh class.

Wire wheel covers were available on most model: Cadillac's customers were ordering optional equip ment in record numbers: 96 per cent opted for viny roofs, 90 per cent specified cabriolet tops, 90 per cer chose tilt-and-telescope steering wheels, 83 per cer ordered cruise control, and 71 per cent the remot trunk lock system.

Sales of 36,800 Fleetwood Broughams also set a ne record. Cadillac built 349,684 cars during th 1978-model production run, down slightly from th previous year but still the second-highest in th division's 76-year history.

The two most popular models in the 1978 Cadillac product lineup were also the least-expensive; the Coupe and Sedan de Ville. The two-door coupe retained its traditional place at the top of the division's sales charts. Next came the companion four-door Sedan de Ville. A total of 88,951 of these four-door sedans were built during the 1978 model year. Manufacturer's suggested retail price was now $10,621 without options. Cadillac customers, however, tended to load up their cars with optional equipment, which added handsomely to the dealer's profit. A luxurious "Sedan de Ville d'Elegance" option was again available, as were new chrome wire wheels.

Cadillac's smart, new "Phaeton" simulated convertible roof option was available on both 1978 de Ville body styles. This is the Sedan de Ville Phaeton. Chrome wire wheels were part of this new exterior appearance group. Inside were luxurious leather-covered seats and a leather-wrapped steering wheel. The tailored top had the authentic appearance of a convertible roof, right down to a convertible-type rear window, stitching and welts. Special "Phaeton" script nameplates appeared on the rear fenders.

The elegant Cadillac Fleetwood Brougham established another new sales record this year. Deliveries of 36,800 of these premium four-door sedans far exceeded the previous record 28,000 sold only the previous year. In fact, sales had virtually doubled in just three years even though the sticker price had climbed to $12,223. New styling touches for 1978 included wreath-and-crest ornamentation on brushed chrome, an Elk-grain vinyl roof with angled opera lamps, and color-keyed, turbine-bladed wheel discs. Standard equipment included new electronic level control and steel-belted radial tires on match-mounted wheels. The ultra-luxurious "Brougham d'Elegance" option was also continued with a choice of Florentine velour or saddle leather upholstery.

1978

Cadillac's most expensive cars were again the stately Fleetwood Limousines, which were built in seven and eight-passenger versions. The 1978 Cadillac Fleetwood Eight-Passenger Limousine carried a base price of $19,285. The companion Fleetwood Formal Limousine for Seven Passengers sold for an even $20,000. Only 1,530 of these limited-production, long-wheelbase limousines were built this year, including 848 Eight-Passenger Limousines and 682 Formal Limousines with standard glass division. A full padded vinyl roof with coach lamps between the "C" pillar and fixed rear quarter window were standard.

This was the eighth and final year for the second-generation front-wheel-drive Fleetwood Eldorado Two-Door Coupe, which had originally been introduced for the 1971 model year. A brand-new "downsized" Eldorado was on the drawing boards, and Cadillac had served notice that the 1978 model year would be the last for this model in its current form. Despite its huge size and somewhat dated appearance, the big Eldo Coupe was still a very strong seller. Sales of 46,816 Eldorado Coupes were only slightly below the record 47,344 delivered the previous year. Sticker price was $11,858. The 1978 Fleetwood Eldorado Coupe shown is equipped with the "Custom Cabriolet" roof option first offered in 1972.

The last of the Big Eldos went out with a real flourish. While just one body style was available, it was offered in no less than four versions. First there was the standard Fleetwood Eldorado Coupe, which could be upgraded with a "Custom Cabriolet" option. Next came the special-edition Eldorado Custom Biarritz, first offered late in the 1976 model year and a very popular 1977 model option. Then, for 1978, Cadillac enhanced the premium Biarritz package with yet another limited edition called the Eldorado Custom Biarritz Classic. The Custom Biarritz Classic featured a distinctive two-tone exterior paint treatment and special interior appointments and trim. The 1978 Eldorado was powered by the same 425 cubic-inch (7.0 litre) V-8 engine used in all other 1978 Cadillacs except the Seville.

For those who insisted on privacy and the epitome of formal elegance, the Cadillac Motor Car Division continued to offer a distinctive landau roof option for its long-wheelbase Fleetwood Limousines. This premium limousine package included closed rear roof quarters, a padded, vinyl-covered rear roof area with smaller, specially-trimmed rear window and ornamental chrome landau irons. Small coach lamps heightened the formal effect. The official name for this rare body style was the Fleetwood Landau Cabriolet Formal Limousine. Some were delivered without the landau irons. Note the extensions of the "B" pillar chrome trim up and over the roof.

The compact Cadillac Seville went into its fourth model year with some important changes. A Diesel engine option was offered in the middle of the 1978 model year and an ultra-luxurious Seville Elegante was added to the top of the line. The standard Seville got new engraved taillight emblems and side and rear accent paint striping. Exterior opera lights were available and Seville purchasers could choose a painted metal or full vinyl covered roof at no additional cost. Chrome wire wheels or wire wheel discs were also optionally available. The Seville's 350 cubic-inch (5.7 litre) V-8 engine with standard electronic fuel injection was equipped with a 31-pound lighter aluminum intake manifold. Seville sales rose to 56,985 during the 1978 model year. Prices started at $14,161. The new Seville Diesel carried a sticker price of $16,447.

An ultra-luxurious new "special edition" broadened the Cadillac Seville offerings for 1978. It was the Seville Elegante, which featured a number of exclusive exterior styling touches. These included a distinctive two-tone exterior color treatment with painted metal roof; brushed chrome side moldings that extended the full length of the car at the fender peak line, and genuine chrome-plated wire wheels. Inside were lush perforated leather seats and a center console. The striking duo-tone exterior color choices included Platinum and Sable Black, or Western Saddle Firemist and Ruidoso Brown. Cadillac Motor Car Division General Manager Edward C. Kennard warned that only 5,000 Seville Elegantes would be built during 1978 model year. The ultimate 1978 Seville was an Elegante with the new Diesel engine option.

1978

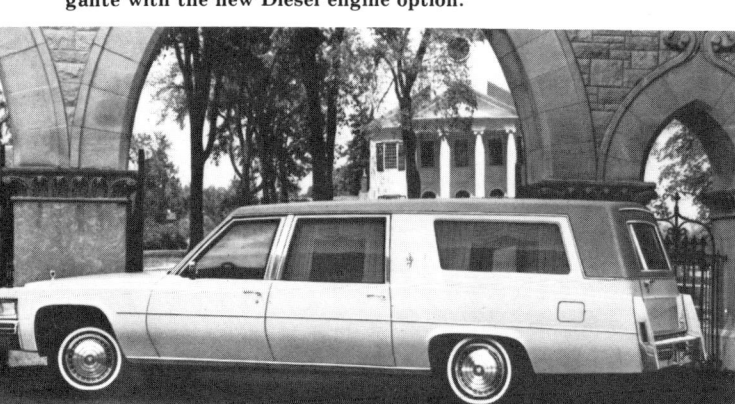

The shrinking, or downsizing, of the Cadillac commercial chassis that began in the 1977 model year had a devastating effect on the U.S. funeral car and ambulance industry. Funeral directors everywhere shied away from the much more expensive, significantly smaller Cadillac hearses. Cadillac shipped only 852 of its special long-wheelbase hearse and ambulance chassis to the three manufacturers during the 1978 model year. Sales would never again reach their traditional annual levels of about 2,000. Within five years, none of these firms would be in business in their original form. This is the 1978 Superior-Cadillac Sovereign Landaulet.

The limousine-style funeral coach or combination was now a rarity. The combination funeral car and ambulance had all but been replaced by the new van-type ambulances, and very few Cadillac-chassised ambulances were being built anymore. The limousine-style hearse had been around since the 1920s. Only Superior Coaches of Lima, Ohio, even offered one, although Miller-Meteor built some high-headroom limousine-type Cadillac ambulances in 1977 and 1978. This is the 1978 Superior-Cadillac Sovereign Limousine, which was offered in both combination funeral car/ambulance and straight hearse configurations.

A new Canadian firm was beginning to make a name for itself in the booming conversion industry. The AHA Manufacturing Co., in the Toronto suburb of Mississauga, Ontario, offered a full range of limousine conversions of everything from Honda Civics on up to Lincolns and Cadillacs. This is a 1978 Cadillac Fleetwood Brougham Town Limousine by A.H.A. Eight extra inches have been added to the wheelbase and rear doors. Note the complementing fillet with Fleetwood emblem in the front of the elongated rear doors. This new company sold most of its production in the U.S.

Armbruster-Stageway Inc. of Fort Smith, Ark., marked its 50th Anniversary in the specialty car industry this year. This company continued to offer six-door limousine conversions of Cadillac Sedan de Villes and Fleetwood Broughams which were now quite popular with funeral directors. Stageway and Moloney six-door Cadillac "stretches" could comfortably seat nine passengers, where Cadillac's downsized Fleetwood Limousines were designed to carry only seven or eight passengers. This is the 1978 Armbruster-Stageway Six-Door Cadillac. Note the opera lights on the rear roof quarters.

Moloney Coachbuilders, in the Chicago area, also offered a wide variety of extended-wheelbase and conventional limousine conversions of 1978 Cadillacs. This is a Moloney long-wheelbase conversion of a 1978 Cadillac Fleetwood Brougham four-door sedan. The modest "stretch" was done between the rear doors and rear wheel openings and in the rear roof area. Note the wide, chrome band that sweeps up and over the roof ahead of the padded vinyl-covered rear roof quarter.

1978

One of the smaller, lesser-known names in the booming limousine conversion business was Breese Custom Limousines Inc. of Dallas, Tex. Breese's nine-passenger, six-door Cadillac Sedan de Ville "stretches" were primarily targeted at the funeral service trade. This is a 1978 Breese-Cadillac limousine conversion. Note the three rows of seats and full chrome wheel discs.

The Wisco Co., in the Detroit suburb of Ferndale, Mich., offered station wagon and utility pickup conversions of standard production Cadillacs on the 121.5-inch wheelbase de Ville and Fleetwood Brougham chassis. This is a very nicely done station wagon conversion of a 1978 Cadillac Sedan de Ville. Conversions like this estate wagon were pretty expensive—about $13,700 over and above the price of the base car.

Here is a Wisco utility pickup conversion of a front-wheel-drive Cadillac Fleetwood Eldorado Coupe. Wisco had been in the conversion business for nearly a decade. The Eldorado made an especially attractive station wagon or pickup conversion, because of the absence of the usual driveline and differential housing, which permitted a low, flat rear floor. This distinctive Eldorado Pickup has relatively high sides above the rear fenderlines. It's doubtful that this dignified "truck" was used to haul trash or boulders.

One of the largest conversion shops in the country, American Custom Coachworks of Beverly Hills, Calif., used the distinctive "Paris" name on a number of its semi-custom products. The "Paris Pickup" was a standard ACC model offering, based on the Cadillac Coupe de Ville. These slick conversions were available only through authorized Cadillac dealers, who drop-shipped "base" cars off at any one of 10 regional ACC plants in the U.S., or chose from units in stock at the L.A. home plant. Note the full upper body vinyl trim on this 1978 Cadillac "Paris" Pickup.

American Custom Coachworks in Beverly Hills, Calif., was now one of the largest and most successful of all U.S. conversion builders. This Los Angeles area firm offered a very wide range of Cadillac conversions, from two and four-door convertibles on up to utility pickups and custom-built limousines. This is a four-door convertible sedan conversion by American Custom Coachworks. The base car was a Sedan de Ville or Fleetwood Brougham. Note how the rear fender has been built up to accommodate the oversized boot required to hold the large convertible top.

American Custom Coachworks Inc. of Beverly Hills, offered full soft-top convertible conversions of standard two and four-door Cadillacs. This is the "Paris Convertible" by ACC, based on the 1978 Cadillac Coupe de Ville. Conversions like this one had become increasingly popular since the demise of the production convertible two years earlier. They were very expensive, however, usually costing at least as much again as the base car.

1978

Armbruster-Stageway, Inc. of Fort Smith, Ark., also offered limited-production conversions of the Cadillac Seville. For those who liked the Seville's compact luxocar concept but who desired something a little longer, Armbruster-Stageway offered this 10-inch Seville stretch. Special styling touches included a deluxe chrome grille surround, full vinyl roof, and coach lamps on the upper body center panel. The price of this conversion was a quite reasonable $3,800 in 1978. Optional equipment available included a drop-leaf table, glass division, and color television set.

Among the most ambitious of all Cadillac Seville conversions attempted thus far was the stunning "Grandeur Opera Coupe", built by the Gandeur Motor Car Corporation of Pompano Beach, Fla. The Grandeur featured a stretched hood and shortened rear body, which resulted in a truly distinctive profile. Dual simulated side-mounted spare tires (last seen in 1940) completed the classic appearance. Public reaction was mixed: you either loved it or hated it. But it sure was different....

The premium Cadillac Seville proved popular with conversion builders from the time it was introduced late in the 1975 model year. The compact Seville seemed to lend itself very well to all kinds of conversions, from subtle "stretches" to two-door coupes and even convertibles. A California enterprise that operated under the name of the Coach Design Group Inc., built a limited number of these exclusive Cadillac Seville "San Remo" Convertible Coupes, which were distributed through Ogner Motors Ltd. of Woodland Hills, Calif. The Seville San Remo Convertible carried a price tag of $46,000. The redesigned rear end did away with the Seville's wraparound taillights, substituting Eldorado horizontal taillights across the lower rear deck.

Here's another Caddy pickup, this one by a San Francisco firm called American Built Cars Inc. This company's Cadillac pickup conversion was called the "Caribou". The upper body was a one-piece fiberglass section that extended from the windshield to the rear of the car. The tailgate was also fiberglass. The spare tire was stored under the Caribou's cargo bed floor.

The versatile Coach Design Group Inc. created this Cadillac Seville Two-Door Coupe for the Hillcrest Motor Co. of Beverly Hills, Calif. Note the premium chrome wire wheels, Double Eagle tires, and fixed quarter windows. This custom creation was also marketed as the "San Remo" by Coach Design Group. Exterior dimensions have not been altered, but the Seville's four doors were replaced with two larger ones and taillight styling was revised a la the companion San Remo Convertible.

At a glance, this looks like any old 1978 Cadillac Seville. But on closer examination, it's a Gucci Seville. International Automotive Design of Miami, Fla., collaborated with fashion designer Aldo Gucci to create this designer Seville, which came with a five-piece set of Gucci matched luggage. Other touches included 24-karat gold-plated Gucci emblems inside and out and Gucci inserts on the chrome wire wheel hubcaps. Had enough? The price was $19,990.

1979

Cadillac's big product story this year was an all new, third-generation Fleetwood Eldorado luxury/personal car. Optional Diesel engine availability was extended to all 1979 Cadillacs, except limousines, and Cadillac Motor Car Division model year production set another new record.

The 1979 Cadillacs were built at three General Motors plants from coast to coast. The division's "home" plant complex on Clark Avenue in Detroit produced Coupe and Sedan de Villes, the Fleetwood Brougham, Seville, all Fleetwood Limousines, and the Cadillac commercial (hearse and ambulance) chassis. The all-new Eldorado was built in the General Motors Assembly Division plant in Linden, N.J. Production of 1979 two and four-door de Villes began in the GMAD plant at South Gate, Calif., near Los Angeles.

The totally-redesigned and restyled Fleetwood Eldorado retained its traditional front-wheel-drive powertrain configuration and contemporary long-hoodline, short rear deck profile. Only one body style was offered, a classically-proportioned two-door hardtop coupe. This "New Breed of Eldorado" rode on the same 114-inch wheelbase as the companion rear-wheel-drive Seville but was a full 20 inches shorter, eight inches narrower, and 1,150 pounds lighter than the 1978 Eldorado Coupe which it replaced. Standard engine in the downsized Eldo was the same 350 cubic inch, electronically fuel-injected V-8 used in the Seville. A Diesel version of this 5.7-litre engine, built to Cadillac specifications by sister GM division Oldsmobile, was available as an extra cost option.

Standard equipment in the new car, which combined superb handling and outstanding fuel economy for a car in its premium price class, included four-wheel independent suspension, four-wheel disc brakes, and electronic level control. The new Eldorado's crisp styling featured a wide, isolated center grille with bold horizontal cross-hatching and a Rolls-Royce-esque gabled header. The Eldorado's very distinctive upper body structure sported a flush-mounted windshield, formal quarters and a chopped-off, nearly vertical rear roofline. Angled taillights in the ends of the rear fenders were reminiscent of those seen on the very first FWD Eldorado back in 1967.

An ultra-luxurious Eldorado Biarritz was also available. The Biarritz had a brushed stainless steel roof cap, padded rear roof area with exterior opera lamps and a bright chrome up-and-over roof molding. Heavy chrome moldings extended from the quarter window sills to the tips of the long front fenders. Accent stripes and new cast aluminum wheels were also part of the Biarritz package.

All standard 1979 Cadillacs were powered by the same 425 cubic inch (7.0 litre) V-8 engine introduced two years earlier. The 350 CID Diesel V-8 was offered as an option in Eldorado and Seville from the beginning of the 1979 model year and was gradually extended to the Fleetwood Brougham and de Ville options lists during the model year. The Diesel had a 22.5-to-1 compression ratio.

The standard 1979 Cadillacs got only minor styling changes. The revised grille design had a new cellular texture and was capped with a rectangular chrome header. De Villes got new standard wheel covers. The compact Seville, now in its fifth and last year in its original notchback sedan configuration, was virtually unchanged. The premium Seville Elegante model introduced the previous year was continued for 1979.

The 1979 Cadillac product lineup consisted of seven basic body styles: Coupe and Sedan de Ville, Fleetwood Brougham Four-Door Sedan; Fleetwood Eldorado Coupe; Seville and the flagship Fleetwood Limousine and Formal Limousine. Cabriolet options were available for the Coupe de Ville and new Eldorado, and luxurious "d'Elegance" upgrade options were offered for Coupe and Sedan de Ville and the Fleetwood

Cadillac's downsized standard models entered their third and final model year with only modest front-end styling changes. All would be extensively restyled for 1980. The modified 1979 Cadillac grille featured a new cellular texture and was capped by a full width chrome header. A thin chrome molding framed the rectangular upper grille opening and extended over the dual headlamp units around the corners of the front fenders, outlining the large cornering lights. The Cadillac name appeared in script on the left side of the prominent grille header. This is the 1979 Cadillac Coupe de Ville, Model 6CD47, which carried a sticker price of $11,728 making it the lowest-priced of the division's seven standard 1979 models. Coupe de Ville d'Elegance and Cabriolet roof options were again available.

Cadillac's least-expensive 1979 cars were again the extremely popular Coupe and Sedan de Ville. These two body styles outsold all other Cadillac models combined and again retained their traditional positions as the top-selling luxury cars in the world. This is the 1979 Cadillac Sedan de Ville, Model 6CD69, which had a manufacturer's suggested retail price of $12,093. A "Sedan de Ville d'Elegance" upgrade package was again offered as a $755 extra-cost option. The optional exterior opera lamps made it difficult to tell the Sedan de Ville apart from the more expensive Fleetwood Brougham. Both four-door sedans shared the same 121.5-inch wheelbase chassis and had an over-all length of 221.2 inches.

1979

Brougham. Four "special edition" Cadillacs were again available this year: Seville Elegante, Eldorado Biarritz, and authentic-looking simulated convertible versions of the de Villes called the Custom Phaeton Coupe and Custom Phaeton Sedan. The sporty Phaetons had been introduced the previous year.

Cadillac's sophisticated Trip Computer was available as an option on the 1979 Seville and Eldorado. This electronic "black box" digitally displayed, at the touch of a button, such useful vehicle operating information as average speed, miles to travel, elapsed time, engine RPM, distance to destination and arrival time. The speedometer and fuel gauge were part of the Trip Computer digital display.

Engineering refinements on the 1979 Cadillacs included retuned suspensions and new body mounts.

Model year production set a new record for the first time in two years. A total of 381,113 Cadillacs were built during the 1979-model production run. These included 260,190 de Villes, Fleetwood Broughams and Limousines; 53,487 Sevilles, and a record 67,436 of the new Fleetwood Eldorado Coupes.

The U.S. auto industry was once more faced with uncertainty when Iran cut off its flow of oil to the United States in the spring of 1979. Lines formed at the gasoline pumps again, and a concerned public began a permanent shift away from traditional full-sized cars to smaller, more fuel-efficient ones. From here on in, the industry would never again be the same.

A pair of sporty simulated convertibles were again offered in the popular de Ville series for 1979. The youthful "Cadillac Phaetons" had been introduced during the 1978 model year, but in 1979 they were called Custom Phaetons. Authentic convertible roof styling was available on both Coupe and Sedan de Ville for an extra $2,029. This attractive appearance option also included chrome wire wheel discs, special striping, luxurious leather seats and upholstery, and a leather-wrapped steering wheel. Soft-top roof detailing was faithful right down to stitching, welting and a convertible-style backlight. All that was missing were the wind and the bugs in your hair, but you could always lower the windows for that top-down feeling. Here are the Custom Phaeton Coupe and Custom Phaeton Sedan of 1979.

Cadillac's most expensive 1979 cars were the quietly impressive long-wheelbase Fleetwood Limousines. Again, this limited-production body style was offered in two versions—an eight-passenger sedan and a seven-passenger formal limousine with fixed partition, adjustable division glass, and a leather-upholstered chauffeur's compartment. The 1979 Cadillac Fleetwood Limousine, Model 6DF23, had a sticker price of $21,869. The Model 6DF33 Formal Limousine was the most expensive car in the 1979 Cadillac line with a base price of $22,640. A landau roof option with closed upper rear quarters was also still available. Except for the revised grille design, exterior styling was 1978 carryover.

The premium Cadillac Fleetwood Brougham Four-Door Sedan was still a strong seller, even at $14,102 a copy, without options. The "Brougham d'Elegance" option with leather seats cost an extra $1,367, or with cloth seats, $997. Exterior opera lights were standard on the Fleetwood Brougham and Limousines and were available on virtually every other 1979 Cadillac body style. The 1979 Cadillac Fleetwood Brougham d'Elegance shown is equipped with optional chrome wire wheel covers. Vaned, turbine wheel discs with color-keyed centers were also available as extra-cost options. The Model 6CB69 Fleetwood Brougham retained its distinctive tapered upper door pillars and broad chrome rocker sill and rear fender extension moldings.

The big news at Cadillac this year was a totally-new Fleetwood Eldorado. The "downsized" third generation Eldorado retained its front-wheel-drive powertrain but was 20 inches shorter, eight inches narrower and more than 1,100 pounds lighter than the 1978 model it replaced. Only one body style, a classically-proportioned two-door coupe, was offered. The all-new 1979 Eldorado was powered by the same 350 cubic inch, electronically fuel-injected V-8 engine used in the Cadillac Seville. A Diesel version of this 5.7-litre engine, still sourced from Oldsmobile, was optionally available. The new Eldo also offered standard four-wheel independent suspension and disc brakes all around. A record 67,435 Eldorado Coupes were built in the car's first year on the market. Model 6EL57 the 1979 Cadillac Fleetwood Eldorado Coupe had a sticker price of $14,668. A Cabriolet roof option was still available.

The all-new 1979 Cadillac Fleetwood Eldorado Coupe was available in two versions—the standard Coupe with or without a Cabriolet roof option, and a premium Eldorado Biarritz. The Biarritz package featured a distinctive stainless steel roof cap and padded vinyl cabriolet-type rear roof treatment. A bright chrome molding framed the leading edge of the rear roof and flowed into heavy upper body moldings that extended to the tips of the front fenders. Cast aluminum wheels and opera lamps were also part of the Eldorado Biarritz package which cost $2,350 with lush cloth upholstery and $2,700 in leather. With Astroroof, this pricey option commanded $3,488 extra in cloth and $3,838 in leather.

1979

The Cadillac Seville got some competition this year from the all-new Eldorado Coupe which shared the same 114.3-inch wheelbase. Both were marketed as "personal" luxury cars. The rear-wheel-drive Seville entered its fifth and final model year in this configuration. A totally-new, front-wheel-drive Seville was planned for 1980. The 1979 Cadillac Seville was virtually unchanged in appearance from the previous year. Two versions, the standard Seville and a premium Seville Elegante, were again offered Model 6KS69, the 1979 Seville Four-Door Sedan had a sticker price of $16,224, still well above that of the companion Eldorado. Sales of 53,487 Sevilles ran slightly below the 56,985 built the previous year.

The ultimate Cadillac Seville was again the exclusive Seville Elegante, which had been introduced the previous year. Distinctive Seville Elegante styling features included a painted metal roof, bold two-tone paint treatment, special upper bodyside moldings and long-laced, chrome wire wheels. The Elegante's interior was upholstered in rich, "breathable" perforated leather. New for 1979 was plush Tangier carpeting which simulated fine fur. The Elegante option was a $2,755 extra. When an astroroof was specified, the package price went up to $3,893. Options like these easily lofted the Seville's price well over the $20,000 mark. The Diesel engine option introduced halfway through the 1978 model year continued for both Sevilles in 1979.

One of the best-known specialty conversion shops in the world, the Hess and Eisenhardt Co. of Cincinnati, Ohio, got into the convertible business during the 1978 model year. Hess and Eisenhardt offered a two-door convertible coupe conversion of the standard Cadillac Coupe de Ville, which it marketed as the "Cadillac Le Cabriolet". Standard equipment included a fully-lined and padded convertible top, premium leather upholstery, a tempered glass rear window with defroster, and a special "Le Cabriolet" hood ornament. The H & E "Le Cabriolet" was distributed through 60 Cadillac dealers across the U.S. Price was an impressive $29,000. More than 300 had been shipped by the end of 1978, including both 1978 and 1979 models. This is the 1979 Cadillac Le Cabriolet.

Even though Cadillac still offered the only assembly line-built long-wheelbase limousines in America, several conversion builders were doing a brisk business in limousine "stretches" of standard Cadillac sedans. The most prolific limo conversion manufacturers at this time were Armbruster/Stageway in Fort Smith, Ark.; the Phaeton Coach Corp. of Dallas; American Custom Coachworks of Beverly Hills, and Moloney Coachbuilders in the Chicago suburb of Schaumburg, Ill. This is a Moloney Professional Six-Door Limousine conversion of a 1979 Sedan de Ville. These elongated sedans were becoming very popular with funeral directors who sorely missed the big nine-passenger Fleetwood Seventy-Fives last offered in 1976.

American Custom Coachworks Ltd. of Beverly Hills, Calif., offered this "extended body" limousine conversion of the 1979 Cadillac Fleetwood Brougham. The four-door sedan was literally cut in two and a 42-inch fixed center section inserted between the front and rear doors. The example shown is equipped with optional chrome wire wheel covers. Note the wide "B" pillar, which is covered with the same padded vinyl applique used on the roof. That's a television antenna on the rear deck. American Custom Coachworks also offered a six-door, nine-passenger limo.

1979

Perhaps the most prolific convertible conversion firm in business in the United States at this time was the American Custom Coachworks Ltd. of Beverly Hills, Calif. This company offered both two and four-door convertible conversions of the Cadillac Coupe and Sedan de Ville and Fleetwood Eldorado. This is ACC's impressive Convertible Sedan de Ville, an exceedingly rare body style. Note how the rear fenders have been built up to accommodate the large folding top. The convertible sedan shown was finished in a dazzling fire engine red with matching red interior and convertible top boot.

Now in its fifth model year, the compact Cadillac Seville had become the darling of the conversion trade. Several conversion shops were doing two-door coupe and convertible coupe conversions of the Seville four-door sedan. This is a 1979 "San Remo" Convertible, which was marketed through Ogner Motors Ltd. in Westlake Village, Calif. Ogner's Coach Design Group also offered an Eldorado Convertible. Well-known automotive photographer Rick Lenz found this 1979 Seville San Remo Soft-Top on a side street in L.A., truly a happy hunting ground for special interest car buffs.

In 1977, M. A. Stein of Toronto acquired Canadian manufacturing rights from Andrew A. Hotton Associates of the Detroit suburb of Belleville Mich. Stein established his new A.H.A. Manufacturing Co. in the Toronto suburb of Mississauga. A.H.A.'s principal products were long-wheelbase Lincoln Continental limousine stretches, but car buff Mel Stein was soon turning out a wonderful variety of customs on a special-order basis. This is A.H.A.'s "Canso," a two-door coupe conversion of a 1979 Cadillac Seville. The Canso nameplates came off a General Motors of Canada A-body compact. With its padded cabriolet roof, fixed quarter windows and chrome roof band, the Seville Canso was one of the nicer Seville specials around.

Anyone can cut and lengthen a Cadillac, but shrinking one is an entirely different matter. This has to be one of the strangest Cadillac Seville conversions that ever actually made it into volume production Called the "Seville Tomaso Coupe," it was marketed by Tomaso of America Inc. of Little Rock, Ark. The two-seater coupe design was by Tom Earnhart of Armbruster/Stageway fame. Dealers shipped stock Sevilles to Custom Car and Detail Inc. of Fort Smith, Ark., which did the conversion in about six weeks. The cost of the conversion was about $11,000 over and above the price of the car. The Tomaso was 20 inches shorter and slightly lower than a standard Seville and had a modified suspension. Distinctive, yes, but decidedly strange-looking . . . . .

Another company doing spectacular Cadillac Seville conversions was the aptly-named Grandeur Motor Car Corp. of Pompano Beach, Fla. Grandeur offered long-nose, close-coupled conversions of the Seville four-door sedan. The four-door in the foreground was marketed as the Grandeur Formal Sedan. The two-door coupe in the background was called the Grandeur Opera Coupe. Both models featured simulated dual sidemounted spare tires on their elongated front fenders. The Seville's doors were shortened, which gave the Grandeurs their very distinctive profiles.

1979

Cadillac's all-new, smaller Fleetwood Eldorado Coupe was an immediate hit with convertible conversion builders. Within a matter of weeks of announcement day, a number of Eldorado Coupes had been shorn of their roofs and like automotive butterflies, emerged from conversion shops as glamorous convertibles. American Custom Coachworks of Beverly Hills, boasted that its new Cadillac Eldorado "Paris" Convertible was America's answer to the Rolls-Royce Corniche. At least two other concerns, Ogner Motors Ltd. of Westlake Village, Calif., and Mark Doyne's Custom Coach of Clearwater, Fla., offered soft-top conversions of the new Eldorado Coupe.

After more than half a century, the prestigious long-wheelbase Cadillac ambulance came to the end of the line. Very few Cadillac ambulances were built in the U.S. after Cadillac downsized its 1977 commercial chassis. Hess and Eisenhardt didn't even offer one after 1976, Miller-Meteor built its last in 1978, and Superior stopped building them at the end of the 1979 model production run. Dr. Roger D. White, a lifelong ambulance buff and staunch advocate of the passenger car-based professional ambulance, personally bought the last Cadillac-chassised ambulance built. Dr. White, also an active member of The Professional Car Society, took delivery of his high-headroom Superior-Cadillac at the Lima, Ohio, plant on February 19, 1980. Superior had built only 10 a year in 1977, 1978 and 1979. This historic ambulance occupies a place of honor in Dr. White's garage in Rochester, Minn.

McClain Leasing of Anderson, Ind., continued to produce small numbers of funeral flower car conversions of Cadillac two-door coupes. McClain's flower cars featured stainless steel flower wells and a stainless steel rear deck boot. The rear roof area was vinyl-covered. This is a 1979 McClain-Cadillac Flower Car, the first one completed. The U.S. funeral car industry had stopped building long-wheelbase flower cars at the end of the 1976 model year, but Superior and Eureka revived this highly-specialized professional car body style in 1981.

American Custom Coachworks offered an impressive range of custom and semi-custom Cadillac body styles, including convertibles, utility pickups and long-wheelbase limousines. All of ACC's Cadillac conversions seemed to carry the "Paris" name. This is an American Custom Coachworks "Paris" Pickup conversion of a Cadillac Coupe de Ville. Its builder boasted that this elegant vehicle was equally at home in town or out on the range. Note the premium chrome wheels and vinyl-covered upper body with fixed backlight recessed between deep sail panels.

The Superior Division of Sheller-Globe added a new low-line Cadillac funeral coach to its product line for 1979. Built for Superior by Armbruster/Stageway of Fort Smith, Ark., it was called the "Royale". This new landau-style hearse was not built on the Cadillac commercial chassis. It was a stretched, much-modified Sedan de Ville. The body was lengthened 16 inches between the side doors and wheels. Armbruster/Stageway also built a Buick station wagon funeral car conversion for Superior on a sub-contract basis.

1980

This was an extremely eventful new product year for the Cadillac Motor Car Division. Cadillac's big news was a totally new, dramatically-restyled front-wheel-drive Seville. In addition to its bold bustle-back styling, the completely redesigned 1980 Cadillac Seville was the first production passenger car built in North America with a standard V-8 Diesel engine.

The standard Cadillacs were extensively restyled for the first time in three years. A two-door Fleetwood Brougham Coupe was added to the line at mid-year, and by the end of the 1980 model run, Cadillac's customers could choose from no fewer than five engines, including a V-6—the first non-V-8 powerplant offered in a Cadillac in 66 years!

The 1980 Cadillac model lineup included eight cars, one more than the previous year, in five series on three different wheelbases. The least expensive were the overwhelmingly popular Coupe and Sedan de Ville. Next came the highline Fleetwood Brougham Four-Door Sedan, which was joined later in the model year by the first two-door body style ever offered in this premium series, the new Fleetwood Brougham Coupe. Cadillac's luxury/personal car entries included the front-wheel-drive Fleetwood Eldorado Coupe and the stunning new Seville Four-Door Sedan. At the top of the heap were the quietly-dignified flagships of the General Motors fleet, the Fleetwood Eight-Passenger Limousine and Seven-Passenger Formal Limousine with glass division.

Cadillac special editions again included an Eldorado Biarritz Coupe and ultra-luxurious Seville Elegante.

Cadillac buyers could upgrade their cars with one of several optional luxury decor packages. These included "De Ville d'Elegance" options for the Coupe and Sedan de Ville and "Brougham d'Elegance" packages for the Fleetwood Brougham Four-Door Sedan and the new Fleetwood Brougham Coupe. Coupe de Ville and Eldorado customers could also specify attractive "Cabriolet" roof options.

The restyled standard Cadillacs—de Villes, Fleetwood Broughams and Fleetwood Limousines—got new aerodynamically-efficient "isolated" grilles flanked by flush-mounted lighting modules. Parking and directional signal lamps, now located below the dual rectangular headlights, had amber lenses. The new grille featured vertical accents and was capped with a bright mitred grille header with the Cadillac name on the left side. De Villes and Broughams got "stiffer", more formal rooflines. The restyled bodies had full-length upper crease lines and higher, straighter rear fenders. New bevelled decklids also had a higher profile. Massive new chrome rear bumper ends with integral side marker lights capped the rear fenders. All standard 1980 Cadillacs sported new deep-dish chrome wheel covers with red center inserts. Wire wheel discs were available for most models. Wheelbases and other major dimensions were generally unchanged.

The 1980 Fleetwood Brougham Sedan got a smaller limousine-style rear window. New electro-luminescent opera lamps were mounted on the door center pillars at the beltline. The elegant Brougham—direct descendent of the 1938 Sixty Special—retained its individually-framed side windows and broad chrome rocker sill moldings.

The new Fleetwood Brougham Coupe, first two-door in this exclusive series since the Sixty Special was introduced 42 years earlier, featured Cabriolet-type upper body styling with a thickly-padded rear roof and fixed vertical coach windows. A bright chrome molding swept up and over the roof at the "B" pillar, and standard opera lamps were mounted on the wide sail panels, behind the quarter windows.

The quietly impressive Fleetwood Limousines for 1980 also got Cadillac's fresh exterior styling. The long-wheelbase limousine's standard coach lamps were moved from the "C" pillar to a new location on the rear roof quarters behind the fixed quarter windows. The 1980 Fleetwood Eldorado Coupe, completely redesigned only the previous year, received minimal styling changes—a bolder eggcrate grille texture and new slotted standard wheel covers.

But Cadillac's (and GM's) styling tour de force this year was the all-new Seville. The Seville's spectacular rear-end styling was every bit as controversial as Cadillac's first tail fins had been in 1948. There was no middle ground: you either loved it or hated it. Designer Wayne Cady bestowed the second-generation Seville with a clifflike, vertical center grille. At the end of an extremely long hood, the windshield was raked back at an extreme angle. The rear roof swept down in a daring

Cadillac's big news this year was an all-new, front-wheel-drive, second-generation Seville. The 1980 Seville four-door sedan was dramatically restyled with a sweeping fastback roofline and sculptured "bustle back" rear deck inspired by the classic Hooper-bodied Rolls-Royces of an earlier era. The 1980 Cadillac Seville was also the first production U.S. car with a standard V-8 Diesel engine. A gasoline V-8 was available as a no-cost option. The 1980 Seville was GM's styling tour de force this year: you either loved it or hated it. The Seville's vertical front end contrasted sharply with the boldly raked rear end. Cast aluminum wheels with brushed chrome centers were standard. Chrome wire wheel covers were optionally available. All 1980 Sevilles had massive chrome rocker sill moldings. This is the standard 1980 Cadillac Seville which sold for $20,477 in base form. A total of 38,344 Sevilles were built this year.

1980

...astback profile broken by a sculptured, bustle-back rear deck. The effect was a visually powerful blend of the old and new. The daring bustle-back trunk was inspired by the classic Hooper-bodied Rolls-Royces of an earlier era.

The new K-body Seville was built in the General Motors Assembly Division plant in Linden, N.J. which also produced the "E"-body Eldorado, Buick Riviera and Oldsmobile Toronado. All previous Sevilles had been built in Cadillac's "home" plant in Detroit.

Mechanically, the 1980 Seville shared the same 114.0-inch wheelbase chassis and front-drive engine and transaxle layout of the companion Fleetwood Eldorado. But the Seville's standard engine this year was a 5.7-litre (350 cubic inch) fuel-injected Diesel V-8. A new 6.0-litre (368 cubic inch) gasoline V-8 with digital electronic fuel injection was available as a no-cost option. The gasoline-engined Seville weighed 300 pounds less than the Diesel. The high-styled Seville was the first U.S.-built sedan to combine front-wheel-drive with fully independent four-wheel suspension, four-wheel disc brakes, electronic level control, electronic climate control air conditioning, and cruise control and a standard Diesel engine.

The Oldsmobile-built 5.7-litre Diesel V-8 was optionally available in all other 1980 Cadillacs except Limousines. All standard 1980 Cadillacs—de Villes, Broughams and Limos—were powered by a new 6.0-litre, 368 cubic-inch-displacement V-8 gasoline engine with four-barrel carburetion. A modified version of this engine with Cadillac's sophisticated new digital electronic fuel injection system was standard in the 1980 Eldorado and was available at no extra cost in the Seville. Eldorados and Sevilles built for California were equipped with a 5.7-litre gasoline V-8 with standard electronic fuel injection. In April, 1980, a V-6 gasoline engine—the first ever offered in a Cadillac—was added as an optional choice in standard series cars. This 4.1-litre, 252 cubic-inch six-cylinder engine was built for Cadillac by Buick.

An "MPG Sentinel" system for monitoring fuel efficiency was standard on Sevilles and Eldorados with the 6.0-litre, DEFI gasoline engine. New Cadillac options this year included heated and electrically-adjustable outside rearview mirrors, a new electronic climate control air conditioning system, three-channel garage door opener; an improved theft-deterrent system that prevented the car from being started, and a Long Distance Cruise Package with 25-gallon fuel tank and electronic cruise control.

In the spring of 1980, General Motors announced plans to build a huge, new Cadillac assembly plant on the former site of the old Dodge Main plant on Detroit's east side, gradually phasing out the 60-year-old Clark Avenue and Fisher-Fleetwood plant complex sometime after 1982. Ford quietly discontinued its compact Lincoln Versailles at the end of the 1980 model year after only a three-year production run.

Triggered by the shutoff of Iranian oil the previous year, the U.S. auto industry was mired in another recession. Passenger car sales went into a prolonged slump. Cadillac built only 231,028 cars during the 1980 model year, 150,000 fewer than the record 381,113 cars built the previous year.

Like its predecessor, the new 1980 Cadillac Seville was offered in two versions—a standard four-door sedan and a premium, ultra-luxurious Seville Elegante. The Elegante was distinguished by a bold two-tone paint treatment split by a heavy chrome accent molding that swept from the tips of the front fenders to the rear of the car, terminating in a plunging French curve. Three two-tone color combinations were available. Elegante script nameplates were mounted on the Seville's broad sail panels next to the Cadillac laurel wreath crest. The interior was upholstered in premium quality leathers throughout. The Seville Elegante package was a $2,934 extra-cost option. Chrome-plated wire wheel covers were offered at no extra cost. Cast aluminum wheels were standard.

The 1980 Cadillacs were extensively restyled. All standard models got new grilles, rooflines and redesigned rear fenders. The Coupe de Ville remained the least-expensive car in the line with a sticker price of $12,899. The smart, new formal rear roof profile added an extra two inches of rear seat legroom. This is the 1980 Cadillac Coupe de Ville with the popular Cabriolet roof option. Opera lamps were standard with the Coupe de Ville d'Elegance upgrade option or could be ordered separately. Note the new deep-dish chrome wheel covers with red centers. Wire wheel covers or special vaned wheel discs exclusive to the de Ville series were also available as extra cost options. The Coupe de Ville was listed as Model 6CD47.

The Sedan de Ville also got a crisp, new formal roofline this year. The rear door was equipped with a large ventilator pane. The redesigned 1980 Cadillac grille featured a series of this vertical bars surmounted by a bright chrome mitered header. The Model 6CD69 Sedan de Ville had a sticker price of $13,282. The d'Elegance upgrade option was priced at $1,005 for both coupe and sedan in Cadillac's lowest-priced series. Opera lamps were part of the d'Elegance option or could be ordered for an extra $71. New de Ville options for 1980 included a heated outside rear view mirror and a three-channel garage door opener. All standard 1980 Cadillacs were powered by a new 6.0 litre (368) cubic-inch) V-8 engine. A 5.7 litre Diesel was optionally available in most models.

1980

The distinctive Fleetwood Brougham Sedan also got extensive styling changes this year. A companion two-door coupe joined this exclusive series midway through the 1980 model year. The Fleetwood Brougham Four-Door Sedan got a new, more formal roofline with a smaller limousine-type rear window. New electro-luminescent opera lamps were mounted on the door pillars at the beltline. The Fleetwood Brougham also sported standard vaned wheel covers with the Fleetwood laurel wreath and crest and broad, chrome rocker sill and rear fender extension moldings. Model 6CB69, the 1980 Cadillac Fleetwood Brougham Sedan had a sticker price of $15,564. The Brougham d'Elegance option with cloth upholstery cost an extra $1,062 and with leather, $1,525. Note the new chrome rear fender caps with built-in side marker lamps.

For the first time since the Sixty Special was introduced away back in 1938, a two-door body style was now available in this distinctive premium series. The new Cadillac Fleetwood Brougham Coupe was added to the line during the 1980 model year. The companion coupe's styling featured a landau-type rear roof treatment with fixed, vertical quarter windows. Opera lights were standard. A bright chrome molding extended over the roof. A Brougham Coupe d'Elegance option was also available. The "1980-½" Fleetwood Brougham Coupe was listed as Model 6CB47 and carried a manufacturer's suggested retail price $260 lower than that of the Fleetwood Brougham Sedan.

The long-wheelbase Fleetwood Limousines were also restyled for 1980. Exterior changes included Cadillac's new aerodynamically-efficient grille, redesigned rear fenders and massive chrome rear bumper ends. The Fleetwood Limousine's standard opera lamps were repositioned from the "C" pillar to a new location on the upper rear roof quarters behind the fixed quarter windows. Deep-dish chrome wheel covers with red inserts were also new. Two models were again offered in Cadillac's most prestigious series—the Model 6DF23 Fleetwood Eight-Passenger Limousine and the Model 6DF33 Formal Limousine for Seven Passengers. The standard limo carried a sticker price of $23,509. The Formal Limousine with glass division was the most expensive car in the 1980 model line with a base price of $24,343.

Totally redesigned the previous year, the front-wheel-drive Fleetwood Eldorado Coupe received only minor exterior styling changes for 1980. The most visible appearance changes were a new "eggcrate" grille texture and slotted standard wheel covers. Standard engine in the 1980 Cadillac Eldorado was a new 6.0 litre (368 cubic inch) V-8 gasoline engine with digital electronic fuel injection. A 5.7 litre (350 cubic inch) V-8 engine was standard in cars built for California. A 5.7 litre, electronically-fuel-injected Diesel V-8 was optionally available in the Eldorado. Model 6EL57, the Fleetwood Eldorado Coupe had a sticker price of $16,492. A total of 52,685 were built during the 1980-model production run. A Cabriolet roof option was again available for an extra $363. The 1980 Eldorado Coupe shown has optional wire wheels.

The 1980 Fleetwood Eldorado Coupe was again offered in three versions—base Coupe, Cabriolet, or the premium Eldorado Biarritz. The very popular Eldorado Biarritz package included a thickly-padded rear roof with standard opera lamps and a distinctive brushed stainless steel front roof cap. Heavy chrome moldings extended from the tips of the front fenders to the base of the Eldorado's narrow quarter windows, then up and over the roof. Special accent striping and "Biarritz" script on the rear roof quarters were also part of the package. Cast aluminum or chrome wire wheel covers were optionally available. The wire wheel covers were equipped with a locking device to deter would-be thieves. Standard equipment on all 1980 Cadillac Eldorados included a new electronic climate control air conditioning system, electronic level control and four-wheel disc brakes.

1980

As quickly as new Sevilles became available, they were customized and restyled by various conversion builders. One of the slickest Cadillac Seville conversions ever seen was the "Seville Renaissance Coupe" designed and built by the A.H.A. Mfg. Co. of Toronto. This cleanly-styled coupe bore its two-door styling extremely well. At a glance, it looked strikingly similar to Chrysler's 1981 Imperial. This is the basic Model R-801, one of seven Renaissance Coupe models offered by A.H.A.

At least eight firms were now actively engaged in the business of building "stretched" Cadillac limousine conversions of standard Sedan de Villes or Fleetwood Broughams. The most prominent of these were American Custom Coachworks of Beverly Hills; Moloney Coachbuilders in the Chicago area; Phaeton Coach Corp. of Dallas; Armbruster/Stageway of Fort Smith, Ark., and the A.H.A. Mfg. Co. in Toronto, Ontario, Canada. This is a 1980 Cadillac Executive Limousine conversion by A.H.A. This model featured a fixed 42-inch center panel extension. Note the chrome wire wheels, attractive two-tone paint treatment and the television antenna affixed to the roof rail. This custom-built limo was photographed in front of the Ontario Legislative Buildings in Queen's Park, Toronto.

Perhaps the most prolific of all the conversion houses in the U.S. at this time was American Custom Coachworks in the trendy L.A. suburb of Beverly Hills. A.C.C. was among the first manufacturers to market a limousine stretch conversion of the new bustle-back 1980 Cadillac Seville. Moloney Coachbuilders of Shaumburg, Ill. also introduced a Seville long-wheelbase limo at about the same time. This is an A.C.C. Seville stretch. A 48-inch center section was inserted between the Seville sedan's doors. The rear roof area is vinyl-covered, and there is a broad chrome crossover roof molding at the "B" pillar. A fixed-glass sun roof has also been added over the rear seat.

This is the "full house" A.H.A. Seville Renaissance Coupe. Styling highlights included flared rear wheel fender skirts and a striking two-tone exterior paint treatment. This style was listed as Model R-806. Six other versions were available. The A.H.A. Mfg. Co. of Toronto offered a full range of Cadillac conversions, ranging from four-inch "executive sedan" extensions to 42-inch custom-built limousine stretches. This company offered Cadillac, Lincoln and Buick conversions and did a few Chryslers as well. One-off A.H.A. limo conversions included a Volvo and at least one Honda Civic!

During 1979, American Custom Coachworks Ltd. introduced an ultra-customized version of the Cadillac Fleetwood Eldorado Coupe. This breathtaking semi-custom boasted just about every styling excess in the book—from dummy sidemounted spare tires to landau irons and rear deck straps. American Custom Coachworks called this very flamboyant limited-edition job the "Eldorado Glamour Coupe". The example shown is also equipped with chrome wire wheels and a sun roof. We can only speculate as to what it looked like inside.

1980

Cadillac's special commercial chassis for 1980 continued on the same 144.5-inch wheelbase as the Fleetwood Limousine. This special hearse chassis was now being supplied to only two funeral car manufacturers—Superior and Hess and Eisenhardt (S & S). Miller-Meteor had gone out of business at the end of the 1979 model year. The 1980 Cadillac commercial chassis got the same fresh exterior styling as Cadillac's standard 1980 passenger cars. This is the 1980 Superior-Cadillac Sovereign Regal Landaulet Funeral Coach. Note the narrow coach window behind the rear side door. Superior was the only manufacturer left that offered a side-loading hearse with movable casket table.

Armbruster/Stageway of Fort Smith, Ark., continued to build an "economy model" funeral coach conversion of the Cadillac Sedan de Ville for Superior. This conversion utilized the Sedan de Ville's standard rear doors, and was marketed as the Superior-Cadillac Envoy Landaulet. The American funeral car industry was never the same after Cadillac downsized its commercial chassis in 1977. One by one, all three of the major manufacturers either threw in the towel or were sold off and reorganized. This would be Superior's last year as a division of Sheller-Globe.

Among the first conversion shops to offer a long-wheelbase limousine conversion of the new bustle-back Cadillac Seville was Moloney Coachbuilders, in the Chicago suburb of Shaumberg, Ill. Moloney's Seville limo stretches were impressive indeed: the Seville's sweeping rear roofline contributed to the car's impression of extreme length. In the background are a pair of Moloney Sedan de Ville and Fleetwood Brougham long-wheelbase limousine conversions.

1981

Cadillac's big news for 1981 was the introduction late in the model year of the division's smallest car in more than 60 years, and an advanced variable-displacement V-8 engine designed to operate on eight, six, or four cylinders.

Actually an early 1982 model, the subcompact "Cimarron by Cadillac" was introduced in May, 1981. Addition of the diminutive (for Cadillac) Cimarron increased Cadillac's model lineup to nine cars, the most the division had offered since 1976. The comprehensive 1981 product array included the best-selling luxury cars built in America, the Coupe and Sedan de Ville; the premium Fleetwood Brougham Coupe and Four-Door Sedan; the luxury/personal Fleetwood Eldorado Coupe and Seville Four-Door Sedan, and the long-wheelbase Fleetwood Seven and Eight-Passenger Limousines. When it arrived near the end of the 1981 model, the four-cylinder Cimarron displaced the Coupe de Ville as Cadillac's least-expensive car.

Cadillac's big engineering story this year, however, was the "modulated displacement" V-8-6-4 engine. This sophisticated, electronically-controlled V-8 gasoline engine was standard in all 1981 Cadillacs except the Cimarron and Seville, but was optionally available in the latter in place of the Seville's standard 5.7-litre Diesel. The 6.0-litre (368 cubic inch) V-8-6-4 engine was equipped with standard digital electronic fuel injection. The DFI system was first offered on 1980 Eldorados and Sevilles but was extended to most Cadillac models for 1981.

The V-8-6-4 engine was equipped with a microprocessor-controlled electromechanical system that automatically actuated only the number of cylinders needed to satisfy driving requirements at any given moment. Eight, six or four-cylinder selection was completely automatic. A digital instrument panel display of the number of cylinders actually in use was part of the standard MPG Sentinel system that also provided a display of average and instantaneous miles per gallon and anticipated fuel range. An Electronic Control Module monitored the engine control system, sensors and actuators. Some problems developed with the complex modulated-displacement engine, however, resulting in complaints about performance. The Cadillac Motor Car Division vigorously defended the V-8-6-4 and offered customers special extended warranty coverage.

The Buick-built 4.1-litre (252 cubic inch) V-6 engine introduced as a Cadillac option late in the 1980 model year was again available in all Cadillacs except Limousines. The V-6 for 1981 was teamed with a four-speed automatic transmission with converter clutch in rear-wheel-drive cars. Front-wheel-drive V-6s employed a three-speed automatic transmission. The V-6 also offered Computer Command Control and closed-loop control to reduce emissions while permitting good fuel economy and performance. The 5.7-litre (350 cubic inch) Diesel V-8 manufactured for Cadillac by Oldsmobile continued as the standard engine in the 1981 Seville and was optionally available in all other models except the Cimarron and Limousines.

Extensively restyled the previous year, the 1981 Cadillacs received only minor appearance changes. DeVilles, Fleetwood Broughams, and Limousines got a modified grille with fine eggcrate texture and a new full-width chrome grille header. The front-wheel-drive Fleetwood Eldorado also got a new crosshatched grille. The "C"-body Cadillacs were the first full-sized GM cars to offer an automatic lap and shoulder belt system. Available as an extra-cost option on 1981 Sedan de Villes equipped with the V-6 engine, this

"A new kind of Cadillac for a new kind of Cadillac owner," is how the division introduced its new Cimarron subcompact on May 21, 1981. Cadillac did not actually call its new, small car a Cadillac: officially, it was the "Cimarron by Cadillac." A derivative of the new General Motors "J" body shared by the Chevrolet Cavalier and Pontiac J-2000, the Cimarron was offered in only a notchback four-door sedan body style. The car took its name from a fast-flowing river in Texas. The Cimarron was initially sold only through exclusive Cadillac dealers—242 of Cadillac's 1,600 U.S. dealers, and just one in Canada. All Cadillac dealers got the Cimarron for the 1982 model year. One of the smallest Cadillacs offered since the Model "D" of 1905, the Cimarron went into production in the GM Assembly Division plant at South Gate, Calif. Introductory price was $12,131 making the Cimarron the lowest-priced car in the 1981 model line—even though it was promoted as an early 1982 model.

Cadillac's new Cimarron subcompact was the first four-cylinder Cadillac built since 1914. It also offered the first manual transmission available in a Cadillac since 1953. The Cimarron by Cadillac was designed to compete with upscale sports sedans like the BMW and Audi. Wheelbase was 101.2 inches and the car had an over-all length of 173.0 inches. Weight was 2,590 pounds. The Cimarron was powered by an 85-horsepower, 1.8-litre (112.4 cubic inch) four-cylinder engine linked to a standard four-speed manual transmission. A three-speed automatic was available as an option. The Cimarron had a transverse-mounted, front-wheel-drive engine/transaxle.

"passive" occupant protection system required no specific action by the driver or front-seat passenger. As the door was opened, lap and shoulder belts extended out across the seat for ease of entry and exit.

The 1981 Eldorado was again available with a Cabriolet roof option or as a premium Eldorado Biarritz with brushed stainless steel roof cap. New wire wheel covers with large center medallions were standard on Biarritz and optionally available on other Eldorados. A new air dam below the front bumper on 1981 Eldorados and Sevilles improved vehicle aerodynamics and contributed to better fuel economy.

The 1981 Seville's vertically-textured grille was capped by a massive chrome header with standup wreath-and-crest hood ornament and windsplit molding. A one-piece bumper with built-in guards cradled the bright grille. The full-length accent molding used exclusively on the premium Seville Elegante in 1980 was now standard on all Cadillac Sevilles. The "ultimate Cadillac," the 1981 Seville Elegante was available in four exclusive two-tone color combinations. New wire wheel covers were optionally available for Seville. Cast aluminum wheels with brushed chrome discs were standard.

A new "Touring Suspension" option offering a firmer ride and handling was introduced during the 1981 model year.

The subcompact Cimarron went into dealer showrooms May 21, 1981. Initially, only exclusive Cadillac dealers got the Cimarron, which was designed to compete with upscale imported sports sedans like BMW, Audi, Volvo, and Saab. Offered in only one body style, a notchback four-door sedan, the Cimarron was Cadillac's version of the new GM "J" body shared by Chevrolet and Pontiac. Oldsmobile and Buick versions came later. The Cimarron was powered by a 1.8-litre (112.4 cubic inch) four-cylinder engine rated at 85 horsepower. The transverse-mounted engine had a front-wheel-drive transaxle and was coupled to a standard four-speed manual transmission—the first stick-shift Cadillac available since 1953. A three-speed automatic was optionally available.

The Cimarron, which took its name from a river in Texas, was built on a 101.2-inch wheelbase and had an over-all length of 173.0 inches, making it the smallest Cadillac built since the four-cylinder Model "D" of 1905. It was also the first four-cylinder Cadillac since the Model Thirty of 1914. The Cimarron weighed 2,590 pounds and carried an introductory price of $12,131 making it the lowest-priced Cadillac of the 1981 model year. The new "baby" Cadillac was built in the General Motors Assembly Division plant in South Gate, Calif. The Cimarron's specially-tuned suspension with McPherson struts, rack-and-pinion steering and semi independent rear suspension provided responsive handling. Steel-belted black sidewall tires were standard. Cadillac marketed the Cimarron as ... "A new kind of Cadillac for a new kind of Cadillac owner." The division was after younger, more affluent car buyers who had never been in a Cadillac showroom.

Chrysler Corporation, thoroughly restructured under the leadership of the dynamic Lee A. Iacocca, re entered the luxury car market this year, taking on Cadillac and Lincoln with a totally-new Imperial—the first since 1975. The high-styled 1981 Imperial produced exclusively in Chrysler's Windsor, Ontario plant, was offered only in a sleek two-door fastback coupe body style with a prominent bustle-type rear deck similar to that on the Cadillac Seville.

The auto industry slump which had begun in 1979 was now headed into its third straight year with no improvement in sight. Imports were taking an alarming share of the U.S. market. Cadillac built 252,256 cars, including 12,376 Cimarrons, during the 1981 model year, an improvement over the 231,028 cars built in 1980 but still far below the record 381,113 Cadillacs built for 1979.

The 1981 Cadillacs, extensively restyled only the previous year, were introduced with only minor styling changes. All standard models got a revised grille design with a fine cellular texture and a new full-width chrome grille header. The Cadillac name again appeared in script on the left-hand side. Until the introduction of the subcompact Cimarron late in the model year, as an early 1982 offering, the lowest-priced Cadillac available was the extremely popular Coupe de Ville. The 1981 Cadillac Coupe de Ville carried a base sticker price of $13,450. The car shown is equipped with the Cabriolet roof option. Other extra-cost equipment on this car includes wire wheels, opera lamps and contrasting two-tone paint treatment. The Coupe de Ville was listed as Model 6CD47.

Cadillac's big engineering innovation this year was a "modulated displacement" version of the division's 6.0-litre V-8 engine designed to run on eight, six, or four cylinders for optimum fuel economy. The fuel-injected "V-8-6-4" engine was standard in all 1981 Cadillac de Ville, Fleetwood Brougham, Limousine and Eldorado models, and was optionally available in the Seville. This is the 1981 Cadillac Sedan de Ville, Model 6CD69, which had a base sticker price of $13,847. Optional equipment on the Sedan de Ville shown includes wire wheels, opera lamps and two-tone paint. Engine options available for de Villes included a V-6 with automatic overdrive or a 5.7-litre Diesel V-8.

1981

The premium Fleetwood Brougham Coupe was now in its first full model year. This two-door companion car to the Fleetwood Brougham four-door sedan had been introduced midway through the 1980 model year. A landau-style rear roof with fixed quarter windows, broad sail panels, a small backlight and gleaming chrome crossover roof molding were standard. Model 6CB47, the lavishly-equipped Fleetwood Brougham Coupe had a sticker price of $15,942. A "Brougham d'Elegance" upgrade option was available for Fleetwood Brougham customers who insisted on even more luxury than the very high level of opulence that was standard in all Broughams. Electro-luminescent opera lamps were standard on both two and four-door Fleetwood Broughams.

Cadillac's most exclusive, and expensive, cars were again the long-wheelbase Fleetwood Limousines, long the favorites of governments, corporate customers, and funeral directors. The 1981 Fleetwood Limousines were powered by Cadillac's new modulated-displacement V-8-6-4 fuel-injected engine. The only significant styling change was a revised grille design with new cross-hatching and a redesigned chrome header. For customers who insisted on rear-seat privacy, a landau roof option was still available on a special-order basis. Two Fleetwood Limousines were offered for 1981: the Model 6DF23 Eight-Passenger Limousine which sold for $24,464, and the Model 6DF33 Formal Limousine for Seven Passengers, the most expensive of all 1981 Cadillacs at $25,323 without options. A sliding glass partition was standard in the Formal Limousine. All Fleetwood Limos had two folding auxiliary seats in the rear passenger compartment.

The biggest change in the 1981 Fleetwood Eldorado Coupe was Cadillac's new variable-displacement 6.0-litre, fuel-injected 368 cubic inch V-8 engine. The "modulated-displacement" V-8-6-4 engine was standard in the Eldorado, along with an "On-Board Computer Diagnostic System" designed to simplify servicing, and an "Econo-minder" to help the driver obtain the highest possible mileage through fuel-efficient driving habits. The 1981 Fleetwood Eldorado got a new, finely-textured cross-hatched grille but retained its classic mitered brushed-chrome grille header. Standard 1981 Eldorado wheel covers had large red centers. The cast aluminum wheels on this car, or new chrome wire wheel covers, were available as extra-cost options. Model 6EL57, the 1981 Fleetwood Eldorado Coupe had a sticker price of $16,492. Production went up to 60,643 this year.

The premium Fleetwood Brougham was successfully marketed as the "Cadillac of Cadillacs". This distinguished series again offered two body styles, a two-door coupe and the traditional Fleetwood Brougham Four-Door Sedan. The 1981 Fleetwood Brougham got Cadillac's new crosshatched grille and variable-displacement V-8-6-4 engine. Model 6CB69, the Fleetwood Brougham Four-Door Sedan retained its own, distinctive styling touches including small limousine-type backlight, individually-framed door uppers, electro-luminescent opera lamps on the center pillars between the doors and broad chrome rocker sill moldings. The 1981 Fleetwood Brougham Sedan carried a manufacturer's suggested retail price of $16,355 without options. Vaned, color-keyed wheel covers were standard. Genuine wire wheels or wire wheel discs were optionally available.

Two Eldorado "special editions" were again offered this year. A Cabriolet roof option was available for the standard Coupe, but the most elegant of all 1981 Eldorados was the distinctive Biarritz. Exclusive Eldorado Biarritz styling features included a brushed stainless steel front roof cap, Cabriolet-type rear roof in thickly-padded Elk grain vinyl, exterior opera lamps, special accent stripes and plush, pillow-type seats. The new wire wheel covers shown were standard on the Biarritz. Among the few options available for this lavishly-equipped and appointed Eldorado was Diesel power and an electrically-operated Astroroof.

The boldly-styled second-generation Seville went into the 1981 model year with minimal appearance changes. The most prominent of these was the chrome accent molding that extended the full length of the car, from the tips of the front fenders and sweeping down to the base of the bustle-back rear decklid. This molding had been exclusive to the premium Seville Elegante in 1980. The 1981 Cadillac Seville Four-Door Sedan, Model 6KS69, had a vertically-textured grille. Cast aluminum wheels were standard. The cross-laced chrome wire wheel covers shown were extra-cost options. Seville production fell by more than 10,000 cars to 28,631 in the recession-plagued 1981 model year.

1981

American Custom Coachworks of Beverly Hills, Cal., introduced a new two-seater coupe conversion of the standard Fleetwood Eldorado. This new ACC custom conversion was called the "Eldorado Paris Coupe II." This very extensive conversion involved stretching the front end 18 inches, shortening the roof 15 inches and adding an extra 12 inches to the rear deck. Needless to say, all this metal-cutting resulted in a much-altered profile. A portion of the front fender behind the wheel opening was also cut away to accommodate a set of chrome-plated exhaust pipes which converged into a single lakes-type exhaust below the rocker sill. Fitted with a refrigerator, crystal, and numerous coats of lacquer, ACC's Eldorado "Paris" Coupes sometimes sold for more than $60,000.

Cadillac marketed its bustle-backed Seville as ... "The Ultimate North American Motor Car." The ultra-luxurious Elegante, then, was the ultimate Seville. The 1981 Seville Elegante was offered in a choice of four contrasting two-tone exterior color combinations—Sheffield Gray Firemist over Sable Black; Superior Blue Metallic over Twilight Blue; Desert Sand Firemist over Briarwood Brown; or Mulberry Gray Firemist over Bordeaux Red. A full-length accent side molding that terminated in a graceful French curve at the rear provided an effective color break. A 5.7-litre Diesel V-8 engine continued as standard in the 1981 Seville. Cadillac's new moduated-displacement V-8-6-4 gasoline engine with electronic fuel injection was optionally available. Wire wheel covers were also optional.

Introduced two years earlier, American Custom Coachworks' flashy Eldorado "Paris" Coupe conversion was again available for 1981. This heavily customized version of the Fleetwood Eldorado Coupe was designed to transport four people in highly-conspicuous automotive luxury. Styling highlights included a lengthened front end, simulated side-mounted spare tires, a broad chrome roof band, ornamental landau irons, and rear deck straps. The rear roof quarters were closed in and six chrome wire wheels provided a racy touch. The Eldo Paris Coupe sports a sunroof, too, of course.

American Standard Coachbuilders offered six and nine-passenger station wagon conversions of the Cadillac Fleetwood Brougham Four-door Sedan. The rear roof was cut away behind the rear doors and a raised roof extension and full rear door added. A chrome roof luggage rack was also part of the package. The rear roof was vinyl-covered. The Fleetwood Brougham Station Wagon shown is equipped with premium chrome wire wheels. Note the wind deflector above the rear window. Retail price was about $54,000.

There was still a strong market for long-wheelbase limousine conversions of production Cadillacs for corporate customers and wealthy individuals. This is an extended-body Cadillac limousine by American Custom Coachbuilders of Los Angeles. A standard 1981 Sedan de Ville was cut in two and a four-foot panel extension inserted between the side doors. With such accessories as television, tape decks, beverage cabinets, mirrored one way glass, and dual heating and air conditioning systems, such custom limousines easily cost twice as much as a production Cadillac limousine.

1981

Among the most attractive Seville customs ever built was this very sporty one-off roadster done by California-based Milan Coach Builders. The base car was an earlier, first-generation Seville, but the two-seater was called the 1981 Milan. The nicely-proportioned Milan Roadster was photographed by well-known automobile photographer Rick Lenz outside the Milan plant. Note the lengthened hood, shortened midsection and Continental Mark-style louvers on the front fenders.

A proud, old name in the U.S. funeral car industry was resurrected this year. The Eureka Co., of Rock Falls, Ill., had ceased funeral car production in 1964. But Thomas A. McPherson, a senior executive of A.H.A. Mfg. Co. in Toronto, a lifelong admirer of Eureka's high-quality products, approached surviving members of the family that had owned Eureka and obtained rights to the name. A new Eureka Coach Co. Ltd. in Toronto, of which Mr. McPherson was President, introduced a line of Eureka-Buick funeral cars and limousines for 1980. Eureka started the industry when it added a line of premium Eureka-Cadillac hearses in mid-1981, using commercial components. This is a 1981 Eureka-Cadillac Concours Landaulet Funeral Coach which Mr. McPherson thoughtfully sent to the author's home in Windsor, Ontario, for photography. A multi-talented person, Mr. McPherson is also the author of two *Crestline Publishing Co.* automotive books—*The Dodge Story*, first published in 1975, and *American Funeral Cars & Ambulances since 1900*, published in 1973. The latter book is the only comprehensive history ever done on this fascinating aspect of automotive history.

After a five-year absence, the classic coupe-style funeral flower car made a dramatic comeback. Superior had built the industry's last long-wheelbase flower cars in the 1976 model year. Up to now, none had been built on Cadillac's downsized commercial chassis. But the new Superior Coaches operation in Lima, Ohio built one during the 1981 model year. The Coupe de Fleur joined Superior's 1982 product line as a standard model offering. This is the prototype 1981 Superior-Cadillac Coupe de Fleur, which was finished in gold with a black vinyl roof. It was delivered to a funeral home in Los Angeles.

After 47 years, the Superior Coach operation on East Kibby St. in Lima, Ohio, was no more. Once the largest and best-known funeral car and ambulance manufacturer in America, Superior had been a component of the Sheller-Globe corporate conglomerate since 1969. Sheller-Globe disposed of its Superior Division late in 1980 and closed the huge Superior plant. What was left of the funeral coach business was sold to Tom Earnhart of Armbruster/Stageway fame who organized a new company called Superior Coaches in a new plant a few miles away in Lima. Production of 1981 Superior-Cadillacs began in January, 1981. This is the 1981 Superior-Cadillac Sovereign Landaulet photographed at the factory. Note the wire wheel covers—rather sporty for a hearse!

1981

Moloney Coachbuilders, in the Chicago suburb of Schaumburg, Ill., also offered a long-wheelbase four-door limousine conversion of the fastback Cadillac Seville. Moloney's Seville stretch featured extra-wide "B" pillars with Cadillac laurel wreath and crest emblems and opera lamps on the "C" post. A standard Seville sedan was literally cut in half and a four-foot extension inserted between the doors. Most such limousine conversions had two rearward-facing fixed seats behind a fixed partition with glass division. There was usually a beverge cabinet between the rear-facing seats.

The newly-formed Eureka Coach Co. Ltd. of Toronto, a spinoff of the A.H.A. Mfg. Co. which was now concentrating on convertible conversions and an Allard J2 reincarnation, offered four and six-door Cadillac limousine "stretches" aimed primarily at the funeral service trade. Eureka's four-door limousine was a greatly-stretched Cadillac Coupe de Ville that revived two dormant General Motors model names—Concours and Talisman. Talisman, of course, was a former premium Fleetwood Brougham option, while the Concours name had been used on a Chevrolet. The 1981 Eureka-Cadillac Concours Talisman Limousine was built on a 155.4-inch wheelbase with an over-all length of 255 inches. Note the huge rear doors which made entry and exit a breeze.

1982

The 1982 Cadillacs went into dealer showrooms on September 24, 1981. Styling changes were minimal, but all 1982 Cadillacs except the subcompact Cimarron and the Fleetwood Limousines were powered by a brand-new, lightweight V-8 engine.

Cadillac's 1982 model lineup included nine cars in six series on four different wheelbases. The smallest and lowest-priced car in the Cadillac family was now the Cimarron, which had been introduced as an early 1982 model the previous May. Next came the top-selling Coupe de Ville and Sedan de Ville. The premium Fleetwood series included the Fleetwood Brougham Coupe and Fleetwood Brougham Four-Door Sedan. Cadillac's luxury/personal specialty cars included the front-wheel-drive Fleetwood Eldorado Coupe and companion Seville Four-Door Sedan. At the very top of Cadillac's comprehensive luxury car offerings were the stately long-wheelbase limousines—the Fleetwood Eight-Passenger Limousine and the Fleetwood Formal Seven-Passenger Limousine with glass division. Wheelbases and principal dimensions were unchanged.

Cadillac's new high-technology engine was called the HT-4100 DFI Power System. It was a totally-new 4.1-litre (252 cubic inch displacement) V-8 gasoline engine with digital electronic fuel injection, coupled to a new four-speed overdrive automatic transmission. The new engine had a die-cast aluminum block with free-standing cast iron cylinders and a separate die-cast aluminum valve lifter carrier. The engine was equipped with a closed-loop digital fuel injection (DFI) system.

A digital microprocessor called the Electronic Control Module controlled electronic spark timing, fuel metering, and idle speed. This extremely efficient, state-of-the-art engine was standard in all 1982 Cadillac de Villes, Fleetwood Broughams, Eldorados, and Sevilles.

For the first time in two years, the Seville did not offer a standard Diesel V-8. The 5.7-litre electronic fuel injection Diesel sourced from Oldsmobile was still available, however, as a Seville option. Other 1982 engine options available in de Villes, Fleetwood Broughams, the Eldorado and Seville included the Buick-built 4.1 litre V-6 gasoline engine and the 5.7-litre EFI Diesel. Cimarron continued to use the GM-built 1.8 litre four-cylinder engine. The modulated-displacement V-8-6-4 engine used across the board in 1981 was now used only in the Fleetwood Limousines.

The 1982 Cadillacs received only minor styling changes. All six "standard" models—de Villes, Fleetwood Broughams and Limousines—got revised vertically-textured grilles with two thin horizontal crossbars and a redesigned full-width chrome header. Tail lamps had new ornamentation and all models got restyled standard wheel covers. The rear window on Fleetwood Broughams was even smaller than before, resulting in a formal "limousine look." The Eldorado Coupe also got a revised grille texture with vertical accents and three fine horizontal cross bars and new taillamp detailing. The Seville was virtually unchanged for the third straight year, retaining its strong vertical grille design and sweeping fastback rear end styling with sculptured, bustle-back trunk.

The subcompact "Cimarron by Cadillac" continued in a single body style, a four-door notchback coupe, but a power-operated Astroroof and Twilight Sentinel automatic headlight system were added to the Cimarron options list. The Cimarron's standard

The subcompact Cimarron, introduced late in the 1981 model year, entered its first full model year with few changes. A new full-width rear seat was added and Cimarron options included a power-operated Astroroof, electronic radio with tape player and Twilight Sentinel automatic headlight system. Cimarron continued in only one body style, a notchback four-door sedan. All Cadillac dealers now sold the Cimarron. Sticker price went up $50 to $12,181. Standard engine was the GM-built 1.8-litre four-cylinder powerplant linked to a four-speed manual gearbox. A three-speed automatic was available as an extra-cost option. Cadillac cautiously promoted its first subcompact as "Cimarron, by Cadillac," never calling it the Cadillac Cimarron. The Cadillac name did not appear anywhere on this car. The only visible clue to its identity was the Cadillac badge in the center of the grille and winged Cadillac emblems in the lenses of the horizontal taillights.

The 1982 Cadillacs received only minor exterior styling changes, including new vertically-textured grilles, new taillight ornamentation, and new standard wheel covers. The big news this year lay under that broad Cadillac hood. All 1982 Cadillacs except the subcompact Cimarron and Fleetwood Limousines were powered by a brand-new lightweight V-8 engine called the HT-4100 Power System. The all-new 4.1-litre (252 cubic inch) V-8 engine had a die-cast aluminum block and digital electronic fuel injection and was coupled to a new four-speed overdrive automatic transmission. The lowest-priced of the six "standard" full-sized Cadillacs was the ever-popular Coupe de Ville which carried a base sticker price of $15,249. The car shown is equipped with the Cabriolet roof option, opera lamps and premium wire wheels.

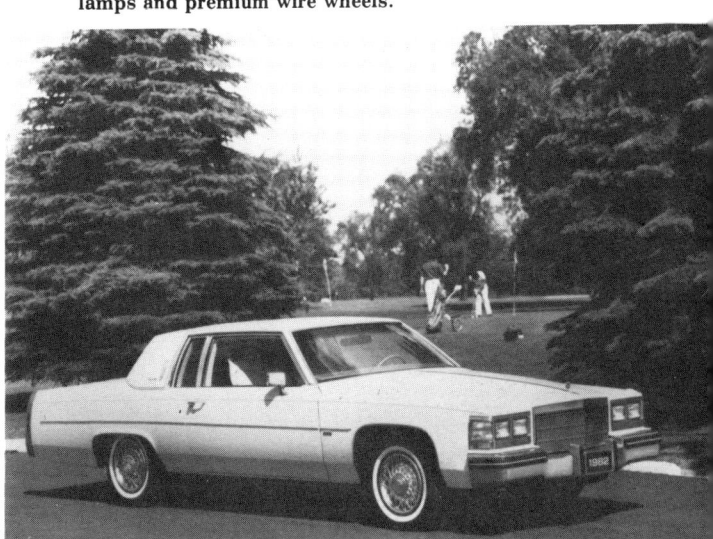

1982

transmission was a four-speed manual with a three-speed automatic available as an extra-cost option. A more powerful 2.0-litre engine and five-speed manual transmission were said to be on the way. All of Cadillac's 1,600 dealers now sold the Cimarron.

Cadillac's customers could again choose from a wide range of luxury upgrade options and "special editions". Luxurious "d'Elegance" decor packages were available for the Coupe and Sedan de Ville and Fleetwood Brougham Coupe and Four Door Sedan. A cabriolet roof option was available for the Coupe de Ville and Eldorado. Premium versions of Cadillac's luxury/personal entries—Eldorado Biarritz and Seville Elegante—were again offered. In March, 1982, an attractive simulated convertible roof styling option was announced for both the Eldorado Coupe and Seville. Called the "Full Cabriolet Roof" option, this convertible-style top treatment was available in three colors and was priced at $995 for each car. The sporty soft-top appearance option evoked memories of the "Phaeton" simulated convertible options available for the Coupe and Sedan de Ville in 1978 and 1979.

But the most intriguing new Cadillac offered this year was a performance-oriented version of the Eldorado called the Eldorado Touring Coupe. Based on the "Touring Suspension" package introduced during the 1981 model year, the Touring Coupe was conceived as the ultimate American driving machine.

The new Touring Coupe included Cadillac's new 4.1-litre HT-4100 DFI Power System, four-speed overdrive automatic transmission and the touring suspension package as standard equipment. The Eldo Touring Coupe rode on steel-belted, radial-ply black sidewall tires mounted on special aluminum wheels with center hubcaps and exposed chrome lug nuts. The usual standup hood ornament was replaced with a cloisonne badge. Headlamp and taillight bezels were finished in body color, and window reveal moldings and windshield wipers were blacked out. The Touring Coupe was available in only one color, Silver Metallic, with large, ribbed gray-finish rocker panel moldings. The interior was trimmed in matching gray leather and the package included a leather wrapped steering wheel.

The Touring Coupe was the closest Cadillac had ventured to producing a sports car in years. The touring suspension delivered increased cornering capability and steering precision, with increased ride and roll rates, greater shock absorber control, increased steering effort and feedback and excellent all-around vehicle performance.

Edward C. Kennard, who had been at the helm of the Cadillac Motor Car Division since he succeeded Robert D. Lund in 1974, projected deliveries of 275,000 Cadillacs during the 1982 model year when he introduced the 1982 cars at the division's annual new model press preview in Detroit. This total included 143,100 standard Cadillacs (De Villes, Fleetwood Broughams and Limousines); 55,300 Eldorados; 34,600 Sevilles and 42,000 Cimarrons.

The North American auto industry was still in recession. Record-high interest rates and the continuing erosion of domestic vehicle markets by imports had reduced 1981 calendar year sales to their lowest level in more than 20 years. The new GM "J" cars had met with an indifferent public reception, and in April, 1982, the South Gate, Calif., plant where the Cimarron was built was closed and Cimarron production was transferred to a GM Assembly Division plant in Janesville, Wisc.

Eighty years after Henry Martyn Leland oversaw the construction and testing of the first car to bear Le Sieur Antoine de la Mothe Cadillac's distinguished name, the Cadillac Motor Car Division maintained a seemingly invincible domination of the fine car field in America. Seven million Cadillacs later, the "Master of Precision" would undoubtedly be pleased!

The 1982 Cadillac's new vertically-textured grille had two thin horizontal cross bars and a mitered, brushed-chrome header. The Cadillac name appeared in script on the left side of this redesigned grille cap. The 1982 Cadillac Sedan de Ville had a manufacturer's suggested retail price, or sticker price, of $15,699—$450 higher than that of the companion Coupe de Ville. The car shown sports Cadillac's new standard wheel discs. Wire wheel covers were available as an option, as were exterior opera lights. This car is also equipped with the "Sedan de Ville d'Elegance" decor option. Note the d'Elegance script on the rear roof quarters.

The premium Fleetwood Brougham Coupe entered its third model year. Basic styling was unchanged, with a distinctive cabriolet rear roof treatment, narrow fixed quarter windows, and a formal rear window. The padded vinyl rear roof was edged with a bright chrome crossover molding. The 1982 Fleetwood Brougham Coupe and Sedan also got new vaned, color-keyed standard wheel covers. All 1982 Cadillacs equipped with Cadillac's new HT-4100 DFI Power System bore prominent "HT-4100 DFI" nameplates on their front fenders. The Fleetwood Brougham Coupe was priced at $18,096 without options. This is the top-line Fleetwood Brougham Coupe d'Elegance.

1982

The 1982 Fleetwood Broughams also got Cadillac's revised grille design and new taillight detailing. The prestigious Fleetwood Brougham Four-Door Sedan got an even smaller, limousine-type rear window. Genuine chrome wire wheels or wire wheel discs were available as extra-cost options. The Fleetwood Brougham name appeared in small script on the ends of the sedan's rear fenders. Note the Cadillac laurel wreath and crest on the padded rear roof quarters. The ultra-luxurious "Brougham d'Elegance" decor option was again available for both the Fleetwood coupe and sedan. The 1982 Cadillac Fleetwood Brougham Sedan had a sticker price of $18,567—$471 higher than the companion two-door coupe.

The Fleetwood Limousines were also given Cadillac's subtle 1982 styling changes—a modified grille design, new wheel covers and new taillamp ornamentation. These regal, long-wheelbase formal cars continued on a 144.5-inch wheelbase. The 6.0-litre, "modulated-displacement" V-8-6-4 engine used in all 1981 Cadillacs except the Seville and Cimarron, was offered only in the Fleetwood Limousines during the 1982 model year. The stately Fleetwood Limousines were the most expensive cars in Cadillac's nine-car 1982 product lineup. The Model 6DF23 Fleetwood Limousine for Eight Passengers carried a sticker price of $27,961. The 6DF33 Seven-Passenger Formal Limousine with glass division was the costliest Cadillac of all, with a base price of $28,941.

The third-generation Fleetwood Eldorado Coupe was now in its fourth model year. Except for minor cosmetic changes, its basic styling was the same as when the downsized Eldo was introduced for the 1979 model year. The 1982 Fleetwood Eldorado got a new, vertically-textured grille with three horizontal crossbars and black bumper rub stips with a white center strip. Cadillac crest ornamentation was also added to the Eldorado's taillight lenses. The Eldorado for 1982 was available in four versions: base Coupe, Cabriolet, premium Eldorado Biarritz and a new Eldorado Touring Coupe. This is the 1982 Cadillac Eldorado Biarritz with its distinctive brushed stainless steel roof cap and padded rear roof with opera lamps and bright chrome crossover molding. The least-expensive Eldorado now carried an $18,716 price tag. Wire wheels were standard on the Biarritz, optional for other Eldorados.

In March, 1982 the Cadillac Motor Car Division introduced a new simulated convertible roof option for the Fleetwood Eldorado Coupe. An imitation soft-top roof option was introduced for the Seville at the same time. Available in a choice of three colors—black, white or dark blue—new "Full Cabriolet Roof" option was priced at $995 for both models. The dummy convertible top is not to be confused with the landau-type rear roof option available for the Eldorado, which was also called a Cabriolet roof.

Cadillac had introduced a special "touring suspension" package during the 1981 model year. This sports/touring concept was expanded into a separate new model for 1982. It was the Eldorado Touring Coupe, a real driver's car with some very un-Cadillac styling and equipment features. The new Eldorado Touring Coupe was available in only one color—a Teutonic silver metalic with blacked-out windshield and window moldings and windshield wipers. Headlamp and taillight bezels were body-colored and the lower sides of the car were covered with ribbed, gray-finished rocker panel moldings. Steel-belted, black-sidewall tires were mounted on aluminum wheels with center hubs and exposed chrome lug nuts. Inside, the Touring Coupe had matching gray leather upholstery and a leather-wrapped steering wheel. The usual standup hood ornament was replaced by a cloisonne hood badge. The touring suspension package and 4.1-litre HT-4100 DFI Power System were also standard. The roof was painted metal.

1982

The fastback, bustle-trunked Cadillac Seville went into its third model year with no significant exterior styling changes. Again, there were two versions—the standard Seville Four-Door Sedan and a premium Seville Elegante. The only way to tell them apart at a glance was by the "Elegante" script on the rear roof panel on the highline job. But there was an important change under the Seville's long hood. A Diesel engine was no longer standard. The 1982 Cadillac Seville got the division's new HT-4100 digital fuel injection Power System with 4.1-litre gasoline V-8 engine and four-speed overdrive automatic. The 5.7-litre Diesel V-8 that had been standard in the 1980 and 1981 Seville was now optional. Inside, the base Seville had a new 45/50 split front seat. This is the 1982 Seville Elegante. Note the vertically-textured grille with massive, mitered chrome cap and chrome wire wheels.

An attractive simulated convertible roof option became available for the Cadillac Seville and Eldorado Coupe midway through the 1982 model year. The sporty "Full Cabriolet Roof" option was priced at $995 for both models. It was available in three colors—black, white and dark blue. The distinctive diamond-grain vinyl cabriolet roof brought back memories of the limited-edition Phaeton option offered on the Coupe and Sedan de Ville in the 1978 and 1979 model years. The new cabriolet top treatment gave the Seville the dashing appearance of a four-door convertible sedan.

One of the nation's largest conversion shops, American Custom Coachworks Ltd. of Beverly Hills, Cal., came out with a jaunty two-seat roadster version of the 1982 Cadillac Seville. The ACC Seville Roadster Convertible had a fully-powered convertible top. Chrome wire wheels and rear decklid luggage straps added to the two-passenger roadster's very sporty appearance. This company also marketed an Eldorado convertible coupe called the Eldorado "Paris" Convertible.

1982

The Eureka Coach Co. Ltd. of Toronto, Ontario, introduced a deck-type flower car at the 1981 National Funeral Directors Association annual convention held in Boston, Mass. Eureka's impressive new coupe-style flower car, built on the company's own commercial chassis, was called the Eureka-Cadillac "Concours Classic". It was built on a 147.5-inch wheelbase. The Concours Classic Flower Car had a notched rear roof and raised, adjustable stainless steel flower deck. A compartment below the flower deck could easily accommodate a casket. Eureka's prototype 1982 Cadillac flower car was painted a lustrous light metallic blue with chrome wire wheels.

The Hess and Eisenhardt Co. got out of the funeral coach business at the end of the 1981 model year. The Cincinnati firm, which had been building hearses for 105 years, sold its S & S funeral car division to former rival Superior Coaches of Lima, Ohio. The only other major funeral car manufacturer left in North America was the Eureka Coach Co. Ltd. of Toronto, Ont., which in only two years had become the second-largest name in the industry, behind Superior. Eureka offered two Cadillac hearses built on its own commercial chassis, the standard Concours Landaulet shown here and a premium Concours d'Elegance Landaulet. This 1982 Eureka-Cadillac was photographed in front of the factory prior to delivery to a funeral director in Minnesota.

The Eureka Coach Co. Ltd. of Toronto continued to offer four and six-door Cadillac limousines. This is the 1982 Eureka-Cadillac Concours Talisman, the four-door model with extra-long rear doors. A nine-passenger car, the Talisman is a lengthened Coupe de Ville with Fleetwood Brougham rocker sill moldings and trim. Model C821541, the 1982 Concours Talisman carried a suggested list price of $31,572.

Eureka Coach Co.'s six-door Cadillac limousine was called the Concours Calais. This nine-passenger car was equipped with three rows of seats. The companion four-door Talisman had two folding auxiliary seats. Model C821615, the 1982 Eureka-Cadillac Concours Calais was Eureka's most expensive limousine with a suggested list price of $33,990. The locking chrome wire wheel covers were a $298 extra-cost option. Eureka's principal competitors in the independent Cadillac limousine business were Hess and Eisenhardt and Armbruster/Stageway.

About the author

A lifelong auto buff, Walter Miller Pearce McCall was born in Simcoe, Ontario, Canada in 1938. He received his elementary education in Simcoe and Toronto schools and graduated from W.D. Lowe Technical School in Windsor, Ont. He went directly into journalism as a reporter-photographer for The St. Thomas (Ont.) Times-Journal. Three years later he joined the editorial staff of another Ontario daily, The Windsor Star.

In 1961 he received Canada's highest journalism award for his first-person coverage of a department store explosion in downtown Windsor. He also won a Western Ontario Newspaper Award for the same spot news story.

He was The Windsor Star's automotive writer for nine years, the only full-time auto writer on a Canadian daily newspaper. His "Autolines" column appeared twice weekly and covered the automotive industry in both Canada and the United States. In 1972 he joined the public relations department of Chrysler Canada Ltd. as Manager of News and Community Relations. He was appointed Manager -Public Relations in 1981.

Walter M. P. McCall comes from a well-known auto writing family. His father, the late Tom C. McCall, was the first Director of Public Relations for the Chrysler Corporation of Canada. A brother, Bruce, is a Contributing Editor of Car and Driver Magazine. A sister, Christine, is a former Car and Driver Managing Editor. Another brother, Hugh, is a former Editor of Canada Track and Traffic, Canadian counterpart to Car & Driver.

For obvious reasons, there were always Chrysler products in the McCall family garage, but Walt's older brothers were irrepressible sports car fans and owned a series of Morgans, Triumphs and a Porsche. Walt McCall developed his interest in Cadillacs during the fabulous fifties. Among his most vivid recollections of "The Standard of the World" was a bright-red Eldorado convertible which led the General Motors "Parade of Progress" through downtown Windsor when this famous mobile industrial exposition, with its fleet of huge "Future liners", came to town in June, 1955.

As a professional auto writer, Walt McCall was finally able to indulge in his hobby — and Cadillacs — for a living. He religiously attended the Cadillac Motor Car Division's annual Press Previews in Detroit between 1963 and 1972 and was a frequent visitor to the Clark Avenue office and plant complex. This working relationship with Cadillac's superb Public Relations Department, headed by the affable and always helpful William J. Knight, was to prove invaluable when the opportunity arose to write this book.

Walt McCall has owned three Cadillacs — a 1953 Fleetwood Seventy-Five Imperial Limousine, and 1956 Series Sixty-Two Hardtop Coupe, and a 1959 Superior-Cadillac Crown Royale Landaulet Funeral Coach.

A prolific writer and author, Walt McCall wrote Crestline's authoratative "American Fire Engines Since 1900" and is currently Editor of two U.S. vintage vehicle quarterly magazines — "Engine!-Engine!", official publication of the Society for the Preservation and Appreciation of Antique Motor Fire Apparatus in America, Inc., and "The Professional Car", the quarterly magazine of The Professional Car Society. He is also Editor of "Third Alarm", a bi-monthly newsletter of the Ontario Fire Buff Associates and is Founding President of the Greenfield Village International Antique Fire Apparatus Association, Great Lakes Chapter of SPAAMFAA. He is a regular contributor to various trade journals, including "Fire Engineering" and "The American Funeral Director", and authored the fire apparatus chapter of Automobile Quarterly's Handbook of Automotive Hobbies. A member of the Society of Automotive Historians and, of course, the Cadillac-LaSalle Club, Walt McCall is a Past Vice-President of the Windsor Branch of the Historic Vehicle Society of Ontario.

He resides with his wife, Denise, son Walter Jr. and daughter, Perri, in Windsor, Ontario, less than five miles from Cadillac's home in neighboring Detroit. The McCall motor fleet, in addition to various current Chrysler products, includes two fire engines — a 1925 Seagrave and a 1925 American LaFrance.